T0142349

Advances in Intelligent Systems and Computing

Volume 365

Series editor

Janusz Kacprzyk, Polish Academy of Sciences, Warsaw, Poland
e-mail: kacprzyk@ibspan.waw.pl

About this Series

The series "Advances in Intelligent Systems and Computing" contains publications on theory, applications, and design methods of Intelligent Systems and Intelligent Computing. Virtually all disciplines such as engineering, natural sciences, computer and information science, ICT, economics, business, e-commerce, environment, healthcare, life science are covered. The list of topics spans all the areas of modern intelligent systems and computing.

The publications within "Advances in Intelligent Systems and Computing" are primarily textbooks and proceedings of important conferences, symposia and congresses. They cover significant recent developments in the field, both of a foundational and applicable character. An important characteristic feature of the series is the short publication time and world-wide distribution. This permits a rapid and broad dissemination of research results.

Advisory Board

More information about this series at http://www.springer.com/series/11156

Wojciech Zamojski · Jacek Mazurkiewicz
Jarosław Sugier · Tomasz Walkowiak
Janusz Kacprzyk
Editors

Theory and Engineering of Complex Systems and Dependability

Proceedings of the Tenth International Conference on Dependability and Complex Systems DepCoS-RELCOMEX, June 29 – July 3 2015, Brunów, Poland

 Springer

Editors
Wojciech Zamojski
Wrocław University of Technology
Department of Computer Engineering
Wrocław
Poland

Jacek Mazurkiewicz
Wrocław University of Technology
Department of Computer Engineering
Wrocław
Poland

Jarosław Sugier
Wrocław University of Technology
Department of Computer Engineering
Wrocław
Poland

Tomasz Walkowiak
Wrocław University of Technology
Department of Computer Engineering
Wrocław
Poland

Janusz Kacprzyk
Polish Academy of Sciences
Systems Research Institute
Warsaw
Poland

ISSN 2194-5357   ISSN 2194-5365   (electronic)
Advances in Intelligent Systems and Computing
ISBN 978-3-319-19215-4   ISBN 978-3-319-19216-1   (eBook)
DOI 10.1007/978-3-319-19216-1

Library of Congress Control Number: 2015939269

Springer Cham Heidelberg New York Dordrecht London

Springer International Publishing AG Switzerland is part of Springer Science+Business Media
(www.springer.com)

10 Years
DepCoS-RELCOMEX
2006 – 2015

Preface

We are pleased to present the proceedings of the Tenth International Conference on Dependability and Complex Systems DepCoS-RELCOMEX which was held in a beautiful Brunów Palace, Poland, from 29th June to 3th July, 2015.

DepCoS - RELCOMEX is an annual conference series organized since 2006 at the Faculty of Electronics, Wrocław University of Technology, formerly by Institute of Computer Engineering, Control and Robotics (CECR) and now by Department of Computer Engineering. Its idea came from the heritage of the other two cycles of events: RELCOMEX (1977 – 89) and Microcomputer School (1985 – 95) which were organized by the Institute of Engineering Cybernetics (the previous name of CECR) under the leadership of prof. Wojciech Zamojski, now also the DepCoS chairman. In this volume of "Advances in Intelligent Systems and Computing" we would like to present results of studies on selected problems of complex systems and their dependability. Effects of the previous DepCoS events were published (in historical order) by IEEE Computer Society (2006-09), PWr Publish House (2010-12) and by Springer in "Advances in Intelligent and Soft Computing" volumes 97 (2011), 170 (2012), 224 (2013) and 286 (2014).

Today's complex systems are integrated unities of technical, information, organization, software and human (users, administrators and management) resources. Complexity of modern systems stems not only from their involved technical and organization structures (hardware and software resources) but mainly from complexity of system information processes (processing, monitoring, management, etc.) realized in their defined environment. System resources are dynamically allocated to ongoing tasks. A rhythm of system events flow (incoming and/or ongoing tasks, decisions of a management system, system faults, "defensive" system reactions, etc.) may be considered as deterministic or/and probabilistic event stream. Security and confidentiality of information processing introduce further complications into the models and evaluation methods. Diversity of the processes being realized, their concurrency and their reliance on in-system intelligence often significantly impedes construction of strict mathematical models and calls for application of intelligent and soft computing.

Dependability is the modern approach to reliability problems of contemporary complex systems. It is worth to underline the difference among the two terms: system dependability and systems reliability. Dependability of systems, especially computer systems and networks, is based on multi-disciplinary approach to theory, technology, and maintenance of systems working in a real (and very often unfriendly) environment. Dependability of systems concentrates on efficient realization of tasks, services and jobs by a system considered as a unity of technical, information and human resources, while "classical" reliability is restrained to analysis of technical system resources (components and structures built from them).

Important research area on contemporary dependable systems and computer networks is focused on critical information systems. These problems were the subject of "Critical Infrastructure Security and Safety (CrISS) - Dependable Systems, Services & Technologies (DESSERT) Workshop" which was prepared by prof. Vyacheslav Kharchenko within the framework of the conference.

In the closing words of this introduction we would like to emphasize the role of all reviewers who took part in the evaluation process this year and whose valuable input helped to refine the contents of this volume. Our thanks go to Ali Al-Dahoud, Andrzej Białas, Dmitriy Bui, Frank Coolen, Manuel Gil Perez, Zbigniew Huzar, Vyacheslav Kharchenko, Dariusz Król, Michał Lower, Jan Magott, István Majzik, Jacek Mazurkiewicz, Marek Młyńczak, Tomasz Nowakowski, Oksana Pomorova, Mirosław Siergiejczyk, Ruslan Smeliansky, Janusz Sosnowski, Jarosław Sugier, Victor Toporkov, Tomasz Walkowiak, Marina Yashina, Irina Yatskiv, Wojciech Zamojski, Wlodek Zuberek.

DepCoS-RELCOMEX 2015 is the 10th conference so we feel obliged and honored to express our sincerest gratitude to all the authors (and there was over 500 of them), participants and Programme Committee members of all the ten events. Their work and dedication helped to establish scientific position of DepCoS-RELCOMEX series and contributed to progress in both basic research and applied sciences dealing with dependability challenges in contemporary systems and networks.

Wojciech Zamojski
Jacek Mazurkiewicz
Jarosław Sugier
Tomasz Walkowiak
Janusz Kacprzyk
(Editors)

Tenth International Conference on Dependability and Complex Systems DepCoS-RELCOMEX

Organized by
Department of Computer Engineering,
Wrocław University of Technology
under the auspices of Prof. Tadeusz Więckowski, Rector

Brunów Palace, Poland, June 29 – July 3, 2015

Program Committee

Wojciech Zamojski (Chairman)	Wrocław University of Technology, Poland
Salem Abdel-Badeeh	Ain Shams University Abbasia, Cairo, Egypt
Ali Al-Dahoud	Al-Zaytoonah University, Amman, Jordan
Artem Adzhemov	Technical University of Communications and Informatics, Moscow, Russia
Włodzimierz M. Barański	Wrocław University of Technology, Poland
Andrzej Białas	Institute of Innovative Technologies EMAG, Katowice, Poland
Ilona Bluemke	Warsaw University of Technology, Poland
Dariusz Caban	Wrocław University of Technology, Poland
Frank Coolen	Durham University, UK
Mieczysław Drabowski	Cracow University of Technology, Poland
Francesco Flammini	University of Naples "Federico II", Napoli, Italy
Manuel Gill Perez	University of Murcia, Spain
Zbigniew Huzar	Wrocław University of Technology, Poland
Igor Kabashkin	Transport and Telecommunication Institute, Riga, Latvia
Janusz Kacprzyk	Polish Academy of Sciences, Warsaw, Poland
Andrzej Kasprzak	Wrocław University of Technology, Poland
Vyacheslav S. Kharchenko	National Aerospace University "KhAI", Kharkov, Ukraine
Mieczysław M. Kokar	Northeastern University, Boston, USA
Krzysztof Kołowrocki	Gdynia Maritime University, Poland
Leszek Kotulski	AGH University of Science and Technology, Krakow, Poland
Henryk Krawczyk	Gdansk University of Technology, Poland
Alexey Lastovetsky	University College Dublin, Ireland
Marek Litwin	ITS Polska, Warsaw, Poland
Jan Magott	Wrocław University of Technology, Poland

Istvan Majzik	Budapest University of Technology and Economics, Hungary
Jacek Mazurkiewicz	Wrocław University of Technology, Poland
Yiannis Papadopoulos	Hull University, UK
Oksana Pomorova	Khmelnitsky National University, Ukraine
Ewaryst Rafajłowicz	Wrocław University of Technology, Poland
Nikolay Rogalev	Moscow Power Engineering Institute (Technical University), Russia
Krzysztof Sacha	Warsaw University of Technology, Poland
Rafał Scherer	Częstochowa University of Technology, Poland
Mirosław Siergiejczyk	Warsaw University of Technology, Poland
Ruslan Smeliansky	Moscow State University, Russia
Czesław Smutnicki	Wrocław University of Technology, Poland
Janusz Sosnowski	Warsaw University of Technology, Poland
Jarosław Sugier	Wrocław University of Technology, Poland
Ryszard Tadeusiewicz	AGH University of Science and Technology, Krakow, Poland
Victor Toporkov	Moscow Power Engineering Institute (Technical University), Russia
Tomasz Walkowiak	Wrocław University of Technology, Poland
Max Walter	Siemens, Germany
Bernd E. Wolfinger	University of Hamburg, Germany
Marina Yashina	Moscow Technical University of Communication and Informatics, Russia
Irina Yatskiv	Transport and Telecommunication Institute, Riga, Latvia
Jan Zarzycki	Wrocław University of Technology, Poland
Włodzimierz Zuberek	Memorial University, St.John's, Canada

Organizing Committee

Wojciech Zamojski (Chairman)
Włodzimierz M. Barański
Monika Bobnis
Jacek Mazurkiewicz
Jarosław Sugier
Tomasz Walkowiak

The 5th CrISS-DESSERT Workshop

Critical Infrastructure Security and Safety (CrISS) –
Dependable Systems, Services & Technologies (DESSERT)

The CrISS-DESSERT Workshop evolved from the conference Dependable Systems, Services & Technologies DESSERT 2006–2014. The 4th CrISS-DESSERT was held in Kiev, May 17, 2014. The CrISS Workshop examines modelling, development, integration, verification, diagnostics and maintenance of computer and communications systems and infrastructures for safety-, mission-, and business-critical applications.

Workshop Committee

Vyacheslav Kharchenko (Chairman)	National Aerospace University KhAI, Ukraine
Jüri Vain (Co-chairman)	Tallinn University of Technology, Estonia
Todor Tagarev (Co-chairman)	Institute of Information and Communication Technologies, Bulgaria
De-Jiu Chen (Co-chairman)	KTH University, Sweden
Iosif Androulidakis	Ioannina University, Greece
Eugene Brezhnev	Centre for Safety Infrastructure-Oriented Research and Analysis, Ukraine
Dmitriy Bui	Taras Shevchenko National University of Kyiv, Ukraine
Vitaly Levashenko	Žilina University, Slovakia
Volodymyr Mokhor	Pukhov Institute for Modeling in Energy Engineering, Ukraine
Simin Nadjm-Tehran	Linköping University, Sweden
Harald Richter	Technical University Clausthal, Germany
Serhiy Shcherbovskikh	Lviv Polytechnic National University, Ukraine
Vladimir Sklyar	RPC Radiy, Ukraine
Sergiy Vilkomir	East Carolina University, USA
Elena Zaitseva	Žilina University, Slovakia

Papers

Contents

Network Anomaly Detection Based on Statistical Models with Long-Memory Dependence

Tomasz Andrysiak and Łukasz Saganowski

Institute of Telecommunications,
Faculty of Telecommunications and Electrical Engineering,
University of Technology and Life Sciences (UTP),
ul. Kaliskiego 7, 85-789 Bydgoszcz, Poland
{andrys,luksag}@utp.edu.pl

Abstract. The paper presents an attempt to anomaly detection in network traffic using statistical models with long memory. Tests with the GPH estimator were used to check if the analysed time series have the long-memory property. The tests were performed for three statistical models known as ARFIMA, FIGARCH and HAR-RV. Optimal selection of model parameters was based on a compromise between the model's coherence and the size of the estimation error.

Keywords: Anomaly detection, long-memory dependence, statistical models.

1   Introduction

Ensuring an adequate level of cyber security is currently the most important challenge for information society. Dynamic development of technologies is gradually opening new means of communication; however, it does also create great threats. More and more often appearing improved penetration and intrusion techniques enforce continuous development of network security systems. They must be able to detect simple attacks such as DoS (Denial of Service) or DDoS (Distributed Denial of Service), but also intelligent network worms up to hybrid attacks, which are a combination of diverse destruction methods.

One of possible solutions to the presented problem is implementation of Anomaly Detection Systems ADS. They are currently used as one of the main mechanisms of supervision over computer network security. Their work consists in monitoring and detecting attacks directed towards resources of computer systems on the basis of abnormal behavior reflected in network traffic parameters.

Undoubtedly, an advantage of the mentioned solutions is their ability to recognize possible unknown attacks.

Their strength lies in the fact that they do not depend on knowledge a priori - attack signatures, but on what does not correspond to given norms, i.e. profiles of the analyzed network traffic.

© Springer International Publishing Switzerland 2015

W. Zamojski et al. (eds.), *Theory and Engineering of Complex Systems and Dependability*,

Advances in Intelligent Systems and Computing 365, DOI: 10.1007/978-3-319-19216-1_1

It is worth noticing that network traffic is represented by time series of the analyzed parameters. Statistical models based on principles of the automatic regression and the moving average are a natural way of the description of such series with reference to differently realized variation of this data.

Currently, the most commonly developed methods of anomaly detection are those based on statistical models describing the analyzed network traffic as time series. The most often employed models are autoregressive ARMA, ARIMA or ARFIMA. They allow estimating the characteristics of the analyzed network traffic [13,16]. In literature, there can also be found signal processing methods based on DWT (Discrete Wavelet Transform [17]) subband analysis, data mining methods [18], hybrid methods that connect pre-processing elements (signal decomposition), and next estimate statistical parameters on so processed signal.

In this article we present the use of statistical estimation of ARFIMA, FIGARCH and HAR-RV models for the analyzed time series describing the given network traffic. Anomaly detection is realized on the basis of estimated models' parameters and comparative analysis of network traffic profiles.

This paper is organized as follows: after the introduction, in section 2 we present the GPH estimator for a test of long-memory dependence. In Section 3 different statistical models for data traffic prediction are described in details. Then, in Section 4 the Anomaly Detection System based on ARFIMA, FIGARH and HAR-RV model estimation is shown. Experimental results and conclusion are given thereafter.

2   Definition and Test Method of Long-Memory

The long-memory matter, also known as property of long-term dependence, is visible in autocorrelation of observations creating time series. Moreover, it is autocorrelation of high order. It means that there is a dependence between observations even if they are distant in time. The phenomenon of long-memory was invented by a British hydrologist Hurst [11].

The properties of the time series are characterized most of all by autocorrelation function (ACF) and partial autocorrelation function (PACF). In case of existence of long-memory property the autocorrelation function ACF falls in hyperbolic pace, i.e. slowly.

The time series holding the property of long-memory have in its spectral domain disintegration with low frequency. Short-memory time series show crucial autocorrelations only of low order, which means the observations that are separated even by a short period of time are not correlated. They can be easily noticed due to the fact that on the one hand, in the time domain ACF they disappear quickly; and on the other hand, in the spectral domain there are disintegrations with high frequency.

2.1   Long-Memory Dependence

It is said that the scholastic process has a long memory with d parameter if its spectral density function $f(\lambda)$ satisfies the condition

$$f(\lambda) \sim c\lambda^{-2d}, \qquad \text{when } \lambda \to 0^+, \tag{1}$$

where c is constant, and $\sim$ symbol means the relation of left and right side is heading towards one. When the process fulfils such a condition and when $d > 0$, then its autocorrelation function disappears in hyperbolic manner [2,3,9,10], i.e.

$$\rho_k \sim c_\rho k^{2d-1}, \qquad \text{when } k \to \infty. \tag{2}$$

As it has already been mentioned, d parameter describes the memory of the process. When $d > 0$, the spectral density function is unlimited surrounding 0. In such a case it is said that the process has a long memory. When $d = 0$, spectral density is limited in 0, and the process is described as short-memory. However, if $d < 0$, then spectral density equals 0 and the process shows negative memory – it is called not persistent. The most popular class of models satisfying condition (1) are ARFIMA, i.e. Autoregressive Fractional Integrated Moving Average [9].

2.2   GPH Test of Long-Memory

The class of estimators of long-memory parameter d that is often applied is semi parametrical estimators, which use approximation of spectral density surrounding 0, which is the result of condition (1). Among them, the most common is estimator based on regression of logperiodogram, proposed by Geweke and Porter-Hudak [8], and it was vastly analyzed by Robinson [12]. Semi parametrical estimators use information included in periodogram, calculated only for very low frequencies. This makes the estimators insensitive to all kinds of short term disorders, as in the case of presented parametrical estimator.

The GPH estimation procedure is a two-step procedure, which begins with the estimation of d and is based on the following regression equation:

$$log\left[I_y(\omega_j)\right] = \beta_0 - dlog\left\{2\sin\left(\tfrac{\omega_j}{2}\right)\right\}^2 + \vartheta_j, \; j = 1,2,\dots,g(n), \tag{3}$$

where $g(n)$ is a function of the sample size n where $g(n) = n^\alpha$ with $0 < \alpha < 1$ and $\beta_0 = logf_U(0) + logf_U(\omega_j)/f_U(0)$, $\vartheta_j = log I_y(\omega_j)/f_y(\omega_j) - \psi(1)$, $\psi(\cdot)$ being the digamma function, i.e. $\psi(x) = dlog\Gamma(x)$ and $\omega_j = 2\pi j/T$ represents the $m = \sqrt{T}$ Fourier Frequencies, $I_y(\omega_j)$ denotes the sample periodogram defined as

$$I_y(\omega_j) = \tfrac{1}{2\pi T}|\Sigma_{t=1}^T y_t\, e^{-\omega_j t}|^2. \tag{4}$$

The GPH estimator is given by

$$d_{GPH} = -\frac{\Sigma_{j=1}^{g(n)}(v_j-\hat{v})log I_y(\omega_j)}{\Sigma_{j=1}^{g(n)}(v_j-\hat{v})^2}, \tag{5}$$

where $v_j = log(2\sin(\omega_j/2))^2$, $g(n)$ being the band within the regression equation. The variance of the GPH estimator is

$$var(d_{GPH}) = \frac{\pi^2}{6\sum_{j=1}^{g(n)}(v_j - \hat{v})^2}. \tag{6}$$

Gewke and Porter-Hudak proved the asymptotic normality of the semiparametric estimator in (5) when $d < 0$ and suggested taking $g(n) = n^\alpha$, $0 < \alpha < 1$. They also suggest that the power of T has to be within $(0.5, 0.6)$ and for the null hypotheses of no long-memory process, the slope of regression d equals zero and the usual t-statistics can be employed to perform the test. A detailed description of the presented algorithm can be found in the work of [4,8].

3   Statistical Models of Long-Memory

An interesting approach towards properties of long-memory time series was applying the autoregression with moving averaging in the process of fractional diversification. As a result, ARFIMA model (Fractional Differenced Noise and Auto Regressive Moving Average) was obtained and implemented by Grange, Joyeux and Hosking [9,10]. ARFIMA is a generalization of ARMA and ARIMA models.

Another approach to describing time series was taking into account the dependence of the conditional variance of the process on its previous values with the use of ARCH model (Autoregressive Conditional Heteroskedastic Model) introduced by Engel [7]. Generalization of this approach was FIGARCH model (Fractionally Integrated GARCH) introduced by Baillie, Bollerslev and Mikkelsen [1]. Its function of autocorrelation of squares of rests of the model is decreasing in the hyperbolical way. Therefore, for small series, the autocorrelation function decreases more quickly than for exponential case. For high series, however, it decreases very slowly. Such a behavior of autocorrelation function enables naming the FIGARCH the model of long-memory in the context of autocorrelation functions of squares of the rests of the models.

A different approach describing the analyzed time series is modelling of their variations within diverse time horizons. Corsi [5] proposes HAR-RV model (Heterogeneous Autoregressive Model of Realized Volatility) as a simple autoregressive model with highlighted additive components of short, medium and long term volatilities. The author proposes that HAR-RV model should be described as approximate model with long-memory, due to the fact that it allows taking into account the observed in data long-memory variation.

3.1   ARFIMA Model

The Autoregressive Fractional Integrated Moving Average model called ARFIMA(p, d, q) is a combination of Fractional Differenced Noise and Auto

Regressive Moving Average which is proposed by Grange, Joyeux and Hosking, in order to analyze the long-memory property [9, 10].

The ARFIMA(p, d, q) model for time series y_t is written as:

$$\Phi(L)(1 - L)^d y_t = \Theta(L)\epsilon_t, \qquad t = 1, 2, \ldots \Omega, \tag{7}$$

where y_t is the time series, $\epsilon_t \sim (0, \sigma^2)$ is the white noise process with zero mean and variance σ^2, $\Phi(L) = 1 - \phi_1 L - \phi_2 L^2 - \cdots - \phi_p L^p$ is the autoregressive polynomial and ` is the moving average polynomial, L is the backward shift operator and $(1 - L)^d$ is the fractional differencing operator given by the following binomial expansion:

$$(1 - L)^d = \sum_{k=0}^{\infty} \binom{d}{k} (-1)^k L^k \tag{8}$$

and

$$\binom{d}{k}(-1)^k = \frac{\Gamma(d+1)(-1)^k}{\Gamma(d-k+1)\Gamma(k+1)} = \frac{\Gamma(-d+k)}{\Gamma(-d)\Gamma(k+1)}. \tag{9}$$

$\Gamma(*)$ denotes the gamma function and d is the number of differences required to give o stationary series and $(1 - L)^d$ is the dth power of the differencing operator. When $d \in (-0.5, 0.5)$, the ARFIMA(p, d, q) process is stationary, and if $d \in (0, 0.5)$ the process presents long-memory behavior [9].

In this paper we assume that $\{y_t\}$ is a linear process without a deterministic term. We now define $U_t = (1 - L)y_t$, so that $\{Y_t\}$ is an ARMA(p, q) process. The process defined in (1) is stationary and invertible (see [10]) and its spectral density function, $f_y(\omega)$, is given by

$$f_y(\omega) = f_U(\omega) \left(2\sin\left(\frac{\omega}{2}\right)\right)^{-2d}, \qquad \omega \in [-\pi, \pi], \tag{10}$$

where $f_U(\omega)$ is the spectral density function of the process $\{Y_t\}$.

3.2 FIGARCH Model

The model enabling description of long-memory in variance series is FIGARCH (p, d, q) (Fractionally Integrated GARCH) introduced by Baillie, Bollerslev and Mikkelsen [1]. The FIGARCH (p, d, q) model for time series y_t can be written as:

$$y_t = \mu + \epsilon_t, \qquad t = 1, 2, \ldots \Omega, \tag{11}$$

$$\epsilon_t = z_t \sqrt{h_t}, \qquad \epsilon_t | \Theta_{t-1} \sim N(0, h_t), \tag{12}$$

$$h_t = \alpha_0 + \beta(L)h_t + [1 - \beta(L) - [1 - \phi(L)](1 - L)^d]\epsilon_t^2, \tag{13}$$

where z_t is a zero-mean and unit variance process, h_t is a positive time dependent conditional variance defined as $h_t = E(\epsilon_t^2 | \Theta_{t-1})$ and Θ_{t-1} is the information set up to time $t - 1$.

The FIGARCH(p, d, q) model of the conditional variance can be motivated as ARFIMA model applied to the squared innovations,

$$\left(1 - \varphi(L)\right)(1 - L)^d \epsilon_t^2 = \alpha + \left(1 - \beta(L)\right)v_t, \quad v_t = \epsilon_t^2 - h_t, \tag{14}$$

where $\varphi(L) = \varphi_1 L - \varphi_2 L^2 - \cdots - \varphi_p L^p$ and $\beta(L) = \beta_1 L + \beta_2 L^2 + \cdots + \beta_q L^q$, $\left(1 - \varphi(L)\right)$ and $\left(1 - \beta(L)\right)$ have all their roots outside the unit circle, L is the lag operator and $0 < d < 1$ is the fractional integration parameter.

If $d = 0$, then FIGARCH model is reduced to GARCH; for $d = 1$ though, it becomes IGARCH model. However, FIGARCH model does not always reduce to GARCH model. If GARCH process is stationary in broader sense, then the influence of current variance on its forecasting values decreases to zero in exponential pace. In IGARCH case the current variance has indefinite influence on the forecast of conditional variance. For FIGARCH process the mentioned influence decreases to zero far more slowly than in GARCH process, i.e. according to the hyperbolic function [1].

3.3   HAR-RV Model

The HAR-RV model (Heterogeneous Autoregressive Model of Realized Volatility) is a simple autoregressive type with highlighted diverse components of volatility realized within few time horizons. In the equation for non-observable partial volatility, on every level of cascade, we enumerate autoregressive and hierarchical component (in relation to propagating asymmetric volatility) stated as expected value of variation within a longer time horizon. Corsi [5] formulated the HAR-RV (Heterogeneous Autoregressive Model of Realized Volatility) model's equation as follows

$$RV_{t+\Delta}^{\Delta} = c + \gamma^d RV_t^{(d)} + \gamma^w RV_t^{(w)} + \gamma^m RV_t^{(m)} + \varepsilon_{t+\Delta}, \tag{15}$$

where Δ is time interval – most often a day (d), a week (w) or a month (m); RV_t^{Δ} is aggregated in the given time period realized volatility $\epsilon_{t+\Delta} \sim N(0, \sigma_t)$.

In order to make realized volatilities RV comparable in analyzed time intervals, Corsi suggests averaging the values. The most common way of model's parameters estimation is the method of least squares. However, the estimated standard errors should be resistant to autocorrelation and heteroskedasticity of the random component.

4   Experimental Result

For experimental results we used traffic from network configuration which we proposed in [13]. Our test network contains SNORT [13, 14] IDS with anomaly detection preprocessor. We used the same subset of network traffic features (see Table 1) as in [13]. SNORT is used as sensor for collecting this traffic features. For evaluating usability of proposed statistical models for anomaly detection we simulated real world attacks by using Kali Linux [15] distribution which consist of many tools to perform attacks on every layer of TCP/IP stack. We simulated attacks that belongs to subsequent groups: application specific DDos, various port scanning, DoS, DDoS, Syn Flooding, packet fragmentation, spoofing, reverse shell and others.

Table 1. Traffic feature description

Feature	Traffic feature description	Feature	Traffic feature description
F1	number of TCP packets	F14	out TCP packets (port 80)
F2	in TCP packets	F15	in TCP packets (port 80)
F3	out TCP packets	F16	out UDP datagrams (port 53)
F4	number of TCP packets in LAN	F17	in UDP datagrams (port 53)
F5	number of UDP datagrams	F18	out IP traffic [kB/s]
F6	in UDP datagrams	F19	in IP traffic [kB/s]
F7	out UDP datagrams	F20	out TCP traffic (port 80) [kB/s]
F8	number of UDP datagrams in LAN	F21	in TCP traffic (port 80) [kB/s]
F9	number of ICMP packets	F22	out UDP traffic [kB/s]
F10	out ICMP packets	F23	in UDP traffic [kB/s]
F11	in ICMP packets	F24	out UDP traffic (port 53) [kB/s]
F12	number of ICMP packets in LAN	F25	in UDP traffic (port 53) [kB/s]
F13	number of TCP packets with SYN and ACK flags	F26	TCP traffic (port 4444)

Table 2. GPH long memory test for time series representing network traffic

Feature	d for $(\alpha = 0.5)$	d for $(\alpha = 0.4)$	d for $(\alpha = 0.6)$
F1	0.34	0.56	0.29
F2	0.41	0.43	0.46
F3	0.12	0.15	0.08
F4	0.30	0.25	0.42
F5	0.38	0.51	0.30
F6	0.15	0.21	0.11
F7	0.43	0.25	0.42
F8	0.11	0.19	0.09
F9	0.40	0.41	0.49
F10	0.35	0.55	0.32
F11	0.13	0.17	0.08
F12	0.37	0.59	0.32
F13	0.26	0.15	0.12

Table 3. Detection rate DR[%] for a given network traffic feature

Feature	HAR-RV	FIGARH	ARFIMA	Feature	HAR-RV	FIGARH	ARFIMA
F1	8.28	5.40	6.26	F14	14.28	10.20	12.24
F2	14.28	10.20	12.24	F15	14.28	10.20	12.24
F3	14.28	10.20	12.24	F16	0.00	0.00	0.00
F4	14.28	10.20	12.24	F17	8.28	5.40	6.26
F5	14.28	10.20	12.24	F18	14.28	10.20	12.24
F6	0.00	0.00	0.00	F19	14.28	10.20	12.24
F7	0.00	0.00	0.00	F20	8.28	5.40	6.26
F8	40.45	32.20	35.64	F21	14.28	10.20	12.24
F9	98.60	90.42	96.52	F22	0.00	0.00	0.00
F10	98.24	90.24	95.45	F23	0.00	0.00	0.00
F11	0.00	0.00	0.00	F24	0.00	0.00	0.00
F12	90.24	80.24	82.24	F25	0.00	0.00	0.00
F13	14.28	10.20	12.24	F26	94.24	78.00	80.00

Table 4. False positive FP[%] for a given network traffic feature

Feature	HAR-RV	FIGARH	ARFIMA	Feature	HAR-RV	FIGARH	ARFIMA
F1	3.24	5.24	4.22	F14	2.42	4.45	3.24
F2	3.22	5.45	4.12	F15	2.24	4.38	3.32
F3	3.24	5.24	4.15	F16	0.02	1.24	0.02
F4	3.42	5.22	4.11	F17	0.24	1.82	0.39
F5	2.54	4.28	3.54	F18	2.54	4.55	3.82
F6	1.84	3.34	2.23	F19	2.20	4.62	3.26
F7	4.64	6.75	5.98	F20	3.55	5.34	4.55
F8	3.48	5.24	4.15	F21	2.62	4.22	3.11
F9	4.24	6.22	5.05	F22	1.24	2.46	1.60
F10	0.34	1.46	0.48	F23	2.45	4.44	3.42
F11	1.52	3.52	2.56	F24	0.00	0.00	0.00
F12	0.05	1.04	0.05	F25	0.02	0.45	0.02
F13	3.54	5.46	4.14	F26	0.02	0.45	0.02

Network traffic is represented as a time series for a given traffic feature. In order to calculate models described in subsections 3.1-3.3 we have to check if time series have long memory properties. We used GHP long-memory test (see subsection 2.2) for feature F1 and F2 (see Table 1). Calculated d parameter showed that we can use models with long memory in order to describe network traffic behavior.

For anomaly detection we compare model parameters calculated for traffic without anomalies (we assume that there is no anomalies in traffic during calculation of model parameters). For ARFIMA based models we calculated prediction interval (30 samples horizon [16]) in order to detect suspicious traffic behavior. In Table 3 and Table 4 we compared DR and FP achieved for HAR-RV, FIGARH and ARFIMA statistical models. HAR-RV gives us the best results in case of DR and FP. DR and FP values depends on given traffic feature. Some traffic features have low values because simulated attacks in our experiment haven't got impact on this feature (e.g. F1, F6, F7).

5   Conclusion

Protection of infrastructures of teleinformatic systems against novel and unknown attacks is currently an intensively studied and developed field. One of possible solutions to the problem is detection and classification of abnormal behaviors reflected in the analyzed network traffic. An advantage of such an approach is no necessity to predetermine and memorize benchmarks of those behaviors. Therefore, in the decision making process, it is only required to determine what is and what is not an abnormal behavior in the network traffic in order to detect a potential unknown attack.

In this article there are presented statistical long-memory models – ARFIMA, FIGARCH and HAR-RV, which were used to estimate behavior of the analyzed network traffic. Time series reflecting the network traffic parameters were also surveyed. In result, with the use of statistical test – GPH estimators, it was stated that time series are characterized by long-memory effect. Parameters estimations and identification of rank of the models are realized as a compromise between the model's coherence and size of its estimation error. As a result of implementation of the described models, satisfactory statistical estimations were obtained within the analyzed signals of the network traffic.

The process of anomaly detection was based on comparison of parameters of a normal behavior estimated with the use of mentioned models and parameters of variation of the real analyzed network traffic. The outcomes explicitly indicate that anomalies included in the signal of a network traffic can be effectively detected by the proposed solutions.

References

1. Baillie, R., Bollerslev, T., Mikkelsen, H.: Fractionally Integrated Generalized Autoregressive Conditional Heteroskedasticity. Journal of Econometrics 74, 3–30 (1996)
2. Beran, J.A.: Statistics for Long-Memory Processes. Chapman and Hall (1994)
3. Box, G.E., Jenkins, M.G.: Time series analysis forecasting and control, 2nd edn. Holden-Day, San Francisco (1976)
4. Box, G., Jenkins, G., Reinsel, G.: Time series analysis. Holden-day, San Francisco (1970)
5. Corsi, F.: A simple approximate long-memory model of realized volatility. Journal of Financial Econometrics 7, 174–196 (2009)
6. Crato, N., Ray, B.K.: Model Selection and Forecasting for Long-range Dependent Processes. Journal of Forecasting 15, 107–125 (1996)
7. Engle, R.: Autoregressive conditional heteroskedasticity with estimates of the variance of UK inflation. Econometrica 50, 987–1008 (1982)
8. Geweke, J., Porter-Hudak, S.: The Estimation and Application of Long Memory Time Series Models. Journal of Time series Analysis (4), 221–238 (1983)

 9. Granger, C.W.J., Joyeux, R.: An introduction to long-memory time series models and fractional differencing. Journal of Time Series Analysis 1, 15–29 (1980)
10. Hosking, J.: Fractional differencing. Biometrika (68), 165–176 (1981)
11. Hurst, H.R.: Long-term storage capacity of reservoirs. Transactions of the American Society of Civil Engineers 1, 519–543 (1951)
12. Robinson, P.M.: Log-periodogram regression of time series with long range dependence. Annals of Statistics 23, 1048–1072 (1995)
13. Saganowski, Ł., Goncerzewicz, M., Andrysiak, T.: Anomaly Detection Preprocessor for SNORT IDS System. In: Choraś, R.S. (ed.) Image Processing and Communications Challenges 4. AISC, vol. 184, pp. 225–232. Springer, Heidelberg (2013)
14. SNORT - Intrusion Detection System, https://www.snort.org/
15. Kali Linux, https://www.kali.org/
16. Andrysiak, T., Saganowski, Ł., Choraś, M., Kozik, R.: Network Traffic Prediction and Anomaly Detection Based on ARFIMA Model. In: de la Puerta, J.G., et al. (eds.) International Joint Conference SOCO'14-CISIS'14-ICEUTE'14. AISC, vol. 299, pp. 545–554. Springer, Heidelberg (2014)
17. Wei, L., Ghorbani, A.: Network Anomaly Detection Based on Wavelet Analysis. EURASIP Journal on Advances in Signal Processing 2009 (2009), doi:10.1155/2009/837601
18. Xie, M., Hu, J., Han, S., Chen, H.-H.: Scalable Hypergrid k-NN-Based Online Anomaly Detection in Wireless Sensor Networks. IEEE Transactions on Parallel & Distributed Systems 24(8), 1661–1670 (2013), doi:10.1109/TPDS.2012.261

Critical Infrastructures Risk Manager – The Basic Requirements Elaboration

Andrzej Bialas

Institute of Innovative Technologies EMAG, 40-189 Katowice, Leopolda 31, Poland
a.bialas@emag.pl

Abstract. The paper concerns the risk assessment and management methodology in critical infrastructures. At the beginning a review is performed of the state of the art, regulations, best practices, EU projects, and other relevant documents. On this basis a set of the most preferable features of a CI risk management tool is identified. These features allow to specify basic requirements for the risk management tool. As the core of the solution is the bow-tie model. A risk register is proposed as an inventory of the hazardous events, along with other data structures for hazards/threats, vulnerabilities, consequences, and barriers. Risk factors and results measures, i.e. likelihood and consequences measures as well as a risk matrix are discussed. Next, a new concept is proposed how to integrate different bow-tie models through internal and external dependencies. These requirements can be implemented on the available software platform for further experiments and validation.

Keywords: critical infrastructure, risk assessment, interdependencies, bow-tie model, tool requirements.

1 Introduction

Effective functioning of today's societies, especially those of well-developed countries, is based on critical infrastructures (CI). These CIs consist of large scale infrastructures whose degradation, disruption or destruction would have a serious impact on health, safety, security or well-being of citizens or effective functioning of governments and/or economies. Typical examples of such infrastructures are energy-, oil-, gas-, finance-, transport-, telecommunications- and health sectors. Critical infrastructures encompass different kinds of assets, services, systems, and networks – very important for citizens, society or even groups of societies. Providing services for the society, CIs support the right relationships between governments and citizens.

All CIs widely use and strongly depend on information and communication technologies (ICT). ICT-supported information processes that are critical infrastructures themselves, or are critical for the operation of other CIs, are called critical information infrastructures (CII). As the functioning of today's societies is based on information, CI protection is a new challenge and has an international dimension.

Infrastructure owners, manufacturers, users, operators, R&D institutions, governments, and regulatory authorities have to keep the performance of CIs in case of

© Springer International Publishing Switzerland 2015
W. Zamojski et al. (eds.), *Theory and Engineering of Complex Systems and Dependability*,
Advances in Intelligent Systems and Computing 365, DOI: 10.1007/978-3-319-19216-1_2

failures, attacks or accidents and minimize the recovery time and damage These institutions have certain programmes and activities which are understood as critical infrastructure protection (CIP). CIs are characterized by complexity, heterogeneity, multidimensional interdependencies, significant risk dealing with natural disasters and catastrophes, technical disasters and failures, espionage, international crime, physical and cyber terrorism, etc. Therefore they require a new, holistic approach to their protection, considering not only information security methods and techniques but also achievements of the safety domain. Co-operation on the international level is needed too. While the key challenge is to improve dependability and survivability of CIs.

Critical infrastructures and their protection programmes can be considered from different perspectives: sectors and its enterprises (organizations), countries, global levels (e.g. EU level).

The right protection programmes can be defined on the basis of risk whose source, character and level should be identified, and the right countermeasures applied. The risk management methodology is the key issue, but it has a specific character with respect to critical infrastructures and applying it properly remains a challenge. The specific CIs issues encompass reciprocal dependencies between CIs, different abstract levels applied to manage them, new effects, such as common cause failures, cascading, escalating effects, etc. The risk management methodology and tools are a subject of current R&D on the national and international levels, including EU level.

The Council Directive [1] refines the critical infrastructures protection issue to the needs of the EU and its member states, especially in CI identification, risk analysis and management programmes. The term ECI (European critical infrastructure) introduced in the Directive [1] means critical infrastructure located in member states, whose disruption or destruction would have a significant impact on at least two member states. The significance of this impact is assessed in a special way with the use of cross-cutting criteria which include three items:

- casualties criterion (the potential number of fatalities or injuries),
- economic effects criterion (the significance of economic loss and/or degradation of products or services, including potential environmental effects),
- public effects criterion (the impact on public confidence, physical suffering and disruption of daily life, including the loss of essential services).

Two ECI sectors, energy and transport, are considered. In the energy sector the three following subsectors are distinguished: electricity (infrastructures and facilities for generation and transmission of electricity), oil (oil production, refining, treatment, storage and transmission by pipelines) and gas (gas production, refining, treatment, storage and transmission by pipelines, LNG terminals). In the transport sector five subsectors are distinguished: road transport, rail transport, air transport, inland waterways transport, ocean and short-sea shipping and ports.

The EU document [2] presents a revised and more practical implementation of the European Programme for Critical Infrastructure Protection (EPCIP). The EC Joint Research Centre (JRC) report [3] features a survey on the existing risk management methodologies on the EU and global level, identifies gaps and prepares the ground for

R&D in this field. Dozens of the EU or worldwide CIP R&D projects have been finalized or are running (FP6, FP7, Horizon2020, CIPS). Most of them concern risk management methodologies and their supporting tools.

The aim of this paper is to specify the requirements for a risk management tool which can be applied in critical infrastructure protection. The paper starts with CI specific issues related to risk management, presents most representative methodologies and tools analyzing their pros and cons, specifies and assesses the requirements for the developed tool.

2 Critical Infrastructures Protection Issues

A given CI can be considered a sociotechnical system, usually heterogeneous, distributed and adaptive. Generally, such systems are composed of processes and assets, including procedures, personnel, facilities, equipment, materials, tools, software. The sociotechnical systems take into account not only hardware, software and liveware, but also environmental, management and organizational aspects. These elements constitute certain symbiosis, as they are harmonized and work together meeting the common objectives, providing important services or products for the society.

These particular critical infrastructures (systems), e.g. electricity, rail transport, gas, port, telecommunications, collaborating with each other, constitute a more complex structure to be analyzed, called a complex system or a system-of-systems (SoS).

Due to the critical infrastructures co-operation, there are many mutual dependencies, called interdependencies, between them. The interdependency is a bidirectional relationship between two infrastructures (systems) through which the state of each infrastructure influences or is correlated to the state of the other [4]. Four types of interdependencies are identified [5]:

- physical interdependency – related to the physical coupling between inputs and outputs, e.g. goods produced/processed by one infrastructure are vital for others;
- cyber interdependency – the state of the infrastructure depends on the information transmitted through the information infrastructure;
- geographical interdependency – one or several elements of infrastructures are in close proximity so that the event in the infrastructure (e.g. fire) may disturb the infrastructures in its neighbourhood;
- logical interdependency – the state of each infrastructure depends on the state of the other by a mechanism that has not a physical, cyber, or geographic character.

The identification of interdependencies helps to assess cascading effects, which are specific for critical infrastructures. A cascading effect is [6] a sequence of component failures: the first failure shifts its load to one or more nearby components which fail and, in turn, shift their load to other components, and so on.

An escalating failure is when an existing disruption in one infrastructure causes an independent disruption of a second infrastructure [4]. The effects of hazardous events may escalate outside the area where they occur and exacerbate the consequences of a given event (generally in the form of increasing the severity or the time for recovery or restoration of the second failure).

The CIs failures can be causally linked. Sometimes the failures have a common cause (failures implied by a single shared cause and coupling to other systems mechanisms) and occur almost concurrently. The common cause failure occurs when two or more CIs are disrupted at the same time. The resilience of the CIs should be understood as an ability of a system to react and recover from unanticipated disturbances and events.CI-specific issues listed in this section are a real challenge during the risk management process.

3   Survey on Representative Methodologies and Tools for Risk Management

The report of the European Institute for the Protection and Security of the Citizen (EC Joint Research Centre) [3] presents a structured review of 21 European and worldwide risk assessment methodologies and identifies their gaps. The report [7] presents a desktop study of 11 risk assessment methodologies related to the energy sector. The ENISA website [8] includes an inventory of risk management/assessment methods, mostly ICT-focused. The [9] characterizes about 30 risk assessment methods for different applications. The book [5] presents a general approach to risk assessment and management in CIs, shows risk case studies for typical infrastructures, and (in Appendix C) compares the features of about 22 commonly used risk analysis methods.

The JRC report [3] points at one of the challenges for risk assessment methodologies, i.e. to develop a framework that would capture interdependencies across infrastructures, cross-sectors and cross-borders with a focus on resilience.

Most risk assessment methodologies cope with the cascading and escalation effects up to a certain extent. Two main approaches have been identified: aggregated impact and scoring [3]. The impact of infrastructure disruption is usually expressed in terms of economic losses. The scoring approach is useful for ordering mitigation measures, e.g. give priority to a certain sector over another one as the latter might bring more severe effects. Some methods are focused on terrorist attacks.

Some methods allow for risk analysis and assessment only, some are more comprehensive and implement a full risk management framework. Not all methods are supported by tools, but sometimes many tools implement the given method. A few representative examples are mentioned as follows.

The Risk and Vulnerability Analysis (RVA) method for critical infrastructures was improved in the Norwegian project DECRIS (Risk and Decision Systems for Critical Infrastructures) [10] – the RVA-DECRIS approach was developed. The RVA is the modified version of PHA (Preliminary Hazard Analysis) [6].

The RVA-DECRIS approach encompasses four main steps:

1. Establish event taxonomy and risk dimensions:
    a. Establish a hierarchy of unwanted events (main event categories are: natural events, technical/human events, and malicious acts).

b. Decide on the consequence dimensions used to analyze the unwanted events (consequence categories: life and health, environment, economy, manageability, political trust, and availability of delivery/supply of infrastructure).
c. Calibrate risk matrices: a probability category and a consequence category.
2. Perform a simple analysis (like a standard RVA/PHA):
   a. Identify all unwanted (hazardous) events.
   b. Assess the risks related to each unwanted event.
3. Select events (with high risk) for further detailed analyses.
4. Perform detailed analysis of selected events. The course of events and various consequences are investigated in more detail. These analyses shall include:
   a. Evaluation of interactions and other couplings in between the infrastructures, and how this affects the consequences of the unwanted events.
   b. Evaluation of vulnerabilities.
   c. Suggesting and evaluating risk and vulnerability reducing measures.

The objective of RVA is to identify hazardous events related to the activity/system as thoroughly as reasonably practicable. Resilience is not directly assessed here.

RAMCAP™ (Risk Analysis and Management for Critical Asset Protection) [11], [12] (currently RAMCAP Plus) was elaborated by the American Society of Mechanical Engineers as an all hazards risk and resilience assessment methodology. It is supported by the US Department of Homeland Security, especially for water providers. RAMCAP has 7 steps: assets identification, threats identification, consequences analysis, vulnerabilities analysis, threats assessment, risk and resilience assessment, risk and resilience management. RAMCAP Plus offers a high-level approach focused on the most critical assets at a given facility. Important advantages are cross-sectoral risk comparisons and resilience management.

The methodical framework EURACOM [13] for hazard identification, including risk assessment and contingency planning and management can be applied in specific energy sectors, yet it emphasizes interconnections of energy networks. The method includes the following steps: set up a holistic team with a holistic view, define the holistic scope, define risk assessment scales, understand the assets, understand the threat context, review security and identify vulnerabilities, and finally evaluate and rank the risks. The favourable feature is that the scope of EURACOM goes beyond the limits of assets of a given CI to the higher level (CI sector, countries). Resilience is not addressed and no supporting software tools are available.

The reviewed methods do not explicitly distinguish CI internal and external causes of hazardous events. They also do not distinguish CI internal non-escalating consequences, consequences generating hazards/threats in the same infrastructure, and consequences generating external hazards/threats for other collaborating infrastructures.

4   Preferred Features of the Risk Management Tools in Critical Infrastructures

A huge number of methods and tools related to critical infrastructure protection risk assessment/management make it impossible to identify one ideal approach. Yet it is possible to specify the most preferable common features of the risk management solution. These features imply the requirements for the risk management tool.

4.1   Conceptual Model of Risk Modeling

Risk assessment in critical infrastructures should consider many causes and many multidirectional impacts of the given hazardous event. The bow-tie conceptual model, described in many publications, e.g. [6], can be useful in risk modeling. The model (Fig. 1) points at relationships within the given hazardous event, its causes and consequences. One diagram presents relationships for one hazardous event.

Assets are placed in the environment, where there are hazards and threats which may cause hazardous events. The triggered hazards or threats exploiting vulnerabilities can overcome proactive barriers (safeguards) existing in the system and may lead to a hazardous event. The consequences of the event may be diversified and multidirectional. They can be mitigated by reactive barriers. In this case vulnerabilities can weaken or even remove these barriers. Barriers identified with different kinds of safeguards are applied with respect to the risk value and are monitored and maintained.

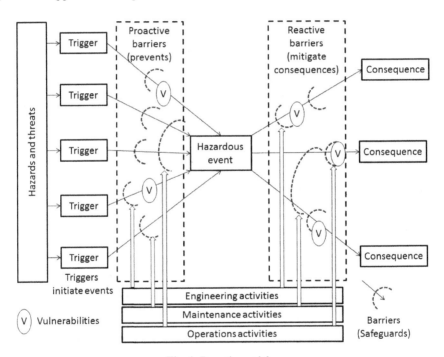

Fig. 1. Bow-tie model

The analysis, provided separately for each event, encompasses the following steps:

1. Specify the hazardous event (what, where, when).
2. Identify hazards, their triggers, and threats.
3. Identify proactive barriers preventing the occurrence of the hazardous event.
4. Identify possible sequences that may follow the considered hazardous event.
5. Identify reactive barriers mitigating event consequences (to stop the event sequence, to decrease consequences).

6. List all potential undesirable consequences.
7. Identify influences from engineering, maintenance and operation activities on the various barriers.

The presented conceptual model and the analysis procedure are very general. The bow-tie model encompasses the cause analysis and the consequences analysis. They can be refined and implemented in less or more complex ways [9], e.g. using FTA (Fault tree analysis) [14] and ETA (Event tree analysis) [15] methods.

FTA can be useful to provide a cause analysis and to identify and analyze potential causes and pathways to an undesired (here: hazardous) event. The causes that may lead to a failure are organized in a tree diagram. The method may be used to calculate the probability of the hazardous event based on the probabilities of each causal event.

ETA can be useful to provide a consequences analysis. It helps to specify the sequence of positive or negative events (here: leading to consequences) implied by the hazardous event that initiates them. Displayed in the form of a tree, ETA is able to present aggravating or mitigating events. It can be used for modeling, calculating, ranking different accident scenarios which follow the initiating event.

There is no analysis of interdependencies in this model – it should be extended in this field. The bow-tie model is focused on risk assessment but it can be used to reassess the risk after new/updated barriers implementation (the risk management aspect).

Proposed requirement: Use the bow-tie risk conceptual model which comprises the cause analysis and the consequences analysis, and considers multilayered barriers, hazards, threats and vulnerabilities. The recommended bow-tie model complies with CIs where hazardous events usually have many sources and generate many diversified and multidirectional consequences. The model considers common cause hazards/threats. It can support risk assessment in interdependent CIs (Section 4.5).

4.2   Risk Register

A CI owner should elaborate a list of hazardous events inherent to the given infrastructure, i.e. the so called risk register. The listed items (data records) should include at a minimum: related hazard/threats, possible corresponding hazardous event, probability of the event and its consequences. The risk management process is performed during the CI life cycle, so the risk register can be continuously updated. It is important for the protection plans.

To make the analyses more structured and traceable, the hazardous events should be grouped in categories, e.g. categories representing main contributors to a hazardous event: natural-, technological-, organizational-, behavioural-, social hazards. The example of a reference list of hazardous events can be found in [5] /Appendix A. Such lists should be elaborated to meet the specific features of the given CI, its environment and stakeholders' expectations.

Proposed requirement: Elaborate and maintain the risk register as the managed inventory of hazardous events. The data structures depend on the applied risk method [9]. The risk management tool should be equipped with proper data structures expressing CI-specific risk issues, and with a functionality to operate on the risk register.

4.3   Risk Related Data – Assets, Societal Critical Functions, Hazards, Threats, Vulnerabilities, and Barriers

To assess the risk, all risk-related data should be gathered and maintained.

The first issue to specify is an asset related to the societal critical functions (SCF) [5] performed by CIs. The SCFs are essential to ensure the basic needs of a society, e.g.: life and health, energy supply, law and order, national security. Some of these functions are provided by CIs. The high level SCFs can be linked to the assets protected in CIs ([5]/Appendix B). The SCFs are identified also as CI services and for this reason the service continuity issue is important. A hazard is "a source of danger that may cause harm to an asset" [6], e.g.: toxic or flammable materials, moving parts, short circuit. A threat is "anything that might exploit vulnerability" [6]. A vulnerability is "a weakness of an asset or group of assets that can be exploited by one or more threat agents" [6]. A threat agent is "a person or a thing that acts, or has the power to act, to cause, carry, transmit, or support a threat" [6]. The threat agent is an analogy to the hazard trigger. Please note that threats, unlike hazards, exploit vulnerabilities. Threats have a broader meaning – a threat is a hazard, but a hazard does not need to be a threat. To decrease the probability of a hazardous event and its consequences, different barriers (safeguards, countermeasures, controls) are applied according to the risk value. A barrier is "a physical or engineered system or human action (based on specific procedure or administrative controls) that is implemented to prevent, control, or impede energy released from reaching the assets and causing harm" [6], e.g.: firewall, evacuation system, isolation of ignition sources, firefighting system. Vulnerabilities express the degree to which a system is likely to experience harm induced by perturbation or stress. Several vulnerability and risk factors may affect the probability of a hazardous event, and later, after the event occurs, the vulnerabilities affect the consequences. The following factors should be identified: temperature, area, duration, level of maintenance or renewal, degree of coupling, mental preparedness, population density, etc. The list of vulnerability and risk factors can be found in [5] /Appendix B. Some are considered before and some after the event occurs.

Proposed requirement: For the given CI prepare and maintain a reference list of records representing assets, hazards, threats, vulnerabilities, barriers and related attributes. The risk management tool should be equipped with CI-specific data structures and a functionality to operate on the risk-related data. The data structures depend on the applied risk method [9]. They represent sources of information for risk management.

4.4   Risk Assessment

Security assessment depends on the applied risk assessment method. Generally, there is a need to define the levels to measures the likelihood (probability, frequency) of a hazardous event and consequence severity (dimensions). Usually, enumerative scales are used, though there are more complex quantitative approaches. Severity is a seriousness of the consequences of an event expressed either as a financial value or as a category [6], e.g.: catastrophic, severe loss, major damage, minor damage.

Consequences have many dimensions (e.g. life and injury, economic losses, unavailability, social impact, etc.) and each of them should be assessed independently and the results aggregated. Table 1 presents examples of categories for four consequences dimensions. Other dimensions, like environment or property dimension, can be defined in the same way. Please note that for the economic losses a simple and useful logarithmic scale was applied. The measures for the social impact dimension can be elaborated in a more precise way with the use of the ValueSec methodology [16]. In this project the special QCA (Qualitative Criteria Assessment) tool was elaborated to assess social impact related to the current security situation.

The problem is that the dimensions cannot overlap and particular levels should be precise, clear-cut and easy to interpret.

Table 1. Examples of the consequences measures in different dimensions

Level of measure	Live and injury dimension	Economic losses dimension (million Euro)	Service unavailability dimension	Social impact dimension
1 (Negligible damage)	< 4 injured/seriously ill	< 0.1	< 6 hours	None or not significant
2 (Minor damage)	4–30 injured/seriously ill	[0.1-1)	6 hours to 1 day	Minor social dissatisfaction
3 (Major damage)	1–2 fatalities, 31–100 injured/seriously ill	[1, 100)	1 day to 1 week	Moderate dissatisfaction, possible episodic demonstrations
4 (Severe loss)	3–20 fatalities, 101–600 injured/seriously ill	[10-100)	1 week to 3 month	Serious dissatisfaction, possible demonstrations, strikes, riots
5 (Catastrophic)	> 20 fatalities, > 600 injured/seriously ill	≥ 100	More than 3 months	Migration from the affected area or country

The second issue is to define the levels to measure the likelihood (probability, frequency), called frequency classes. A good example [6] is placed in Table 2.

Table 2. Levels to measure likelihood of the event (frequency classes)

Level of measure	Frequency per year	Description
5 (Fairly normal)	10 – 1	Event that is expected to occur frequently
4 (Occasional)	1 – 0.1	Event that may happen now and then and will normally be experienced by personnel
3 (Possible)	$10^{-1} - 10^{-3}$	Rare event, but will be possibly experienced by personnel
2 (Remote)	$10^{-3} - 10^{-5}$	Very rare event that will not necessarily be experienced in a similar plant
1 (Improbable)	$0 - 10^{-5}$	Extremely rare event

A common and simple way is to present the risk in a form of a risk matrix. It is a tabular illustration of the consequences severity (horizontally) and frequency (vertically) of hazardous events (Table 3). The number of rows and columns depends on analytical needs. The aggregated risk value can be a sum/product of values coming from the cross-cutting rows and columns. In Table 3 a sum is used – risk can be assessed from 2 to 10 (in the case of a product – in the range: 1 to 25). Sometimes non-linear, "flattening" scales are applied. The risk matrix gives a picture of the risk and helps to choose a risk management strategy according to the assumed risk acceptance level. For the sociotechnical systems the ALARP (As Low As Reasonably Practicable) approach [5], [6] is applied.

Risk range is divided into three areas to prepare further risk management actions:

1. Unacceptable area, where risk is intolerable except in extraordinary conditions – the risk reduction countermeasures should be applied.
2. ALARP area, where risk reduction countermeasures are desirable, but may not be implemented if their cost is grossly disproportionate to the benefits gained.
3. Broadly acceptable area, where no further risk reduction countermeasures are needed – further reduction is uneconomical and resources could be spent better elsewhere to reduce the total risk.

Usually these three areas are marked by colours: red, yellow and green.

Table 3. Example of a risk matrix (5x5)

		Likelihood				
		1 Improbable	2 Remote	3 Possible	4 Occasional	5 Fairly normal
Consequences	1 Negligible	2	3	4	5	6
	2 Minor damage	3	4	5	6	7
	3 Major damage	4	5	6	7	8
	4 Severe loss	5	6	7	8	9
	5 Catastrophic	6	7	8	9	10

In the example presented in Table 3:

1. Unacceptable area: 8 to 10; red – it will require risk reducing measures and/or additional, more detailed analyses to be carried out.
2. Acceptable (ALARP area): 6 to 7; yellow – risk-reducing measures will typically be considered based on their efficiency; use the ALARP principle and consider the further analysis.
3. Broadly acceptable area: 2 to 5; green – no reduction needed.

Proposed requirement: Select a risk assessment method [9], right risk factors measures and a calculation method. In the simplified variant define the likelihood levels, consequences dimensions and their severity levels. The risk management tool should be flexible with respect to the number of dimensions and levels of the measure. Reporting functionality, based on different diagrams useful for operators and decision makers, is very important.

4.5   Critical Infrastructure Specific Phenomena

A typical risk assessment/management method applied to a sociotechnical system is focused on this systems and not on other systems working in its neighbourhood. Such methods are not prepared to accurately capture (physical, cyber, logical, geographical) interdependencies across infrastructures, cross-sectors and cross-borders (Section 3).

Here a simple mechanism, i.e. an interdependency diagram [5], is proposed to supplement the methodology presented in the paper. The diagram, which allows to consider interdependencies, will be elaborated as an auxiliary tool for a set of collaborating infrastructures. This diagram will be used to identify threats/hazards coming from external infrastructures to the given infrastructure and check if the consequences identified in the given infrastructure appear as the hazards or threats in the neighbouring infrastructures. It requires to build a separate risk management system (bow-tie based) for each CI, to use common risk measures and to distinguish their internal and external data (CI risk management related – section 4.2-4.3). This approach is able to consider common cause failures spreading across critical infrastructures.

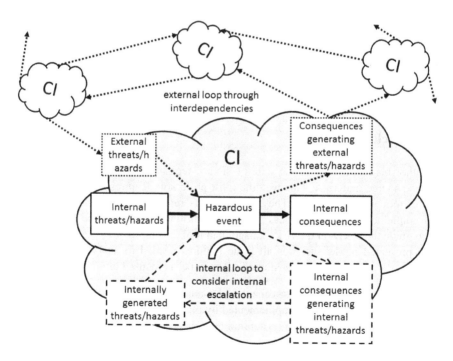

Fig. 2. Conceptual model for risk assessment in interdependent infrastructures

Figure 2 presents a CI (marked grey) collaborating with some interdependent infra-structures. This CI has its own risk assessment tool based on the bow-tie concept. The tool distinguishes internal threats/hazards which invoke hazardous events and internal consequences caused by these events – this complies with Fig. 1. It was assumed that such consequences do not invoke any other hazards or threats which cause different impacts described above. Let us call them internal consequences.

Some consequences can invoke internal hazards/threats leading to new hazardous events which escalate harm in the given CI. These are specially distinguished (dashed lines). To consider internal escalation and cascading effects, the analytical bow-tie process should be repeated until no new hazards/threats are generated. Additionally, the internally generated threats/hazards should be watched on the model input. These consequences can be called consequences generating internal threats/hazards.

Due to interdependencies, certain consequences can spread through CIs couplings outside and generate hazards/threats in the neighbouring infrastructures and cause harms there. In addition, the external CIs may be sources of hazards/threats for the considered CI. The external pathways are marked by dot lines. To consider external escalation and cascading effects, the analytical bow-tie process should be repeated until no new hazards/threats are generated. Additionally, the externally generated threats/hazards should be watched on the model input. For this reason the third conse-quences category is introduced – consequences generating external threats/hazards.

There is an analogy to asynchronous digital circuits. Such issues are rather compli-cated and CIs co-operation requires certain synchronization – checking in a certain time interval if the internally and/or externally generated hazards/threats are invoked.

Proposed requirement: Prepare an interdependency diagram (or better implement it in the tool). Categorize hazards/threats and extend the analytical possibilities of pre-viously developed bow-tie model implementations (data structures, functionality). Organize information exchange between risk management systems of CIs.

5   Conclusions

The author made a review of the state of the art, regulations, best practices, EU pro-jects, and source materials. This way the most preferable features of the CI risk man-agement tool were identified. They can serve as reference requirements for a tool which implements most usefulness features. Additionally, they can be discussed and used to guide the R&D on the risk management tool development.

The presented model includes all basic features needed to analyze critical infra-structures risk and its specific phenomena. The paper is focused on:

- conceptual model of risk assessment and management – the bow-tie approach is preferred, though it may be supplemented by the FTA and ETA methods;
- key data structures, i.e.: risk register and the related hazards/threats, vulnerabilities, consequences, barriers registers;
- risk analysis procedure and assessment, including likelihood and consequences measures, and the results presentation;

- critical infrastructures specific phenomena, including these implied by interdependencies – the new concept is proposed to integrate the different bow-tie models through internal and external dependencies (categorization of the causes and consequences of hazards/threats, generation of internal causes and causes for external CIs), though further experimentation and validation are needed.

The paper does not consider other issues important for CIs, like: resilience analysis, human and management aspects, compliance with EU regulations, communications in risk management, real-time risk management aspects, and many others – because the approach presented here is limited to the key risk management issues.

The presented concept will be a subject of experimentation. The OSCAD [17] software will be used as a platform for its implementation and validation on 2 selected critical infrastructures: railway transport and electricity. This will enable to acquire experience to formulate the requirements for the CIs dedicated solution.

This work is related to the CIRAS [1]project [18] launched by the organization of the paper author.

References

1. Council Directive 2008/114/EC on the identification and designation of European critical infrastructures and the assessment of the need to improve their protection (2008)
2. Commission Staff Working Document on a new approach to the European Programme for Critical Infrastructure Protection Making European Critical Infrastructures more secure. European Commission. Brussels, SWD, 318 final (August 28, 2013)
3. Giannopoulos, G., Filippini, R., Schimmer, M.: Risk assessment methodologies for Critical Infrastructure Protection. Part I: A state of the art. European Union (2012)
4. Rinaldi, S.M., Peerenboom, J.P., Kelly, T.K.: Identifying, Understanding and Analyzing Critical Infrastructure Interdependencies. IEEE Control Systems Magazine, 11–25 (2001)
5. Hokstad, P., Utne, I.B., Vatn, J. (eds.): Risk and Interdependencies in Critical Infrastructures: A Guideline for Analysis, Reliability Engineering. Springer-Verlag London (2012)
6. Rausand, M.: Risk Assessment: Theory, Methods, and Applications. Series: Statistics in Practice (Book 86). Wiley (2011)
7. Deliverable D2.1: Common areas of Risk Assessment Methodologies. Euracom (2007)
8. ENISA: http://rm-inv.enisa.europa.eu/methods (access date: January 2015)
9. ISO/IEC 31010:2009 - Risk Management - Risk Assessment Techniques
10. Utne, I.B., Hokstad, P., Kjolle, G., Vatn, J., et al.: Risk and vulnerability analysis of critical infrastructures – the DECRIS approach. SAMRISK, Oslo (2008)
11. RAMCAP[TM] Executive Summary. ASME Innovative Technologies Institute, LLC (2005)
12. All-Hazards Risk and Resilience: Prioritizing Critical Infrastructures Using the RAMCAP Plus Approach. ASME Innovative Technologies Institute, LLC (2009)

---

[1] This project has been funded with support from the European Commission. This publication reflects the views only of the author, and the European Commission Carnot be held responsible for any use which may be made of the information contained therein.

13. EURACOM Deliverable D20: Final Publishable Summary, Version: D20.1 (March 2011), http://cordis.europa.eu/result/rcn/57042_en.html (access date: January 2015)
14. EN 61025 Fault tree analysis (FTA) (IEC 61025:2006), CENELEC (2007)
15. EN 62502 Event tree analysis (ETA) (IEC 62502:2010), CENELEC (2010)
16. ValueSec FP7: http://www.valuesec.eu (access date: January 2015)
17. OSCAD project: http://www.oscad.eu/index.php/en/ (access date: January 2015)
18. CIRAS project: http://cirasproject.eu/content/project-topic (access date: January 2015)

Experiment on Defect Prediction

Ilona Bluemke and Anna Stepień

Institute of Computer Science, Nowowiejska 15/19, 00-665 Warsaw, Poland
I.Bluemke@ii.pw.edu.pl

Abstract. It is important to be able to predict if a module or a class or a method is faulty, or not. Such predictions can be used to target improvement efforts to those modules or classes that need it the most. We investigated the classification process (deciding if an element is faulty or not) in which the set of software metrics is used and examined several data mining algorithms. We conducted an experiment in which ten open source projects were evaluated by ten chosen metrics. The data concerning defects were extracted from the repository of the control version system. For each project two versions of code were used in the classification process. In this study the results of two algorithms i.e. *k*- NN and decision trees used in the classification process are presented.

Keywords: defect prediction, object metrics, data mining, prediction model.

1 Introduction

Testing of software systems is time and resources consuming. Applying the same testing effort to all parts of a system is not the optimal approach, because the distribution of defects among parts of a software system is not balanced. Therefore, it would be useful to identify fault-prone parts of the system and with such knowledge to concentrate testing effort on these parts. According to Weyuker et al. [1,2] typically 20% of modules contain 80% of defects. Testers with a tool enabling for defect prediction may be able to concentrate on testing only 20% of system modules and still will find up to 80% of the software defects. The models and tools which can predict the modules to be faulty, or not, based on certain data, historic or current ones, can be used to target the improvement efforts to those modules that need it the most.

Considerable research has been performed on the defect prediction methods e.g.[3-22] also many surveys on this subject are available e.g. [3-4,8,19]. In the defect prediction models often software or/and product metrics are used. The application of metrics to build models can assist to focus quality improvement efforts to modules that are likely to be faulty during operations, thereby cost-effectively utilizing the testing and enhance the resources. There is abundant literature on software and product metrics e.g. [23-25].

Our goal was to investigate the classification process (deciding if an element is faulty or not) in which the set of software metrics is used and examine several data mining algorithms. We also had to build a tool extracting information about the defects from the repository of the version control system.

© Springer International Publishing Switzerland 2015

W. Zamojski et al. (eds.), *Theory and Engineering of Complex Systems and Dependability,*

Advances in Intelligent Systems and Computing 365, DOI: 10.1007/978-3-319-19216-1_3

The paper is organized as follows: In section 2 the defect prediction is introduced and our prediction model is outlined. The experiment is described in section 3. This includes the presentation of investigated projects and methods for data acquiring. In section 3.1 the steps of experiment are listed as well as the values calculated in the experiment. In section 3.2 metrics used in the prediction process are given. Subsequent sections contain the results of the experiment: in section 3.3 the results of classification with k-NN algorithm, while in section 3.4 with decision trees, are presented. Conclusions and future research are given in section 4.

2   Defect Prediction

The defect prediction has been conducted so far by many researchers [3-22] and many prediction models has been proposed. Catal and Diri in [8] divided the fault prediction methods into four groups and also calculated the percentage of the usage of these methods (given in brackets):

1. statistical methods (14%),
2. machine learning based methods (66%),
3. statistical methods with an expert opinion (3%),
4. statistical methods with machine learning based methods (17%).

The defect predictor builds a model based on data:

- from old version of a project (often data from version control system are used) or
- current system measures.

The data from previous system versions to predict defects were used e.g. in [9-14]. In defect predictor using current system version e.g. [6,15], different metrics are typically used. These metrics could be calculated at different levels: process, component, files, class, method. According to Catal and Diri [8] the method level metrics are used in 60% of defect predictors and class level in 24% while process level only in 4%. Depending on the metric level appropriate code level can be identified as faulty. When method level metrics are used the predicted fault-prone modules are methods, if class level then classes.

In our experiment with the defect prediction we decided to use machine learning methods and metrics. We conducted the experiment with several machine learners but in this study the k-NN algorithm and decision trees are used. These algorithms are briefly described below. The set of metrics we used in the experiment is shown in section 3.3

2.1   *k* – NN Algorithm

The k-Nearest Neighbors algorithm (k-NN for short) [26] is a simple algorithm used for classification and regression. The algorithm stores all available samples and classifies new ones based on a similarity measure which is the distance functions.

A sample is classified by a majority vote of its neighbors, with the sample being assigned to the class most common amongst its k nearest neighbors measured by a distance function. If k=1, then the sample is simply assigned to the class of its nearest neighbor. A commonly used distance metric for continuous variables is Euclidean distance [26], also other metrics like Manhattan [26] can be used. For discrete variables another metric are used e.g. Hamming distance [26].

One major drawback in calculating distance measures directly from the training set is in the case where variables have different measurement scales or there is a mixture of numerical and categorical variables. The solution of this problem could be to standardize the training set.

The best choice of k depends upon the data: generally, larger values of k reduce the effect of noise on the classification, but make boundaries between classes less distinct. A good k can be selected by various heuristic techniques.

2.2   Decision Trees

Decision tree learning is a method commonly used in data mining [27]. The goal is to create a model that predicts the value of a target variable based on several input variables. A decision tree or a classification tree is a tree in which each internal (non-leaf) node is labeled with an input feature (e.g. height, age). The arcs coming from a node labeled with a feature are labeled with each of the possible values of the feature. Each leaf of the tree is labeled with a class or a probability distribution over the classes.

A tree can be "learned" by splitting the source set into subsets based on the feature value test. This process is repeated on each derived subset in a recursive manner called recursive partitioning. The recursion is completed when the subset at a node has all the same value of the target variable, or when splitting no longer adds value to the predictions.

Algorithms for constructing decision trees usually work top-down, by choosing a variable at each step that best splits the set of items. Different algorithms use different metrics for measuring "best". These generally measure the homogeneity of the target variable within the subsets. Examples of such metrics are: Gini impurity, information gain (entropy) [26], test χ^2 .

3   Experiment

In defect prediction experiments many researchers are using NASA projects from the Promise repository [28] e.g.[20-21], others are using "private" or open source projects e.g. [5-6,22]. We decided to use in our experiment ten open source Java projects. The fault prediction was made for two successive releases with the usage of object metrics and learning algorithms. Object metrics were calculated by Ckjm tool [5] which enables to calculate 19 object oriented metrics (WMC, DIT, NOC, CBO, RFC, LCOM, CA, CE,NPM, LCOM3, LOC, DAM, MOA, MFA, CAM, IC, CMB, AMC, CC) from Java code.

The information about errors were obtained by specially designed and implemented application – Bug Miner [29]. Given a bug pattern (a regular expression describing either a bug identifier e.g. *PROJECT-1234* or a process of fixing e.g. *fixed, bugfix*), the Bug Miner finds in the repository of the control version system comments associated with error repairs. As bug patterns are project specific, they were obtained by analyzing each of the projects naming conventions. The *git* version control system was used [30]. Bug Miner also uses the metrics values calculated by *Ckjm* and generates statistics for the project, its main operation is presented in Fig.1. Statistics are generated as a csv file, in which each line consists of a class name, metric values and information about the presence of bugs.

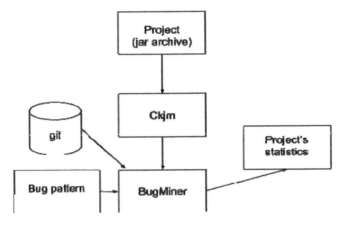

Fig. 1. Bug Miner operation

In Table 1 projects which were used in the experiment are listed.

Table 1. Analyzed projects

project	Reference	Number of classes	Number of classes containing errors	Percentage of errors
Apache Lucene	[31]	624	65	10,56%
Apache Hadoop	[32]	701	59	8.40%
Apache Tika	[33]	174	5	2.86%
MyBatis	[34]	361	31	8.56%
Apache Commons Lang	[35]	84	30	35.29%
Hibernate	[36]	2123	174	8.19%
Jsoup	[37]	45	3	6.52%
Apache Velocity	[38]	228	5	2.18%
Elasticsearch	[39]	2464	27	1.09%
Zookeeper	[40]	246	37	15.04%

In majority of projects the number of classes containing errors is small. Especially in projects Elasticsearch and in Apache Velocity the ratio of classes with errors is only 1,09% and 2,18% respectively. The project Apache Commons Lang differs significantly from other projects because in subsequent release the ratio of classes containing errors was 35%. As it can be seen in Table 1 data which are used in the fault prediction process (classes containing errors) constitute small ratio of all data. Such imbalanced distribution makes a problem for the classification process as it can assign all elements to the dominant group and in the classification for the fault prediction the less numerous group is of higher interest. Solution to this problem is resampling:

- over-sampling the minority class,
- under–sampling the majority class or
- combined approach.

We used SMOTE (Synthetic Minority Oversampling Technique) algorithm proposed by Chawla et.al [41]. The minority class is over-sampled by creating "synthetic" examples. The minority class is over-sampled by taking each minority class sample and introducing synthetic examples along the line segments joining any/all of the k minority class nearest neighbors. Neighbors from the k nearest neighbors are randomly chosen. Synthetic samples are generated in the following way: the difference between the feature vector (sample) under consideration and its nearest neighbor is taken. This difference is multiplied by a random number between 0 and 1, and added to the feature vector under consideration. This approach effectively forces the decision region of the minority class to become more general.

3.1   Steps of Experiment

The datasets used in the experiments were generated from the statistics files obtained by Bug Miner application. A single sample was a n-dimensional vector of metrics values. The testing procedure consisted of the following steps performed for each classification algorithm and was repeated N = 1000 times:

- Preparation of data with the SMOTE algorithm,
- 10-fold cross-validation [26],
- Results collection.

In the k-fold cross validation (we used k=10) the data set is divided into k subsets, and the method is repeated k times. Each time, one of the k subsets is used as the test set and the other k-1 subsets are put together to form a training set. The advantage of this method is that it matters less how the data get divided. Every data point gets to be in a test set exactly once, and gets to be in a training set k-1 times. The disadvantage of this method is that the training algorithm has to be rerun k times, which means it takes k times as much to make the evaluation.

The result of the classification algorithm could be :

- **TN** (True Negatives) - the number of negative examples correctly classified - the unit do not contains errors and was classified into group of units with no errors.
- **FP** (False Positives) - the number of negative examples incorrectly classified as positive - the unit do not contains errors and was classified into group of units with errors,
- **FN** (False Negatives) - the number of positive examples incorrectly classified as negative - the unit contains errors and was classified into group of units with no errors,
- **TP** (True Positives) - the number of positive examples correctly classified – the unit contains errors and was classified into group of units with errors.

In each iteration following performance measures were calculated:

- **Precision** – the ratio of the units with errors correctly classified to all units in this class: TP / TP + FP
- **Recall** – the ratio of the units with errors correctly classified, to all units which should appear in that class: TP / TP+FN
- **Specificity** – the ratio of the units without errors correctly classified to all units in this class: TN / TN + FP ,
- **F1** measure – the harmonic mean of precision and recall: 2* precision*recall / (precision + recall).
- **ROC** (Receiver Operating Characteristic) [26] curves can be thought of as representing the family of best decision boundaries for relative costs of TP and FP. On an ROC curve the X-axis represents recall and the Y-axis represents 1- Specificity. The ideal point on the ROC curve would be (0,1), that is all positive examples are classified correctly and no negative examples are misclassified as positive.
- **AUC** - area under the ROC curve – used for model comparison, a reliable and valid AUC estimate (AUC value is in [0,1]) can be interpreted as the probability that the classifier will assign a higher score to a randomly chosen positive example than to a randomly chosen negative example (AUC=0.5)..

3.2   Object Metrics

The Ckjm tool is able to calculate nineteen metrics. Among them are widely known Chidamber Kemerer metrics (CK metrics) [23], QMOOD metrics proposed by Bansiya and Davis [7], Martin's metrics [24] and some Henderson-Sellers metrics [25].

We decided to use ten metrics (out of nineteen available) and to choose the best ones we performed an analysis by random forests [26] for all projects. Random forests are a combination of tree predictors such that each tree depends on the values of a random vector sampled independently and with the same distribution for all trees in the forest. As features used in the splitting we used two attributes: mean decrease in accuracy and mean decrease of Gini index (probability that randomly chosen two samples are in the same class).

The most suitable ten metrics obtained after the random forests analysis are as follows:

1. LOC – number of lines of code,
2. AMC - Average Method Complexity [42]- the average method size for each class: the size of pure virtual methods and inherited methods are not counted, large method, which contains more code, tends to introduce more faults than a small method,
3. RFC - Response For a Class (CK metric) - the cardinality of the set of all methods that can be potentially executed in the response to the arrival of a message to an object, a measure of the potential communication between the class and other classes,
4. CE - Efferent Coupling (Martin metric) – the number of other classes that the class depends upon, an indicator of the dependence on externalities, efferent means outgoing ,
5. CBO - Coupling Between Objects (CK metric) - the number of couplings with other classes (where calling a method or using an instance variable from another class constitutes coupling),
6. CAM - Cohesion Among Methods of Class (QMOOD metrics) - the relatedness among methods of a class based upon the parameter list of the methods, is computed using the summation of number of different types of method parameters in every method divided by a multiplication of number of different method parameter types in whole class and number of methods
7. WMC - (CK metric) - the number of methods in the class (assuming unity weights for all methods)
8. LCOM - Lack of Cohesion in Methods (CK metric) - counts the sets of methods in a class that are not related through the sharing of some of the class fields. The lack of cohesion in methods is calculated by subtracting from the number of method pairs that do not share a field access the number of method pairs that do.
9. NPM -Number of Public Methods (QMOOD metrics) - counts all the methods in a class that are declared as public, the metric is known also as Class Interface Size (CIS)
10. LCOM3 - Lack of Cohesion in Methods (Henderson-Sellers metric) [25].

3.3   Results for k-NN Algorithm

The experiments with k-NN algorithm used in the classification process were executed for $k \in \{3,5,7\}$. For searching the nearest neighbor the kd tree [26] and cover tree [43] algorithms were used, both were operating on the Euclidean distance. The mean values for all projects are shown in Table 2. The values of precision and F1 are rather high, close to 0.8 but the value of AUC (area under the ROC curve) is low, close to 0.5 so it is similar as for the random classifier. It can be noticed that the algorithm used for searching the nearest neighbor (kd tree and cover tree) has no influence on the results of precision, recall, F1 and AUC as well. The values of precision, recall,

F1 and AUC decrease with the increase of k. The best results were obtained for k = 3, the worst for k = 7. This phenomena can be explained analyzing the operation of the k-NN classifier. Classifier makes the decision by majority voting of k nearest neighbors. In the examined problem the data were imbalanced as classes containing errors constitute only small ratio of all data (Table 1). If k increases, in the neighborhood of an element from the minority class there are more elements from the dominant class so the results of classification can be incorrect.

Table 2. Results for k-NN   k = 3, 5, 7

Measure	Kd tree k=3	Cover tree k=3	Kd tree k=5	Cover tree k=5	Kd tree k=7	Cover tree k=7
Precision	0.799192	0.7991514	0.7783662	0.7783623	0.766027	0.7660199
Recall	0.7088503	0.7088663	0.663491	0.6635084	0.6372275	0.6372375
F1	0.7483749	0.7483661	0.7130383	0.713046	0.6921973	0.6922009
AUC	0.544651	0.5446305	0.5385112	0.5384918	0.5279101	0.527888

3.4   Results for the Decision Tree

Prediction models based of decision trees produced significantly better results than the *k*-NN algorithm: precision (0.8143766), recall (0.8083666), F1 (0.8087378) and AUC (0.8816798). The value of AUC is high so the accuracy of decision tree classifier is very high.

4   Conclusions

We investigated the classification process (deciding if an element is faulty or not) in which the set of software metrics was used and examined several data mining algorithms. In this study we presented the results of defect prediction by *k*-NN and decision trees algorithms for ten open source projects, of different sizes and from different domains. We also developed a tool extracting information about the defects from the repository of the *git* version control system [30].

In our experiment the decision tree algorithm produced significantly better results than the *k*-NN algorithm, mean values of AUC are 0.88 and 0.54 accordingly. Our results did not confirm the observations of Lessmann et. al presented in [20]. They compared 22 classifiers for ten projects from NASA repository [28] and stated that differences in the values of AUC obtained by competing classifiers are not significant. For *k*-NN algorithm they obtained AUC ≈ 0.7 while we only 0.54. Possible explanations could be the different set of metrics used in the classification process (LOC, Halstead, McCabe metrics) and the specificity of NASA project.

However our analysis of the set of suitable metrics (by random forests) (section 3.2) was different than the analysis of e.g.: Singh et. al [21], Jureczko et. al [5-6], Gyimothy et. al [22] we obtained very similar set of metrics. It can be claimed that this set of metrics is very efficient and useful in the defect predictors.

The results of defect predictors could be influenced by the set of chosen metrics, specificity of projects being the subject of prediction and learning algorithms applied. It should be mentioned that these results also strongly depend on the data concerning defects. We were using the inscriptions in the repository of the version control system concerning defects, in fact some of them could be false.

References

1. Weyuker, E., Ostrand, T., Bell, R.: Adapting a Fault Prediction Model to Allow Widespread Usage. In: Proc. of the Int. Workshop on Predictive Models in Soft. Eng., PROMISE 2008 (Leipzig, Germany, May 12-13) (2008)
2. Weyuker, E., Ostrand, T., Bell, R.: Do too many cooks spoil the broth? Using the number of developers to enhance defect prediction models. Empirical Soft. Eng. 13(5), 539–559 (2008)
3. D'Ambros, M., Robbes, R.: An Extensive Comparison of Bug Prediction Approaches, http://inf.unisi.ch/faculty/lanza/Downloads/DAmb2010c.pdf (access date: January 2015)
4. D'Ambros, M., Lanza, M., Robbes, R.: Evaluating Defect Prediction Approaches: A Benchmark and an Extensive Comparison. Empirical Softw. Eng. 17(4-5), 531–577 (2012)
5. Jureczko, M., Spinellis, D.: Using Object-Oriented Design Metrics to Predict Software Defects. In: Models and Methodology of System Dependability, pp. 69–81. Oficyna Wydawnicza Politechniki Wroclawskiej, Wroclaw (2010)
6. Jureczko, M., Madeyski, L.: Towards identifying software project clusters with regard to defect prediction. In: Menzies, T., Koru, G. (eds.) PROMISE, p. 9. ACM (2010)
7. Bansiya, J., Davis, C.G.: A hierarchical model for object-oriented design quality assessment. IEEE Trans. Softw. Eng. 28(1), 4–17 (2002)
8. Catal, C., Diri, B.: Review: A systematic review of software fault prediction studies. Expert Syst. Appl. 36(4), 7346–7354 (2009)
9. Nagappan, N., Ball, T.: Use of relative code churn measures to predict system defect density. In: Proc. 27th Int. Conf. on Soft. Eng., pp. 284–292 (2005)
10. Graves, T.L., Karr, A.F., Marron, J.S., Siy, H.: Predicting fault incidence using software change history. IEEE Trans. Softw. Eng. 26(7), 653–661 (2000)
11. Moser, R., Pedrycz, W., Succi, G.: A comparative analysis of the efficiency of change metrics and static code attributes for defect prediction. In: Proc. of the 30th Int. Conf. on Soft. Eng., ICSE 2008, New York, NY, USA, pp. 181–190 (2008)
12. Hassan, A.E.: Predicting faults using the complexity of code changes. In: Proc. of the 31st Int. Conf. on Soft. Eng., ICSE 2009, Washington, DC, USA, pp. 78–88 (2009)
13. Hassan, A.E., Holt, R.C.: The top ten list: Dynamic fault prediction. In: Proc. of the 21st IEEE Int. Conf. on Soft. Maintenance, ICSM 2005, pp. 263–272 (2005)
14. Kim, S., Zimmermann, T., Whitehead Jr., E.J., Zeller, A.: Predicting faults from cached history. In: Proc. of the 29th Int. Conf. on Soft. Eng., ICSE 2007, pp. 489–498 (2007)
15. Elish, M.O., Al-Yafei, A.H., Al-Mulhem, M.: Empirical comparison of three metrics suites for fault prediction in packages of object-oriented systems: A case study of eclipse. Adv. Eng. Softw. 42(10), 852–859 (2011)
16. Pinzger, M., Nagappan, N., Murphy, B.: Can developer-module networks predict failures? In: Proc. of the 16th ACM SIGSOFT Int. Symp. on Found. of Soft. Eng., pp. 2–12 (2008)
17. Zimmermann, T., Nagappan, N.: Predicting defects using network analysis on dependency graphs. In: Proc. of the 30th Int. Conf. on Soft. Eng., pp. 531–540 (2008)

18. Shin, Y., Bell, R., Ostrand, T., Weyuker, E.: Does calling structure information improve the accuracy of fault prediction? In: 6th IEEE Int. Working Conf. on Mining Soft. Repositories, pp. 61–70 (2009)
19. Mende, T., Koschke, R.: Revisiting the evaluation of defect prediction models. In: Proc. of the 5th Int. Conf. on Predictor Models in Soft. Eng., PROMISE 2009, pp. 7:1–7:10 (2009)
20. Lessmann, S., Baesens, B., Mues, C., Pietsch, S.: Benchmarking classification models for software defect prediction: A proposed framework and novel findings. IEEE Trans. on Soft. Eng. 34(4), 485–496 (2008)
21. Singh, Y., Kaur, A., Malhotra, R.: Predicting software fault proneness model using neural network. In: Jedlitschka, A., Salo, O. (eds.) PROFES 2008. LNCS, vol. 5089, pp. 204–214. Springer, Heidelberg (2008)
22. Gyimothy, T., Ferenc, R., Siket, I.: Empirical validation of object-oriented metrics on open source software for fault prediction. IEEE Trans. on Soft. Eng. 31(10), 897–910 (2005)
23. Chidamber, S.R., Kemerer, C.F.: A metrics suite for object oriented design. IEEE Trans. on Soft. Eng. 20(6), 476–492 (1994)
24. Martin, R.: OO Design Quality Metrics - An Analysis of Dependencies. In: Proc. of Workshop Pragmatic and Theo. Directions in Object-Oriented Soft. Metrics, OOPSLA 1994 (1994)
25. Henderson-Sellers, B.: Object-Oriented Metrics, Measures of Complexity. Prentice Hall (1996)
26. Witten, I.H., Frank, E.: Data Mining: Practical Machine Learning Tools and Techniques. Elsevier (2005) ISBN-13: 978-0-12-088407-0
27. Rokach, L., Maimon, O.: Data mining with decision trees: theory and applications. World Scientific Pub. Co. Inc. (2008) ISBN 978-9812771711
28. Promise: http://promise.site.uottawa.ca/SERepository/datasets-page.html (access date: January 2015)
29. Stępień, A.: Fault prediction with object metrics (in Polish). MSc thesis to appear, Institute of Computer Science (2015)
30. Git: http://git-scm.com (access date: January 2015)
31. Apache Lucene: http://lucene.apache.org (access date: January 2015)
32. Apache Hadoop: http://hadoop.apache.org (access date: January 2015)
33. Apache Tika: http://tika.apache.org (access date: January 2015)
34. MyBatis: http://mybatis.github.io/mybatis-3 (access date: January 2015)
35. Apache Commons Lang, http://commons.apache.org/proper/commons-lang (access date: January 2015)
36. Hibernate: http://hibernate.org (access date: January 2015)
37. Jsoup: http://jsoup.org (access date: January 2015)
38. Apache Velocity: http://velocity.apache.org (access date: January 2015)
39. Elasticsearch: http://www.elasticsearch.org (access date: January 2015)
40. Apache Zookeeper: http://zookeeper.apache.org/ (access date: January 2015)
41. Chawla, N.V., Bowyer, K.W., Hall, L.O., Kegelmeyer, P.: Smote: Synthetic minority oversampling technique. J. Artif. Int. Res. 16(1), 321–357 (2002)
42. Tang, M.-H., Kao, M.-H., Chen, M.-H.: An Empirical Study on Object-Oriented Metrics. In: Proc. of te Soft. Metrics Symp., pp. 242–249 (1999)
43. Beygelzimer, A., Kakade, S., Langford, J.: Cover Trees for Nearest Neighbor. In: Proc. Int. Conf. on Machine Learning (ICML), pp. 97–104 (2006), doi:10.1145/1143844.1143857

The Spatial Interactions Using the Gravity Model: Application at the Evaluation of Transport Efficiency at Constantine City, Algeria

Salim Boukebbab and Mohamed Salah Boulahlib

Laboratoire Ingénierie des Transports et Environnent,
Faculté des Sciences de la Technologie,
Université Mentouri Constantine 1, Campus Universitaire Zarzara,
25017 Constantine, Alegria
boukebbab@yahoo.fr, msboulahlib@gmail.com

Abstract. The spatial interactions by definition represent a movement of people, freight or information between an origin and a destination. It is a relationship between transport demand and supply expressed over a geographical space. This method has a very particular importance in the transport geography, and relates how to estimate flows between locations, since these flows, known as spatial interactions, aim to evaluate the demand for transport services existing or potential. In this paper, the case study concerns the economic exchange in the Constantine city (which is a commune of north-eastern of Algeria) between the main towns (El-Khroub, Ain-Smara and Hamma-Bouziane) geographically belonging to the Constantine city. The gravity model is used, and it is the most common formulation of the spatial interaction method; it uses a similar formulation of Newton's law of gravity.

Keywords: Spatial interaction, transport, gravity model, accessibility, matrix origin/destination.

1 Introduction

The importance of transport systems in these different forms continues to grow in Algeria, because they are the key factor in the development of modern economies. However, the Algerian transport system must cope with the request for a young population increasingly more demanding of mobility and a public opinion which don't support the level of services quality offered.

Land transports are the most usable means in Algeria. Be it for passengers or merchandises, 85 % of populations take the road per day, what causes many problems of congestion. To solve this problematic, the Algerian state has made great effort to develop public transportation (subways, trams, cable cars, trains, buses, planes, etc.) [1]. Nevertheless, individual transports such as car, bicycle and motorcycle, present an evident flexibility comparatively to the public transport and took these last year's an significant market share.

© Springer International Publishing Switzerland 2015
W. Zamojski et al. (eds.), *Theory and Engineering of Complex Systems and Dependability*,
Advances in Intelligent Systems and Computing 365, DOI: 10.1007/978-3-319-19216-1_4

Currently, due to a continuous increase of the vehicle fleet estimated at 5.6 million in 2013 [2], the great Algerian's cities have experienced: congestion problems, air pollution, noise and road safety, which are generating the dysfunctions in the management of their utilities, roads and means of transport. For this reason, the public authorities encourage transport companies to increase efforts to maintain and improve the quality of services proposed [3]. This development is expected to be continued to meet the common objectives of decongestion of the cities and the asphyxiated effect by traffic, to open up the outlying areas and improving environmental quality.

The main goal of this paper is to help reorganize the routes of the Public Transport Company and Merchandizes of the Constantine city (ETC, east). This will allow him to develop new service strategies. Because the company wants to know the land transport demand potential as well as within inside Constantine city and between the main neighboring towns. Once this data is available, it will be able to improve efficiently transport planning and see where are the best market prospects [4].

Many methods have been developed in the literature who to treat this problem. In our case study, we will use the gravity model. This method has a particular importance in the treatment of problem transport geography refers to how to estimate flows between locations, known as spatial interactions, and enable to evaluate the demand for transport services [9-10].

A principal challenge in the usage of the gravity model is their calibration. Because, the principal constraint is to estimated the results according to the observation flows. For that, the calibration model is necessary and consists in finding the value of each parameters of the model (constant and exponents) to insure the previous constraint [4].

2   Presentation of the Constantine City

The Constantine city is the capital of east of Algeria, despite the competition from other cities. She occupies a central location in the region, being a pivotal city between the crossroads of major routes north-south and west-east. She is also the metropolis of the East and the largest inland city in Algeria [1].

The Constantine metropolitan area extends over 15 to 20 km of radius (figure 1), which includes, apart from the city center, two new cities and three important areas, such as [2]:

- The zone of El Khroub, based on an open site, near a major cros-sroad. It benefited from the installation of a large market and two industrial zones that are Hammimine and Oued Tarf
- The area of Ain Smara, ancient village, she has a large area of mechanical manufacturing
- The area of Hamma Bouziane, former colonial village, has a cement plant and several brick manufacturer.

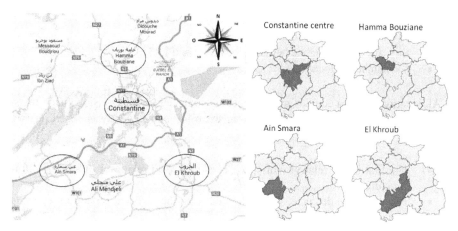

Fig. 1. Constantine metropolitan area

The Constantine city has nine suburban areas. A suburban agglomeration is a nearby area of habitat, representing the extension in terms of habitat and sometimes activities of the Constantine city [3]. Like most major Algerian cities, the Constantine agglomeration is known a strong urbanization over the last decade. For this rapid urbanization result an increase of population movement. However the travel need and the means related to transport infrastructure observe a delayed development.

This paper proposes a procedure to examine the economic activities that generate and attract movements which deploy in the space of the metropolitan area of Constantine. For that, the spatial interaction by using the gravitational model is used. The aim is to highlight the most important economic activity like: job commerce, industrial production, business or leisure. It is therefore a transport request made through a geographical area.

3   Geographical Accessibility

The humans have always the reflex to minimize displacements. They opt at all the time for take the shortest way to get from one location to another. For this, the transport system, as an economic activity, has much to gain from the minimization of distances between origins and destinations [5].

In this context; accessibility is the measure of the capacity of a location to be reached by, or to reach different locations. She represents the capacity of a location to be reached from other locations which have a different geographical location. The accessibility is a good indicator of the spatial structure since it takes into consideration location as well as the inequality conferred by distance. The following formulation is used to calculate the geographical accessibility matrix.

$$A(Geo)_i = \sum_{j=1}^{n} L_{ij} \qquad (1)$$

With:

- A(Geo) = geographical accessibility matrix.
- Lij = the short distance between location i and j.
- n = number of locations.

All locations are not equal (figure 2) because some are more accessible than others, which imply inequalities. The notion of accessibility consequently is based on two concepts:

- The first is the relativity of space is estimated in relation to transport infra-structures,
- The second is distance, which derived from the physical separation between locations.

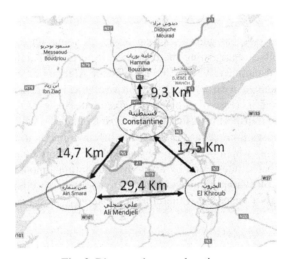

Fig. 2. Distances between locations

Table 1. The matrix of geographical accessibility

Lij (km)	Constantine	Ain Smara	El Khroub	Hamma Bouziane	A(Geo)j
Constantine	1	14,7	17,5	9,3	42,5
Ain Smara	14,7	1	29,4	24	69,1
El Khroub	17,5	29,4	1	26,8	74,7
Hamma Bouziane	9,3	24	26,8	1	61,1
A(Geo)i	42,5	69,1	74,7	61,1	247,4

The summation values are the same for columns and rows since this is a transposable matrix [4]. The most accessible place is Constantine center, since it has the lowest summation of distances (42,5). After this step, we can determinate the potential accessibility using the attribute commercial surface.

Table 2. Data input of potential accessibility

	Commercial Surface in Km²	Population	Total Superficies in Km²
Constantine	0,93	448374	231,63
Ain Smara	1,58	36988	175
El Khroub	0,98	90122	244,65
Hamma Bouziane	0,28	79952	71,18

4   Potential Accessibility

The potential accessibility is more complex measure than geographic accessibility. Potential accessibility includes simultaneously the concept of distance associated to the location attributes. It's important to point out here that all locations are not equal and thus some are more important than others. Potential accessibility can be measured as follows formulation:

$$A(Pot)_i = \sum_{j=1}^{n} \frac{Pot_j}{L_{ij}} \qquad (2)$$

With:
- $A(Pot)$ = potential accessibility matrix,
- L_{ij} = distance between location i and j (derived from valued graph matrix),
- Pot_j = attributes of location j,
- n = number of locations.

The particularity of the potential accessibility matrix, she is not transposable since locations do not have the same attributes, the same for rows and columns represent respectively notions of emissiveness and attractiveness:

- Emissiveness (λ) is the capacity to leave a location,
- Attractiveness (α) is the capacity to reach a location.

In our case the attribute is considered the commercial area, we obtain the following result:

Table 3. The potential accessibility matrix

Poti / Lij	Constantine	Ain Smara	El Khroub	Hamma Bouziane	A(Pot)j
Constantine	0,93	0,11	0,06	0,03	1,12
Ain Smara	0,06	1,58	0,03	0,01	1,68
El Khroub	0,05	0,05	0,9786	0,01	1,10
Hamma Bouziane	0,10	0,07	0,04	0,28472	0,49
A(Pot)i	1,14	1,80	1,10	0,34	4,39

We can then summarize the results of the calculation of the potential accessibility matrix:

Table 4. The potential accessibility matrix

	Population	λ	α
Constantine	448374	1,12	1,14
Ain Smara	36999	1,68	1,80
El Khroub	179033	1,10	1,10
Hamma Bouziane	79952	0,49	0,34

Now calculate the spatial interactions can started with the determination of the origin / destination matrix.

5   The Gravity Model

The formulation of gravity model is the most common spatial interactions method; he uses a similar formulation at gravity Newton's law; which states that, the attraction between two objects is proportional to their mass and inversely proportional to their respective distance [4]. Accordingly, the spatial interactions model is given by the following formulation:

$$IT_{ij} = kc \times \frac{Pop_i \times Pop_j}{L_{ij}} \tag{3}$$

With:
- Popi and Popj : The population of the locations origin and destination.
- Lij : Distance between origin and destination.
- kc is a proportionality constant related to the rate of the event   year or one week.

A data-processing model is realised in Excel software and provides after computation the results that are represented in table 5. In this last all the possible interactions between location pairs are given:

Table 5. Elementary spatial interaction matrix

	Constantine	Ain Smara	El Khroub	Hamma Bouziane	ITi
Constantine		112,85	**458,71**	**385,47**	957,03
Ain Smara	112,85		22,53		135,38
El Khroub	458,71	22,53			481,24
Hamma Bouziane	385,47				385,47
ITj	957,03	135,38	481,24	385,47	1959,11

After this, we identify the two most important origin-destination pairs in the matrix. Figure 3 illustrates the two important pairs.

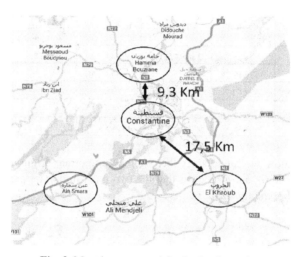

Fig. 3. Most important origin-destination pairs

It should be noted that the calibration can also be considered for different O/D matrices according to type of merchandise and modal choice. A significant challenge related to the usage of spatial interactions gravity model, is their calibration. This calibration is of crucial importance and consists in finding the value of each parameters of the model to insure that the estimated results are similar to the observed flows. Because we can easily notice that the location of Hamma-Bouziane has a very low emissiveness (λ=0,49) and attractiveness (α=0,34) compared to the others localities. For that, the model is almost useless as it predicts or explains little. Therefore, spatial interactions between locations i and j are proportional to their respective importance divided by their distance. The gravity model can be calibrated to include several parameters like emissiveness (λ) and attractiveness (α), and others variables as follows:

$$IT_{ij} = kc \times \frac{Pop_i^{\lambda} \times Pop_j^{\alpha}}{L_{ij}^{\beta}} \tag{4}$$

With:
- **Pop**, **L** and **kc** refer to the variables previously discussed.
- β (beta) : The efficiency of the transport system between two locations.
- λ (lambda) : Potential to generate movements (emissiveness).
- α (alpha) : Potential to attract movements (attractiveness).

The value of 1 is generally given to the parameters (λ, α, β), and then they are progressively altered until the estimated results are similar to observed results [4]. The value of k will be higher if interactions were considered for a year comparatively to the value of **kc** for one week [6]. In our case the value of **kc** = 0,0000001 and β=1.

Table 6. Calibration Spatial Interactions matrix

	Constantine	Ain Smara	El Khroub	Hamma Bouziane	ITi
Constantine		**2468871,83**	7741,77	1,04	2476614,65
Ain Smara	947545,78		105111,50		1052657,28
El Khroub	9282,17	**328365,88**			337648,06
Hamma Bouziane	7,44				7,44
ITj	956835,39	2797237,72	112853,27	1,04	3866927,42

After computation, we can identify the two most important origin-destination pairs in the matrix (Table 6), and it is clear that the two most important pairs are changed, such as presented in the figure 4.

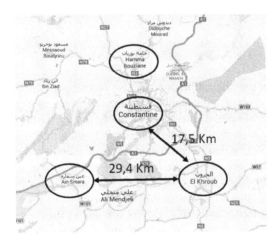

Fig. 4. Most important origin-destination pairs a calibration model

A part of the scientific research in transport aims at finding accurate parameters for spatial interactions model [7]. Once a spatial interactions model is validated for a city or a region, it can then be used for simulation and prediction of transport system.

In our study case, the Public Transport Company of Constantine city (ETC, east) will be able to improve efficiently his transport market if it's taking into account the new potential prospects market between El khroub and Ain smara locations; and it's necessary to take an account that the value of the parameters can change in time due to factors such as technological innovations, new transport infrastructure and economic development [11].

6   Conclusion

Historically the transport organization and planning are adopted in the aim to resolve the transport problems. By this work, the main goal is to help reorganize the routes of the Public Transport Company of Constantine city (ETC, east). This will allow him to develop new service strategies. Once data is available, it will be able to improve efficiently transport planning and see where are the best market prospects. For that the spatial interactions model is used. This last, is a methodology which has a particular importance for transport geography in the aim to estimate flows between locations, since these flows, known as spatial interactions, enable to evaluate the demand (existing or potential) for transport services [4]. However, it is difficult to empirically isolate impacts of locations use on transport and vice versa because of the multitude of concurrent changes of attributes [12].

In this study, we shown that the gravity model un-calibrated diverges from the reality observed compared to the calibrated model that takes into account the attributes of locations for better management of transport flows. In addition, the spatial resolution of present models is coarse because it not take account the integration of environmental sub-models for air quality, traffic noise, are likely to play an important role [8].

References

1. Guechi, F.-Z.: Constantine: une ville, des héritages, 231 p., pp. 16–231. Média Plus (2004) ISBN 9961-922-14-X
2. ONS, Armature urbaine: Les principaux résultats de l'exploitation exhaustive, Alger, Office National des Statistiques, 213 p., p. 109 (September 2011) ISBN 978-9961-792-74-2
3. Babo, D.: Algérie, Éditions le Sureau, collection "Des hommes et des lieux" (2010) ISBN 978-2-911328-25-1, "Le Constantinois"
4. Rodrigue, J.-P.: The geography of transport systems, 3rd edn., 416 p. Routledge, New York (2013) ISBN 978-0-415-82254-1
5. Goldman, T., Gorham, R.: Sustainable Urban Transport: Four Innovative Directions. Technology in Society 28, 261–273 (2006)
6. Roy, J.R., Thill, J.-C.: Spatial interaction modelling. In: Fifty Years of Regional Science, pp. 339–361 (2004) ISBN: 978-3-642-06111-0
7. Geurs, K.T., van Wee, B.: Accessibility evaluation of land-use and transport strategies: review and research directions. Journal of Transport Geography 12, 127–140 (2004)
8. Wegener, M.: Overview of land-use transport models. In: Henscher, D.A., Button, K. (eds.) Handbook in Transport. Transport Geography and Spatial Systems, vol. 5, pp. 127–146. Pergamon/Elsevier Science, Kidlington, UK (2005)

9. Fotheringham, A.S., O'Kelly, M.E.: Spatial Interaction Models: Formulations and Applications. Kluwer Academic Publishers, Dordrecht (1989)

10. Goodchild, M.: Geographic information science. International Journal of Geographical Information Systems 6, 31–45 (1992)

11. Burger, M., van Oort, F., Linders, G.-J.: On the Specification of the Gravity Model of Trade: Zeros, Excess Zeros and Zero-inflated Estimation. In: Spatial Economic Analysis, pp. 167–190. Taylor & Francis (2009), doi:10.1080/17421770902834327

12. Fotheringham, A.S.: Spatial flows and spatial patterns. Environment and Planning A 16(4), 529–543 (1984), doi:10.1068/a160529

Axiomatics for Multivalued Dependencies in Table Databases: Correctness, Completeness, Completeness Criteria

Dmitriy Bui and Anna Puzikova

Taras Shevchenko National University of Kyiv, 64/13, Volodymyrska Street, City of Kyiv, Ukraine
buy@unicyb.kiev.ua, anna_inf@mail.ru

Abstract. Axiomatics for multivalued dependencies in table databases and axiomatics for functional and multivalued dependencies are reviewed; the completeness of these axiomatics is established in terms of coincidence of syntactic and semantic consequence relations; the completeness criteria for these axiomatic systems are formulated in terms of cardinalities (1) of the universal domain D, which is considering in interpretations, and (2) the scheme R, which is a parameter of all constructions, because only the tables which attributes belong to this scheme R are considering.

The results obtained in this paper and developed mathematical technique can be used for algorithmic support of normalization in table databases.

Keywords: table databases, functional dependencies, multivalued dependencies, completeness of axiomatic system.

1 Introduction

Data integrity of relational (table) databases is dependent on their logical design. To eliminate the known anomalies (update, insertion, deletion) the database normalization is required. Analysis of the current state of normalization theory in relation databases indicates that the accumulated theoretical researches not enough to satisfy the needs of database developers (see, survey [1] based on 54 sources). This is evidenced works devoted to the ways of solving existing problems of designing database schemas (see, for example, [2]) and improvement of algorithmic systems for normalization (see, for example, [3]).

The process of normalization is based, in particular, upon functional and multivalued dependencies theory the foundation of which is made by corresponding axiomatics and their completeness. The overview of research sources has shown that these axiomatic systems lack the proof of completeness that would comply with mathematical rigor; in fact, these vitally important results have a seemingly announcement status.

© Springer International Publishing Switzerland 2015
W. Zamojski et al. (eds.), *Theory and Engineering of Complex Systems and Dependability*,
Advances in Intelligent Systems and Computing 365, DOI: 10.1007/978-3-319-19216-1_5

Known CASE-tools (Computer-Aided Software Engineering tools) (for example, ERwin[1], Vantage Team Builder (Cadre), Silverrun[2]) perform normalization to 3NF (Third Normal Form). The results obtained in this paper can be used for development of the CASE-tools which support normalization. Since without proper mathematical results for the correctness and completeness of axiomatics the algorithms don't have foundation and keep heuristic in nature.

Features of this paper is clear separation of syntactic and semantic aspects.

All undefined concepts and notation are used in understanding of monograph [4], in particular, $s \mid X$ – restriction the row s to the set of attributes X. Symbol $\square$ means the end of statement or proof, symbol $\circ$ – end of logical part of proof.

2   Axiomatic for Multivalued Dependencies

Let t – a table, R – the scheme of the table t (finite set of attributes); X, Y, W, Z – subsets of scheme R; s, s_1, s_2 – the rows of table t. Henceforth we shall assume that set R and *universal domain* D *(the set, from witch attributes take on values in interpretations)* are fixed.

A multivalued dependence (MVD) $X \rightarrow\rightarrow Y$ is valid on the table t of the scheme R (see, for example, [4]), if for two arbitrary rows s_1, s_2 of table t which coincide on the set of attributes X, exists row $s_3 \in t$ which is equal to the union of restrictions of the rows s_1, s_2 to the sets of attributes $X \cup Y$ and $R \setminus (X \cup Y)$ respectively:

$$(X \rightarrow\rightarrow Y)(t) = true \overset{def}{\Leftrightarrow} \forall s_1, s_2 \in t(s_1 \mid X = s_2 \mid X \Rightarrow$$

$$\Rightarrow \exists s_3 \in t(s_3 = s_1 \mid (X \cup Y) \cup s_2 \mid R \setminus (X \cup Y))).$$

Thus, from the semantic point of view MVD – parametric predicate on tables of the scheme R defined by two (finite) parameter-sets of attributes X, Y.

Structure of table t, which complies with MVD $X \rightarrow\rightarrow Y$, can be represented using the following relation. We say that rows s_1, s_2 of table t are in the relation $=_X$, if they coincide on the set of attributes X:

$$s_1 =_X s_2 \overset{def}{\Leftrightarrow} s_1 \mid X = s_2 \mid X.$$

It is obvious that relation $=_X$ is equivalence relation and therefore it partitions the table s into equivalence classes, which are as follows:

$$[s]_{=_X} = \{s \mid X\} \otimes \pi_Y([s]_{=_X}) \otimes \pi_{R \setminus (X \cup Y)}([s]_{=_X}),$$

where s – arbitrary representative of the class.

---

[1] http://erwin.com/products/detail/ca_erwin_process_modeler/
[2] http://www.silverrun.com/

A table $t(R)$ is the model of a set of MVDs G, if each MVD $X \twoheadrightarrow Y \in G$ is valid on table $t(R)$:

$$t(R) \text{ is the } model \text{ of } G \overset{def}{\Leftrightarrow} \forall(X \twoheadrightarrow Y)(X \twoheadrightarrow Y \in G \Rightarrow (X \twoheadrightarrow Y)(t) = true).$$

Here and in the sequel the notion $t(R)$ stand for the table t of the scheme R. The next axioms and inference rules are valid [5].

Axiom of reflexivity: $\forall t(X \twoheadrightarrow Y)(t) = true$, where $Y \subseteq X$.

Axiom: $\forall t(X \twoheadrightarrow Y)(t) = true$, where $X \cup Y = R$.

Rule of *complementation*: $(X \twoheadrightarrow Y)(t) = true \Rightarrow (X \twoheadrightarrow R \setminus (X \cup Y))(t) = true$.

Rule of *augmentation*:
$(X \twoheadrightarrow Y)(t) = true \& Z \subseteq W \Rightarrow (X \cup W \twoheadrightarrow Y \cup Z)(t) = true$.

Rule of *transitivity*:
$(X \twoheadrightarrow Y)(t) = true \& (Y \twoheadrightarrow Z)(t) = true \Rightarrow (X \twoheadrightarrow Z \setminus Y)(t) = true$.

As an example we give the proof of the axiom of reflexivity.

Proof. Let s_1 and s_2 be the rows of table t for which $s_1 \mid X = s_2 \mid X$ is carried out. We show that the row $s_1 \mid (X \cup Y) \cup s_2 \mid R \setminus (X \cup Y)$ belongs to the table t. Restrict both parts of equality $s_1 \mid X = s_2 \mid X$ to the set Y: $(s_1 \mid X) \mid Y = (s_2 \mid X) \mid Y$. According to the property of restriction operator $((U \mid Y) \mid Z = U \mid (Y \cap Z))$ [1, p. 24]) it follows $s_1 \mid (X \cap Y) = s_2 \mid (X \cap Y)$. Consequently, and from the condition $Y \subseteq X$ we get $s_1 \mid Y = s_2 \mid Y$. According to the distributive property of restriction operator $(U \mid \bigcup_i X_i = \bigcup_i (U \mid X_i))$ [1, p. 24] it follows: $s_1 \mid (X \cup Y) \cup s_2 \mid R \setminus (X \cup Y) =$
$= s_1 \mid X \cup s_1 \mid Y \cup s_2 \mid R \setminus (X \cup Y) = s_2 \mid X \cup s_2 \mid Y \cup s_2 \mid R \setminus (X \cup Y) = s_2 \mid (X \cup Y \cup R \setminus (X \cup Y)) = s_2 \mid R = s_2$. $\quad\square$

The proof of other axiom and rules is given similarly.
A MVD $X \twoheadrightarrow Y$ is semantically deduced from the set of MVD's G, if at each table $t(R)$, which is the model of set G, MVD $X \twoheadrightarrow Y$ is valid too:

$$G \models X \twoheadrightarrow Y \overset{def}{\Leftrightarrow} \forall t(R)(t \text{ is the model of the } G \Rightarrow (X \twoheadrightarrow Y)(t) = true).$$

From above-mentioned axioms and inference rules follow corollaries.

Lemma 1. The next properties of the semantic consequence relation are valid:
1) $\varnothing \models X \twoheadrightarrow Y$ for $Y \subseteq X$;
2) $\varnothing \models X \twoheadrightarrow Y$ for $X \cup Y = R$;
3) $G \models X \twoheadrightarrow Y \Rightarrow G \models X \twoheadrightarrow R \setminus (X \cup Y)$;

4) $G \models X \rightarrow\rightarrow Y \& Z \subseteq W \Rightarrow G \models X \cup W \rightarrow\rightarrow Y \cup Z$;

5) $G \models X \rightarrow\rightarrow Y \& G \models Y \rightarrow\rightarrow Z \Rightarrow G \models X \rightarrow\rightarrow Z \setminus Y$;

6) $G \models X \rightarrow\rightarrow Y \& G \models Y \rightarrow\rightarrow Z \& Z \cap Y = \varnothing \Rightarrow G \models X \rightarrow\rightarrow Z$.   □

A MVD $X \rightarrow\rightarrow Y$ is syntactically derived from the set of MVD's G with respect to the scheme R ($G \models_R X \rightarrow\rightarrow Y$), if there is a finite sequence of MVD's $\varphi_1, \varphi_2, ..., \varphi_{m-1}, \varphi_m$ where $\varphi_m = X \rightarrow\rightarrow Y$ and for all $\forall i = \overline{1, m-1}$ each φ_i is either the axiom of reflexive or belongs to G, or is derived with some inference rule for MVD's (complementation, augmentation, transitivity) from the previous in this sequence φ_j, φ_k, $j, k < i$.

Let sequence $\varphi_1, \varphi_2, ..., \varphi_{m-1}, \varphi_m$ be called proof, following the tradition of mathematical logic [6].

Let there be given certain set of MVD's G. *Closure* $[G]_R$ is a set of all MVD's, that are syntactically derived from G:

$$[G]_R \overset{def}{=} \{X \rightarrow\rightarrow Y \mid G \models_R X \rightarrow\rightarrow Y\}.$$

For notational convenience, we write $\models$ for $\models_R$.

Lemma 2. Next properties are valid:
1) $G \subseteq [G]$ (*increase*);
2) $[[G]] = [G]$ (*idempotency*);
3) $G \subseteq H \Rightarrow [G] \subseteq [H]$ (*monotonicity*).   □

Proof. Let's prove proposition 1. Let MVD's $X \rightarrow\rightarrow Y \in G$, then $G \models X \rightarrow\rightarrow Y$ with one number of steps of proving, hence, $X \rightarrow\rightarrow Y \in [G]$.   □

Let's prove proposition 2. According to property 1, we have $[G] \subseteq [[G]]$. Let us prove the reverse inclusion $[[G]] \subseteq [G]$. Let $X \rightarrow\rightarrow Y$ – arbitrary MVD, such that $X \rightarrow\rightarrow Y \in [[G]]$. Then there is a finite sequence MVD's $\varphi_1, \varphi_2, ..., \varphi_{m-1}, \varphi_m$, such that $\varphi_m = X \rightarrow\rightarrow Y$ and for all $\forall i = \overline{1, m-1}$ each φ_i is either the axiom of reflexive property or belongs to $[G]$, or is derived with the help of some inference rule for MVD's from the previous in this sequence φ_j, φ_k, $j, k < i$. Let us make a new sequence according to such rules:

— if φ_i is the axiom of reflexivity, then we write down this MVD without any changes;

— if $\varphi_i \in [G]$, then according to the definition of closure this MVD has a finite proof $\psi_1, ..., \psi_{l-1}, \psi_l$ from G. Instead of MVD φ_i let's insert this proof;

— if φ_i is derived according to any inference rule from the previous in this sequence MVD's φ_j, φ_k, $j, k < i$, then also we write down φ_i without any changes.

Clearly, created in such a way the sequence is a proof of MVD $X \twoheadrightarrow Y$ from G that is $G \vdash X \twoheadrightarrow Y$, hence, $X \twoheadrightarrow Y \in [G]$.   □

Thereby, operator $G \mapsto [G]$ is closure operator in terms of [7].

Observe that properties of operator $G \mapsto [G]$ listed in Lemma 2, are carried out in axiomatic systems (see, for example, [8]).

From reflexivity axiom and inference rules indicated above is possible to get other inference rules for MVD's [2, 6].

Rule of *pseudo-transitivity*:
$$\{X \twoheadrightarrow Y, Y \cup W \twoheadrightarrow Z\} \vdash X \cup W \twoheadrightarrow Z \setminus (Y \cup W).$$

Rules of *difference*:
a) $\{X \twoheadrightarrow Y\} \vdash X \twoheadrightarrow Y \setminus X$;
b) $\{X \twoheadrightarrow Y \setminus X\} \vdash X \twoheadrightarrow Y$;
c) $\{X \twoheadrightarrow Y\} \vdash X \twoheadrightarrow R \setminus Y$.

Rule of union: $\{X \twoheadrightarrow Y_1, X \twoheadrightarrow Y_2\} \vdash X \twoheadrightarrow Y_1 \cup Y_2$.

Rules of decomposition:
a) $\{X \twoheadrightarrow Y_1, X \twoheadrightarrow Y_2\} \vdash X \twoheadrightarrow Y_1 \cap Y_2$;
b) $\{X \twoheadrightarrow Y_1, X \twoheadrightarrow Y_2\} \vdash X \twoheadrightarrow Y_1 \setminus Y_2$.   □

Lemma 3. The next properties are valid for $n = 2, 3, \ldots$:
1) $\{X \twoheadrightarrow Y_1, \ldots, X \twoheadrightarrow Y_n\} \vdash X \twoheadrightarrow Y_1 \cup \ldots \cup Y_n$;
2) $\{X \twoheadrightarrow Y_1, \ldots, X \twoheadrightarrow Y_n\} \vdash X \twoheadrightarrow Y_1 \cap \ldots \cap Y_n$.   □

The proofs of this lemma constructed by the induction in the n, according the rules of augmentation and transitivity.

3   Axiomatic for Multivalued Dependencies and Functional Dependencies

It will be recalled that a functional dependence $X \to Y$ is valid on the table t, if for two arbitrary rows s_1, s_2 of table t which coincide on the set of attributes X, their equality on the set of attributes Y is fulfilled (see, for example [4]), that is:

$$(X \to Y)(t) = true \overset{def}{\Leftrightarrow} \forall s_1, s_2 \in t \big(s_1 | X = s_2 | X \Rightarrow s_1 | Y = s_2 | Y \big).$$

Let there be given a sets F and G of FD's and MVD's respectively. A table $t(R)$ is the model of a set $F \cup G$, if each dependency $\varphi \in F \cup G$ is valid at table t :

$$t(R) \text{ is } model \text{ of } F \cup G \overset{def}{\Leftrightarrow} \forall \varphi \, (\varphi \in F \cup G \Rightarrow \varphi(t) = true).$$

Mixed inference rules for FD's and MVD's are valid [5].

1. Rule of *extension* FD to MVD

$(X \to Y)(t) = true \Rightarrow (X \to\to Y)(t) = true$.

2. $(X \to\to Z)(t) = true$ & $(Y \to Z')(t) = true$ & $Z' \subseteq Z$ & $Y \cap Z = \varnothing \Rightarrow$

$\Rightarrow (X \to Z')(t) = true$.

The proof of *extension* rule is presented, for example, in the monograph [1, p. 73]). Let's prove mixed inference rules for FD's and MVD's (proposition 2).

Proof. Let s_1 and s_2 be the rows of table t for which $s_1 \mid X = s_2 \mid X$ is carried out and MVD $X \to\to Z$ holds for table t . Therefore, there is row $s_3 \in t$ that $s_3 = s_1 \mid X \cup s_1 \mid Z \cup s_2 \mid R \setminus (X \cup Z)$. Let FD $Y \to Z'$ holds on table t , where $Z' \subseteq Z$ and $Y \cap Z = \varnothing$.

First we show that equality $s_2 \mid Y = s_3 \mid Y$ for rows s_2 i s_3 is fulfilled. From equalities $s_2 \mid X = s_3 \mid X$ (by assumption, $s_1 \mid X = s_2 \mid X$ and by construction of row s_3 , $s_1 \mid X = s_3 \mid X$) and $s_2 \mid R \setminus (X \cup Z) = s_3 \mid R \setminus (X \cup Z)$ (by construction of row s_3) we have $s_2 \mid (R \setminus (X \cup Z) \cup X) = s_3 \mid (R \setminus (X \cup Z) \cup X)$, that is [3] $s_2 \mid (R \setminus Z \cup X) = \ = s_3 \mid (R \setminus Z \cup X)$; hence and from inclusion $R \setminus Z \subseteq R \setminus Z \cup X$ we have equality $s_2 \mid (R \setminus Z) = s_3 \mid (R \setminus Z)$. By condition, we have $Y \cap Z = \varnothing$ therefore $Y \subseteq R \setminus Z$, hence, $s_2 \mid Y = s_3 \mid Y$.

By condition, we have $(Y \to Z')(t) = true$, hence $s_2 \mid Z' = s_3 \mid Z'$. Since $s_1 \mid Z = s_3 \mid Z$ (by construction of row s_3), then from inclusion $Z' \subseteq Z$ it follows $s_1 \mid Z' = s_3 \mid Z'$. Thus, for rows s_1 and s_2 which coincide on the set of attributes X , equality $s_1 \mid Z' = s_2 \mid Z'$ is fulfilled. Thus, FD $X \to Z'$ holds for table t . □

FD or MVD φ is semantically deduced from the set of dependencies $F \cup G$, if at each table $t(R)$, which is the model of a set of dependencies $F \cup G$, dependency φ is valid too:

$$F \cup G \models \varphi \overset{def}{\Leftrightarrow} \forall t(R)(t(R) - \text{model of } F \cup G \Rightarrow \varphi(t) = true) .$$

From above-mentioned mixed inference rules for FD's and MVD's follow corollaries (the properties of semantic consequence relation):

1. $F \models X \to Y \Rightarrow F \models X \to\to Y$;

2. $G \models X \to\to Z$ & $F \models Y \to Z'$ & $Z' \subseteq Z$ & $Y \cap Z = \varnothing \Rightarrow F \cup G \models X \to Z'$.

---

[3] It is required to take into account the succession of set-theoretic equalities $R \setminus (X \cup Z) \cup Z = (R \setminus X \cap R \setminus Z) \cup Z = (R \setminus X \cup Z) \cap \ (R \setminus X \cup Z) \cap \ (R \setminus Z \cup Z) = = (R \setminus X \cup Z) \cap R = R \setminus X \cup Z$.

Lemma 4. Let H_1 i H_2 – the sets of dependencies (FD's or MVD's) and T_1, T_2 – the sets of all their models respectively. Then implication $H_1 \subseteq H_2 \Rightarrow T_1 \supseteq T_2$ is carried out. □

Corollary 1. The next properties of the semantic consequence relation are valid:
1) $F \models \varphi \Rightarrow F \cup G \models \varphi$;
2) $G \models \varphi \Rightarrow F \cup G \models \varphi$. □

Lemma 5. The next properties of semantic consequence relation are valid:
1) $F \models X \rightarrow Y \Rightarrow F \cup G \models X \cup Z \rightarrow Y \cup Z$ for $Z \subseteq R$;
   $F \cup G \models X \rightarrow Y \Rightarrow F \cup G \models X \cup Z \rightarrow Y \cup Z$ for $Z \subseteq R$;
2) $F \models X \rightarrow Y \& F \models Y \rightarrow Z \Rightarrow F \cup G \models X \rightarrow Z$;
   $F \cup G \models X \rightarrow Y \& F \cup G \models Y \rightarrow Z \Rightarrow F \cup G \models X \rightarrow Z$;
3) $G \models X \rightarrow\rightarrow Y \Rightarrow F \cup G \models X \rightarrow\rightarrow R \setminus (X \cup Y)$;
   $F \cup G \models X \rightarrow\rightarrow Y \Rightarrow F \cup G \models X \rightarrow\rightarrow R \setminus (X \cup Y)$;
4) $G \models X \rightarrow\rightarrow Y \& Z \subseteq W \Rightarrow F \cup G \models X \cup W \rightarrow\rightarrow Y \cup Z$;
   $F \cup G \models X \rightarrow\rightarrow Y \& Z \subseteq W \Rightarrow F \cup G \models X \cup W \rightarrow\rightarrow Y \cup Z$;
5) $G \models X \rightarrow\rightarrow Y \& G \models Y \rightarrow\rightarrow Z \Rightarrow F \cup G \models X \rightarrow\rightarrow Z \setminus Y$;
   $F \cup G \models X \rightarrow\rightarrow Y \& F \cup G \models Y \rightarrow\rightarrow Z \Rightarrow F \cup G \models X \rightarrow\rightarrow Z \setminus Y$;
6) $F \models X \rightarrow Y \Rightarrow F \cup G \models X \rightarrow\rightarrow Y$;
   $F \cup G \models X \rightarrow Y \Rightarrow F \cup G \models X \rightarrow\rightarrow Y$;
7) $F \cup G \models X \rightarrow\rightarrow Z \& F \cup G \models Y \rightarrow Z' \& Z' \subseteq Z \& Y \cap Z = \emptyset \Rightarrow F \cup G \models X \rightarrow Z'$. □

FD or MVD φ is syntactically derived from the set of dependencies ($F \cup G \vdash_R \varphi$), if there is a finite sequence of FD or MVD $\varphi_1, \varphi_2, ..., \varphi_{m-1}, \varphi_m$ where $\varphi_m = \varphi$ and for all $\forall i = \overline{1, m-1}$ each φ_i is either the axiom of reflexivity (FD's or MVD's) or belongs to $F \cup G$ or is derived with some inference rule (complementation for MVD's, augmentation (for FD's or MVD's), transitivity (for FD's or MVD's), mixed inference rules for FD's and MVD's) from the previous in this sequence φ_j, φ_k, $j, k < i$.

As has been started above, let sequence $\varphi_1, \varphi_2, ..., \varphi_{m-1}, \varphi_m$ be called proof of φ from set of dependencies $F \cup G$.

Let there be given certain sets F and G of FD's and MVD's respectively. *Closure* $[F \cup G]_R$ – is a set of all FD's and MVD's that are syntactically derived from $F \cup G$:

$$[F \cup G]_R \overset{def}{=} \{\varphi \mid F \cup G \vdash \varphi\}.$$

Lemma 6. Next properties are valid:

1) $F \cup G \subseteq [F \cup G]$ (*increase*);
2) $[[F \cup G]] = [F \cup G]$ (*idempotency*);
3) $F' \cup G' \subseteq F \cup G \Rightarrow [F' \cup G'] \subseteq [F \cup G]$ [4] (*monotonicity*);
4) $[F] \subseteq [F \cup G]$, $[G] \subseteq [F \cup G]$;
5) $[F] \cup [G] \subseteq [F \cup G]$. □

From the propositions 1-3 it follows that operator $F \cup G \mapsto [F \cup G]_R$ is closure operator.

To be mentioned one more mixed rule for FD's and MVD's [5]:

$$\{X \rightarrow\rightarrow Y, X \cup Y \rightarrow Z\} \vdash X \rightarrow Z \setminus Y.$$

Closure $[X]_{F \cup G, R}$ of a set X (with respect to the set of dependencies $F \cup G$ and scheme R) is the family of all right parts MVD's which are syntactically derived from the set $F \cup G$:

$$[X]_{F \cup G, R} \overset{def}{=} \{Y \mid X \rightarrow\rightarrow Y \in [F \cup G]_R\}.$$

Obviously, $[X]_{F \cup G, R} \neq \varnothing$ since, for example, $X \in [X]_{F \cup G, R}$, ($X \rightarrow\rightarrow X$, $X \rightarrow X$ are axioms of reflexivity); the latter statement can be strengthened: actually performed inclusion $2^X \subseteq [X]_{F \cup G, R}$, where 2^X – Boolean of a set X.

Let $[X]_F$ – closure of a set X with respect to the set of FD's F [9]. Note that by definition $[X]_F \subseteq R$.

Lemma 7. Next properties are valid:

1) $Y \subseteq [X]_F \Rightarrow Y \in [X]_{F \cup G, R}$;
2) $[X]_{F \cup G, R} = [[X]_F]_{F \cup G, R}$. □

Observe that operator $X \mapsto [X]_{F \cup G, R}$ is not closure operator; it is based on the fact that this operator has no idempotency property (notion $[[X]_{F \cup G}]_{F \cup G, R}$ has no sense).

Basis $[X]_{F \cup G, R}^{bas}$ of a set X with respect to the set of dependencies $F \cup G$ and scheme R is subset of closure $[X]_{F \cup G, R}$, such that:

1) $\forall W (W \in [X]_{F \cup G, R}^{bas} \Rightarrow W \neq \varnothing)$ (i.e., basis contains only nonempty sets of attributes);
2) $\forall W_i W_j (W_i W_j \in [X]_{F \cup G, R}^{bas} \& W_i \neq W_j \Rightarrow W_i \cap W_j = \varnothing)$ (i.e., sets of basis are pairwise disjoint);

---

[4] From the fact that sets FD's and MVD's are disjoint it follows that inclusion $F' \cup G' \subseteq F \cup G$ is equivalent to the conjunction of inclusions $F' \subseteq F$, $G' \subseteq G$.

3) $\forall Y (Y \in [X]_{F \cup G, R} \Rightarrow \exists \mathfrak{I} (\mathfrak{I} \subseteq [X]_{F \cup G, R}^{bas} \ \& \ \mathfrak{I} - \text{finite} \ \& \ Y = \bigcup_{W \in \mathfrak{I}} W)$ (i.e., each
set of attributes from closure $[X]_{F \cup G, R}$ is equal to finite union of some sets from
basis).

Lemma 8. Next properties are valid:
1) $\bigcup_{W \in [X]_{F \cup G, R}^{bas}} W = R$ for $X \subseteq R$ (i.e. basic is partition R);

2) $A \in [X]_F \Rightarrow \{A\} \in [X]_{F \cup G, R}^{bas}$.   □

These lemmas are needed to establish the following main results.

4   Correctness and Completeness of Axiomatic for FD's and MFD's

Let φ – FD or MVD.

Statement 1 (Correctness of axiomatic for FD's and MFD's). If dependency φ is
syntactically derived from the set of dependencies $F \cup G$, then φ is derived semantically from $F \cup G$:

$$F \cup G \vdash \varphi \Rightarrow F \cup G \models \varphi .$$ □

The proof is carried out by induction in the length of proving.

Statement 2 (Completeness of axiomatic for FD's and MFD's). If dependency φ is
derived semantically from the set of dependencies $F \cup G$, then φ is syntactically
derived from $F \cup G$ under the assumption $|R| \geq 2$ and $|D| \geq 2$ [5]:

$$F \cup G \models \varphi \Rightarrow F \cup G \vdash \varphi .$$ □

Condition $|R| \geq 2$ and $|D| \geq 2$ is obtained through a detailed analysis of the
proofs.

Theorem 1. The relations of semantic and syntactic succession coincide for axiomat-
ic of FD's and MFD's under the assumption $|D| \geq 2$ and $|R| \geq 2$:

$$F \cup G \models \varphi \Leftrightarrow F \cup G \vdash \varphi .$$ □

The proof follows directly from statements 1 and 2.
Analogous theorem holds for axiomatic of MFD's (for axiomatic of FD's see [9]).

---

[5] For details see further.

5   Completeness Criteria for Axiomatic of FD's and MFD's

Analysis of the proof of the main result (Theorem 1) shows that it is constructed under the assumption $|D| \geq 2$ and $|R| \geq 2$.

The dependence of coincidence of relations of syntactic and semantic succession for different values of cardinalities of the sets R and D is indicated respectively in the tables 1 and 2. The symbol "+" (respectively "-") in the cell means that these relations coincide (do not coincide respectively) under specified assumptions.

Table 1. All variants of cardinalities of the sets R and D for axiomatic of MFD's

D \ R	$|R|=0$	$|R|=1$	$|R| \geq 2$
$|D|=0$	+	+	−
$|D|=1$	+	+	−
$|D| \geq 2$	+	+	+

Table 2. All variants of cardinalities of the sets R and D for axiomatic of FD's and MFD's

D \ R	$|R|=0$	$|R|=1$	$|R| \geq 2$
$|D|=0$	+	−	−
$|D|=1$	+	−	−
$|D| \geq 2$	+	+	+

The above table shows the following main results.

Theorem 2. The relations of semantic and syntactic succession coincide for axiomatic MVD's if and only if $|R| \leq 1$ or ($|R| \geq 2$ & $|D| \geq 2$).   □

Theorem 3. The relations of semantic and syntactic succession coincide for axiomatic of FD's and MVD's if and only if $|D| \geq 2$ or $|R| = 0$.   □

6   Conclusion

In this paper we construct a fragment of the mathematical theory of normalization in relational (table) databases – considered axiomatic for multivalued dependencies and axiomatic for functional and multivalued dependencies. For each axiomatic relations of syntactic and semantic succession are considered and the conditions under which these relations coincide (do not coincide) found.

In particular, it is shown that known in the literature proof of the completeness of these axiomatics constructed under the assumption for scheme R and universal domain D: $|R| \geq 2$, $|D| \geq 2$.

The authors believe that the demonstrated approach and developed mathematical apparatus can be successfully used for other tasks of data modelling.

The next challenge – research of independence of axiomatic' components (axioms and inference rules).

References

1. Bui, D.B., Puzikova, A.V.: Theory of Normalization in Relational Databases (A Survey), `http://www.khai.edu/csp/nauchportal/Arhiv/REKS/2014/REKS514/BuyPuz.pdf`

2. Darwen, H., Date, C., Fagin, R.: A Normal Form for Preventing Redundant Tuples in Relational Databases. In: Proceedings of the 15th International Conference on Database Theory – ICDT 2012, Berlin, Germany, March 26-30, pp. 114–126 (2012)

3. Bahmani, A., Naghibzadeh, M., Bahmani, B.: Automatic database normalization and primary key generation. In: CCECE/CCGEI, Niagara Falls, Canada, May 5-7, pp. 11–16 (2008)

4. Redko, V.N., et al.: Reliatsiini bazy danykh: tablychni alhebry ta SQL-podibni movy. Akademperiodyka, Kyiv (2001)

5. Beeri, C., Fagin, R., Howard, J.: A complete axiomatization for functional and multivalued dependencies. In: Proceedings of the ACM-SIGMOD Conference, Toronto, Canada, August 3-5, pp. 47–61 (1977)

6. Lyndon, R.: Notes on Logic. D. Van Nostrand Company, Inc., Princeton (1966)

7. Skornyakov, L.A.: Elementyi teorii struktur. Nauka, Moskva (1982)

8. Shoenfield, J.: Mathematical logic. Addison-Wesley (1967)

9. Bui, D.B., Puzikova, A.V.: Completeness of Armstrong's axiomatic. Bulletin of Taras Shevchenko National University of Kyiv. Series Physics & Mathematics 3, 103–108 (2011)

On Irrational Oscillations of a Bipendulum

Alexander P. Buslaev[1], Alexander G. Tatashev[2], and Marina V. Yashina[2]

[1] Moscow State Automobile and Road Technical University (MADI) and MTUCI,
125319, Leningradskii pr. 64, Moscow, Russia
apal2006@yandex.ru
[2] Moscow Technical University of Communications and Informatics (MTUCI) and MADI,
111124, 8-a Aviamotornaya Str., Moscow, Russia
yash-marina@yandex.ru

Abstract. In this paper there are discussed the problems associated with yearly introduced by authors the class of dynamical systems, which have occurred from traffic. But now these dynamical systems become one of indicators of rational and irrational numbers connected with computer sciences basics.

Some results of computer experiments are presented. Also hypothesis and future works are provided.

Keywords: Transport-logistic problem, dynamical system, multi-dimensional Turing tape, positional numeral system.

1 Introduction. Pre-Planning Dynamical System (PPDS)

We consider the algebraic interpretation of the dynamical system that was introduced in [1], [2]. There are *two vertices* V_0, V_1 *two particles* P_0, P_1 and *discrete time* $T = 0, 1$, Each particle is at one of two vertices at each time instant, Fig. 1.

Suppose the following binary representations of two numbers a_0, a_1, $0 \leq a_0, a_1 < 1$,

$$a_0 = 0, a_{01} a_{02} \dots a_{0k} \dots, \tag{a}$$

$$a_1 = 0, a_{11} a_{12} \dots a_{1k} \dots, \tag{b}$$

determine the plan of particles relocations, which is called *a multidimensional Turing tape.*

Each particle reads one of digits on its Turing tape TTP_i at every time instant. The particle P_i must be located, at successive time instants, in the vertices $V_{a_i(T)}$ in accordance with the plan of this particle, $i = 0, 1, T = 1,2, \dots$. If the particle P_i is in the vertex $V_{a_i(T)}$ at the instant T, then, at the *instant $T + 1$, $T = 0,1,2, \dots$* this particle will be in the vertex $V_{a_i(T+1)}$ except the cases of competitions. *A competition takes place if two particles try to move in the opposite directions,* Fig. 1. In the case of a competition, only one of two competing particles wins the competition. This particle changes its state of its Turing tape on one positions to the left. The other particle does not change its state and its tape does not move. Thus, the implementation of the plan is delayed at least for one step. We assume that some initial conditions are given. These conditions determine the location of particles at the instant $T = 0$.

W. Zamojski et al. (eds.), *Theory and Engineering of Complex Systems and Dependability*,
Advances in Intelligent Systems and Computing 365, DOI: 10.1007/978-3-319-19216-1_6

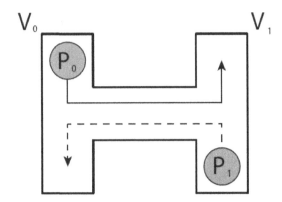

Fig. 1. Movement in the opposite directions through the channel is forbidden

2 Characteristics of Bipendulums

Denote by $D_i(T)$ the number of delays for the plan a_i of relocation of the particle P_i over the time interval $(0,T]$, $i = 0,1$, $T = 1, 2, \ldots$ Let

$$c_i(T) = D_i(T)/T \tag{1}$$

be the average conflict rate of P_i and

$$w_i(T) = (T - D_i(T))/T \tag{2}$$

be the average velocity of TTP_i. The limit

$$c_i = \lim_{T \to \infty} c_i(T) \tag{3}$$

is called *the conflict rate of the particle P_i tape* if this limit exists.
The limit

$$w_i = \lim_{T \to \infty} w_i(T) \tag{4}$$

is called *the velocity of the TTP_i*, if this limit exists. This limit characterizes the measure of the particle P_i plan implementation.

The value $w_i = (w_0 + w_1)/2$ is called the tapes velocity of the system, and $c_i = (c + c_1)/2$ is called the conflict rate of the system.

Since we assume that each particle wins every conflict with probability $1/2$, then $w_0 = w_1 = w$.

The system is in the state of *synergy* if no conflicts takes places after an instant T_{syn}. If the system comes to the state of synergy, then the velocities of tapes are equal to one. The problem is *to study the tapes velocity and the conditions of synergy*. The next problem is *to study conditions of existence of the limits* (3), (4).

3   The Classification of Bipendulums

(3.1) Rational bipendilum   **(RBP)**

A bipendulum is called   *rational* if both the plans are rational numbers.

(3.2) Irrational bipendilum **(IBP)**

A bipendulum is called   *irrational* if at least one of two plans is an irrational number.

(3.3) Phase bipendilum   **(PhBP)**

A bipendulumis called *a   phase bipendulum* if we get one of the plans, shifting the representation of the other bipendulum onto a number of positions.

(3.4) Random walking bipendilum   **(RWBP)**

We consider also the dynamical system with stochastic planning.

The definition of this system suggests that the digit of   *(T+1)* - plan of every particle is determined independently at time *T* and it is equal to 0 or 1 with probability *0.5*.

4   Some Properties of Rational Bipendulums (RBP)

We have obtained the following result, [3],[4].

(4.1) *If a_0 and a_1 are rational numbers, then there exist the limits (3)-(4).*
From this fact, it follows that the conflict rate of a rational bipendulum is well defined problem.

(4.2) *The minimum possible value of the tape velocity of   RBP equals to $\frac{4}{5}$, and the maximum possible tape velocity of RBP equals to 1. The maximum possible conflict rate of RBP equals $\frac{2}{5}$.*

(4.3) *For other conflict resolution rules, the tape velocity of RBP can be equal to $\frac{1}{2}$, but not less than $\frac{1}{2}$.*

5   Random Walking Bipendilum (RWBP)

We have obtained the following, [3].

(5.1) *The tapes velocity of   RWBP   is equal to $w = \frac{19}{20}$.*

(5.2) *The conflict rate of   RWBP   is equal to $c = \frac{1}{10}$.*

6   Irrational Bipendilum (IBP)

6.1   Problems

Some examples, [3], show that limits (3) - (4) do not exist. We cannot represent the process of the system work for irrational plans as a finite Markov chain. The numerical characteristics of bipendulums are discontinuous, because the significance of digits of the position representation of a real number is exponentially decrease, but the significance of digits of plan representation of IBR do not decrease. This is a reason of difficulties.

The problem of constructive determination also exists. It is not always possible to know whether two different algorithms give representations of the same number.

The problem also relates to the concept of normal number, i.e., number with equiprobable distribution of digit values.

6.2   The Tape Velocity Simulations of IBP for Famous Irrational Plans

We investigated, by simulation, the tapes velocity for famous irrational numbers. We assume that each of numbers a_0 and a_1 are equal to one of the numbers $\sqrt{2} \bmod 1$, $\sqrt{3} \bmod 1$, $\pi \bmod 1$, $5 \bmod 1$, $\frac{1}{\varphi} = \frac{\sqrt{5} \bmod 1}{2}$, and $a_0 \neq a_1$. The results of simulation are shown in Fig. 2.

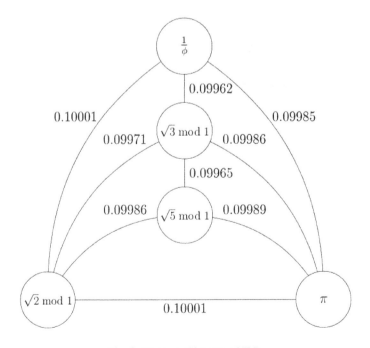

Fig. 2. Mean conflict rate of IBP

Binary representation of the number π with *30000* digits was found at [5]. Binary representations of other numbers with *500000* digits were computed using square root and basic arithmetic functions from arbitrary size float math package [6].

Then they're proved to be of the claimed precision by digit-by-digit squaring and comparison with the expected exact square. The length of the simulation interval corresponds to *30000* digits for simulations with π , and to *500000* digits for simulations with other numbers.

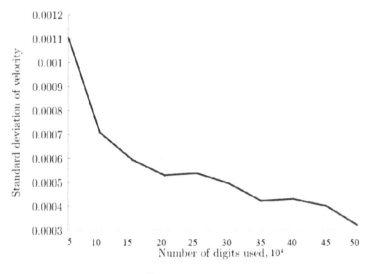

Fig. 3. $a_0 = \sqrt{5} \bmod 1$, $a_0 = \varphi \bmod 1$

We conclude from this experiments the following.

(6.2.1) *For the other pairs* a_0 *and* a_1 *the velocity w(T) converges to* $\frac{19}{20}$ *as* $T \to \infty$, *Fig. 2, 3 .*

(6.2.2) The conflict rate $C(T)$ converges to $\frac{1}{10}$.

(6.2.3) *If one of the number* a_0 *and* a_1 *equals* $\sqrt{5} \bmod 1$ *and the other number equals* $\frac{1}{\varphi}$, *then the value* c(T) *does not converge numerically, Fig. 4. Probably, it is not enough the volume of calculation for phase bipendulums in this case.*

(6.2.4)   *Thus IBP simulation gives the same tape velocity as for the RWBP.*

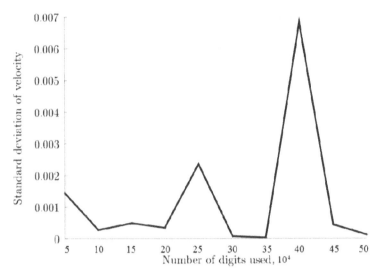

Fig. 4. $a_0 = \sqrt{2} \; mod \; 1$, $a_1 = 3 \; mod \; 1$

6.3   An Approach of Plans Determination

We can use a constructive approach of plans determination. For example, consider the square root of real numbers. Let the plans a_0 and a_1 be irrational numbers such that $x_i^{(n)}$ tends to a_i, $(i=0,1)$.

$$a_i = \lim_{n \to \infty} x_n^{(i)}, \; i = 0,1,$$

where $x_n^{(i)}$ is a rational number,

$$x_{n+1}^{(i)} = f\big(x_n^{(i)}, m_i\big), \; i = 0,1,$$

$f(x,m)$ is a given function, m is a parameter. A conflict rate is defined as

$$\lim_{n \to \infty} C\big(x_n^{(i)}, x_n^{(1)}\big) = \; C(a_0, a_1),$$

where $C(a,b)$ is conflict rate of the bipendulum with plans (a) and (b). For example,

$$x_{n+1}^{(i)} = \frac{1}{2}\Big(x_n^{(i)} + \frac{m_i}{x_n^{(i)}}\Big), \tag{5}$$

$$x_0^{(i)} \approx \sqrt{m_i}, \tag{6}$$

$i = 0, 1.$

7   Future Works

(7.1) The quality analysis for *PPDS* system with rational plans (a) and (b) was investigated in [3].

What can we say about irrational plans in common case?

(7.2) We plan to investigate several versions and generalizations of the *PPDS* system. One of these generalization is the system with vertices and M particles, where N and M are arbitrary numbers. It is interesting to consider an asymptotic of the conflict rate if $N \to \infty$.

(7.3) Some problems relate to approach, described in Subsection 6.3. We must choice an appropriate metric in definition of convergence of plans and investigate the continuity of conflict rate in this metric.

Acknowledgements. The computer simulations was preferred by students Kuchelev D.A. and Schepshelevich S.S.

References

1. Kozlov, V.V., Buslaev, A.P., Tatashev, A.G.: Monotonic walks on a necklace and coloured dynamic vector. International Journal of Computer Mathematics (2014), doi:1080/00207150.2014.915964
2. Kozlov, V.V., Buslaev, A.P., Tatashev, A.G.: A dynamical communication system on a network. Journal of Computational and Applied Mathematics 275, 247–261 (2015), doi:10.1016/j.cam.2014.07.026
3. Kozlov, V.V., Buslaev, A.P., Tatashev, A.G.: On real-valued oscillations of bipendulum. Applied Mathematical Letters, 1–6 (2015), doi:10.10.1016/j.aml.2015.02.003
4. Buslaev, A.P., Tatashev, A.G.: Bernoulli algebra on common fractions. International Journal of Computer Mathematics, 1–6 (2015) (in press)
5. A binary representation of the number Pi, http://www.befria.nu/elias/pi/binpi.html
6. The Perl 5 language interpreter, http://perldoc.perl.org/Math/BigInt.html

Algorithmic and Software Aspects of Information System Implementation for Road Maintenance Management

Alexander P. Buslaev[1,2], Marina V. Yashina[2,1], and Mikhail Volkov[2]

[1] MADI, 64, Leningradskiy Pr., Moscow, 125319, Russia
apal2006@yandex.ru
[2] MTUCI, 8a, Aviamotornaya Street, Moscow, 111024, Russia
yash-marina@yandex.ru, mmvolkov@gmail.com

Abstract. A distributed information system for business processes of road maintenance is presented in the paper. The architecture of developed system is the type of client-server system with terminal devices such as smart phones with the Android and IOS Operating Systems. The system has been called "Server – Smartphone – Student – Receiver (SSSR) – Road", and it gives possibility to fulfill the optimal scheduling, local routing for services movements and maintenance organization on the road traffic network by communications technology. Problems of dependability and system reliability are essentially developing on the accuracy of objects positioning. So methods of accuracy improving are provided in the paper.

Keywords: distributed information systems, scheduling and management, road networks maintenance, GPS\GLONASS, navigation, iOS, Android.

1 Introduction

Intelligent Transport Systems of megalopolis have been constantly improved by the development of electronics and Internet technology. Software design for transport is widely interested by specialists in Computer science.

Analyzing scientific literature of this scope we can formulate the following main areas of applications of intelligent systems for transport:

— traffic management of motorways;
— vehicles avoidance collision and the safety of their movement;
— electronic payment system of transport services;
— traffic management of main street networks;
— consequences elimination of accidents and safety providing;
— navigation.

In [1], a new approach of mathematical modeling of particle flows in the one-dimensional one-way supports was proposed. It allows designing distributed systems of monitoring and management for transport flows on complex networks by using modern information and communication technologies. Thereat, in [1], the Server – Smartphone – Student – Receiver (SSSR) system concept was formulated. The system with mobile

W. Zamojski et al. (eds.), *Theory and Engineering of Complex Systems and Dependability*,
Advances in Intelligent Systems and Computing 365, DOI: 10.1007/978-3-319-19216-1_7

phones, as terminal devices, collects structured information, distributed over the whole network, processes it, analyzes and makes decisions. These ideas were developed for analysis of real-time characteristics of saturated flows, [2], and [3].

It is necessary to have cartographic navigation data in many applications. At present, it is important to get exact connection between coordinates received from the GPS or GLONASS devices and real objects with the development of mobile and embedded systems for monitoring and traffic control.

Now great attention is given to problems of transport. One of urgent problems is optimization model of transportation in a region. In distributed systems, the response services have to get necessary link between a moving object position and the edges of the road network graph, [4]. There are relevant models for other kinds of transport, such as sea transport, [5]. It is important to get real traffic data and to implement this type of optimization models, preferably arranged in the flow of information. In the paper, we discuss methods allowed to extract such information flows.

For the adequacy of integrated models for traffic control in server applications, it is necessary to use theoretical results of objects movement modeling on networks with complex geometry. In this regard, it should be noted, [5].

2   Information System for Monitoring and Management of Enterprise Organization on Distributed Network Using Terminal Devices

Starting in 2011, Server – Smartphone – Student – Receiver(SSSR)-AN system, [1], was developed for monitoring of road pavement defects by road maintenance services. The system works with client devices was based on the operating system Windows Mobile, where the data were sent to the FTP-server, [6].

Currently, we develop a new platform for the implementation of the SSSR-Road system. A system of monitoring and maintenance distributed problems and elimination of defects of the road pavement and infrastructure has been created for the company carrying out maintenance and repair of road network of federal highways.

The system consists of the components and methods of collecting information processing. The collecting of information by mobile devices is fulfilled with using specially installed software and the server software stores and processes of received information.

The system is implemented in accordance with the client-server architecture. Information is stored in a database that is located on a remote server. The problem of filling and correction is the main database, especially if the monitoring objects are distributed by distance. The modern approach is to use smart phones as endpoints.

For example, in [1] how to use the system formed SSSR in problems of transport. SSSR system is used to collect information in real-time, [2], and may also be used to control a group of clients on network from a central location, [8]. In this paper we discuss the application of the system SSSR for scheduling processes. System feature is data distribution in space and time.

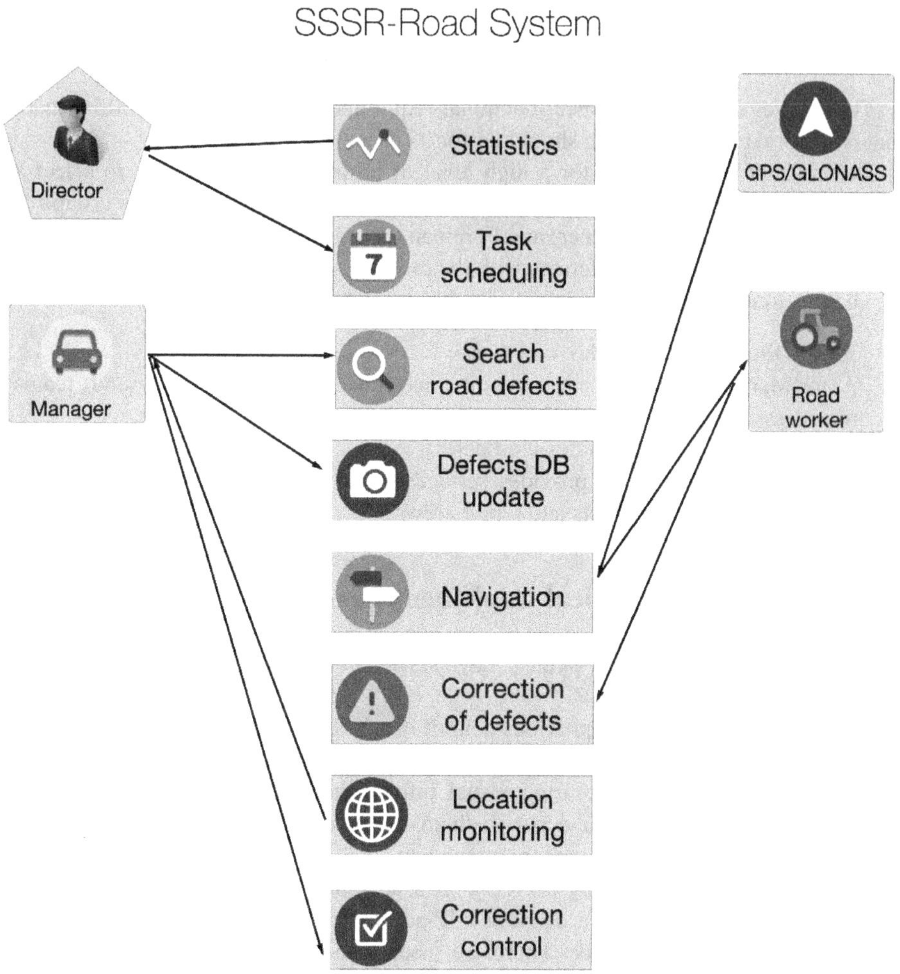

SSSR-Road System

Fig. 1. Concept of SSSR-Road system

The system of monitoring the state of the road infrastructure has two main functions: state monitoring of road management and road maintenance. The close functions include planning of repair road, repair crews monitoring locations, monitoring of service implementation, the search of road defect location, navigation road crews to a point defect. These functions define the following use cases:

— work scheduling;
— view info graphics and statistics for the road works;
— addition of defects to the base;
— control of the job;
— search for a defect location;
— navigate to the point defect;

— renovate of road defect;
— select a task to perform.

These cases establish the basic functional requirements of the system SSSR-Road, that is, reported that the system should do for its users.

Defining the requirements for a high level of abstraction is necessary to select a group of people interacting with the system. With a system interacting three groups of people: Director, Master (Worker) and Foreman (Manager) (Figure 1).

Additionally, the system interacts with the external system GPS / GLONASS.

These actors play the following roles in the system:

- Actor Director monitors the road works
- Actor Manager finds defects, enters the coordinates of the database, sends teams to address them and controls the removal of defects
- Actor Road worker eliminates road defect
- Actor GPS / GLONASS provides services in certain geographic coordinates to track the location of defects and repair crews

3   Features of Technical Implementation

Modern mobile devices, smart phones, have sufficient capacity to perform basic tasks of the system.

Module GPS / GLONASS provides data of the geographical location of the device with special satellites, speed of movement in space and other statistical information.

Built-in camera allows capturing visual information about the state of roads or of the different situations that occur to them. Software features allow sending this information to the server, accompanying geodata.

The system uses a stationary server. It is a personal computer running Windows 2008 Server. The server has a dedicated IP-address and installed services database servers and HTTP Web server. Therefore, users can connect external devices and exchange information with the server.

Web service has been built in accordance with the methodology of RESTfull service. RESTfull web services uses Persistence API to communicate with databases. More specifically, RESTfull web services integrates the entity classes and the persistence unit as defined in the Persistence API. Entity classes are classes associated with objects in a relational database.

The server performing the function is storing and processing information, also making summary reports and publishing   them on the Internet (Figure 2).

System uses JSON method for data output to external devices. JSON-text is the (encrypted) set of pairs <key: value>. In various software languages, the process is realized as an object, record, structure, dictionary, hash table, a list of key or an associative array. The key can only be a string with value in any form. They can be an ordered set of values. In many programming languages, data structure is realized by array, vector, list, or sequence.

Fig. 2. Web-interface of distributed system

The system uses the GPS and GLONASS navigation for getting link between the data and the domain. Typical accuracy for autonomous civilian GPS receivers in the horizontal plane is about 15 meters (about 5 in good visibility of satellites).

The client part of the system consists of 2 program products: "Master" app and "foreman" app. Application "Master" collects graphics (photos from the camera of the smartphone) and statistics (data from the GPS) information in special interface. In addition, the data block contains information about the user who fulfills the acts and place the data on the type of fixed faults, defects, etc.

Mobile operators send all captured information to an external server via any available Internet connection: 3G / 4G networks, Wi-Fi via Ethernet.

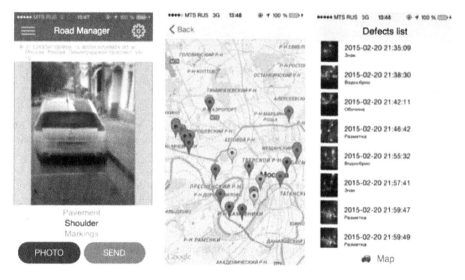

Fig. 3. The interface of the mobile client for iOS

The mobile client is implemented as a native application for the two most popular mobile platforms - Apple iOS (Figure 3) and Android. Android Application written in JAVA using the Android SDK and Google services (Figure 4).

IOS app is written in a new language swift and uses the framework Cocoa Touch. Library Cocoa Touch provides an abstraction layer for iOS. Cocoa Touch is based on the framework classes Cocoa, used in Mac OS X, and similarly it allows you to use Objective-C language or swift. Cocoa Touch should design pattern Model-View-Controller. Swift is the result of the latest research in the field of programming languages combined with years of experience in the development of Apple platforms. Named arguments from Objective-C code is transferred to net Swift, making the API on Swift easier to read and maintain. Type inference makes the code cleaner and eliminates errors at the same time get rid of the header modules and use namespaces (namespaces). Memory management is automatic.

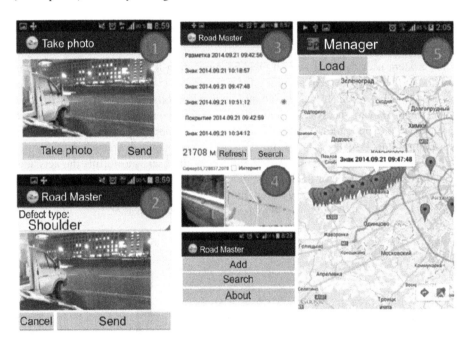

Fig. 4. Android Application Interface, 1 - screen photo of fixing the defect, 2 - the screen to send a photo of the defect, 3 - selection screen of the defect base, 4 - the start screen of the program, 5 - defect map

Information is sending to the server where information is cataloged and processed: photos flipped for easy perception, statistical and text, information are displayed as a table.

The results can be viewed on any computer or mobile phone with access to the Internet. Information can be sorted by type of data plots using browser.

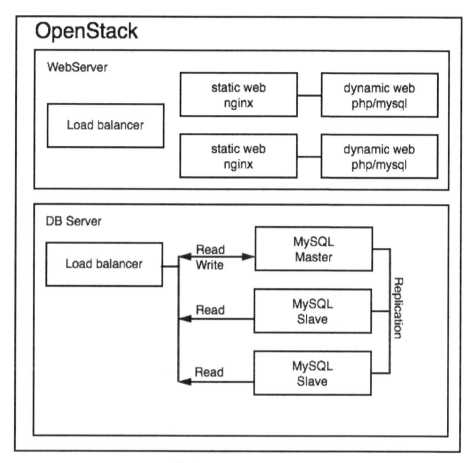

Fig. 5. Backend replication

So new system provides several new features and advantages:

1. Interactive user notification of new defects;
2. Better positioning accuracy;
3. Modern UI of iOS and Android applications for modern devices;
4. Increased server performance;
5. Optimization of transmitting/receiving traffic (improved protocols);
6. Increased server stability (new system supports data replication and web/application/data base server redundancy).

4   Methods for Increment of GPS-Data Accuracy

The accuracy of the GPS receiver depends on different factors. Main factors impact on the radio path for its passage from the satellite to the receiver. They are described below. Another cause depends on small errors in the satellite atomic clocks, and control the flight path of satellites on orbit.

Positioning accuracy is of great importance for ensuring the reliability and stability of the system SSSR-Road as a whole.

In the monitoring system of the road sector [5,6] describing the point with defect data are linked both to the map (graph of the road network), site-specific edges of the graph (indicating kilometers and meters from the beginning of the edge (roads, trails)) and a predetermined area of the partition area of monitoring.

In many cases, it is useful to determine not only an edge of the road networks graph, but also exact number of road lane on which the point is located. This problem can be particularly difficult in the case of insufficient accuracy of navigation data and on the site of the road network of complex configuration (multi-level interchanges, conventions and so on. D.).

Radio signal is distorted with the passage of charged particles through ionosphere (upper atmosphere at altitudes of 60 to 1,000 miles above the earth's surface), and then through the vapor cloud in a downstream troposphere. Because of this, there are small errors in the synchronization, so as the signal passes a distance slightly greater than the distance in a straight line. Ionospheric disturbance can be compensated by two methods:

1. modeling of errors;
2. measuring at two frequencies.

In the modelling of error, mathematical model is used, that allows to predict the typical error of the impact of the ionosphere and the troposphere on a particular day, then makes adjustments. This method is used to counter inaccuracies in civilian receivers GPS.

Method of measuring the two frequencies is more accurate and used for military receivers. It is based on particular radio signals of different frequencies, like the ability to a variety of largest refractive distortions. By comparing the differences in the timing of the two signals of different frequencies, current value refraction can be calculated.

GPS satellites transmit signals on two frequencies denoted L1 and L2. Civilian receivers can receive only signals at a frequency of L1, while working with both military frequencies, allowing them to perform appropriate correction.

In the above table it shows the contribution of various sources of error in a typical margin of error for GPS guidance location:

Table 1. Contribution of error sources

Source of error	Value of error
Watch satellite	1.5 m
Deflection	2.5 m
Ionosphere	5.0 m
Troposphere	0.5 m
Multipath error	0.5 m
Typical error reading	10 m

Method of positioning accuracy improving is based on the account of these factors. It is integrated into the SSSR-Road system with the incremental static correlation processing geo-data object in the previous steps to ensure their continuity.

5 Conclusion

The control system has been tested in actual operation on the road operating company serving the complex distribution of the transport network.

An efficient client-server system for receiving, processing and receiving information in real time and distributed over the territory.

Embedded applications for all authors of the system, i.e., road masters, supervisors and directors.

Methods to improve the accuracy of GPS data are presented that allow to detail information about the objects with an area not exceeding $1 \, m^2$. The proposed methods get possibilities to classify geo-objects processed by road services, up to a strip of the road.

Thus, routing navigation to the specified object, built-in system enables optimal solutions to determine the shortest path to the goal. A series of experiments showed that errors identification lanes decreases on order, and reached 7%.

References

1. Bugaev, A.S., Buslaev, A.P., Kozlov, V.V., Yashina, M.V.: Distributed Problems of Monitoring and Modern Approaches to Traffic Modeling. In: 14th International IEEE Conference on Intelligent Transportation Systems (ITSC 2011), Washington, USA, October 5-7, pp. 477–481 (2011), doi:10.1109/ITSC.20116082805
2. Yashina, M.V., Provorov, A.V.: Verification of Infocommunication System Components for Modeling and Control of Saturated Traffic in Megalopolis. In: Zamojski, W., Mazurkiewicz, J., Sugier, J., Walkowiak, T., Kacprzyk, J. (eds.) New Results in Dependability & Comput. Syst. AISC, vol. 224, pp. 531–542. Springer, Heidelberg (2013), doi:10.1007/978-3-319-00945-2_49
3. Buslaev, A.P., Strusinskiy, P.M.: Computer Simulation Analysis of Cluster Model of Totally-Connected Flows on the Chain Mail. In: Zamojski, W., Mazurkiewicz, J., Sugier, J., Walkowiak, T., Kacprzyk, J. (eds.) New Results in Dependability & Comput. Syst. AISC, vol. 224, pp. 63–73. Springer, Heidelberg (2013), doi:10.1007/978-3-319-00945-2_6
4. Gorodnichev, M.G., Nigmatulin, A.N.: Technical and Program Aspects on Monitoring of Highway Flows (Case Study of Moscow City). In: Zamojski, W., Mazurkiewicz, J., Sugier, J., Walkowiak, T., Kacprzyk, J. (eds.) New Results in Dependability & Comput. Syst. AISC, vol. 224, pp. 195–204. Springer, Heidelberg (2013), doi:10.1007/978-3-319-00945-2_17
5. Caban, D., Walkowiak, T.: Reliability Analysis of Discrete Transportation Systems Using Critical States. In: Zamojski, W., Mazurkiewicz, J., Sugier, J., Walkowiak, T., Kacprzyk, J. (eds.) New Results in Dependability & Comput. Syst. AISC, vol. 224, pp. 83–92. Springer, Heidelberg (2013), doi:10.1007/978-3-319-00945-2_8

6. Buslaev, A.P., Abishov, R.G., Volkov, M.M., Yashina, M.V.: Distributed Computing and intelligent monitoring of complex systems. Part 1: Basic programming technology pro-client applications on smartphones. Textbook. Publisher MTUCI, Moscow (2011)
7. Buslaev, A.P., Yashina, M.V., Provorov, A.V., Volkov, M.M.: SSSR - system in Nondestructive Measurement. In: Proceedings of the 9th Int. Conf. TGF 2011. T-Comm - Telecommunications and Transport. Publishing House Media publisher, Moscow (2011)
8. López, J.A.M., Ruiz-Aguilar, J.J., Turias, I., Cerbán, M., Jiménez-Come, M.J.: A Comparison of Forecasting Methods for Ro-Ro Traffic: A Case Study in the Strait of Gibraltar. In: Zamojski, W., Mazurkiewicz, J., Sugier, J., Walkowiak, T., Kacprzyk, J. (eds.) Proceedings of the Ninth International Conference on DepCoS-RELCOMEX. AISC, vol. 286, pp. 345–353. Springer, Heidelberg (2014), doi:10.1007/978-3-319-07013-1_33
9. Buslaev, A., Volkov, M.: Optimization and control of transport processes in the distributed systems. In: Zamojski, W., Mazurkiewicz, J., Sugier, J., Walkowiak, T., Kacprzyk, J. (eds.) Proceedings of the Ninth International Conference on DepCoS-RELCOMEX. AISC, vol. 286, pp. 123–132. Springer, Heidelberg (2014), doi:10.1007/978-3-319-07013-1_12

Climate Changes Prediction System Based on Weather Big Data Visualisation

Antoni Buszta[1] and Jacek Mazurkiewicz[2]

[1] Dolby Poland Sp. z o.o.
ul. Srubowa 1, 53-611 Wroclaw, Poland
Antoni.Buszta@dolby.com
[2] Department of Computer Engineering, Wroclaw University of Technology
ul. Wybrzeze Wyspianskiego 27, 50-370 Wroclaw, Poland
Jacek.Mazurkiewicz@pwr.edu.pl

Abstract. The paper introduces a new approach to weather forecasting. Overall prediction process consisted of processing big data, turning processed data to visualization, and later this visualization has been used for enhancing forecasting methods using artificial neural networks. The following assumptions are proved: data visualization gives additional interpretation possibilities, it is possible to enhance weather forecasting by data visualization, neural networks can be used for visual weather data analysis, neural networks can be used for climate changes prediction.

Keywords: weather forecasting, big data system, neural networks.

1 Introduction

Weather forecasting is nowadays relying mainly on computer-based models which takes into consideration multiple atmospheric factors [2]. However there is still needed human input to choose which prediction model is best for forecasting, to recognize weather impacting patterns, and to know what models are performing better or worse. Big data which is collected, processed and used for weather forecasting results in extensive computational needs. More than 15% of supercomputers, which has its application area specified, from the *Top 500 Supercomputers* list are dedicated only to weather and climate research [15]. Such high demand on processing speed and performance leads to searching for solutions which can improve already developed software. When data is too complicated to be understand directly, people are often using some approximation techniques, presenting data indirectly in multiple dimensions using tables or visualizing data [6]. One of the data presentation methods used most in meteorology is by using visualization [4]. Visual presentation is often more readable for humans than any other data presentation techniques. In this paper there will be created multiple weather visualizations to check how meteorological data presented in such way can enhance local weather changes prediction. Strong emphasis in this work will be put on creating automatic and generic algorithms which can reduce human

© Springer International Publishing Switzerland 2015
W. Zamojski et al. (eds.), *Theory and Engineering of Complex Systems and Dependability*,
Advances in Intelligent Systems and Computing 365, DOI: 10.1007/978-3-319-19216-1_8

interaction needed for forecasting. An attempt has been made for using neural networks in solving problem of visualizations interpretation and weather prediction. Artificial neural networks has been used for weather forecasting before [1], however researchers have not taken bigger effort to use visual input as part of the weather prediction. Visualizations may lead to many meaningful conclusions and this topic will be further analysed. Finally, processing huge amounts of data needed for visualization requires developing rules which can be easily followed when designing and implementing algorithms [7].

2   Classic Weather Data Analysis

The classic weather forecasting began near 650 BC, when the Babylonians started to predict weather from cloud patterns. Even early civilizations used reoccurring events to monitor changes in weather and seasons. There are five main weather forecasting methods [20]. *Persistence* - This is the most simple and primitive method of weather forecasting. The persistence method takes as main input information that the conditions will not change during whole time of the forecast. Despite looking wrong this way of predicting weather is well performing for regions where weather does not change so often like sunny and dry regions of California in the USA. *Trends* - Basing on speed and directions of weather changes simple trend model can be calculated. Atmospheric fronts, location of clouds, precipitation and high and low pressure centers are main factors of this method. This model is accurate for forecast weather only few hours ahead, especially for detecting storms or upcoming weather improvement. *Climatology* - This method bases on statistics and uses long term data accumulated over many years of observations. It is first method that can take into account wider spectrum of changes occurring in some region and thus can provide more accurate local predictions. The main disadvantage is ignoring extremes, as whole model output is averaged. It also does not recalculate on radical climate changes (like more and more warm winters) so it can provide year by year incorrect output. *Analog* - Analog method is more complicated than previous ones. Involves collecting data about current weather and trying to find analogous conditions in the past. This method assumes that in current situation the forecast is same as this one found in archival data. *Numerical Weather Prediction* - Most common method of forecasting nowadays is called Numerical Weather Prediction. First attempt to predict weather in this way was in 1920 but due to insufficient computer power realistic results was not provided until 1950. This method uses mathematical models of the atmosphere to calculate possible forecast. It can be used for both - short and long term weather predictions [12]. However despite supercomputer power increase weather calculations are so complicated that this model provide valid forecasts for up to one week only. Due to chaotic nature of atmosphere simulations is almost impossible to solve multiple equations considering them without any error. What is more even small error in the beginning of calculation yields in huge mistake in further forecast.

3   Weather Visualization Methods

One of the most common weather visualization techniques are satellite photos (example shown in Fig. 1.). This approach does not involve any preprocessing of the data. Images collected by International Space Station or satellites can provide information about current sky state above some region. This technique is often used for manual weather forecasting and for such tasks like storm detection or disaster warnings (tsunami, earthquakes, etc.). Wind can be visualized with consideration of speed, intensity, direction or other parameters therefore it can be a good source of interesting and meaningful images or animations. Fig. 1. shows how wind weather data can be presented. Other technique of data presentation in meteorology is by isopleths. *Isopleth* (also known as contour line or isoline) is a curve along which some parameter has constant value. There are multiple isopleths used in meteorology: *Isobar* - a line which connects regions of constant pressure [16]. (Fig. 2a.), *Isotherm* - a line of equal temperature, often used to find regions with warm or cold air advection as well as waves, fronts and temperature gradient boundaries [17]. (Fig. 2b.), *Isotachs* - a line of equal wind speed, used often for locating jet streams [18]. (Fig. 2c.), *Isodrosotherm* - a line of equal dewpoint [19]. (Fig. 2d.) *Isopleths* are most flexible method of visualization. Application which was developed for the purposes of this paper had to provide easy access to multiple weather data parameters therefore isopleths were chosen as main visualization tool. Using contour drawing it was possible to create generic drawing mechanism. What is more presentation layer supplied by contour plots is undemanding for interpretation by humans and it is possible to be analyzed by automated algorithms.

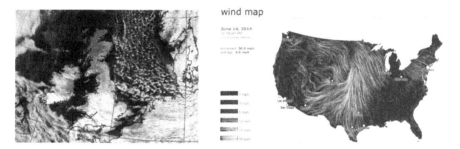

Fig. 1. Satellite photo of sky above Great Britain and live wind speed and directions over USA

Data used for visualization is extracted using processing application. Algorithm is configured to collect all measurements about specific region and sort them out chronologically. This allows to provide easy access to even oldest weather reports. Density of isopleths has been chosen experimentally to provide most informative visualizations. Threshold has been set for each parameter on such level that every image contains 20 equally distributed levels.

Collecting data sets which are large and complex is very difficult in laboratory scale environments. Weather data is known to be one of the most memory consuming data sources. Main two reviewed worldwide weather sources were used:

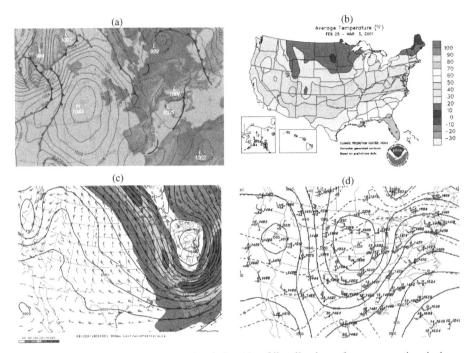

Fig. 2. Weather visualization using isopleths (a) - Visualization of pressure using isobars (b) - Visualization of temperature using isotherms (c) - Visualization of wind using isotachs (d) - Visualization of dew point using isodrosotherms

HadCRUT3 records of the global temperature from the Intergovernmental Panel on Climate Change (IPCC). It is data that has been designated by the World Meteorological Organization (WMO) for use in climate monitoring. It contains monthly temperature values for more that 3000 land stations. National Climatic Data Center from Asheville, NC, USA reports collecting global summary of the day data (GSOD). Data which creates the Integrated Surface Data (ISD) daily summaries is obtained from USAF Climatology Center. Beginning from 1929, over 9000 land and sea stations are monitored. Both of this data sources are available freely for non-commercial use. Daily reports from GSOD contain multiple collectible data types which can serve as perfect input to big data processing. Data was provided in Imperial units but in later processing it have been converted to Metric system.

4   Forecasting Using Neural Networks

4.1   State of Art

Neural networks were successfully used in meteorology before. In *Temperature forecasting based on neural network approach* [5], authors try to utilize artificial neural networks for one day ahead prediction of only one weather parameter - temperature. Model was build for city of Kermanshah which is located in west of Iran. Researchers decided to use Multilayer Perceptron [8]. Datasets were constructed of ten years of meteorological data (1996-2006). For results improvements there has been created four different neural networks. They were used to forecast weather for each season. Multilayer Perceptron has provided good results in temperature forecasting, giving average difference between predicted and expected temperature less than 2°C. Similar work has been done by researchers in *Ensemble of neural networks for weather forecasting* [9]. Authors have chosen Multilayer Perceptron, Elman Recurrent Neural Network, Radial Basis Function Network and Hopfield model [10][13][14] to create and compare temperature, wind speed and humidity forecasting. All predictions has been assumed for city in southern Saskatchewan, Canada. Similarly to previous papers researchers choose to create four models for each season and to produce one day look ahead forecast. Most accurate forecast were provided by ensemble of multiple neural networks. Andrew Culclasure in *Using neural networks to provide local weather forecasts* [3] presents another application of using Artificial Intelligence for weather forecasting. Three different neural networks were created and trained with data collected in Stetesboro in Georgia, USA every 15 minutes over a year in personal weather collection station. Researcher uses small scale and imperfect datasets for creating forecast for 15-minute, 1-hour, 3-hour, 6-hour, 12-hour and 24-hour look ahead forecasts. Despite using neural network for weather forecasting, it has been also used to nonlinear image processing [11]. It has been proven by authors that Artificial Intelligence can be used for object recognition and feature extraction [10][13][14].

4.2   Adaptive Processing

Artificial Intelligence seemed to be natural choice for interpreting data which is provided in "human" way [10][13][14]. From extensive research [11] it was proven that neural networks can be used for object recognition and feature extraction. Using neural networks for visual analysis of images delivered by created application appeared to be hard and demanding task to achieve. Due to immense research material, unsatisfactory results and insufficient time for continuing the paper research it was decided to use neural network in another way. Visualization itself gave chances for manual pattern recognition. Neural networks on the other hand have been proven to be successful tool for forecasting weather. Main idea which has evolved from these two statements was to use visualization for pre-filtering the data and passing it to neural network to achieve even better results of forecasts [2]. Weather forecasting is almost always related to phenomenons occurring in some specific region. Especially models build for persistence or climatology prediction models do not provide any reasonable results to other place in the world. One model which can take advantage from different locations is forecasting by analogies. Because analog method uses data to find analogous conditions in the past, it can be used with data from other (but similar) location to help predicting weather state for situations when no analogies can be found. It is crucial to thoroughly adjust other location model. When location model with nearly constant temperature variance is applied to location where temperatures are changing very quickly it is not possible to achieve good results. Good results for other forecasting model can be   achieved by averaging models build from surround-ing locations. These locations cannot be far away and cannot include extreme loca-tions. Often weather model is very similar for whole regions, in which there are multiple data collecting stations. They can complement each other data and provide better, and more complete results. However sometimes there are weather stations which location is on top of the hill or mountain and data collected there contains more raw temperature, pressure or wind speed measurements. Whole process of finding locations with analogous weather models or finding locations which are close enough to provide good results can be called adaptive processing. That means that resulting model for location of interest is actually build as the union of similar models, adapted to provided best results. The model is build based on visual analysis and is later used as neural network input. Visual analysis allowed to choose which locations are related and which places data should be used to train, validate and test neural network. Pre-dicting weather changes using neural network can be compared to two forecasting methods: *Climatology* - Neural network can predict weather conditions according to many years of data collection [1]. Result of forecast will be an average state of tem-perature, pressure, wind speed or other parameters. Network will be also aware for which part of the year every forecasting is about. Such forecasting is done in clima-tology model. *Analog* - Without consideration which part of the year it is, forecasting using neural network is similar to forecasting using analogies. When network is fed using parameters which are similar to this passed earlier it will provide similar output as before.

4.3   Neural Network Design

Based on [5], [9] and theoretical research Multilayer Perceptron was chosen as neural network model [8]. Application created as a result of this work allowed to create net-works with configurable number of inputs and outputs. It did not have constant archi-tecture, but could be easily parametrized instead. Possibility to forecast multiple meteorological events and multiplicity of parameters which can be chosen as most significant in process of forecasting were main reasons for such design. Sample con-figurations were provided in Tab. 1.

Table 1. Sample neural network configurations

Configuration description	Input parameters	Input neurons	Output parameter	Output neurons
Seven days forecast based temperature only. This method is very similar to analog forecasting	Temperature	7	Temperature	1
Seven days forecast based on temperature and number of day in the year. Similar to previous one, but considers also transience of seasons by knowing which part of the year it is	Temperature, day of the year	8	Temperature	1
One day forecast based on temperature, mean station pressure and wind speed for that day. Analog forecasting aware of multiple weather parameters	Temperature, mean station pressure for the day and mean wind speed	3	Temperature	1

We assumed the input temperature from previous two days, pressure and wind speed from one previous day should provide one day temperature forecast. The algorithm needs to keep history of three days of measurements. Two days will be used as neural network input, and last day will be used for validation or testing if provided answer is correct. The neural network build for the purposes of forecasting application was trained using most classic method of training – backpropagation [8][10]. What is important here, neural network is not being trained until convergence. It is impossible to find optimal solution and create model which will perfectly correspond weather conditions in any situation.

4.4   Learning Benchmarks

All developed functions combined together create neural network which can be used for weather forecasting. Configuration which has been used for benchmarking network learning performance consisted of 7 days of temperature measurements and one look ahead temperature prediction. This resulted in 7 input neurons and 1 output neurons. Three places in the world were chosen to benchmark neural network performance and effectiveness. These cities were Opole, Helsinki and Mexico City. Choice has been made based on how fast is weather changing within one year in each place. What is more in these cities there has been collected enough meteorological data for creating representative dataset and later neural network.

Opole is a city in southern Poland on the Oder River (Odra). It can be characterized by moderate winters and summers. Neural network model build for this city achieves error rate around 0.01 after 5 epochs of learning (Fig. 3.).

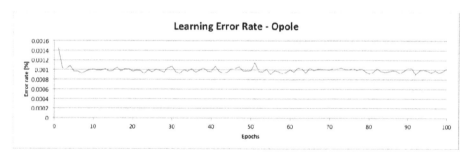

Fig. 3. Neural network learning error rate for creating prediction model for Opole, Poland

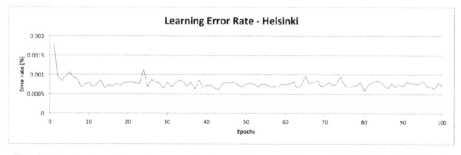

Fig. 4. Neural network learning error rate for creating prediction model for Helsinki, Finland

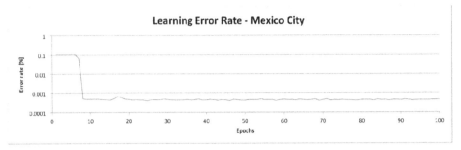

Fig. 5. Neural network learning error rate for creating prediction model for Mexico City

Helsinki is the capital and largest city of Finland located in southern Finland, on the shore of the Gulf of Finland, an arm of the Baltic Sea. This city has a humid continental climate and it has lower minimal temperature than previous one. Neural network has less stable but lower error rate (Fig. 4.).

Mexico City is capital of Mexico. It has a subtropical highland climate, and due to high altitude of 2,240 m temperature is rarely below 3°C and above 30°C. Due to almost constant temperature along the year network has very low error rate (Fig. 5.).

Neural network learning benchmarks have revealed that model build for specific city correlates with weather and climate from that city. Neural networks are learning to almost constant error rate after maximum of 10 epochs. Another assumption is also that if city has more constant weather conditions like Mexico City Neural network achieves lower error rates (Fig. 5.). More fluctuating temperature in Helsinki and Opole provide slightly higher error rates. Helsinki however has a lot colder winters (greater difference between lowest and highest temperature) which impacts model, which gives more uncertain answers.

5   Climate Change Prediction

Whole processing chain has been performed for cities used for testing learning capabilities of neural network. These cities were Opole, Helsinki and Mexico City. Main experiment aim was to create weather forecast for chosen regions based on different atmospheric parameters. Weather prediction was targeted to whole year 2013, because it was the last year for which dataset with complete input to output relations could be acquired. First forecast was created based on 7 day temperature input. Additionally neural network had information about which part of the year it was. This was assured by providing number of the day in year as one of the network inputs. Model created for Opole city consisted of data beginning with year 1973 and ended with year 2012. Year 2013 was used for testing forecasting possibilities. Forecasting results are presented in Fig. 6. Average error is estimated to 2.06°C, and standard deviation was equal 1.59°C. From presented figure it can be deduced that most wrong predictions were made around 200th day of the year, in July. This month is in summer and it can be foreseen that neural network may have problems with detecting temperature extremes. Better results are in months where temperature is closer to yearly average.

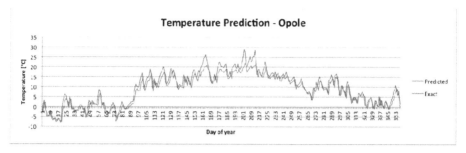

Fig. 6. Weather forecasting results based on day of the year value and 7 days look behind prediction for Opole, Poland

Fig. 7. Weather forecasting results based on day of the year value and 7 days look behind prediction for Helsinki, Finland

Similar experiment has been performed for Helsinki in Finland. Results are presented in Fig. 7. Results for this city does not correspond with results for Opole. Neural network better performed in summer, where atmospheric conditions were moderate. Achieved algorithms were not able to correctly detect temperature drop in end of February. This can be explained with expected neural network behavior. Model has been trained using representative data sets of measurements collected through many years. In consequence predicted network output is most likely to be average of such measurements. When some year weather stands out from the average, neural network will probably forecast wrong temperature.

Best results from this three cities were achieved for Mexico City. Average difference between predicted and observed temperature was equal 1.49°C and standard deviation equal 1.21°C. Such great results were achieved mostly because of nearly constant temperature in Mexico City. Training neural network for providing similar output data is faster than in Opole and Helsinki (no more that 5 training epochs were needed). Only one major wrong temperature forecast was provided. Similarly to predictions in Helsinki most probable reason for such application behavior was unusual observed weather conditions. Due to weather partial uncertainty this situation could not be prevented.

Opole was chosen for further research. Configuration with 3 day and 20 day temperature based forecast has been provided, however no result improvement has been noticed (Fig. 9.) and (Fig. 10.) respectively.

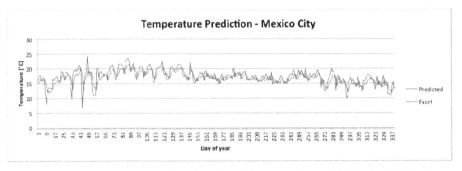

Fig. 8. Weather forecasting results based on day of the year value and 7 days look behind prediction for Mexico City, Mexico

Average difference between predicted and exact value for equal 3.37°C for 3 day forecast, 4.14°C for 20 day forecast. Most probable reason for such behavior is presenting too few or too many data to neural network. Even ensemble of parameters does not provide better results that most classic 7 day forecast. Configuration consisted of input based on 3 day of collected mean temperature, station pressure, sea level pressure dew point and wind gust. Results of using this configuration is presented in (Fig. 11.).

Similar tests have been performed for other parameters like: mean dew point, mean sea level pressure, mean station pressure, maximum and minimum temperature.

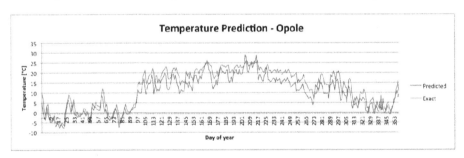

Fig. 9. Weather forecasting results based on day of the year value and 3 days look behind prediction for Opole, Poland

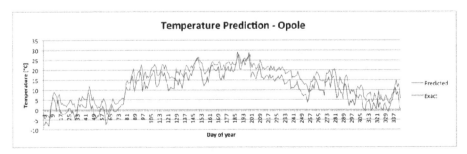

Fig. 10. Weather forecasting results based on day of the year value and 20 days look behind prediction for Opole, Poland

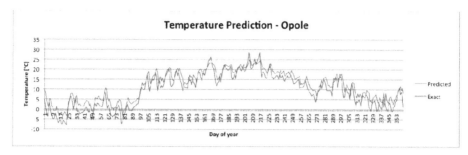

Fig. 11. Weather forecasting results based on ensemble of multiple atmospheric parameters prediction for Opole, Poland

As long as weather reports for this parameters contain correct data creating prediction of this variables is possible and provides correct and surprising results. Example of prediction of mean sea level pressure for Opole is presented in Fig. 12.

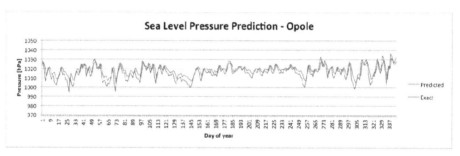

Fig. 12. Mean sea level pressure forecasting results based on 3 daily pressure reports for Opole, Poland

6   Conclusions

Achieved results were satisfactory and proved that it is possible to use neural network for weather forecasting. Big data visualization on the other hand provided easy access to data which in another way would be hard to process. Processing of big data was found to be complex task. After research few guidelines have been introduced and this allowed to create processing algorithm which was able to parse even infinite data sources. Analysis showed that generative algorithms should be used in favor of iterative ones due to better processing speed and lower memory consumption. After easy access to big weather data has been guaranteed two tasks has been done. First one was to create application which can visualize weather in any world region, and second one was to forecast the weather using neural networks. Executed test and benchmarks, and further analysis have shown that developed application creates valid forecasting reports and it can be successfully used for prediction of multiple weather parameters. Minor drawback of created application was that location choice had to be done manually and weather reports are not being generated without human interaction.

References

1. Bonarini, A., Masulli, F., Pasi, G.: Advances in Soft Computing, Soft Computing Applications. Springer (2003)
2. Bell, I., Wilson, J.: Visualising the Atmosphere in Motion. Bureau of Meteorology Training Centre in Melbourne (1995)
3. Culclasure A. Using Neural Networks to Provide Local Weather Forecasts. Electronic Theses & Dissertations. Paper 32, (2013)
4. Francis, M.: Future telescope array drives development of exabyte processing, http://arstechnica.com/science/2012/04/future-telescope-array-drives-development-of-exabyte-processing/
5. Hayati, M., Mohebi, Z.: Temperature Forecasting Based on Neural Network Approach. World Applied Sciences Journal 2(6), 613–620 (2007)
6. Hoffer, D.: What does big data look like? visualization is key for humans, http://www.wired.com/2014/01/big-data-look-like-visualization-key-humans/
7. Katsov, I.: In-stream big data processing, http://highlyscalable.wordpress.com/2013/08/20/in-stream-big-data-processing/
8. Kung, S.Y.: Digital Neural Networks. Prentice-Hall (1993)
9. Maqsood, I., Riaz Khan, M.: An ensemble of neural networks for weather forecasting. Neural Comput & Applic. (2014)
10. Pratihar, D.K.: Soft Computing. Science Press (2009)
11. de Ridder, D., Duin, R., Egmont-Petersen, M., van Vliet, L., Verbeek, P.: Nonlinear image processing using artificial neural networks (2003)
12. Sandu, D.: Without stream processing, there's no big data and no internet of things, http://venturebeat.com/2014/03/19/without-stream-processing-theres-no-big-data-and-no-internet-of-things/
13. Sivanandam, S.N., Deepa, S.N.: Principles of Soft Computing. Wiley (2011)
14. Srivastava, A.K.: Soft Computing. Narosa PH (2008)
15. Top 500 supercomputers list, http://www.top500.org
16. Isobar visualization, http://www.daviddarling.info/encyclopedia/I/isobar.html
17. Isotherm visualization, http://www.nc-climate.ncsu.edu
18. Isotachs visualization, http://www.erh.noaa.gov
19. Isodrosotherms visualization, http://ww2010.atmos.uiuc.edu
20. Guideline for developing an ozone forecasting program. U.S. Environmental Protection Agency, Office of Air Quality Planning and Standards (1999)

Practical Problems of Internet Threats Analyses

Krzysztof Cabaj, Konrad Grochowski, and Piotr Gawkowski

Institute of Computer Science, Warsaw University of Technology (WUT)
Nowowiejska 15/19, 00-665 Warszawa, Poland
{K.Cabaj,K.Grochowski,P.Gawkowski}@ii.pw.edu.pl

Abstract. As the functional complexity of the malicious software increases, their analyses faces new problems. The paper presents these aspects in the context of automatic analyses of Internet threats observed with the HoneyPot technology. The problems were identified based on the experience gained from the analyses of exploits and malware using the dedicated infrastructure deployed in the network of the Institute of Computer Science at Warsaw University of Technology. They are discussed on the background of the real-life case of a recent worm targeting Network Attached Storage (NAS) devices vulnerability. The paper describes the methodology and data analysis supporting systems as well as the concept of general and custom HoneyPots used in the research.

Keywords: Network Security, HoneyPot Systems, Network Attacks, Exploits, Malware.

1 Introduction

The growing number of devices connected to the Internet pushed forward the necessity of secure communication and data storage. In this light the security is one of the most important issues in the IT systems. Cloud-based and cloud-related solutions has become a reality introducing new challenges for security and personal privacy. As the high speed broadband is more and more popular [1], it is observed that software manufactures push users towards their cloud-based services. At the same time, mostly because of the high quality multimedia materials, home users require more and more storage space. Moreover, users want to share them with their family and friends as well as to stream to their TV sets, etc. That means the necessity to use Network Attached Storage devices at home or even to provide them the Internet connectivity. So, many security problems (well-known in companies) are now faced by average home users, which are not IT specialists aware of recent security threats.

Several types of threats can be pointed out in the above context. One of them is related with the confidentiality of the communication channels (e.g. robustness of routers in public Wi-Fi installations) and application protocols (e.g. in plain text). In [2] it is showed that single-bit faults can lead to major security issues. However, most of the problems are related to the quality of software. It may contain (or rather contains) some yet-undiscovered bugs that can be exploited to take over the target system. In the security field this bugs are called *vulnerabilities*. They can be exploited

W. Zamojski et al. (eds.), *Theory and Engineering of Complex Systems and Dependability*,
Advances in Intelligent Systems and Computing 365, DOI: 10.1007/978-3-319-19216-1_9

through malicious documents sent by e-mail, accessing the infected web page or direct attack on the services served over the network (e.g. remote code injection through malicious network request). In this last scenario the infection does not require any user actions on the target machine to activate malicious code. Thus, it is more dangerous as without the proper system monitoring the infection could be undiscovered for a long time. System user might be unaware of the problems. Here arises the problem of abnormal system behaviour – how to recognise any anomalies from normal operation [3].

Home network attached systems (e.g. routers, NAS) that provide some services for the Internet (like data sharing, web servers, peer-to-peer clients) are very attractive to hackers. They are typically not (or not regularly) monitored by their users. So, the attacks and eventually the overtaking is unlikely to be discovered. At the other hand, these devices are effective enough to handle some malicious tasks along with their normal operation. The threat of private information leakage is just one side of the coin: compromised device can be used to infiltrate home network but also is a good base for further attacks over the Internet. Thus, the hacked device is possibly becoming a source of attacks or a malicious code repository. In this context it is worth noting, that source of attacks might be not the hacker's IP but in fact an innocent victim. Nevertheless, it is important to identify all the sources of a violent activities.

One of the main problems of Internet security is the number of new, so far unknown threats due to existing security holes in the software – called *zero day threats*. Because attackers are usually first to know and exploit vulnerabilities, HoneyPot systems were created to monitor behaviour of attackers in real environment, providing 'traps' to capture potentially malicious activities. Following and understanding them (and mechanisms behind) allow identifying software bugs that lead to the successful attack. Any system or resource can be treated as a HoneyPot, as long as it used only as monitor (not for any production purpose). All access attempts to that resource are unexpected and thus can indicate malicious activity. A good survey on HoneyPots can be found in [4, 5]. Some of our real-life cases identified with our own installations are given in [8]. Using these systems and analysing the collected data we have identified some problems that aggravate the analyses.

Most of the papers in the literature describe HoneyPot systems itself or a malware analyses as a separate tasks [13, 14]. Obviously, such approach is not practical as the whole process depends on iterative improvements of HoneyPots to allow deeper analysis of multistage attack scenarios that are faced in the Internet nowadays. Each stage of the attack require some specific actions to be made by the HoneyPot, data classifiers, dynamic analyses of malware, etc. Based on real-life Internet threat we discuss the problems faced in detailed analysis starting from initial detection of vulnerability scans, attack attempts, and finally, to understand the behaviour of compromised network device. In the next section we describe the considered (in our opinion the most typical) scenario of a takeover attack on network attached system together with real-life example observed and analysed using our systems.. Based on that, practical problems are discussed in Section 3, including existing software capabilities. Paper is concluded in section 4.

2   Machine Takeover Scenario

Contemporary network worms targeted at network attached devices realize quite complex scenarios consisting several stages. Hereafter we try to give a general description of it. Next, the real-life QNAP worm is analysed to illustrate further discussion over the analysis problems.

2.1   Attack Stages

Recon Phase. Typical remote attacks exploiting vulnerability start with the recon phase – vulnerable machines accessible over the Internet are sought. This process could be performed in advance as massive scanning of large ranges of IP addresses. It can be seconds or months before the real attack. Monitoring of all incoming connections is a valuable source of data to trace potentially dangerous activities. However, additional analysis is needed to distinguish real attacker from, for instance, scanning made by service discovery software. Moreover, at the connection instant the one (security supervisor) will hardly know if it is a malicious connection or not. The only thing the supervisor can do is to keep track of widely-known addresses of attackers to limit their access to the supervised resources. Moreover, he can observe and react on noticeable changes in communication trends (e.g. more frequent connection attempts).

However, the recon phase may not require any physical connection from the attacker to the target system as services like Shodan [6] provide all the necessary information. Shodan sequentially scans all machines connected to the Internet and records all enabled services and used software versions – attacker can search for the potential victim based on the vulnerability he is looking for or victim's geographical location.

Exploit. When potential victim is found the attacker sends so called *exploit*. The exploit is specially crafted packet, communication session or user data that performs some actions (using a bug or imperfection in target software) not intended by a programmer. Two most popular ways of attacks are remote code execution and command injection [7]. In the first case a bug allows the attacker to execute provided machine code. This can be done for example by buffer overflow vulnerability (leading to overwritten return address). The other kind of software errors allows direct execution of some (restricted) commands. In both cases the attacker's code or commands are very limited. In effect, this phase of the attack is a first stage, used only to allow further downloading and execution of a larger, main malicious software.

Malware. After successful exploitation of vulnerable machine, the attacker gains enough control to download and execute more sophisticated software on it. Attacker could be a person or automate (script, software, etc.). In this phase he starts using victim's machine for his purposes. In most observed cases downloaded software is a kind of backdoor or bot. The first one allows easy and stealth access to the machine without subsequent exploitation (like logging into the machine, using it for finding other vulnerable machines, hosting malware or performing further attacks). During our research we identified a machine used for storing backups of legitimate user and at the same time to store some malware. The machine was used as a malware repository during attacks at vulnerable phpMyAdmin software. The second group turns the machine to be a part of a botnet: a group of compromised systems controlled by attacker. Executed malware connects to the command and control (C&C) server, and waits for further commands (to download new module, perform IP scanning or take a part in DDoS attack, etc.). During our research, we observed MS Windows malware which connects to C&C server, downloads new IP address range and starts finding VNC-enabled Linux machines. In the last year a new trend was observed: exploited machine downloads e-currency mining software, joins to the pool and starts mining for the attacker [8] (due to lower network activity it is harder to detect).

2.2   QNAP NAS Example

QNAP attack uses two-level exploit to download and run malware on a target machine. First level uses Bash shell interpreter vulnerability – Shellshock: some special values of environment variables are treated as procedures and executed by Bash [9]. To set the required value of a variable the attack uses User-Agent HTTP header (it is passed by HTTP server to underlying CGI script via environment variable). Instead of standard data (describing user's HTTP client), the HTTP request header provides a Bash code, as seen below.

```
() { :; }; /bin/rm -rf /tmp/S0.php && /bin/mkdir -p
/share/HDB_DATA/.../ && /usr/bin/wget -c
http://XXX.VVV.YYY.ZZZ:9090/gH/S0.php -O /tmp/S0.sh &&
/bin/sh /tmp/S0.sh 0<&1 2>&1
```

Starting "() { :; }; " part is Shellshock specific and results in execution of further commands. They were slightly customized for QNAP NAS (i.e. folders paths, target URLs), however, the way of their activation can be easily identified as a typical Shellshock. The first-level script forces the device to download and execute second-level exploit in a form of another Bash script. This script is even more dedicated for QNAP NAS devices as it uses specific folders and configuration files. It is worth noting that the exploit is prepared to work with multiple architectures and downloads different binaries depending on attacked system specification. One of the first downloaded binaries is a fix for used vulnerability, so it ensures the attacker for being the only one on the machine. At the same time it makes the device to look as secured despite being compromised.

In the next step the second-level script downloads and installs SSH daemon. Then it opens SSH access to the NAS on non-standard port. It also adds new machine's user with hashed password hardcoded in script, creating backdoor access to the machine. Password can be easily found on the Internet using its hashed version, allowing anyone to access infected machines.

At the end the second-level script downloads some malware binaries and a script to propagate itself over the network. Moreover, QNAP specific configuration files are modified ensuring automatic execution of malicious software when machine boots.

Self-propagation script uses a free tool (binary of which is also downloaded) to send HTTP requests with malformed HTTP header to random range of IP addresses. No preliminary scans etc. are performed by the exploit, making it easy to spot on Honey-Pot systems. Below a part of propagation script is shown.

```
rand=`echo $((RANDOM%255+2))`
url="http://XXX.VVV.YYY.ZZZ:9090/gH/S0.php"

download="/bin/rm -rf /tmp/S0.php && /bin/mkdir -p
/share/HDB_DATA/.../ && /usr/bin/wget -c $url -O
/tmp/S0.sh && /bin/sh /tmp/S0.sh 0<&1 2>&1 \n\n\n"

get="GET /cgi-bin/authLogin.cgi HTTP/1.1\nHost:
127.0.0.1\nUser-Agent: () { :; }; $download \n\n\n"

./pnscan -rQDoc -w"$get" -t500 -n100
$rand.0.0.0:255.0.0.0 8080 > /dev/null &
```

3   Identified Problems

To observe threats activities the HoneyPot systems are used. We use various types and instances of them in our Institute. We have started with open source low-interaction HoneyPots, Nepenthes, and its successor Dionaea. However, both of them have very limited functionalities associated with simulating behaviour of a web server and web applications - simple scans can just be registered. Although in some attacks a scan is made before the real exploit to detect and omit some basic HoneyPots. In such cases the whole service might need to be simulated (some sequence of proper request-response pairs might be needed to observe complete attack behaviour). Due to this fact we have developed WebHP - a HoneyPot dedicated for observing all scanning activities and obtaining details of all attacks performed with HTTP protocol [8]. Moreover, some application specific systems were created to cover in more detail particular attacks (e.g. HeartBleed, phpMyAdmin). Emulating with greater accuracy the desired target for the attacker allows to collect longer scenario of attack and application level data.

Due to fact that exploits usually utilize undocumented functions or nuance of vulnerable software, HoneyPots and analytical systems have to reproduce original one with all details. Any, even minor, deviation can lead to failure in observing whole attack scenario. Attackers can also use some specific behaviour as a countermeasure. Moon worm, attacking Linksys routers, is a good example: exploit set compression flag in HTTP communication, yet it rejected all compressed data. In effect, standard Apache server (which supports compression) could not work as a HoneyPot and required reconfiguration – disabling of compression at server level. Only after that manual intervention HoneyPot was able to capture next stages of the worm.

Collected scanning activity (see Section 2) is a starting point for the analyses. The growing activity of a specific connection type rivet our attention as new (probably unknown) vulnerability may be identified. However, some more advanced data analysis could utilize data mining pattern discovery techniques (like frequent sets and jumping emerging patterns). Scanning, complex attack or worm outbreak generate repeated activities, which could be detected using these methods. To illustrate the problem of the amount of complex data it is worth to note that WebHP sensors during two years of operation (1st January 2013 to the end of December 2014) observed over half of million connections. Acquired data contains various types of activities, even completely innocent, like access from various web crawlers used by search engines.

To allow analysis of gathered data a HoneyPot Management System (HPMS) was developed. Basic HPMS functions allow convenient access to raw data gathered by WebHP sensor and plots generation for interesting time ranges. HPMS user can manually investigate collected data. In-depth manual analysis of such volumes of data is almost impossible. So, some supporting software is needed to easy identification, grouping, visualisation, etc. Still, the expert knowledge and manual analysis is required in this process to be used on the automatically detected sets of data.

In case of QNAP the WebHP system was able to register only the first step of the attack as an increased amount of HTTP requests targeting the same URL with one of the headers matching the Shellshock behaviour marker (Fig 1). It was possible, because Shellshock is a well-known attack and can be easily recognized by specific sequence of characters. It gets a lot more complex with majority of attacks, which are directed to specific applications – like mentioned phpMyAdmin attack which used application internal structure (serialized as part of request) to inject malicious code. In scan only specific URL appears and request contents – expert knowledge is needed to associate *PMA_Config* structure used in attack with targeted application. Only after recognizing attack and its target further analysis can start.

After identification step, in HPMS any known (identified) activity can be described with appropriate rules and added to HPMS (like in QNAP example). Using these rules the system automatically marks all existing and new data using configured labels [10]. Preliminary experiments shows that detected patterns describe all known threats, previously discovered manually. Moreover, manual analysis of discovered patterns reveals suspicious machines and activity. Without such support, most of them will be hinder in vast amounts of other data.

Even with data labelling new attacks has to be noticed manually, as in QNAP example: WebHP system has been registering Shellshock attack attempts since 12 September 2014, which was the day when that vulnerability was first publicly announced. Fig. 1 shows how attacks intensified on December 2014, when QNAP worm targeted WebHP system. To distinguish this attack, manual analysis was required and new label created, separating it from Shellshock. Since then, worm activity is approximately constant, higher peaks indicate scans. It is noticeable, that QNAP provided firmware patch for its devices on 5 October 2014, more than a month before attacks became frequent.

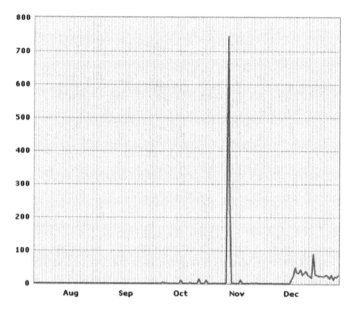

Fig. 1. Overall Shellshock activity (attacks attempts count over time) as seen in HPMS system

In our work various data analysis supporting methods were used. One of them util-ize visualization and simple data filtering. Applied visualization techniques use sim-ple plots (like this presented in the Fig. 1) as well as more complex graphs joining the attacker, used malware filename and server hosting it. It is worth noting, that in this graph two interesting kinds of relations can be pointed. We found a group of attackers which directs their exploit to download the same malware file from a bunch of differ-ent hosts. At the same time multiple attackers use a single malware from single host. More details are provided in our paper [11].

Another problem related with threat analysis concerns its mutations. When new at-tack appears in the Internet, soon its mutations are active. Frequently the mutated attack differs only in C&C server or malware hosting servers. However, this aggra-vates the analysis (it is harder to identify two different attacks as the same type). Fig-ure 2 shows how different variants of QNAP worm were registered by WebHP over the time and suggests that this vulnerability is actively utilized, enhanced and tailored by attackers, despite being quite a few months old. Seven mutations of QNAP worm were identified. This number of versions could be analysed manually, despite the fact that precious time of security specialist is needed. However, when attack uses hun-dredths of various exploits or malware names automatic support for detecting simi-larities is needed.

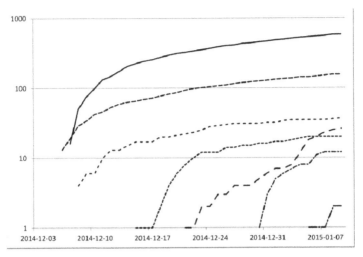

Fig. 2. Different QNAP worm variants activity (attack count over time, logarithmic scale)

Analysing the worm the most of the detailed knowledge can be obtained by analysis of subsequent stages of the attack and malware which is downloaded after successful exploitation of vulnerability. For example, described QNAP worm behaviour was investigated not in detected by HoneyPot systems exploit but from analysis of subsequent (manually) downloaded files which locations were enclosed in following exploit stages (see Section 2.2). It is obvious that manual downloading of files have some drawbacks. The first disadvantage is associated with huge number of data. For example, our HoneyPot in the 2014 year observed more than 283 unique URLs, which was used for attacking phpMyAdmin management software. The second disadvantage is even more important. Some modern threats have a protection against delayed downloading of malware. URL used in the exploit, which points to the second stage, is generated randomly for each victim and becomes invalid after a few minutes. In normal circumstances, just seconds after exploit execution the malware should be downloaded. If not, later attempts (performed manually by administrators or security officers) will fail. Such behaviour poses, for example, Linksys worm observed in the 2014 at our HoneyPot.

Due to this fact deployed HoneyPot system should automatically downloads all data pointed by URLs enclosed in observed exploits. For well-known threats this functionality is easy to implement (as the dedicated HoneyPot can handle the worm). However, problems occur with unknown, zero day threats. As was described earlier, from observed seven URLs used by QNAP worm, only from two of them we succeeded to download copy of a malware for further analysis. To solve this problem some heuristics can be used, for example finding all strings which looks like URL or attempts to execute unknown data [12]. To overcome these issues the HoneyPot should operate together with the malware analysing system.

Malware can be analysed using two methods: static and dynamic. In static analysis it is investigated without execution. However, even if the malware is a plain text script, such analysis can be hard to proceed. In QNAP example, some scripts use a lot of nested variables, a values of which depend on machine state and configuration (see Section 2.2). Contrary, in dynamic analysis a malware is executed in a sandbox environment and its behaviour is observed. For this purpose a Maltester system was developed and deployed in our Institute. It analyses MS Windows targeted malware. The system consist of supervisor software and specially crafted virtual machine currently using Windows XP as a guest. The whole environment is secured to allow limited access to the Internet and to protect Internet users from hostile activity. Supervisor software works in Linux and Xen hypervisor is used for managing virtual machines. The only part of the Maltester in the guest OS system is simple server for receiving malware samples from the supervisor. Prepared guest machine is hibernated and awaits for sample for analysis. When a new sample is added to Maltester, the guest machine is woken up from hibernation and executes the sample. After configured amount of time the guest machine is hibernated once again. Then the supervisor performs various analysis, for example, checks which files are changed or added and what new process are executed. After the analysis the guest machine is restored to clean state and awaits for next sample. Performed experiments proves usefulness of such approach. This is very fast and efficient way to gather initial data concerning analysed malware. For example, we use this initial analysis for detecting C&C servers IP addresses and automatic gathering of sample concerning scanning and attacking tools used by attackers. Using such dynamic approach can also show the potential scope of damage made by the exploit and malicious software on the compromised systems. Moreover, it can help to identify the other web resources engaged in spreading the malicious software over the Internet.

4   Conclusions

Analysis of malicious software is much more complicated than several years before. Any attack consists of several steps – each require sophisticated (and safe) environments to be analysed. Worms consist of many mechanisms hindering the analysis. During this process the time window might be opened only for a short period – after that the next step of the analysis might be interrupted. Moreover, the binary of a malware may be encrypted or obfuscated, so, the dynamic evaluation is needed in order to discover its behaviour and further stages of the attack.

Our systems proved to be a significant assist in detection and analysis of observed threats (using jumping emerging and frequent sets patterns). Configurable HoneyPot system (WebHP) allowed capturing details about new web attacks. These data was then used for manual malware downloading. Some dynamic analysis of the malware samples were made with Maltester system. The conducted experiments showed that analytical software is required to support security specialist. However, more work on process automation is still required. Without that, rapid analysis of gathered data is almost impossible. Our further work will focus on building and integrating different kinds of complex malware analytical systems with the network monitoring targeted at anomaly detection.

References

1. Akamai Releases Third Quarter, 2013 'State of the Internet' Report,
   `http://www.akamai.com/html/about/press/releases/`
   `2014/press_012814.html` (access date: January 2015)
2. Nazimek, P., Sosnowski, J., Gawkowski, P.: Checking fault susceptibility of cryptographic algorithms. Pomiary-Automatyka-Kontrola (10), 827–830 (2009)
3. Sosnowski, J., Gawkowski, P., Cabaj, K.: Exploring the Space of System Monitoring. In: Bembenik, R., Skonieczny, Ł., Rybiński, H., Kryszkiewicz, M., Niezgódka, M. (eds.) Intell. Tools for Building a Scientific Information. SCI, vol. 467, pp. 501–517. Springer, Heidelberg (2013)
4. Provos, N., Holz, T.: Virtual Honeypots: From Botnet Tracking to Intrusion Detection. Addison-Wesley Professional (2007)
5. Bringer, M.L., Chelmecki, C.A., Fujinoki, H.: A Survey: Recent Advances and Future Trends in Honeypot Research. I. J. Computer Network and Information Security 10, 63–75 (2012)
6. Bodenheim, R., Butts, J., Dunlap, S., Mullins, B.: Evaluation of the ability of the Shodan search engine to identify Internet-facing industrial control devices. International Journal of Critical Infrastructure Protection 7(2), 114–123 (2014)
7. Anderson, R.J.: Security Engineering: A Guide to Building Dependable Distributed Systems. John Wiley & Sons, Inc., New York (2001)
8. Cabaj, K., Gawkowski, P.: HoneyPot systems in practice. Przeglad Elektrotechniczny, Sigma NOT 91(2), 63–67 (2015), doi:10.15199/48.2015.02.16
9. Ullrich, J.: Update on CVE-2014-6271: Vulnerability in bash (shellshock) InfoSec Handlers Diary Blog, `https://isc.sans.edu/diary/18707` (access data: January 2015)
10. Cabaj, K., Denis, M., Buda, M.: Management and Analytical Software for Data Gathered from HoneyPot System. Information Systems in Management 2, 182–193 (2013)
11. Cabaj, K.: Visualization As Support For Data Analysis. To appear in Information Systems in Management
12. Koetter M.: libemu: Detecting selfencrypted shellcode in network streams. The Honeynet Project (access date: January 2015)
13. Baecher, P., Koetter, M., Holz, T., Dornseif, M., Freiling, F.: The nepenthes platform: An efficient approach to collect malware. In: Zamboni, D., Kruegel, C. (eds.) RAID 2006. LNCS, vol. 4219, pp. 165–184. Springer, Heidelberg (2006)
14. Xu, M., Wu, L., Qi, S., Xu, J., Zhang, H., Ren, Y., Zheng, N.: A similarity metric method of obfuscated malware using function-call graph. Journal in Computer Virology Archive 9(1), 35–47 (2013)

Risk Assessment of Web Based Services

Dariusz Caban and Tomasz Walkowiak

Wrocław University of Technology, Wybrzeże Wyspiańskiego 27, 50-320 Wrocław, Poland
{dariusz.caban,tomasz.walkowiak}@pwr.edu.pl

Abstract. Web based information systems are exposed to various faults during their lifetime (originating in the hardware, in the software or stemming from security vulnerabilities). Service reconfiguration strategies are used to improve their resilience. A risk assessment approach is proposed to analyze the vulnerabilities of the system with reconfiguration. The proposed technique involves assessment of likelihood and consequences of occurrence of various fault combinations handled by the reconfiguration strategies.

Keywords: Web based systems, risk assessment, security, reconfiguration.

1 Introduction

Whenever a fault manifests itself in a Web based system, whether it is a hardware failure, a software error or a security attack, it impacts the ability of a business to provide service. Isolation of the affected hardware and software is usually the first reaction (to prevent propagation of the problem to yet unaffected parts of the system). It then follows that the most important services have to be moved from the affected hosts/servers to those that are still operational. This redeployment of services [1, 6] is further called system reconfiguration.

Reconfiguration limits the adverse business consequences of a fault occurrence, but it does not eliminate them. The system usually operates with a reduced effectiveness until its full operational readiness is restored [2]. The problem of contingency planning requires a method of assessing the risks connected with the various faults occurrence, both in case of single faults and coincidence of multiple faults.

Risk analysis is a method of assessing the importance of the various possible faults, so that the maintenance policies may be better focused on improvement. In case of Web based systems, the usual approach is to eliminate or limit the number of the single points of failure, i.e. avoiding single faults that lead to the denial of service. When reconfiguration is employed, single points of failure are very unlikely. A more sophisticated analysis is required to pinpoint the weakest points of the system. Risk analysis can be used: based on assessing the likelihood of faults and their combinations, and then on confronting this likelihood with the gravity of the faults impact.

© Springer International Publishing Switzerland 2015
W. Zamojski et al. (eds.), *Theory and Engineering of Complex Systems and Dependability*,
Advances in Intelligent Systems and Computing 365, DOI: 10.1007/978-3-319-19216-1_10

97

2   Approaches to Risk Assessment

Risk analysis is used to ascertain the consequences of an event or decision, when these cannot be a priori determined. It is now used in all major planning activities. The term "risk" is used in decision-making, when the results of the decisions cannot be predicted [9]. The analysis ensures that the risks are identified, likelihood of their occurrence is established and the consequences are evaluated.

There are various alternative approaches to risk assessment [5]. Most popular is the qualitative approach, based on expert opinions [7]. It describes the risk in generic terms of 'high', 'medium' and 'low'. Quantitative analysis uses numeric values to signify similar notions [8]. The basis of risk analysis is to classify the risks in accordance to the likelihood of their occurrence and the gravity of their consequences. The risk is proportional to the product of likelihood and consequences:

$$R = LC. \tag{1}$$

We will use similar range of values to assess both the likelihood and the consequence gravity (in case of consequences, 0 indicates no impact) : very low (1), low (2), medium (3), high (4), very high (5).

The assessment of consequences of an event (fault) is very application specific. It will be considered later, after introducing the model of Web based services. Currently, it is sufficient to remark that the consequences have to be mapped to digits in the range of 1 to 5.

Likelihood is very closely related to probability of events occurrence. Web based systems are exposed to multiple faults: hardware failures, software errors, human mistakes, security issues. Some of them (especially hardware malfunctions) are customarily characterized by using probability measures. In this approach, the exponential scale of likelihood is assumed. This means that the likelihood is proportional to the rank of probability. Let's assume that an event occurs with a known constant intensity λ_i and its effects prevail for a mean renewal time T_i. Then, the probability that at a specific time instant the system is affected by the event is approximately equal to the product of them, and the likelihood is given by equation:

$$L_i = \alpha + \beta \log \lambda_i \cdot T_i, \quad \text{assuming } \lambda_i \cdot T_i \ll 1. \tag{2}$$

The constants α and β are used to ensure the required range of likelihood values.

By deploying reconfiguration techniques, the consequences of single faults in a Web based system are very low. This implies that risk analysis cannot be restricted to single faults only. Thus, it is required to consider coincidence of multiple faults at the same time. It is necessary to assess likelihood of multiple events at the same time. In case of independent events, the probabilities of single faults multiply, and the likelihood is proportional to the sum of single faults likelihood, i.e.

$$L_{i,j} = L_i + L_j - \alpha. \tag{3}$$

The above is not applicable if the events occurrence is not probabilistic or if the events are not independent. An approximate technique for dealing with likelihood of multiple events is proposed in [4]. This approach was adopted in the reported considerations. It is based on the following Table 1.

Table 1. Likelihood of coincidence of two events

	very low (1)	low (2)	moderate (3)	high (4)	very high (5)
very low (1)	1	1	1	1	1
low (2)	1	1	1	2	2
moderate (3)	1	1	2	2	3
high (4)	1	2	2	3	4
very high (5)	1	2	3	4	4

3   Web Based Systems Model

3.1   Services

The Web based systems provide some business service(s), useful to the end-users as a result of interaction between communicating component services, which are transparent to the end-user. In Fig. 1, the system is represented by the interacting service components, which are deployed on various hosts (networked computer nodes). The services make use of the hosts to provide the required processing capabilities, and of the network resources to ensure visibility and data exchange.

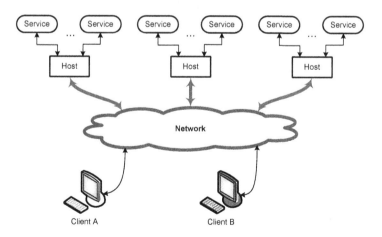

Fig. 1. Infrastructure of a Web based system

Each host is characterized by its computing resources: processing power, memory and external storage, installed software, etc. The services determine the demand for these resources. If the cumulated demand for a resource of all the services deployed in a node exceeds the available level, then all the services will be degraded. Similarly, the logical connections between the services determine the demand on the communication resources at both end nodes of the connection. The network throughputs or SLAs (Service Level Agreements) determine the limits placed on the cumulated communication demands in any single node or group of nodes. Thus, any change in the placement of services onto hosts affects both the time of processing requests by the services and the time of transmitting requests and responses.

3.2   System Reconfiguration

System configuration is determined by the deployment of service components onto the system hosts. A configuration ensures system operability if the services are so deployed that the hosts are not overloaded and the demand for communication between them does not violate the SLA limits. The set of all possible configurations that satisfy these conditions is denoted by C_{up}. This set is referred to as the set of permissible configurations. It should be noted that some deployments will not be possible due to conflicting requirements of the services regarding the host resources, such as the versions of installed software. The corresponding configurations will also be excluded from the set of permissible ones.

One of the most promising techniques to improve dependability and to avoid risks of faults occurrence is based on reconfiguration of services when failures occur. Reconfiguration is used to improve the availability of systems, invoked as a reaction to a fault occurrence.

Reconfiguration (change of system configuration) takes place when service deployment is changed. If we reconfigure the system to any configuration from the set C_{up}, then its operability will be preserved. Of course, this does not mean that the quality of the service will not be affected. The various permissible configurations may differ in the efficiency of generating the responses to client requests. This leads to the degraded operation after some reconfigurations. The permissible and degraded-operation configurations can be found using standard combinatorial techniques and simulation. Due to the size of the problem, it is almost never feasible to compute the full sets, though.

The reconfiguration graph [3] is built to define the changes in the configuration that tolerate the various faults. Set C_{up} is at the root of the graph, since any admissible configuration ensures system being up. The branches leaving the root correspond to the various faults affecting hosts or services. They point at subsets of C_{up} obtained by eliminating the configurations which do not ensure system operation in presence of the specified faults, i.e. if a host is down as the effect of the fault occurrence, then all the configurations that assume deployment of a service to that host are eliminated.

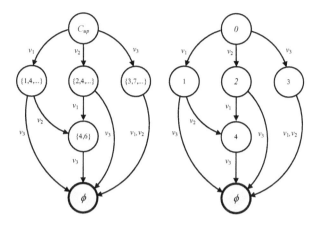

Fig. 2. An example of a simple reconfiguration graph (the numbers in the nodes correspond to the arbitrary numbering of permissible configurations) and one of the possible strategies

Further branches of the graph, corresponding to subsequent faults, are produced by eliminating configurations from those listed in higher graph node. The procedure is continued until the elimination produce empty sets ϕ that correspond to combinations of failures that cannot be tolerated by any reconfiguration. This approach to the reconfiguration graph construction ensures that all the possible configurations are taken into account. An example of such a reconfiguration graph is presented in Fig. 2.

It should be noted that the reconfiguration graph illustrates all the possible changes in the service deployment that will preserve the system operability. Reconfiguration strategy is developed by choosing a single configuration from the set at each vertex of the reconfiguration graph. There are various approaches to making this choice, as discussed in [10].

For risk assessment, a specific reconfiguration strategy is assumed. Thus, the assessment highlights the operational risks in a Web based system realizing this specific reconfiguration strategy.

4   Faults and Their Likelihood

When considering the operational risks of a Web based system, it is necessary to analyse a very diverse set of faults. It encompasses hardware faults, errors in the software, security vulnerabilities. The most suitable for the proposed analysis is the classification of faults that is based on the effects they have on the Web system. The proposed taxonomy of faults, as described in Fig. 3, is addressed to the detected faults only. Undetected faults proliferate through the system, eventually causing detected data inconsistencies.

It should also be noted that the communication faults are usually handled at the infrastructure level (by retransmission, error correction techniques, rerouting, etc.). Thus, even though they are indicated in the taxonomy, we will not consider them as events requiring risk analysis.

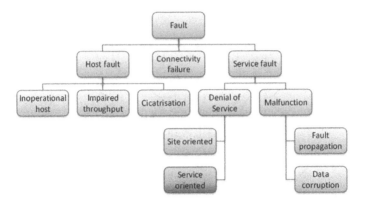

Fig. 3. A classification of system faults reflecting their impact on reconfiguration

The faults can either affect a host or only a service running on it. We distinguish the following classes of faults that affect the host:

Host crash – the host cannot process services that are located on it, these in turn do not produce any responses to queries from the services located on other hosts. The likelihood of this type of faults is low to medium (the rate of occurrence is extremely small in modern hardware, but renewal times are significant once the fault occurs).

Performance fault – the host can operate, but it cannot provide the full computational resources, causing some services to fail or increasing their response time above the acceptable limits. The likelihood of this fault is also low to medium.

Host infection – caused by the proliferation of software errors, effects of transient malfunctions, exploitation of vulnerabilities, malware propagation. The operation of services located on the host becomes unpredictable and potentially dangerous to services at other nodes (service corruption fault). Due to the potential damage that the host may cause, it is usually isolated from the system. This is equivalent to a crash fault with potential service corruption. Nowadays, likelihood of these faults is high to very high.

The faults that affect a single service include:

Inaccessible service – the service component becomes incapable of responding to requests, due to exploitation of vulnerabilities or a DOS attack. This fault can be location dependent (**location locked fault**), in which case relocation may be a fast and effective remedy. On the other hand, it may be **service locked**, in which case relocation will be ineffective and potentially dangerous to the new location. Likelihood of these faults is medium to high (they occur as often as host infection, but renewal is usually shorter).

Corrupted service – the service commences to produce incorrect or inconsistent responses due to software errors or vulnerabilities. Usually, this is a propagated fault that can be simply eliminated by restarting the affected software. It should be noted, though, that the effects of a corrupted service propagate to other service components, possibly located on other hosts. The likelihood is low due to the short time of recovery once the fault is detected.

Data inconsistency – propagating errors and malware may cause more persistent effects, by corrupting the system database. This type of faults can be very costly to recover. Technically, though, they are also remedied by service restart from the last valid backup point. The likelihood is medium to high.

It should be noted that all the faults lead to system failure if left unhandled. System reconfiguration may preserve the system functionality, though in some cases it might be an over-reaction. In case of service oriented DOS attacks, relocation is insufficient, requiring additional handling.

5   Operational Risks in Web Based Systems

5.1   Performance Prediction Using Network Simulation Techniques

Web based systems provide some business services for the users. The quality of these services depends on the configuration used. Reconfiguration causes performance loss which represents the consequences of faults occurrence, used in risk assessment.

There are two important characteristics of a Web service: the response time and the service availability (or accessibility).

The response time is defined as the time that elapses from the moment a client starts sending a request until the response is complete transmitted back to it. The average response time is computed over a mixture of user requests, characteristic for the system workload. The average response time strongly depends on the rate of service requests, as illustrated in Fig. 4a.

Service availability is defined as the probability that a request is correctly responded to by the service. It is computed as the number of properly handled requests expressed as a percentage of all the requests over a short time period $(t, t+\Delta t)$:

$$a(t) = \frac{n_{ok}(t,t+\Delta t)}{n(t,t+\Delta t)} .$$

$$(4)$$

The service availability also changes with the rate of requests sent to the system. The number of error responses should be negligible unless the system becomes overloaded. The service availability needs to be assessed for a typical workload, similarly to the response time. Fig. 4b presents a typical characteristic of availability dependent on the rate of service requests.

The typical characteristics of response times and availability have a distinct range of requests rate, where they rapidly deteriorate. This determines the maximum rate of input requests that is properly handled by the system, while using a fixed configuration. The value of this threshold requests rate decreases when the system is reconfigured. It is proposed that this value be used as the measure of system degradation, i.e. the consequence of having to react to a fault.

A simple technique, sacrificing some precision, is used to fix the maximum requests rate as midpoint in the range between under- and over-utilization. As shown in Fig. 4 this value aligns well both for response times and service characteristics.

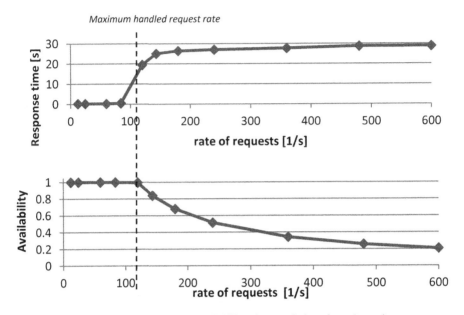

Fig. 4. Response time and availability characteristics of a web service

5.2   Performance Prediction Using Network Simulation Techniques

To predict the consequences of a combination of faults, it is necessary to determine the performance and availability characteristics of the Web system, for each configuration used in the reconfiguration strategy. As proposed in [2], this can be simplified by using a service level simulator. It is based on a comprehensive model of the computational demands placed on the service hosts. The simulator uses special purpose queuing models for predicting tasks processing times, based on the models of the end-user clients, service components, processing hosts (servers), network resources. The client models generate the traffic, which is transmitted by the network models to the various service components. The components react to the requests by doing some processing locally, and by querying other components for the necessary data (this is determined by the system choreography, which parameterizes both the client models and the service component models). The request processing time at the service components depends on the number of other requests being handled concurrently and on the loading of other components deployed on the same hosts.

The results of simulation provide the predicted characteristics of response time and availability, ensured by each simulated systems configuration. These can be used to determine the maximum requests rate for the configuration, as discussed in 5.1.

5.3   Identification of the Risks in a Reconfiguration Strategy

Risk analysis is performed to determine the weakest points in a Web system implementing a specific reconfiguration strategy. The analysis starts off with the reconfiguration strategy graph discussed in Section 2. Simulation is performed for the configuration at each vertex of the graph. The predicted maximum requests rate is associated with each vertex.

The requests rate is then normalized to the range appropriate for describing gravity of consequences (0 .. 5). It is assumed that the original configuration (associated with root vertex) has 0 consequences, while a completely inoperational system has the maximum consequence equal to 5. Decreased rates are mapped to consequences using a nonlinear function:

$$C(r) = \left(-\log\left(\frac{r}{r_0} + e^{-w}\right) - w\right)\frac{5}{w + \log(1 + e^{-w})} + 5, \qquad (5)$$

where

- r denotes the maximum handled requests rate,
- r_0 is the maximum requests rate associated with the root vertex,
- w is an arbitrarily chosen shape coefficient (set to 5 in our considerations).

Once the consequences are determined, the likelihood of the corresponding combinations of faults must be computed. The likelihood of single faults is determined using guidelines from Section 4 or formula (2). Likelihood of multiple faults is then determined from Table 1. Finally, risk is determined using equation (1).

Let's consider a simple example. A Web based system consists of three hosts, on top of this hardware, there are deployed three services. Normally, one host runs the front-end web server. This server uses services located on the other hosts. If any of the hosts becomes inoperational (or it is isolated after a security issue), then the affected service is relocated. Fig. 2b shows the reconfiguration strategy being analyzed. The results of the analysis are given in Table 2.

Table 2. Risk assessment of the example Web system

Configuration i	Consequences C_i	Likelihood L_i	Risk R_i
0	0	-	0
1	0.40	4	1.6
2	0.68	3	2.0
3	0.46	2	0.9
4	1.76	2	3.5
5	5	2	10

6   Conclusions

As shown in the case study example, risk analysis identifies the most problematic situations in a reconfiguration strategy. In this case configuration 5, i.e. system failure due to a combination of faults that cannot be remedied by reconfiguration, poses the highest risk. Also, configuration 4 is associated with much higher risk than any other configuration preserving functionality. By comparison, the operational risk in the same system without configuration is assessed at the level of 20 (likelihood 4, consequences 5).

Risk analysis is demonstrated as a useful tool to assess the weak points in a strategy. It uses a lot of approximations and thus it is inadequate for comparison of strategies and their optimization.

The presented work was supported with statutory funds of Wroclaw University of Technology.

References

1. Caban, D., Walkowiak, T.: Dependability oriented reconfiguration of SOA systems. In: Grzech, A. (ed.) Information Systems Architecture and Technology: Networks and Networks' Services, pp. 15–25. Oficyna Wydawnicza Politechniki Wrocławskiej, Wroclaw (2010)
2. Caban, D., Walkowiak, T.: Prediction of the performance of Web based systems. In: Zamojski, W., Sugier, J. (eds.) Dependability Problems of Complex Information Systems. AISC, vol. 307, pp. 1–18. Springer, Heidelberg (2014)
3. Caban, D., Zamojski, W.: Dependability analysis of information systems with hierarchical reconfiguration of services. In: Second International Conference on Emerging Security Information, Systems and Technologies, SECURWARE, pp. 350–355. IEEE Press (2008)
4. Center for Chemical Process Safety: Guidelines for Enabling Conditions and Conditional Modifiers in Layer of Protection Analysis. John Wiley & Sons, Inc. (2013)
5. Cox, L.A.: What's Wrong with Risk Matrices? Risk Analysis 28(2) (2008), doi:10.1111/j.1539-6924.2008.01030.x
6. Flyvbjerg, B.: From Nobel Prize to Project Management: Getting Risks Right. Project Management Journal 37(3), 5–15 (2006)
7. Kaplan, S., Garrick, B.J.: On the Quantitative Definition of Risk. Risk Analysis 1(1), 11–27 (1981)
8. Pérez, P., Bruyère, B.: DESEREC: Dependability and Security by Enhanced Reconfigurability. European CIIP Newsletter 3(1) (2007)
9. Talbot, J., Jakeman, M.: Security Risk Management Body of Knowledge. Wiley Interscience (2009)
10. Walkowiak, T., Caban, D.: Improvement of dependability of complex web based systems by service reconfiguration. In: Zamojski, W., Sugier, J. (eds.) Dependability Problems of Complex Information Systems. AISC, vol. 307, pp. 149–166. Springer, Heidelberg (2014)

Towards an Ontology-Based Approach to Safety Management in Cooperative Intelligent Transportation Systems

DeJiu Chen[1], Fredrik Asplund[1], Kenneth Östberg[2],
Eugene Brezhniev[3], and Vyacheslav Kharchenko[3]

[1] KTH Royal Institute of Technology, Department of Machine Design,
Division of Mechatronics, Sweden
chen@md.kth.se, fasplund@kth.se
[2] Electronics / Software, SP Technical Research Institute of Sweden, Borås, Sweden
kenneth.ostberg@sp.se
[3] National Aerospace University KhAI, Kharkiv, Centre for Safety Infrastructure
Oriented Research and Analysis, Ukraine
milestone@list.ru, V.Kharchenko@khai.edu

Abstract. The expected increase in transports of people and goods across Europe will aggravate the problems related to traffic congestion, accidents and pollution. As new road infrastructure alone would not solve such problems, Intelligent Transportation Systems (ITS) has been considered as new initiatives. Due to the complexity of behaviors, novel methods and tools for the requirements engineering, correct-by-construction design, dependability, product variability and lifecycle management become also necessary. This chapter presents an ontology-based approach to safety management in Cooperative ITS (C-ITS), primarily in an automotive context. This approach is supposed to lay the way for all aspects of ITS safety management, from simulation and design, over run-time risk assessment and diagnostics. It provides the support for ontology driven ITS development and its formal information model. Results of approach validation in *CarMaker* are also given in this Chapter. The approach is a result of research activities made in the framework of Swedish research initiative, referred to as SARMITS (Systematic Approach to Risk Management in ITS Context).

Keywords: Cooperative Intelligent Transportation System, Safety, Ontology, knowledge, safety loop, vehicle.

1 Introduction

The expected increase in transports of people and goods across Europe will aggravate the problems related to traffic congestion, accidents and pollution. As new road infrastructure alone would not solve such problems, ITS (*Intelligent Transportation System*) has been considered as a necessary initiative. In essence, the ITS-based approach emphasizes the provisions of new services for advanced collaborative and cooperative

© Springer International Publishing Switzerland 2015
W. Zamojski et al. (eds.), *Theory and Engineering of Complex Systems and Dependability*,
Advances in Intelligent Systems and Computing 365, DOI: 10.1007/978-3-319-19216-1_11

behaviors through information and communication technologies. In such a context, the perception of operational situations is supported both by in-vehicle sensors and through the V2V (vehicle-to-vehicle) and V2I (vehicle-to-infrastructure) communication channels [1, 2].

While traditionally focusing on traffic efficiency [3], ITS provides many new opportunities for promoting traffic safety. One main innovation would be the provision of system wide safety services by integrating the existing local sensing and safety features of individual traffic objects. This constitutes an important basis for reaching the goal of *Vision Zero* [4], i.e. that no one will be killed or seriously injured within the road transport system. On the other hand, for the automotive industry, the transition into ITS represents many technology and culture leaps. For example, an ITS-based service for traffic safety requires not only a functional conformity of traffic objects, but also a guarantee of the performance and dependability of their coordinated behaviors. This in turn implies both design-time measures (e.g. safety process) and run-time features (e.g. quality-of-service), where a consideration of multiple traffic objects beyond the traditional automotive vehicle centered view becomes necessary. In particular, due to the complexity of cooperative operational situations, novel methods and tools for the requirements engineering, correct-by-construction guarantee, variability and lifecycle management become also important. For the design of safety functions, there is a need of capturing the operational behaviors of all related traffic objects under dynamically changing conditions for behavior control and anomaly detection. Such a specification of operational behaviors is often not supported by current approaches to the design of safety functions, which rest on worst-case analyses [5].

This chapter presents an ontology-based approach to safety management in C-ITS (*Cooperative Intelligent Transportation Systems*), primarily in an automotive context. It describes the methodology where formal models play a key role both for supporting the perception of operational situations and for dynamically assessing the safety risks and planning the behaviors. The rest of this chapter is structured into the following sections: Section 2 discusses the state-of-the-art approaches to safety management through ITS. Section 3 introduces the envisioned ontology-based approach, including the minimal version of this ontology, description of design and deployment stages.

2   Related Work

There is a wide range of approaches to the management of transport system safety using communication and information technologies. While some of these approaches focus on the integration of *strategic and tactical decisions* (dealing with long-time goals), the others deal with the integration of *operational actions* (dealing with decisions seconds into the future). In this section, we also compare some related safety management approaches in other domains.

For road transport, the integration of strategic and tactical level decisions has been studied in particular for the transport of hazardous material. For example, *dynamic risk assessment* has been suggested as a way to minimize the risk during transport routing and emergency response. One type of systems that provides such services today is ATMS (*Advanced Traffic Management Systems*). These systems typically attempt to use available traffic information to develop optimal traffic control strategies. The focus is usually on a centralized solution for monitoring and controlling traffic behavior at *macroscopic* and *mesoscopic* levels (i.e. with whole cities and large city blocks) [6]. The *microscopic* level (i.e. separate road segments) behavior is often left uncontrolled. An ATMS normally includes methods and tools to support incident detection, incident verification, etc. However, static models developed for larger geographical road systems are often inefficient when dealing with unforeseen dynamic factors such as peak hours. Tying together models at the macro- and mesoscopic levels with simulation at the microscopic level has therefore been suggested as a way of addressing these dynamic factors better. However, to avoid a too high demand on computational resources by simulations the microscopic level is typically only considered in particular "problem areas".

An elaboration of safety-relevant requirements on ITS based rail crossings can be found in [7]. The focus is however on the support for operational (vehicle) level safety in predefined static situations. The dynamical aspect, i.e. runtime risk assessment, is not discussed. In [8], a simulation-based approach to autonomous vehicle safety assessment is presented. The core is an ontology that stipulates the concerns in situation analysis and task planning. It is suggested that *dynamic risk assessment* (continuous evaluation of the risk of possible actions and the selection of the "best" action) can be necessary in complex and less controlled environments where *predetermined risk assessment* (analysis of possible accidents and the inclusion of protection mechanisms) is problematic to apply. An approach to offline risk analysis with Bayesian Networks for the modeling traffic accident data is given in [9]. The factors of concern include the characteristics of the road, traffic flow, time/season and the people involved in an accident. The work presented in [10] also considers the factors relating to driver behavior and vehicle dynamics including sensor uncertainty and vehicle state. In this approach, the risk level is assessed at run time by combining traffic rules, vehicle dynamics, and environment prediction. However, it does not cover cooperative behaviors and lifecycle perspective of ITS. In [11], an evaluation and testing of the two demonstrator vehicles developed for intersection driver assistance is described. The dynamic risk assessment (DRA) is supported through object tracking and classification and the communication of traffic management and driver intention.

The overall dependability of industrial installations in the nuclear, automotive, chemical and energy domains is centered on *functional safety* and *risk management* provided by IEC61508 [12] and associated standards (e.g. ISO 26262 [13]). According to these standards, a risk management lifecycle typically goes across the design, the deployment, and the post-accident analysis stages. During the design, the risk management is focused on eliminating known hazards by safety measures that keep the system in a safe state. The success can however be restricted because of a high degree of uncertainty due to a lack of knowledge, insufficient model accuracy, etc. Post-deployment risk management refers to constantly conducted safety evaluations because of changes in the system configuration, component reliability, maintenances, etc.

Such measures are implemented in many industrial installations to prevent accidents. Post-accident risk assessment produces a detailed description of the incident/accident and its consequences, proposes additional prevention and protection barriers, etc.

All of these risk assessment methods are based on some assumed accident scenarios [14]. For automotive vehicles, such accident scenarios are often given by some estimated *worst case scenarios*. From a safety engineering perspective, such deterministic estimations often abstract away details about the combinatorial effects of environmental events and system anomalies in spatial and temporal domains for the efficiency of assessment. In nuclear domain, after the Fukushima Dai-Ichi accident, when deterministic safety approaches failed, these have been complemented by probabilistic safety assessment/probabilistic risk assessment (PSA/PRA). Currently both activities are implemented during Nuclear Power Plant Instrumentation&Control System development stages. The current PSA framework has some limitations in handling the timing of automatic and personnel actions. The conventional PSA techniques (Event Tree / Fault Tree) methodology may not yet yield satisfactory results, but they open the way for use of new dynamic techniques to accurately describe system dynamics while considering e.g. state uncertainties. These new methodological approaches to risk assessment include statistical analysis of near-miss and incident data using Bayesian theory to estimate operational risk value and the dynamic probabilities of accidents sequences having different severity levels [15]; and the application of simulation models to analyze scenarios using dynamic fault trees [16]. ITS can potentially be benefited by the approaches in other safety critical industrial processes. Here, one challenge is related to the complexity of ITS where unknown emergent behaviors dominate. For example, the risk assessment, e.g. in the nuclear domain, is performed to build risk profiles for critical processes. The application of dynamic risk assessment just allows one to make these profiles dynamical.

3   An Ontology-Based Approach

By *system ontology*, we refer to the formalization of system-wide concerns in terms of models. Such concerns typically range from the definitions of system constituent units (i.e. the traffic objects and traffic environments) in regard to their boundaries, compositions, technological preferences, to the specifications of the interactions of such objects in regard to the functionalities and extra-functional constraints. As a generic support for knowledge formalization, the ontology-based approach aims to promote not only the quality management at design time across engineering disciplines and teams, but also the data treatment and decision making at run time across traffic objects.

3.1   An Overview

For the functional safety of road vehicles, ISO26262 represents the domain consensus on the state-of-the-art approaches [12]. It is centered on a reference safety lifecycle through which the work tasks as well as the information to be generated and communicated for risk management and requirements engineering are stipulated. In the case of ITS, the connectivity across multiple traffic objects implies that a system-of-systems perspective on functional safety becomes necessary. This means in practice that the *safety-loop*, i.e. the loop of system safety lifecycle, now needs to cut across the safety lifecycles of involved traffic objects and infrastructure [6]. As outlined earlier, the complexity of operational situations also makes dynamic risk assessment services necessary for enabling optimized control in *a priori* unknown situations or simply for guaranteeing more qualified specification of safety goals.

The ontology-based approach described here emphasizes the provision of an integrated knowledge model both for safety engineering and for the design of advanced safety features. It in particular considers *coordinated driving* when two (or more) vehicles coordinate their behaviors either based on predefined traffic rules (i.e. *choreographed*), or through active communication and consolidation of intents (i.e. *cooperative*), or by active negotiation for consensus (i.e. *collaborative*). An overview of the target lifecycle phases, work tasks, and run-time services, all centered on a common ontology, is given in Fig. 1.

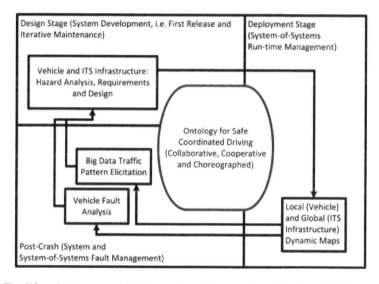

Fig. 1. The lifecycle phases, work tasks and run-time services to be benefited by a common ontology

Post-accident analysis is supposed to be one of main stage of the ITS safety management process. The information accumulated during system runtime would be stored and processed during vehicle fault analysis through big-data and pattern elicitation. Detailed tracking and analysis of vehicle malfunctions and failures will allow the updates of failure probabilities used during design stage. Besides, it decreases uncertainties in risk and hazard analysis.

3.2   The Modeling Framework

There are different roles of the ontological core. During system development, the focus is on a model-based approach to risk assessment and elicitation of safety requirements. It would provide a well-structured and standardized specification of the operational situations, the system architecture in terms of functional and technical design, as well as the related safety requirements and constraints. Beyond the development time, the support is focused on the provision of a knowledge model for dynamically consolidating the monitored operational situations, coordinating the control behaviors, and handling possible anomalies. For instance, during the deployment stage, the ontology allows for transformations of information during V2V and V2I to provide a common basis for logging, shared perception and decision making. For post-accident analysis, it assures that the factors of importance for the analysis are available from the data logged by the ITS infrastructure.

The key base technology to support the deployment of such an ontology-based approach is the *EAST-ADL* (Electronics Architecture and Software Technology - Architecture Description Language). As a modeling framework, EAST-ADL represents a key European initiative towards a standardized multi-viewed description of automotive electrical and electronics systems [17]. It integrates many existing frameworks (e.g. SysML, RIF/ReqIF, ISO26262) while allowing a wide range of functional safety related concerns (e.g. hazards, faults/failures, safety requirements) to be declared and structured seamlessly along with the lifecycle of nominal system development. Based on such a structured description, EAST-ADL also provides necessary modeling support for functional safety [18]. Moreover, through its support for behavior description, the modeling framework also allows the developers to precisely capture various behavioral concerns in requirements engineering, system design, and safety engineering [19]. However, although constituting a very good basis for capturing and formalizing various aspects of ITS, current EAST-ADL does not provide any explicit methodology on the modeling and analysis of ITS systems in regard to the emergent properties and safety issues. Therefore, language extensions and specializations in regard to cooperative ITS (C-ITS) are being developed. One key modeling package being extended is the *Environment Model*, shown in Fig 2. In particular, the following additional concepts are introduced to support the specification of operational situations:

- *Scene* - a description of characteristics and objects that are of interest and "static" at a strategic or tactical level. Typical static scene data include: 1. Weather conditions, i.e. air density, humidity and pressure, solar radiation, temperature, etc. 2. The terms defined by the WGS 84, OpenDrive and OpenCRG standards; 3. Regions of interest, which are defined by a polygon set, a type (e.g. boundary) and an object identified (e.g. fence)
- *Situation* - a scene populated with dynamic objects, which are defined by: 1. WGS 84; 2. Mass and Speed; 3. Behavior, which is defined by a type (e.g. choreographed) and a trajectory (i.e. an intent); 4. An associated region of interest; 5.A probability distribution tied to each of these terms.
- *Scenario* - a set of situations linked in time.

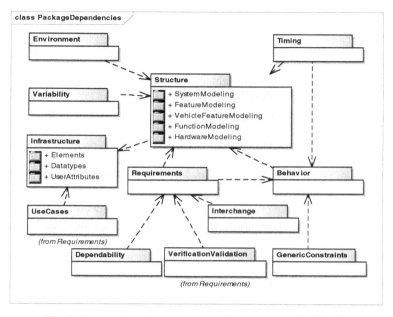

Fig. 2. An overview of the meta-model packages of EAST-ADL

3.3   The Case Study with Virual Depolyment

As a first step towards analyzing the use of our ontology during the deployment stage we used *IPG CarMaker* to simulate interactions between an ego-vehicle and both uncoordinated and all types of coordinated drivers on a four lane circular track (see Fig. 3). If all information defined by the ontology is available, then an ego-vehicle can make informed control decisions by sharing data across all ITS levels. In real implementation, there would of course be many challenges since there is a high probability for the information being lost or corrupted (e.g. the sheer size of the data, failing sensors, transmission issues, etc.). As a fundamental requirement, such problems would not result in risk, but rather in a lower level of certainty of the dynamic assessment outcome that would in turn imply the activation of ADAS function (for involving the human driver in the loop as the fail-safe).

Fig. 3. CarMaker Simulation

A related problem of concern is that each ego vehicle will have to evaluate the *trustworthiness* of each piece of information provided via ITS. For legal reasons it is unreasonable to assume that the manufacturer of a vehicle will be able to completely trust and coordinate with all *other* manufactures of vehicles and infrastructure. This implies two things. Firstly, that the probability distribution tied to each piece of information will have to be subjective for each ego-vehicle in the ITS system. Secondly, that the probabilities of most importance to dynamic risk assessment in an ITS systems are not those related to the outcomes an ego-vehicle set of possible actions, but rather those related to the validity of the provided decision support. Support for reasoning about such uncertainty, such as that provided by *Dempster-Shafer Theory* and *Belief Propagation*, across ITS levels is therefore likely to be a well-motivated research direction. It is also likely that those objects in ITS on which a specific vehicle can trust completely are going to become important as evaluators of other entities. For example, a trusted vehicle driving in front a less trusted vehicle can be used to evaluate the accuracy of the latter vehicle´s sensors; trusted infrastructure can estimate the speed of a vehicle, which can then be compared with the broadcasted value.

4   Conclusions

Deployment of ITS is expected to bring the many benefits for all traffic participants. ITS will help to improve the transport efficiency, passengers comfort, decrease environmental contamination, etc. Due to the inherent complexity of ITS, it is impossible to cover all of the possible traffic scenarios during the design stages. The traditional approaches to safety management in ITS would not be sufficient due to their heavy reliance on worst case assumptions. The approach proposed by this paper is based on an ontology that allows one to formally capture the concerns in temporal and spatial domains and thereby constitutes a basis for a novel safety lifecycle with *knowledge-in-the-loop*. The approach emphasizes the interplay of model-based system development and the design of advanced system services, through which meta-knowledge for robust perception and safe operation will be deployed and maintained at system run-time.

References

1. Sussman, J.: Perspectives on Intelligent Transportation Systems. Springer, New York (2005)
2. Vision for ITS. Proceedings of the National Workshop on Intelligent Vehicle/Highway systems sponsored by Mobility 2000, Dallas, TX (1990),
   http://ntl.bts.gov/lib/jpodocs/repts_te/9063.pdf
   (accessed June 2014)
3. Papageorgiou, M.: ITS and Traffic Management. In: Handbook in OR&MS, vol. 14 (2007), doi:10.1016/S0927-0507(06)14011-6
4. Tingvall, C.: Vision Zero - An ethical approach to safety and mobility. In: Proceedings of the 6th ITE International Conference Road Safety & Traffic Enforcement: Beyond 2000, Melbourne (1999)

5. Chen, D., et al.: A systematic approach to risk management in ITS context – challenges and research issues. Радіоелектронні і комп'ютерні системі 5(69), 11 p. (2014) ISSN 1814-4225

6. Östberg, K., Törngren, M., Asplund, F., Bengtsson, M.: Intelligent Transport Systems - The Role of a Safety Loop for Holistic Safety Management. In: Bondavalli, A., Ceccarelli, A., Ortmeier, F. (eds.) SAFECOMP 2014. LNCS, vol. 8696, pp. 3–10. Springer, Heidelberg (2014)

7. Larue, G.S., et al.: Methodology to assess safety effects of future Intelligent Transport Systems on railway level crossings. In: Proceedings of Australasian Road Safety Research, Policing and Education Conference, Wellington, New Zealand, October 4-6 (2012)

8. Wardzinski, A.: Dynamic Risk Assessment in Autonomous Vehicles Motion Planning. In: Proceedings of the First International Conference on Information Technology, Gdansk, Poland (2008), doi:10.1109/INFTECH.2008.4621607

9. Simoncic, M.: A Bayesian Network Model of Two-Car Accidents. Journal of Transportation and Statistics 7(2/3) (2004)

10. Cheng, H., et al.: Interactive Road Situation Analysis for Driver Assistance and Safety Warning Systems: Framework and Algorithms. IEEE Transaction on Intelligent Transportation System 8(1), 157–167 (2007)

11. Fuerstenberg, K., et al.: Results of the EC-Project INTERSAFE. Advanced Microsystems for Automotive Applications VDI-Buch, pp. 91–102 (2008)

12. International Electrotechnical Commission. IEC 61508:2010, Functional safety of electrical/electronic/programmable electronic safety-related systems (2010)

13. International Organization for Standardization. ISO 26262:2011, Road vehicles - Functional safety (2011)

14. Swaminathan, S., et al.: The Event Sequence Diagram framework for dynamic Probabilistic Risk Assessment. Reliability Engineering and Systems Safety 63, 73–90 (1999)

15. Anjana Meel, A.: Plant-specific dynamic failure assessment using Bayesian theory. Chemical Engineering Science 61, 7036–7056 (2006)

16. Hong, X.: Combining Dynamic Fault Trees and Event Trees for Probabilistic Risk Assessment Reliability and Maintainability. In: Annual Symposium - RAMS, pp. 214–219 (2004) ISBN: 0-7803-8215-3

17. EAST-ADL. EAST-ADL Domain Model Specification, Version M.2.1.12 (2014), http://www.east-adl.info/

18. Chen, D., Johansson, R., et al.: Integrated Safety and Architecture Modeling for Automotive Embedded Systems. e&i 128(6) (2011)

19. Chen, D., Feng, L., et al.: An Architectural Approach to the Analysis, Verification and Validation of Software Intensive Embedded Systems. Computing (2013), doi:10.1007/s00607-013-0314-4

Optimization of Couriers' Routes by Determining Best Route for Distributing Parcels in Urban Traffic

Piotr Chrzczonowicz and Marek Woda

Department of Computer Engineering, Wroclaw University of Technology
Janiszewskiego 11-17, 50-372 Wroclaw, Poland
{172800,marek.woda}@pwr.edu.pl

Abstract. The aim of this paper was to propose an efficient algorithm for determining the best route for couriers' in urban traffic. The best route is defined as the order in which the parcels should be delivered. The specification of couriers' work may result in different criteria being considered optimal, including but not limited to: travelled distance, elapsed time, combination of preceding.

For this reason, the optimality criteria is defined and proposed solution is verified against those optimization targets.

The scope of this paper consists of researching modern methods of solving TSP. In addition to that, a simulation environment that allows to conduct experiments was developed. Secondly, a selected portion of methods described in literature was implemented to form a baseline for benchmarking proposed solution. Furthermore, authorial solutions are implemented and the results of running those are compared against baseline results. A best route is considered in terms of parcel delivery order and not finding the shortest path between subsequent delivery addresses. Consequently the routing problem (finding the shortest path) is beyond the scope of this paper.

Keywords: TSP, routes optimization, parcels distribution.

1   Introduction

Let's define the problem of finding the best route for a courier in a city traffic. The courier starts work at a distribution center with N packages to deliver. Let's assume, without loss in generality, that every package needs to be delivered to different recipient and that everyone is at home, so that the courier don't have to visit any address more than once. The distance between any two addresses could be read from the map. The goal is to find the order in which the courier should visit the recipients so that the travel time or travelled distance is minimal.

In general, the courier can travel between the recipients in any order. This could be represented as a complete graph where the addresses to deliver packages and a starting location are the vertices $V = \{v_0, v_1, ..., v_n\}, |V| = N + 1$. Let the $E = \{v_a, v_b\}: v_a, v_b \in V\}$ be the edge set of this graph. Let the distance d between any two points be the weight of the edge connecting those vertices $w_{v_a, v_b} = w_e = d_{ab}, e \in E$. Let $W = \{d_{ab}\}_{N+1 \times N+1}\}$ be the matrix of costs. Then the goal is to find a

© Springer International Publishing Switzerland 2015   117
W. Zamojski et al. (eds.), *Theory and Engineering of Complex Systems and Dependability,*
Advances in Intelligent Systems and Computing 365, DOI: 10.1007/978-3-319-19216-1_12

Hamiltonian, i.e. such cycle $(v_0, v_a, v_b, ..., v_0)$ where every vertex is used exactly once, where the sum of the weights of the edges connecting subsequent vertices is minimal.

According to Ore's theorem, since the graph is complete, there always would be a *Hamiltonian* cycle, assuming that the $|V| > 2$ [13].

This matches the definition of so called *Travelling Salesman Problem* abbreviated *TSP*, which is a very interesting problem that is still being actively researched. The methods of solving this problem, and its variations, are very useful for optimizing processes in different areas, such as robotics control, manufacturing, scheduling, routing many vehicles, etc. [9, 5].

While *TSP* term could be used to describe a certain class of problems, it usually refers to so called *symmetric TSP* or *sTSP* in short. The symmetry refers to the distances between vertices, i.e. $w_{v_a, v_b} = w_{v_b, v_a}$ as opposed to the asymmetric variant \emph{aTSP}, where $w_{v_a, v_b} \neq w_{v_b, v_a}$ [5].

Other variant of *TSP* covered in this paper is a *dynamic TSP*. The difference from the classical problem is that the number of vertices and edge weights is variable in time, so $d_{ab}(t) = w_{v_a, v_b}(t)$ and the cost matrix is defined by $W(t) = d_{ab}(t)_{N(t) \times N(t)}$. In this definition the time is continuous, but in real life calculation it is common to discretize time changes [10]. The definition assumes that the number of vertices change over time, but in case described in this paper, the number of vertices would be constant $N(t) = N + 1$, which is equal to the number of recipients and the distribution centre.

1.1   Past and Recent Works Related to Subject

The first research around *TSP* is believed to begin around XVIII century in studies of Sir William R. Hamilton and Thomas P. Kirkman, but the problem in currently known form was introduced much later [5, 10]. [12] showed that finding a *Hamiltonian* path in a graph could be reduced to *Satisfiability* problem hence it is *NP-complete*. Finding the shortest of all *Hamiltonian* cycles in a graph, hence solving *TSP*, is at least as hard as finding a *Hamiltonian* cycle, and this matches the informal definition of being *NP-hard*.

The *TSP* finds many usages in many engineering and science disciplines, so many algorithms allowing to solve the problem were developed over the years. Next the most notable ones are reviewed, divided into two groups: *exact* and *approximate* solutions.

Exact solutions. Due to the complexity of *TSP*, even for modern computers, it is hard to find an optimal solution for every instance of the problem. The naive (*Brute force*) solution is to find all possible *Hamiltonian cycles* and return the one with the lowest cost as a result. Let's introduce a fact:

Lemma 1. *In a complete graph $G(V, E)$ every permutation of its vertices defines a Hamiltonian cycle.*

Proof. (1) Let the $\sigma(V)$ be the permutation of vertices of complete graph G. Due to completeness of the graph $\forall\sigma(V)_i, \sigma(V)_{i+1} \in \sigma(V)), \exists e \in E,$ so $\sigma(V)$ defines a cycle.

(2) $\sigma(V)$ is a permutation of V so every vertex appears only one time.

(1) $\wedge$ (2) $\rightarrow \sigma(V)$ defines a *Hamiltonian cycle*.

$c \leftarrow maxval$
for all *route* $\leftarrow \sigma(V)$ **do**
    if $cost(route) < c$ **then**
        $c \leftarrow cost(route)$
        $best \leftarrow route$
    end if
end for
return best

Fig. 1. Algorithm 1

As a consequence of lemma 1 a naive algorithm could be introduced - Fig. 1. Since the loop is executed for every possible permutation of vertices, the algorithm is $O(n!)$. Note that assuming that cycle should start (and end) in arbitrary chosen vertex and that the instance of the problem is symmetric, the number of tested permutations could be limited to $\frac{(n-1)!}{2}$, which is still inefficient due to the rate of growth of the factorial [9].

The operation time of this algorithm could be strongly influenced by the algorithm used to generate permutations, but still research conducted shows that this algorithm useful only to find the solution in small graphs, i.e. consisting of less than thirteen vertices. Because of that fact, more sophisticated algorithms were introduced.

Other approach on finding optimal solution is to use *Dynamic programming* methods. Let's choose, without loss in generality, a starting point v_o, then a way to find a solution for TSP would be to find a path e $(v_0, v_i, ..., v_n)$ starting from v_0 and visiting every $v_i \in V \setminus v_o$ vertex, this yields a permutation of the set V, which as noted earlier according to *lemma 1* defines the *Hamiltonian* cycle. It is important to remember that besides the cost of the path, the cost of returning back from v_n to v_o must be included while calculating the cost of the *Hamiltonian* cycle being the solution. Thus let's define:

$$S_n \subseteq V, where \; |S_n| = n \; and \; v_o \in S_n \tag{1}$$

Now let the function $f_c(S_n, v_i)$ be the cost of the shortest path containing every $v_j \in S_n$, starting at v_o and ending in v_i. This function could be defined as:

$$f_c(S_n, v_i) = \begin{cases} 0 & when \; S_n = \{v_i\} \\ \infty & when \; i = 0 \\ min\{d_{v_i v_j} + f_c(S_n \setminus v_i, v_j)\} & otherwise \end{cases} \tag{2}$$

The second case is needed, because the path could not start and finish in v_o. Given this, the cost of minimal *Hamiltonian* cycle is defined as:

$$min\left\{f_c(V, v_j) + d_{v_j v_o}\right\} \tag{3}$$

The cycle with this cost is the solution for given instance of *TSP* [1, 4]. Based on that an algorithm could be produced as shown on Fig. 2.

---

initialize *costDictionary*
for $x \leftarrow 2$ to n **do**
    for all $S_x : S_x \subseteq V$ **do**
        for all $v_i \in S_x : v_i \neq v_0$ **do**
            store value for $f_c(S_x, v_i)$) in *costDictionary*
        end for
    end for
end for
return $\min\left\{f_c(V, v_j) + d_{v_j v_0}\right\}$

Fig. 2. Algorithm 2

The algorithm returns the cost of optimal path, the optimal path itself could be retrieved from *costDictionary* as $(v_0, v_i, ..., v_j)$ where subsequent v_i are those with the minimal value that were actually used during the cost calculation. The count of sub problems is less or equal to $n * 2^n$ and each is solved in linear time, what yields a computational complexity of $O(n^2 2^n)$[4].

Approximate solutions. While exact algorithms are guaranteed to produce optimal solution for *TSP*, due to the computation complexity, it is not always feasible to compute the optimal solution, usually a "*good enough*" solution may be sufficient. Many approximate or heuristic algorithms have been proposed over the years this section covers some most notable of them. Let's define the optimality of a solution as:

$$\eta = \frac{cost\ of\ approximate\ path}{cost\ of\ optimal\ path} \tag{4}$$

---

append v_0 to *route*
vertices $\leftarrow V \setminus v_0$
lastSelected $\leftarrow 0$
while $|vertices| > 0$ **do**
    $v_j \leftarrow v_i \in vertices$ such that $d_{v_{lastSelected} v_i}$ is minimal
    append v_j to route
    vertices $\leftarrow vertices \setminus v_j$
    lastSelected $\leftarrow j$
end while
return *route*

Fig. 3. Algorithm 3

Some of the algorithms has been proven to produce solution η always within certain bounds. *Nearest neighbor* is one of the simplest approximate algorithms for solving *TSP*. The operation of algorithm is greedy, meaning that in every stage a locally optimal choice is being made. This solution was proven to produce solution with $\eta \leq \frac{1}{2} \lceil \lg(n) \rceil + \frac{1}{2}$ [6]. Since the loop executes $\eta - 1$ times and finding the nearest vertex is $O(n)$, the whole algorithm is $O(n^2)$.

Another algorithm worth mentioning was introduced in 1976 [3]. The instance of *TSP* must be *Euclidean*, i.e. the triangle inequality must be satisfied, to be properly solved by this algorithm. The algorithm itself consists of few simple steps including computing Minimal Spanning Tree (MST) and Minimum Weight Matching (MWM), as shown on Fig. 4. The last two steps might need a comment. MWM is calculated on vertices with odd rank and added to the MST of graph G to ensure that the *Eulerian* cycle exists. The *Eulerian* cycle could be converted into *Hamiltonian* cycle by "taking shortcuts", i.e. removing the edges leading to vertices visited earlier and inserting the edge to next unvisited edge.

$mst \leftarrow$ compute MSTG
$mwm \leftarrow$ compute MWM for $\{v : v \in G \wedge \text{rank}(v) \text{ is odd}\}$
find Euler cycle in $mst \cup mwm$.
convert Euler cycle into Hamiltonian cycle in G.

Fig. 4. Algorithm 4

This algorithm is proven to produce solutions within $\eta \leq \frac{3}{2}$ of optimal [3], but usually it produces solution within $\leq 1 + \frac{1}{10}$, and the computational complexity is $O(n^3)$ [5]. Another popular approximate algorithm is the *Insertion heuristic*. The algorithm is straightforward, and consists of loop of 2 steps [5]: select next vertex, insert selected vertex to the cycle. It is presented in pseudo-code on Fig. 5.

---

append v_0 to *route*
$v_n \leftarrow$ select next $v \in V$
append v_n to *route*
$vertices \leftarrow V \setminus \{v_0, v_n\}$
while $|vertices| > 0$ **do**
    $v_n \leftarrow$ select next $v \in vertices$
    insert v_n into *route*
    $vertices \leftarrow vertices \setminus v_n$
end while
return *route*

Fig. 5. Algorithm 5

The insertion step is always the same, and involves computing the costs of all possible cycles that could be derived from inserting a newly selected vertex into already created cycle, and selecting the one with minimal cost as a result. Selection step on the other hand is more interesting subject, since many ways of selecting next vertex was proposed, such as: farthest, nearest, cheapest, random, and others. The random selection is straightforward. The farthest selection involves selecting a vertex v_j from remaining vertices that has the $max\left\{d_{v_j v_i}\right\}$ for any v_i in remaining vertices. The nearest is similar, but the minimum cost is searched for. The cheapest selection is slightly more complicated, since it involves computing costs after all possible insertions for all remaining vertices and selecting the one where the resulting cost is the smallest. The farthest selection is considered to produce the best results [9]. The computational complexity of the algorithm is really dependent of the selection method that is being used, but for nearest/farthest selection it is $O(n^2)$ [5].

One of most frequently used heuristics are the *k-opt heuristics*. In general this method is used to improve already computed approximate solutions of the *TSP*. Single step of K-opt algorithm involves separating existing *Hamiltonian* cycle into $k \geq 2$ paths and rejoining them together in such way that the new route is a Hamiltonian cycle and is shorter than the previous one. A 2-opt heuristic is often producing a results within $\eta \leq 1 + \frac{1}{20}$ of the optimal solution, while 3-opt often results with $\eta \leq 1 + \frac{3}{100}$ [5]. In general, the greater the k the better results are possible, but due to fast growing number of possible combinations that needs to be computed, and $k > 3$ is usually not feasible. In general there are $\binom{n}{k}$ ways to choose edges to remove, assuming k that at least one reconnection is tested, the algorithm ends up being at least $O(n^k)$ [9]. Fig. 6 shows the pseudo-code of k-opt procedure. Due to the 2-opt's simplicity and acceptably good results, it is used as a local search optimization heuristic, in conjunction with some algorithms described in state of the art section of this paper, in order to form hybrid approaches.

$route \leftarrow$ a valid solution
$bestC \leftarrow$ cost of $route$
$newC \leftarrow 0$
while $newC < bestC$ **do**
    $bestC \leftarrow$ cost of $route$
    $newC \leftarrow \infty$
    chose k edges to be removed $e_1, \ldots, e_k$
    $paths \leftarrow$ remove $\{e_1, \ldots, e_k\}$ from $route$
    for all $newRoute \in paths$ **do**
        $cost \leftarrow$ cost of $newRoute$
        if $cost <= bestC$ and $cost < newC$ **then**
            $route \leftarrow newRoute$
            $newC \leftarrow$ cost of $newRoute$
        end if
    end for
end while
return $route$

Fig. 6. Algorithm 6

Genetic Algorithm and *hybrid* approach are the algorithms that are receiving much research interest nowadays. What is worth noticing, is that those methods, together with *k-opt* are *Tour improvement* methods [5], i.e. are operating on valid *Hamiltonian* cycles, instead of building one. This allows not only to solve static *TSP* but also a dynamic instances of the problem. Very popular method of solving optimization problem is to use the *Simulated annealing* (*SA*). This is the local search algorithm and a variant of threshold algorithm. The algorithm is presented on Fig.7.

$currentTemp \leftarrow$ initial temperature
$currentSolution \leftarrow$ a valid solution
while $currentTemp > endTemp$ **do**
    $nextSolution \leftarrow$ Select next valid solution
    if $F_{accept}(currentSolution, nextSolution)$ **then**
        $currentSolution \leftarrow nextSolution$
    end if
    $currentTemp \leftarrow currentTemp * annealRate$
end while
return $currentSolution$

Fig. 7. Algorithm 7

It starts with a one of possible solutions and in each iteration the solution from the neighborhood is selected. What is different from standard threshold algorithm, is that not only better solutions are accepted. The function deciding whether selected solution will be accepted in any iteration is given as:

$$f_c(S_n, v_i) = \begin{cases} 1 & when\ fit(s_s) \leq fit(s_1) \\ exp\left(\frac{fit(s_s)-fit(s_1)}{currentTemp}\right) & when\ fit(s_s) > fit(s_1) \end{cases} \qquad (5)$$

The state of the algorithm is also characterized by the initial temperature that is decreased with every iteration. The algorithm stops when certain temperature is achieved [1]. Another popular approach of solving NP-hard problems is the Genetic Algorithm. This method is based on natural process of evolution. First work related to this topic is dated at 1957, and since then it is receiving growing interest in research communities. As of 2007 it was described around 170 crossover routines [6], and published a second volume just for mutation routines. The canonical structure of *Genetic algorithm* is presented on Fig. 8.

Constructing an evolutionary algorithm for TSP involves defining a fitness function as well as crossover, together with mutation, operators [5]. The fitness function for the algorithm to minimize is straightforward - cycle cost. For the mutation routine there are many possibilities: random swapping order of some points in cycle, perform a step of k-opt operation, any other routine transforming one Hamiltonian cycle into another. The main difficulty is proper crossover operator. First an appropriate encoding must be defined, and only then a crossover routine could be matched to the encoding, to name a few: binary matrix encoding with matrix crossover [7], priority encoding with greedy algorithm [15], no crossover [7], permutation encoding with appropriate operator and others.

$pNumber \leftarrow 1$
$population \leftarrow$ generate initial population   ɪ
while $stop\ condition = False$ **do**
    calculate fitness of $population$
    select parents set $P \subseteq population$
    use crossover and mutation to create $descendants$ from P
    $population \leftarrow$ create population($pNumber$) from $descendants$
    $pNumber \leftarrow populationNumber + 1$
end while

Fig. 8. *Algorithm 8*

Other method popular in recent research is to create so called hybrid algorithms. Such approach is taken to produce algorithms presented in this paper. General idea of this method is to use one or more existing approaches and use those to improve another method, usually a state of the art solutions are used to get good preliminary result that could be further improved. Since genetic algorithms tend to find local optimum, it is possible to modify it in such way, that it will be more likely to find global optimum. Genetic algorithm could be as well used to reinforce other methods, such as *Ant Colony Optimization* [15]. Usually, the methods chosen are similar in case of computational complexity thus one ends up with better results at little expense of complexity.

2   Proposed Approach

2.1   Simulated Annealing and Local Search Heuristic Hybrid Solution

A baseline implementation of *simulated annealing* for solving *TSP* is using a random permutation as a starting point and a simple random shift of single vertex to get new arrangement in each iteration. While there might exist more sophisticated variants of the *SA* algorithm for solving *TSP*, this method appears to produce optimally looking paths rapidly when it is a subject to visual audition. The produced results tend to have crossing sections. To further improve the path a *2-opt* operation could be used. Based on this, a hybrid algorithm is proposed *SA_2OPT* as shown on Fig. 9. To further improve the operation, the initial path generation could be altered.

$initial \leftarrow$ Run the SA on the problem instance
$result \leftarrow$ Run $2 - opt$ algorithm using $initial$ as initial path
return $result$

Fig. 9. Algorithm 9

Instead of random permutation, a good starting point could be achieved by using one of the Insertion heuristics described in introduction. So another variant of the *SA* method is proposed as shown in Fig. 10.

$initial \leftarrow$ Run the *Farthest Insertion* on the *problem*
$result \leftarrow SA(problem)$
$result \leftarrow 2OPT(result)$
return *result*

Fig. 10. Algorithm 10

2.2   Genetic and Local Search Heuristic Hybrid Solution

A baseline implementation of the *GA* for *TSP*, using simple permutation encoding for crossover and mutation, while runs very quick, tends to produce worse solutions for bigger graphs as found out during visual auditing. To address this issue a variation of the *GA* is proposed. A hybrid approach using Insertion heuristic in order to improve initial population for genetic algorithm is presented on Fig. 11. Rest of the algorithm remains unchanged.

procedure GENERATEINITIALPOPULATION($size$)
    $result \leftarrow$ initialize empty collection
    $permutation \leftarrow$ random permutation
    while $result.size < size$ **do**
        if $random\ value\ from[0,1] < 0.5$ **then**
            $newPermuation \leftarrow$ Run *Random Insertion* on the problem ins
            append $newPermutation$ to $result$
        else
            $newPermuation \leftarrow$ Shuffle the *permutation*
            append $newPermutation$ to $result$
        end if
    end while
    return $result$
end procedure

Fig. 11. Algorithm 11

2.3   Genetic and Simulated Annealing Hybrid Solution

*GA*s are well suited for optimization tasks and are usually able to find local optimum in small number of iterations. The one drawback is that, when a population is focused around local optimum it might be hard to achieve global optimum of the fitness function, even with high mutation rate. In order to address this problem, a new variant of *GA* is proposed, as presented on Fig. 12. The use of the *simulated annealing* algorithm, which allows to accept non optimal moves, may allow to further improve the produced solution.

---

initial ← Run the GA_LS on the problem instance
result ← Run SA algorithm using *initial* as initial path
return *result*

Fig. 12. Algorithm 12

3   Research and Results

To evaluate the performance of proposed approaches, a systematized experiments were conducted. All experiments were conducted using custom simulation software, according to mentioned methodology. A specialized software suite was created that allowed for: visually representing the problems and solutions, saving and restoring problem instances, performing multiple experiments, outputting results in comma separated values (.csv) format. All algorithms were first tested using graphical user interface to validate if those generate valid solution, i.e. *Hamiltonian cycle*. This allowed to find bugs in implementation and helped to select algorithms to benchmark proposed approaches against. Additionally a visual audit allowed to identify what could be improved in existing solutions, for instance a crossed paths appearing in solutions produced by *SA*. In order to achieve best performance of each algorithm, every parameter was tested within given range to determine its best value.

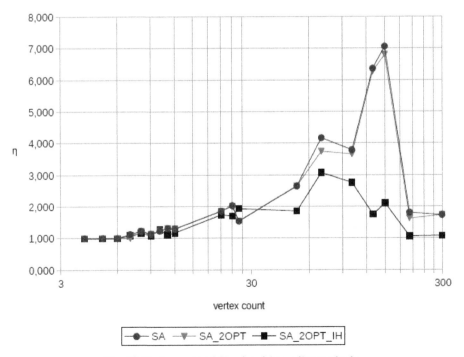

Fig. 13. Performance of *Simulated Annealing* methods

Since testing all combinations of parameters would be difficult, a following algorithm was used: (1) initialize all parameters with reasonable values,   (2) choose one parameter to optimize (3) perform experiments in order to determine best value for parameter,   (4) fix parameter to given value, (5) if any parameters left to optimize go to (2). All parameter optimization was made using *gr202* data set from *TSPLIB* [12].

The Fig. 13 presents the performance of different variants of *SA* algorithm as described in 2.1. The *SA* is baseline implementation of Simulated annealing method, *SA_2OPT* is the baseline algorithm followed by *2-opt*, and *SA_2OPT_IH* is the variant with additional *Farthest insertion* stage. As it can be seen, the *2-opt* gives little to none improvement, whereas a proper initial cycle can produce significant improvement in the solution. It is worth noticing that for small graphs, a near optimal solution is found by every variant.

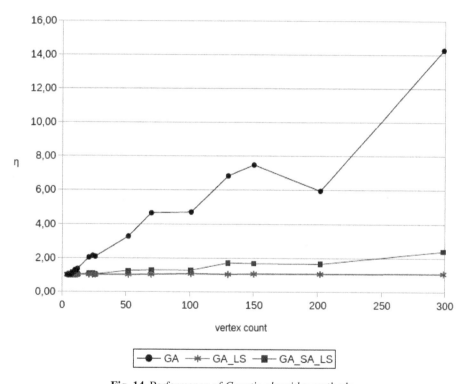

Fig. 14. Performance of *Genetic algorithm* methods

The Fig. 14 shows the performance of the different *Genetic algorithm* variants. The data series marked as *GA* in the chart is the performance of the baseline implementation of the *Genetic algorithm*. The *GA_LS* data series presents the performance of the *GA* with local search heuristics method, as described in 2.2. The *GA_SA _LS* data series presents the performance of the hybrid Simulated annealing and *GA_LS* method as described in 2.3. Despite the assumption, that *SA* on top of the *GA* method could produce better results, almost every time a *GA* with *Insertion heuristics* variant produced much better results.

But still, both modifications yielded a results that were far superior to baseline *GA* implementation. This leads to the conclusion that even simple crossover operator are efficient if the initial population is properly constructed.

4   Conclusions

The main aim of this paper has been achieved, which is to propose an efficient algorithm for determining the best route for couriers' in urban traffic. From the courier's point of view it is sufficient to solve a problem with $|V| < 50$, and for this task the hybrid *Genetic algorithm* with *Insertion heuristic* method as proposed in 2.2 is performing extremely well. It produced near optimal or optimal solution, for every researched static *TSP* instance, as well as outperform other tested algorithms while solving *DTSP*. The optimality η along with the cost, which refers to total distance or total travel time of the salesman, were described in 1.1. From the researcher's point of view, a *SA* with the *Insertion heuristic* hybrid looks promising, since it lacks the main disadvantage of the proposed *GA* with *Insertion heuristic* method, i.e. it runs fast even for very large graphs. Further work should focus on improving performance of *SA* with *Local search optimization* hybrid method, what would allow to efficiently solve small as well as large *DTSP* instances. The resulting performance, while solving small instances of problem, achieved by proposed hybrid variant of *GA* exceeded the expectations, and surely could be used as a base algorithm for a hypothetical mobile application to aid courier at work. While all around solution was expected to be produced, the research could still be considered success, since whole range of problems could be assessed using proposed methods. Introduced research methodology, proved itself to be very useful for testing and assessing the performance of different algorithms against each other. It could easily be used in any future work similar in concept to the research conducted by authors. Backed by the research data presented in this paper, it is possible that, even better results in terms of execution time as well as performance of the algorithm could be achieved by using for instance *Nearest neighbors* method instead of *Insertion heuristic*, and by researching other cycle encoding schemas together with better cross-over routines.

References

1. Aarts, E., van Laarhoven, P.: Simulated annealing. In: Aarts, E., Lenstra, J.K. (eds.) Local Search in Combinatorial Optimization, pp. 91–96. John Wiley & Sons Ltd. (1997)
2. Bellman, R.: Dynamic Programming Treatment of the Travelling Salesman Problem. RAND Corporation, Santa Monica, California (1961)
3. Christofides, N.: Worst-Case Analysis of a New Heuristic for the Travelling Salesman Problem. Defense Technical Information Center (1976)
4. Dasgupta, S., Papadimitriou, C., Vazirani, U.: Algorytmy. Wydawnictwo Naukowe PWN SA, Warszawa (2012)

5. Davendra, D. (ed.): Traveling Salesman Problem, Theory and Applications. InTech (2010), http://www.intechopen.com/books/travelingsalesman-problem-theory-and-applications
6. Gwiazda, T.D.: Algorytmy Genetyczne kompendium, vol. 1. Wydawnictwo Naukowe PWN SA, Warszawa (2007)
7. Homaifar, A., Guan, S., Liepins, G.E.: Schema analysis of the traveling salesman problem using genetic algorithms. Complex Systems 6 (1992)
8. Rosenkrantz, D.J., Stearns, R.E., Lewis, P.M.: An analysis of several heuristics for the traveling salesman problem. In: Ravi, S., Shukla, S.K. (eds.) Fundamental Problems in Computing, pp. 45–69. Springer, Netherlands (2009)
9. Sysło, M.M., Deo, N., Kowalik, J.S.: Algorytmy optymalizacji dyskretnej z programami w jezyku Pascal. Wydawnictwo Naukowe PWN SA, Warszawa (1999)
10. Laporte, G.: A Short History of the Traveling Salesman Problem. Canada Research Chair in Distribution Management and Centre for Research on Transportation (CRT) and GERAD HEC Montreal, Montreal, Canada (2006)
11. Li, C., Yang, M., Kang, L.: A new approach to solving dynamic traveling salesman problems. In: Wang, T.-D., Li, X., Chen, S.-H., Wang, X., Abbass, H.A., Iba, H., Chen, G.-L., Yao, X. (eds.) SEAL 2006. LNCS, vol. 4247, pp. 236–243. Springer, Heidelberg (2006)
12. Karp, R.M.: Reducibility among combinatorial problems. In: Miller, R.E., Thatcher, J.W., Bohlinger, J.D. (eds.) Complexity of Computer Computations. The IBM Research Symposia Series, pp. 85–103. Springer US (1972)
13. Ore, O.: Note on Hamilton circuits. The American Mathematical Monthly 67(1) (January 1960), http://www.jstor.org/stable/2308928
14. Universitat Heidelberg. Tsplib, http://www.iwr.uni-heidelberg.de/groups/comopt/software/TSPLIB95/ (accessed on: December 20, 2014)
15. Wei, J.-D., Lee, D.T.: A new approach to the traveling salesman problem using genetic algorithms with priority encoding. In: Congress on Evolutionary Computation, CEC 2004, vol. 2, pp. 1457–1464 (June 2004)
16. Zukhri, Z., Paputungan, I.V.: A hybrid optimization algorithm base on genetic algorithm & ant colony optimization. International Journal of Artificial Intelligence and Applications (IJAIA) 4(5) (September 2013)

Mutation Testing Process Combined with Test-Driven Development in .NET Environment

Anna Derezińska and Piotr Trzpil

Institute of Computer Science, Warsaw University of Technology,
Nowowiejska 15/19, 00-665 Warsaw, Poland
A.Derezinska@ii.pw.edu.pl

Abstract. In test-driven development, basic tests are prepared for a piece of program before its coding. Many short development cycles are repeated within the process, requiring a quick response of the prepared tests and the tested code extract. Mutation testing, used for evaluation and development of test cases, usually takes a considerable time to obtain a valuable test assessment. We discuss combination of these techniques in one development process. The presented ideas are implemented in VisualMutator – an extension of Visual Studio. The tool supports selected standard and object-oriented mutations of C# programs. Mutations are introduced at the level of the Common Intermediate Language of .NET. A program or its selected methods can be mutated in a batch process or during interactive sessions.

Keywords: mutation testing, test-driven development, TDD, C#, Common Intermediate Language, Visual Studio.

1   Introduction

In the agile methodology of software development [1,2], the process should adopt to changes in software requirements and technology. A program development process is interactive, easy to modify and tests are also used for documentation purposes. Developers are dealing with testing of an uncompleted program and it is often combined with the Test-Driven Development (TDD) approach [3,4].

Mutation testing process is focused on a test suite quality evaluation and development, usually dealing with a complete program under test [5-7]. A typical mutation testing process is sequential and can be summarized as follows:

1. the process configuration is established, e.g. mutation operators are selected,
2. *mutants* of the whole program, i.e., program variants with injected changes specified by mutation operators, are generated,
3. mutants are executed with tests; mutants are said to be *killed* if they behave differently to the original program,
4. optionally, *equivalent* mutants are recognized (mutants with equivalent behavior),
5. *mutation score* is calculated, as the number of killed mutants over the number of all non-equivalent mutants; it denotes an ability to kill mutants of the test suite,
6. optionally, additional tests are developed to kill more mutants.

© Springer International Publishing Switzerland 2015
W. Zamojski et al. (eds.), *Theory and Engineering of Complex Systems and Dependability*,
Advances in Intelligent Systems and Computing 365, DOI: 10.1007/978-3-319-19216-1_13

One of obstacles in the mutation testing process is a long execution time and a high computational complexity. The long delay before obtaining mutation results is especially hardly acceptable in the TDD method. There are different approaches to cope with the performance of mutation testing. One of possibilities is application of various methods reducing the number of generated and tested mutants and the number of test runs, such as selective mutation, mutant sampling or clustering [5], [8,9]. However, those methods can be associated with lowering of a mutation result accuracy.

There are also different directions toward the problem that could be considered. The first one is shortening of the waiting time. This can be accomplished by division of a process and by speeding up the execution. The second direction is another arrangement of a user activity within a process. Therefore, more information can be delivered to a user on demand, a user can control the selected process steps, and partial results of mutation testing can be provided on the fly.

Both directions were considered in the context of a mutation testing combining with the TDD approach realized in the .NET environment. It needs an effective and user friendly tool support. The proposed solutions were implemented in VisualMutator [10] – the mutation testing tool for C# programs fully integrated with the Visual Studio (VS in short). Is supports selected standard and object-oriented (OO in short) mutation operators that introduce changes at the level of the Common Intermediate Language of .NET (CIL).

In the next Section an approach to interactive mutation testing will be discussed. Different features of an automated support to mutation testing are surveyed in Sect. 3. Section 4 presents VisualMutator and Sect. 5 concludes the paper.

2   Interactive Mutation Testing Process

The basic scenario of Test-Driven Development can be described in the following steps [3,4]:

1. A unit test is created that specifies required activity of a small program part (module, class, method). In this stage, sometimes various test doubles (i.e., dummy objects, fakes, stubs, mocks) are used to substitute the missing program elements [11].
2. Running a test that should fail in order to eliminate situation when a test always succeeds, even for a non-existent functionality.
3. Development of a simple, but correct program unit with a desired functionality.
4. Running the test, until succeeded.
5. Code evaluation, refactoring, if necessary.
6. Re-running the test. Test success is expected.

In TDD we focus on the locally expected results that are specified before the code is implemented. It should be noted, that even if we do not follow a pure TDD paradigm, in the incremental and interactive code development the preparation of tests is similar. A developer works at one time mainly on a limited functionality, preparing a small number of tests for its verification. The mutation testing approach could be incorporated in this process to support this limited scope iterative process. Therefore, we need techniques for local interactive mutation testing.

The global mutation causes generation of many mutants and execution of a whole program. This approach is still valuable, especially for the later stages of a program development. But it can be inconvenient for TDD.

Performing the mutation testing process for an extract of a program (locally) can decrease the overall time and lead to collecting valuable results in the respect to the desired scope. However, the number of mutants to be tested can still be significant – it depends on the size of a selected code and the number of active mutation operators. Two general directions are applied to cope with this problem: an interactive organization of the mutation testing process and performance optimizations.

The interactive process is divided into shorter phases/steps. A mutation session can be easily started and re-launched, the activities are visible for a user that can influence their course. A user might be interested in a piece of information about killing a given mutant by any test, its mutated code, or its equivalence. The user needs it as soon as possible in order to create additional tests for the considered program part if necessary. Requirements of an interactive process can comprise assignment of high priority for a selected mutant, mutant code preview at any time of the process, as well as incremental delivery of partial results of mutant creation and test execution.

Performance optimization are applied by parallel processing of the source code, parallel execution of many mutants, application of a cache of the source code and a mutant cache.

An interactive process of test-driven software development with mutation testing can be described by a following use case scenario:

1. A user selects a mutation scope, it could be a class or a method of a developed program. The selection should be supported by a convenient GUI.
2. A mutation system searches all test cases that might refer to the considered program extract and presents them to the user. The user has the opportunity to discard chosen test cases, as not appropriate, or not desired at this stage of development. Hence, the user approves the whole proposed set of tests or its subset.
3. The mutation system prepares the mutation testing process for the selected program part. Mutants are created and mutant testing is launched against tests from the approved test subset.
4. The user can observe the progress and partial results during testing. A code change of a mutant can be viewed if desired, i.e., the high level code can be reconstructed if necessary.
5. The user can mark a selected mutant (tested or not) as an equivalent mutant.
6. The final results are given.
7. The user can create more tests if the results are not satisfactory and perform another mutation testing process with new tests.

The management of a program code within the process can be based on the producer-consumer pattern. A producer creates copies of the original code. This creation can be completed independently of other activities and started even before a mutation testing session. The spare copies can be kept in some storage, e.g. in a code cache.

Consumers are tasks that are requested to create mutants and execute them. A consumer takes a spare code copy and transforms it into a mutant. This transformation is easy to perform, assuming that all information about the mutant (location of a change, kind of a code substitution) is delivered in a mutant descriptor. Mutant descriptors are prepared by a mutation engine, which also verifies all necessary conditions of mutation applicability of selected mutation operators. We proposed a lifecycle of a mutant that can be described by the following steps:

1. A mutant is created based on its descriptor. One code copy, taken from the code cache, is used for the mutant construction. The mutant is stored in a mutant cache.
2. Just before the mutant is to be tested, it is stored on a disk together with other assemblies required to run it. The non-modified assemblies are copied from the source locations if necessary.
3. Selected tests are run with the mutant. The unit tests are executed under a supervision of a given test environment. Program results are analyzed on the fly, determining a mutant status (live, killed by tests or by timeout).
4. The mutant stays in the mutant cache as long as the memory space does not need to be reclaimed. A mutant from the cache is ready to be used immediately if its code preview or tests are demanded. If a demand occurs after the memory release, the mutant has to be created once again, as in point 1.

3   Mutation Process Support – Related Work

There are several tools dealing with mutation testing of C# programs. However, some projects were abandoned, and others are not suitable for TDD.

The first tool for mutation of C# programs was Nester [12]. It was based on simple transformations of C# expressions specified in an XML configuration file. The code changes were inserted in the run time. Different colors of a mutated code presented results such as: all mutants killed, at least one mutant alive or not covered code. Nester provided only structural mutations (also invalid), and since .NET 2.0 is not developed.

Object-oriented mutations of C# were implemented for the first time in the CREAM tool [13,14]. Its current version [9] supports 18 OO and 8 standard mutation operators. Mutations are introduced at the syntax tree level of a C# code, and a program recompilation is necessary. The tool is integrated with the compiler and unit test frameworks, but is not combined with the VS development environment.

Mutation of C# programs at the CIL level were introduced in IlMutator [15]. Selected 11 OO and 4 standard mutations are mapped into changes of an Intermediate Language code and processed in a stand-alone tool. Moreover, the tool has not been updated and .NET 4.0 and later versions are not supported.

Few standard mutations were also provided at the CIL level in NinjaTurtles [16]. Tests that are dealing with a mutated class or method are executed. A mutated code cannot be observed, but summary mutation results for each source line are given. A program can be mutated with the stand-alone tool, or the tool library can be called in specially prepared tests. The project has been suspended since 2012.

Another goals are established in the specification-based Mutation Engine of C# and Visual Basic code [17]. Mutants that meet code contracts requirements are selected. Though, it is not given how mutants invalidating the contracts are distinguished, whether it is done in a static or dynamic analysis. The tool is integrated with VS 2010, therefore information about current projects, libraries, and other resources are easily accessible.

The similar problems of mutation testing process can be also considered in the tools for other languages, especially Java. Jumble tool [18] is associated with the continuous integration system. It supports a limited number of standard operators, but can be accommodated to TDD as it is called from a command line, and works incrementally on new classes and JUnit tests.

The most comprehensive set of mutation operators for Java, including 29 object-oriented operators, is available in MuJava [19,20]. A new version of the tool appeared in 2013. After selecting mutation operators and a code to be mutated, a given test can be run. One of obstacles in the system application is a special test format, different from the typical JUnit tests. Based on the mutation engine of MuJava – the MuClipse tool - an extension for Eclipse was created. Using MuClipse a program can be mutated, and the results of mutants can be viewed in a tree form together with the mutated code. In spite of integration with the development environment, a manual configuration of a source code, compiled files, etc., is necessary.

Advantages of the Javalanche tool [21] are assessment of equivalent mutants, parallel mutant execution, and running only tests covering the mutated code. It is combined with the Ant tool for automating software build but supports only 4 standard operators. Another Java tool similar to the above is Judy [22,23]. It is suitable for JUnit tests and has still extending functionality, including second-order mutation.

The idea of a fast mutation process was central for the PIT tool for Java [24]. Many JUnit tests can be executed in parallel, a mutated class is substituted in a project stored in the operational memory, and tests are run for the covered code. An important feature is the possibility of incremental analysis, useful when new mutations are introduced into the code that had partially been changed. However, impact of code changes on the mutation result can be ambiguous. Several solutions of the PIT engine, e.g. two different phases of mutation place recognition and mutation introduction, are similar to those of VisualMutator.

Apart from Java, a mutation tool integrated with the development environment and speed up options is MutPy for Python programs [25,26]. Yet, it lacks convenient GUI.

An inspiration of the architecture for interactive testing in the .NET environment were also Pex and CodeDigger tools [27]. These extensions to Visual Studio are used for finding input data for a selected method based on the dynamic symbolic execution. Additionally, Pex helps in generating of parametrized unit tests that cover the test data found.

4   VisualMutator

VisualMutator is a mutation testing tool designed as a Visual Studio extension [10]. It can be used to verify quality of a test suite in an active C# solution. The first version of VisualMutator (v.1) was mainly devoted to mutation testing of ASP .NET MVC applications [28]. Therefore, most of the mutation operators were specialized for ASP .NET MVC to modify action parameters, action methods and their results. Both unit tests and functional tests can be used in the mutation testing process. Although the performed experiments confirmed the proof of concept, the practical utilization of the mutation testing in this area was not effective. Therefore, in the next version of the tool the operators of ASP .NET MVC are no more supported.

The second version of VisualMutator followed the framework of the first one and is tightly coupled with the Visual Studio environment, which makes the mutation process considerably convenient to a developer. In comparison to the first version, the tool has a new mutation engine, enables different ways of mutation testing processes (interactive and global) and supports selected general purpose mutation operators (standard and object-oriented). The main features of the tool are the following:

— First-order mutants can be created using built-in and custom mutation operators.
— The modified code fragments can be viewed in C# and in the intermediate code (CIL).
— Generated mutants can be run with NUnit or xUnit tests.
— Details of any mutants can be observed just after the start of a mutation testing process.
— Mutation score and numbers of passed and failed tests are instantly calculated and updated.
— Detailed results can be optionally reported in an XML file.

Interactive interface can be used for mutation testing of program fragments or the whole complete program. In order to perform mutation of a complete program in an automatic way, a command line interface can also be used.

4.1   Mutation Operators

Object-oriented mutation operators in VisualMutator 2.0 were selected after research and experiments on the operators of C# used the CREAM tool [9], based on their specification [29], as well as other experiences in C#/CIL [15] and Java - mainly with MuJava [19], [30,31]. Standard operators cover functionality of selective mutation [32]. The tool can be easily extended with other operators, due to its architecture. The current choice of operators was founded on the following premises:

— avoiding generation of many equivalent mutants,
— popularity and usefulness of a language feature to which an operator refers,
— assessment of a real error occurrence that is mimicked by an operator,
— possibility of unambiguously reflecting a C# code change at CIL level.

In Table 1, the set of object-oriented (ISD...EHC) and standard (AOR...ABS) mutation operators implemented in the tool is shown. Some operators (MCI, PRV, EHR, UOI) have the restricted functionality. Selected standard operators (AOR, LOR, ROR and ABS) have extended functionality in comparison to other typical tools. The changes defined by UOR (*Unary Operator Replacement*) are incorporated in mutants generated by the ASR operator, because at the CIL level there are no special ++ or – arithmetic operators, but the regular addition and subtraction are used.

Table 1. Mutation operators of VisualMutator v2.0

Operator Id	Mutation operator (scope restriction or extension)
ISD	*Super/Base Keyword Deletion* (ISK)
OMR	*Overloading Method Contents Change*
DMC	*Method Delegated for Event Handling Change*
DEH	*Method Delegated for Event Handling Change*
JID	*Member Variable Initialization Deletion*
EAM/EMM	*Accessor/ Modifier Method Change*
MCI	*Member Call from Another Inherited Class* (restriction - used variables are fields of the current class, as they are initialized and invalid mutants are avoided)
PRV	*Reference Assignment with Other Compatible Type* (restriction - the reference is changed only on a class field, as they are initialized and invalid mutants are avoided)
EHR	*Exception Handler Removal* (restriction – a *catch* block is deleted only if there is more than one *catch* block, due to implementation at the CIL level)
EXS	*Exception Swallowing*
EHC	*Exception Handling Change*
AOR	Arithmetic Operator Replacement (arithmetic operator +, -, *, /, % swapped with another, or substitution of the operation with a value of its right or left operand)
ASR	*Assignment Operator Replacement* (assignment operator =+, =/, etc. swapped with another one)
LOR	*Logical Operator Replacement* (logical operator &, |, ^ swapped with another one, or substitution of the operation by a negation of a right or left operand)
LCR	*Logical Connector Replacement* (&&, || swapped with another)
ROR	*Relational Operator Replacement* (relation operator >,<,<+,>=,++,!= swapped with another one, or the whole relation substituted by constant *true* or *false*)
SOR	*Shift Operator Replacement* (shift operator >>, << swapped with another)
UOI	*Unary Operator Replacement* (expression is proceeded by the logical negation '!' or arithmetic negation '-' in accordance to the expression type) The operator was restricted, as no '+' was added to expression due to possibility of an equivalent mutation.
ABS	*Absolute Value Insertion* (each numerical expression is mutated by three functions: *Abs* that returns an absolute value, *NegAbs* returning its negation, *ThrowOnZero* returns the argument or kills the mutant for the 0 value).

4.2   Mutation Testing Session

VisualMutator 2.0 can be installed as an extension in Visual Studio v. 2012 or 2013 IDE. Its window can be opened in the parent environment and all solution components can be processed in parallel to other development activities. A basic entity of a mutation testing process is a mutation session. It can be run on the whole code or in a context of one chosen method. A user can configure a session by selection of test cases, mutation operators, and other parameters. One of methods to start a limited-scope mutation session is selection of one option to *mutate and test* a method with the tests using it. In this quick mode, more choices are made automatically without involving a user, and it is convenient for TDD.

During a mutation testing session the mutants are created on the fly and executed in parallel. The progress of execution of mutants and tests, status of mutants, and a mutation score calculated based on the partial results can be observed by a user.

A code of each mutant can be viewed on demand. It is shown at the CIL level, or in the C# source code. The changes of the code, added and deleted statements, are marked with appropriate signs and colors. If a code of a mutant is not available at the moment, it is reconstructed from the intermediate tree form if necessary.

In the cache of the source code, several copies of the original code are prepared beforehand the proper mutation occurred and in parallel to it. They are used during mutant generation. The code cache also delivers quickly a code copy if a user demands preview of a mutant that is not currently processed or has not been used recently.

The interaction and code preview is speeded up by a mutant cache. In the mutant cache a limited number of recently applied mutants is stored. If one of such mutants is demanded by a user it can be delivered immediately. Otherwise, the mutant code can be prepared using a mutant descriptor and one of the copy from the source code cache. Therefore, the interactive process can be effectively realized.

Mutation process can be controlled and tuned by a user by adjusting different parameters, e.g. a number of threads to prepare program copies, a number of mutants run in parallel, management of mutant cache, etc.

4.3   Application of VisualMutator

The VisualMutator tool was, among others, used by participants of "Diagnostics of Computes Systems" - an advanced course for post-graduate students run in the Institute of Computer Science WUT. During practical classes, test sets of given programs were evaluated and extended, and, on the other hand, the tool utility was verified.

As an example, we discuss experiments carried out on the DSA (Data Structures and Algorithms) program available at http://dsa.codeplex.com. Considered tests of the program had a satisfactory code coverage (96 %). Code overage is treated as a preliminary but not sufficient criterion of the test quality.

The mutation results for mutants generated by object-oriented operators are summarized in Table 2. No mutants were created for some OO operators, which are not included in the table. This is a typical experience of advanced operators and can be treated as an implicit selective mutation [9].

Table 2. Mutation results for object-oriented mutation operators

Operator	ISD	DMC	JID	EAM	PRV	EHC	JTD
Mutation score [%]	25	100	83	88	87	100	83

Mutants generated by standard operators are in general easer to be killed and mutation results are not so diverse as for object-oriented operators. Most of such mutants generated were killed by the tests, giving an average mutation score of 98 %.

The primary goal of the experiment was identification of omissions in the test set. The test evaluation helped in recognizing several situations that were not handled by the test cases. For example, the following omission types were pointed out in selected program areas in reference to the given mutation operators: a remove operation is tested only for a list having one element or the number of elements is not verified after this operation (AOR); an algorithm is checked only for one direction of a tree traversal instead of two - left and right (EAM); tests assumed a minimal/maximal value stored only in the tree root instead of any tree node (EAM); values returned by methods are not verified (EAM); a case of an empty list is not verified (ISD, PRV); a domain value of a variable is not satisfied (JID, JDT). Additional tests could be created to cover those situations.

5   Conclusions

We proposed a new interactive mutation testing process that can assists the TDD approach. The process is supported by the tool. VisualMuator is the first tool implementing object-oriented mutation of C# that is fully integrated with VS. Its new architecture allows to perform an interactive mutation testing of a selected scope. Each mutant can be viewed during the process on demand. Experiences gathered till now are promising. However, further development of VisualMutator is planned, assuming different paradigm. Evaluation of the tool facilities requires a bigger amount of controlled experiments to assess their impact on the development process and a user convenience.

References

1. Agile Alliance, The Manifesto for Agile Software Development, vol. 2003. Agile Alliance (2000)
2. Zhong, S., Liping, C., Tian-en, C.: Agile Planning and Development Methods. In: 3rd International Conference on Computer Research and Development, pp. 488–491. IEEE Press (2011)
3. Beck, K.: Test Driven Development: By Example. Addison-Wesley Professional (2002)
4. Cauevic, A., Punnekkat, S., Sundmark, D.: Quality of Testing in Test Driven Development. In: 8th International Conference on the Quality Information and Communications Technology, pp. 266–271. IEEE CPS (2012)
5. Jia, Y., Harman, M.: An Analysis and Survey of the Development of Mutation Testing. IEEE Trans. Softw. Eng. 37(5), 649–678 (2011)
6. Vincenzi, A.M.R., Simao, A.S., Delamro, M.E., Maldonado, J.C.: Muta-Pro: Towards the Definition of a Mutation testing Process. J. of the Brazilian Comp. Soc. 12(2), 49–61 (2006)
7. Mateo, P.R., Usaola, M.P., Offutt, J.: Mutation at the Multi-Class and System Levels. Science of Computer Programming 78, 364–387 (2013)

8. Usaola, M.P., Mateo, P.R.: Mutation Testing Cost Reduction Techniques: a Survey. IEEE Software 27(3), 80–86 (2010)
9. Derezińska, A., Rudnik, M.: Quality Evaluation of Object-Oriented and Standard Mutation Operators Applied to C# Programs. In: Furia, C.A., Nanz, S. (eds.) TOOLS Europe 2012. LNCS, vol. 7304, pp. 42–57. Springer, Heidelberg (2012)
10. VisualMuator, http://visualmutator.github.io/web/
11. Meszaros, G.: xUnit Test Patterns: Refactoring Test Code. Addison Wesley Professional (2007)
12. Nester, http://nester.sourceforge.net (access date: January 2015)
13. Derezińska, A., Szustek, A.: CREAM - a System for Object-Oriented Mutation of C# Programs. In: Szczepański, S., Kłosowski, M., Felendzer, Z. (eds.) Annals Gdańsk Univ. of Techn. Faculty of ETI, No. 5, Inform. Technologies, Gdańsk, vol. 13, pp. 389–406 (2007)
14. CREAM - Creator of Mutants, http://galera.ii.pw.edu.pl/~adr/CREAM/
15. Derezińska, A., Kowalski, K.: Object-Oriented Mutation Applied in Common Intermediate Language Programs Originated from C#. In: 4th International Conference on Software Testing Verification and Validation Workshops, pp. 342–350. IEEE Comp. Soc. (2011)
16. NinjaTurtles - .NET mutation testing, http://www.mutation-testing.net/ (access: January 2015)
17. Yiasemis, P.S., Andreou, A.: Testing Object-Oriented Code through a Specification-Based Mutation Engine. Inter. Journal on Advances in Software 5(3), 179–190 (2012)
18. Jumble: http://jumble.sourceforge.net/ (access: January 2015)
19. Ma, Y.-S., Offutt, J., Kwon, Y.-R.: MuJava: an Automated Class Mutation System. Softw. Testing, Verif. and Reliab. 15(2), 97–133 (2005)
20. Mujava, http://cs.gmu.edu/~offutt/mujava/ (access date: January 2015)
21. Javalanche – Mutation Testing, http://javalanche.org/ (access date: January 2015)
22. Madeyski, L., Radyk, N.: Judy – a mutation testing tool for Java. IET Software 4(1), 32–42 (2010)
23. Judy – Java mutation tester: http://mutationtest.com/ (access date: January 2015)
24. PIT: http://pitest.org/ (access date: January 2015)
25. Derezinska, A., Hałas, K.: Improving Mutation Testing Process of Python Programs. In: Silhavy, R., Senkerik, R., Oplatkova, Z.K., Prokopova, Z., Silhavy, P. (eds.) Software Engineering in Intelligent Systems. AISC, vol. 349, pp. 233–242. Springer, Heidelberg (2015)
26. MutPy, https://bitbucket.org/khalas/mutpy (access date: January 2015)
27. Pex, http://research.microsoft.com/en-us/projects/pex/ (access date January 2015)
28. Derezińska, A., Trzpil, P.: Mutation Testing of ASP.NET MVC. In: Swacha, J. (ed.) Advances in Software Development, pp. 127–136. Scientific Papers of the Polish Information Processing Society Scientific Council, Warsaw (2013)
29. Derezińska, A.: Specification of Mutation Operators Specialized for C# code, ICS Res. Report2/05, Warsaw University of Technology (2005)
30. Amman, P., Offut, J.: Introduction to Software Testing. Cambridge Univ. Press (2008)
31. Hu, J., Li, N., Offutt, J.: An analysis of OO mutation operators. In: 4th International Conference on Software Testing, Verification and Validation Workshops, pp. 334–341 (2011)
32. Offut, J., Rothermel, G., Zapf, C.: An Experimental Evaluation of Selective Mutation. In: 15th International Conference on Software Engineering, pp. 100–107 (1993)

Boltzmann Tournaments in Evolutionary Algorithm for CAD of Complex Systems with Higher Degree of Dependability

Mieczyslaw Drabowski

Cracow University of Technology, Faculty of Electrical and Computer Engineering
drabowski@pk.edu.pl

Abstract. The paper includes a proposal of a new algorithm for Computer Aided Design (CAD) of complex system with higher degree of dependability. Optimal scheduling of processes and optimal resources partition are basic problems in this algorithm. The following criteria of optimality are considered: costs of system implementation, its operating speed and power consumption. Presented the CAD algorithms may have a practical application in developing tools for rapid prototyping of such systems.

Keywords: complex system, scheduling, partition, allocation, dependable, optimization, genetic, simulated annealing, Boltzmann tournament, CAD tools.

1 Introduction

The aim of computer aided design of complex systems (i.e. systems, which contain many resources and operations) is to find an optimum solution consistent with the requirements and constraints enforced by the given specification of the system. A specification describing a computer system may be provided as a set of interactive processes.

The partitioning of resources between various implementation techniques is the basic matter of automatic design. Such partitioning is significant, because every complex system must be realized as result of hardware implementation for its certain operations. The problems of processes scheduling are one of the most significant issues occurring in design of operating procedures responsible for controlling the allocations of operations and resources in complex systems.

In the design methods, which were presented and implementation so far [1], the software and hardware parts were developed separately (and concurrent) and then connected together, which increased the costs and decreased the speed and the dependability of the final solution. The resources distribution is to specify, what hardware and software are in system and to allocate theirs to specific processes, before designing execution details.

Another important issue that occurs in designing complex systems is assuring their fault-free operation. Such designing concentrates on developing dependable and fault-tolerant architectures and constructing dedicated operating procedures for them.

© Springer International Publishing Switzerland 2015
W. Zamojski et al. (eds.), *Theory and Engineering of Complex Systems and Dependability,*
Advances in Intelligent Systems and Computing 365, DOI: 10.1007/978-3-319-19216-1_14

In this system an appropriate strategy of self-testing during regular exploitation must be provided.

The general model and concept of parallel to processes scheduling and resources partition for complex systems with higher degree of dependability was presented in [2]. We proposed the following schematic diagram of a coherent process of fault tolerant systems synthesis (Figure 1).

The suggested parallel analysis consists of the following steps:

1. specification of requirements for the system,
2. specification of processes,
3. assuming the initial values of resource set,
4. defining testing processes and the structure of system, testing strategy selection,
5. scheduling of processes,
6. evaluating the operating speed and system cost, multi-criteria optimization,
7. the evaluation should be followed by a modification of the resource set, a new system partitioning into hardware and software parts and an update of test processes and test structure (step 4).

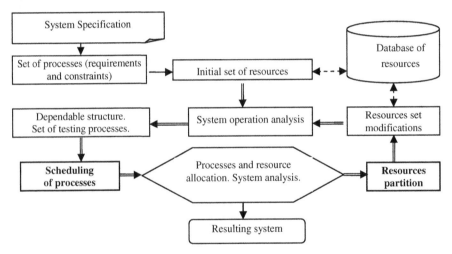

Fig. 1. The process parallel design of dependable computer system

Modeling fault tolerant systems consists of resource identification and processes scheduling problems that are both hard NP-complete [3]. Algorithms for solving such problems are usually based on heuristic approaches. The objective of this paper is to present the approach of combined approach to the problem of fault tolerant systems design, i.e. a parallel solution to processes scheduling and resource assignment problems. We suggest in this paper meta-heuristic and hybrid algorithm: evolutionary with simulated annealing, in which there are Boltzmann tournaments [4].

Synergic design methodology for partition of redundant structures with higher degree of dependability and processes scheduling witch self testing strategy may have practical application in developing the tools for automatic aided for rapid prototyping of such systems.

2   Evolutionary with Boltzmann Tournaments Algorithm

In this chapter we describe the algorithm realizations aimed to optimize resource partition and processes scheduling, as well as the adaptation of those algorithms for design of complex systems realization.

In order to eliminate solution convergence in genetic algorithms, we use data structures which ensure locality preservation of features occurring in chromosomes and represented by a value vector [5]. Locality is interpreted as the inverse of the distance between vectors in an n-dimension hyper-sphere. Then, crossing and mutation operators are data exchange operations not between one-dimensional vectors but between fragments of hyper-spheres. Thanks to such an approach, small changes in a chromosome correspond to small changes in the solution defined by the chromosome. The presented solution features two hyper-spheres: processes hyper-sphere and resource hyper-sphere.

The solutions sharing the same allocations form the so-called clusters. The introduction of solution clusters separates solutions with different allocations from one another. Such solutions evolve separately, which protects the crossing operation from generating defective solutions. There are no situations in which a process is being allocated to a non-allocated resource. Solution clusters define the structures of the system under construction (in the form of resources for task allocation). Solutions are the mapping of tasks allocated to resources and scheduling of processes. During evolution, two types of genetic operations (crossing and mutation) take place on two different levels (clusters and solutions).

A population is created whose parameters are: the number of clusters, the number of solutions in the clusters, the digraph of processes and pool (database) of resources. For the design purposes, the following criteria and values are defined: optimization criteria and algorithm iteration annealing criterion if solution improvement has not taken place, maximum number of generations of evolving solutions within clusters, as well as the limitations - number of resources, their overall cost, total time for the realization of all processes, power consumption of the designed system and, optionally, the size of the list of the best and non-dominated individuals.

Algorithm contains information about the parameters of global temperature:

- Current temperature.
- Ratio of cooling.
- Step of temperature.

During the performance of algorithm, the temperature will diminish with accordance to the function $f(x) = e^{-a \cdot x}$, where a - the ratio of cooling. The workings about step the algorithm of reducing the temperature the argument x be reduced in time – Fig. 2.

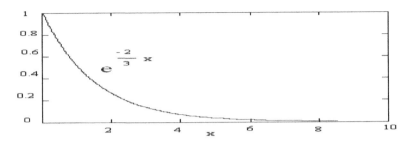

Fig. 2. Chart of reducing the temperature of algorithm

Algorithm keeps two hyper sphere:

- Processes hyper sphere is two-dimensional, representing processes digraph structure. Each of the nods is defined by two coordinates: an indicator obtained through topological sorting (the processes are "closest" if one of them is direct successor of the other), and an indicator calculated from the BFS algorithm parallel processes are equally distant from the beginning of the digraph).
- Resources hyper sphere is three-dimensions representing the dependence of resource features. Each of the resources may be defined by the following coordinates: cost, speed and power consumption.

2.1   Partition of Resources

It is the data the digraph of processes, pool of resources as well as criterions of optimality. The algorithm of partition of resources has determine resources, which have execute all processes with all criterions – Fig. 3.

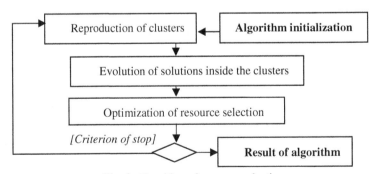

Fig. 3. Algorithm of resource selection

The cluster mutation operator consists in mutating allocation vectors in the following way: a cluster with identical likelihood is picked at random and copied. The number of the resource which will be mutated in a new cluster is picked randomly then, a number in the 0-1 range is picked - if the number is smaller than the **global temperature**, the resource is added, otherwise it is subtracted. Adding resources is limited by the maximum resource number parameter. At the beginning of the algorithm operation, resources will be added to the structure. As the algorithm approaches the end of the run defined by the cooling process, resources will be subtracted. This is aimed at creating a cost-effective structure. The cluster crossing operator consists in randomly picking two clusters and copying them. Crossing is achieved through cutting the resource hyper sphere with a hyper plane. The information contained on "one side" of the hyper plane is exchanged between clusters – Fig. 4.

The algorithm for cutting the hyper-sphere with a hyper-plane:

- Determining the cutting hyper-plane by picking n points inside an n-dimensional hyper-sphere.
- Creating a random permutation.
- Constructing the point displacement vector in respect to the hyper-sphere center; square coordinates are picked consistent with dimension permutations, e.g. for three dimensions with the permutation (2, 1, 3): $y2 = rand() \% r2$, $x2 = rand() \%$ $(r2 - y2)$, $z2 = rand() \% (r2 - (y2 + x2))$, where: r – hyper-sphere radius, and (x, y, z) are the coordinates of the constructed point in a three-dimensional space.
- The roots of square coordinates are calculated.
- A coordinate radical sign is picked.
- The hyper sphere center coordinates are added to the new point resulting in obtaining a new point inside the n-dimensional hyper-sphere.
- The equation of the hyper-plane cutting the hyper-sphere is calculated and the obtained system of equations is solved.

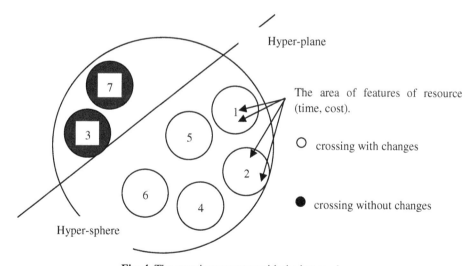

Fig. 4. The crossing operator with the hyper-plane

After solution reproduction, a new procedure is called to save the globally non-dominated solutions generated during evolution. This procedure executes:

- Searches for non-dominated solutions in the present generation.
- Creates the ranking of the best solutions saved so far and in the present generation.

- Saves the non-dominated solutions from both the "old" and the "new" solutions.
- Deletes the solutions saved in the past if they were dominated by new solutions; if there are more than one solution whose all optimized criteria values are identical, only one of those solutions is saved (the "newest" one).
- If the new solutions dominated none of the ones saved in the past, the population was not improved.
- The number of non-dominated solutions that the algorithm can save is defined by an algorithm parameter.

At this stage of the algorithm, half the individuals are removed from the population. The initial number of individuals is restored. The elimination of individuals is carried out using Boltzmann tournament selection strategy.

2.2   Boltzmann Tournament

The calculations of following equation the winner of tournament be appeared on basis of result:

$$\left[1 + e^{\frac{(r1-r2)}{T}} \right]^{-1} \tag{2.1}$$

where: r1 - ranking of first solution, r2 - ranking of second solution, T - global temperature

They are values of this function the number from compartment from $< 0,1 >$. We draw in aim delimitations the winner of tournament number from compartment $(0,1)$. If she is larger from enumerated number with example then individual about ranking is winner *r1*. Second individual in opposite incident winner is (about ranking *r2*) [4].

It the analysis of results of tournament was it been possible was to conduct on basis of graph of function of x (Fig. 5.):

$$\left(1 + e^x \right)^{-1} \tag{2.2}$$

where:

$$x = (r1 - r2) / T \tag{2.3}$$

If $r1 < r2$ this x is negative and for high temperature larger probability exists, that individual about rank *r1* *will* win tournament than for lower temperatures. For low temperatures winner the most often will be individual about rank *r2*.

If $r1 > r2$ this x is positive and for high temperature larger probability exists, that individual *r2* *will* win tournament than for lower temperatures. For low temperatures winner the most often will be individual about rank *r1*.

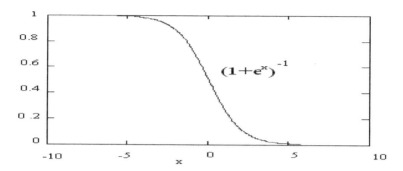

Fig. 5. The chart of probability of victory Boltzmann tournament in depending on global temperature

2.3   Scheduling of Processes

Processes scheduling is aimed at minimizing the schedule length (the total processes **completion time**) [6]. The diagram of the algorithm of processes scheduling is showed on Fig. 6.

Solutions are reproduced using the genetic operators: crossing and mutation. Solutions are reproduced until their number doubles.

The mutation operator produces one and the crossing operator two new solutions. The likelihood of using either of the genetic operators is defined by the algorithm parameters.

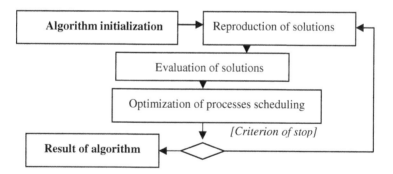

Fig. 6. Algorithm of processes scheduling

The mutation operator of processes allocation to resources acts in the following manner: a solution is randomly selected and copied. Then, the number of processes in the system is multiplied by the **global temperature**. When the global temperature is high, the number of processes changed in the allocation to resources will be greater than that in later stages of the evolution. Processes are picked at random and allocated to resources. The schedule mutation operator acts in the following manner: if due to the mutation operation of process allocation to resources, the resource the process had been running on was changed, then the process is removed from the schedule for the "old" resource and boundaries are set on the new resource schedule between which the process may be allocated. A location within the boundaries is picked and the process is allocated. The crossing operator of process allocation to resources resembles cluster crossing, however, the processes digraph hyper-sphere is used for that purpose. Schedule crossing operator acts in the following way – after the allocations have been crossed, a map is created defining which parent a given feature of an offspring comes from. The offspring stores the allocation vector (obtained after crossing process allocations to resources) and the empty vector of lists with schedules of processes on available resources. The algorithm analyzes the processes by checking their position on the digraph. For all processes in one position, the resources on which the processes will be performed (defined by the vector of allocation to resources) are put on the list. If in a position there is only one process ran on a given resource, the process is entered into the resource schedule, otherwise the processes are sorted according to the starting time they had in the parent and are placed in the schedule in ascending order.

2.4   Parallel Resource Partition and Processes Scheduling

The diagram of the algorithm of the parallel resources selection and processes scheduling according to genetic approach, is showed on Fig. 7.

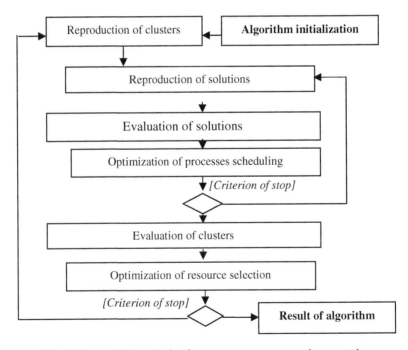

Fig. 7. The parallel synthesis of computer system – genetic approach

The input parameters are the number of clusters in the population and the number of solutions in clusters. Solution clusters represent the structures sharing the same resource allocation, but with different processes allocation to resources and different schedules.

The outer loop of the algorithm (realizes resource selection) is ran until the number of generations without population improvement is exceeded. This value is defined by the annealing criterion parameter. There are few outer loops at the beginning of the algorithm operation.

The number of iteration of internal loop algorithm be definite:

$$f(x) := -k \cdot \left(e^{-a \cdot x}\right)^3 + k \qquad (2.4.)$$

where: k – the parameter of algorithm, a - the annealing parameter. Argument x in every generation is enlarged by the step of temperature. At the beginning of the performance of algorithm, internal loops are scarce – Fig. 8.

Their number grows until it reaches the value of k with the falling of the temperature. Fewer task allocations and scheduling processes are performed at the beginning. When the temperature falls sufficiently low, each inner loop has k iterations. The number of iterations may be regulated with the temperature step parameter. The greater the step, the faster the number of inner iterations reaches the k value.

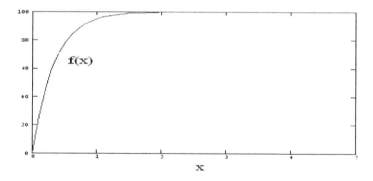

Fig. 8. The chart of function f(x) from formula 2.4

3   Computational Experiments – Multi-criteria Optimization

Experiments were conducted for non-preemptive and dependent tasks. Parameters of constraints: the maximum number of processors, maximum cost, maximum time. It the area of optimum solutions in result was received was in sense Pareto [1]. The following charts (3.1 – 3.3) presented solutions (in Pareto area) for the cost, time and power consumption and solution "compromising" – balancing the values of optimizing criteria.

The above presented charts shows of multi-criteria optimization for parallel design of complex system. The designer in result of working of algorithm receives in sense the gathering of optimum solutions Pareto. It stays with the designer's processes the selection the most answering his requirements of solution. In dependence from this what are for system requirements it was it been possible to lean on one of got results. To get to know for given authority of problem the specific of space of solutions well, important the use is long the list of remembering the best solutions (in tests the parameter of algorithm "it quantity *the best"* it was established was value 50). Important the settlement of slow refreshing the algorithm is equally (the parameters "the *step of temperature"* 0.1 and "the *coefficient of cooling"* - in dependent on from quantity of tasks in system; generally smaller than 0,05). We prevent thanks this sale large convergence in population. The algorithm searches near smaller temperature, the larger area in space of solutions. It has also been noticed that the bigger probability of mutation helps o look for a better system structure, whereas a bigger probability of crossing improves the optimization of time criterion.

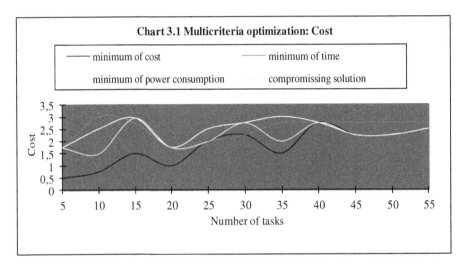

Chart 3.1 Multicriteria optimization: Cost

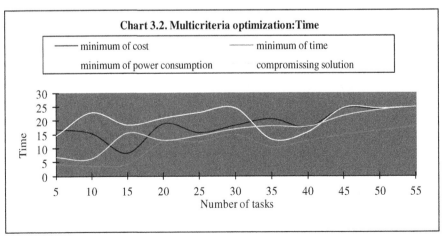

Chart 3.2. Multicriteria optimization:Time

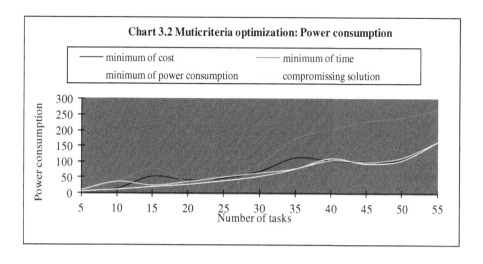

Chart 3.2 Muticriteria optimization: Power consumption

4   Conclusions

This paper is about the problems of parallel design of complex systems. Such a design is carried out on a high level of abstraction in which system specification consists of a set of operations, which should be implemented by a series of resources and these are listed in the database (container, pool, or a catalogue) and are available (do exist or can be created). Resources possess certain constraints and characteristics, including speed, power, cost and dependability parameters. Thus such a design concerns systems of the following type – resource and operation complex and the problems of resource partitioning (selection) as well as scheduling (sequencing) of operations performed on these resources are determined on this level. Optimization of aforementioned design actions occurs on the same level.

When one possesses operations (system specifications) and selected resources (software and hardware which can perform these operations) as well as defined control which allocates operations and resources and schedules operations, then one possesses general knowledge necessary for the elaboration of design details – to define standard, with dedicated parameters, specific, physical processors and software modules, to apply available components and, when it is necessary to prepare special hardware subcomponents and software modules for the implementation of the whole system of greater efficiency (cheaper, faster, consuming less power, a higher degree of dependable).

Problems of design in parallel approach are solved simultaneously and globally and as it is confirmed by calculation experiments, the solutions of these problems are more efficient than non-parallel (eg. concurrent) ones. Of course these problems, as mentioned earlier are computationally NP-complete, as a result there is a lack of effective and accurate algorithms, thus one has to use heuristics in solutions. In the thesis one proposed the so called artificial intelligence implementation methods. Obviously these methods were chosen out of many and proposed adaptations of these methods can be different. However, for methods and adaptations presented herein calculations clearly point out the advantage of parallel design, i.e. joint optimization of resource partitioning and, then on selected resources, processes scheduling over non-parallel design, where resource partitions and processes scheduling are optimized separately. Among presented results of computational experiments the best solutions were obtained with the evolutionary and simulated annealing algorithms with Boltzmann tournaments.

The issues for other methods are now studied.

References

1. Drabowski, M.: A genetic method for hardware-software par-synthesis. International Journal of Computer Science and Network Security 5(22), 90–96 (2006)
2. Drabowski, M., Wantuch, E.: Deterministic schedule of task in multiprocessor computer systems with higher degree of dependability. In: Zamojski, W., Mazurkiewicz, J., Sugier, J., Walkowiak, T., Kacprzyk, J. (eds.) Proceedings of the Ninth International Conference on DepCoS-RELCOMEX. AISC, vol. 286, pp. 165–175. Springer, Heidelberg (2014)
3. Garey, M., Johnson, D.: Computers and intractability: A guide to the theory of NP-completeness. Freeman, San Francisco (1979)
4. Aarts, E.H.L., Korst, J.: Simulated Annealing and Boltzmann Machines. J. Wiley, Chichester (1989)
5. Eiben, A.E., Aarts, E.H.L., Van Hee, K.M.: Global convergence of genetic algorithms: A Markov chain analysis. In: Schwefel, H.-P., Männer, R. (eds.) PPSN 1990. LNCS, vol. 496, pp. 3–12. Springer, Heidelberg (1991)
6. Błażewicz, J., Drabowski, M., Węglarz, J.: Scheduling multiprocessor tasks to minimize schedule length. IEEE Trans. Computers C-35(5), 389–393 (1986)
7. Dick, R.P., Jha, N.K.: COWLS: Hardware-Software Co-Synthesis of Distributed Wireless Low-Power Client-Server Systems. IEEE Trans. on Computer-Aided Design of Integrated Circuits and Systems 23(1), 2–16 (2004)
8. Kirkpatrick, S., Gelatt Jr., C.D., Vecchi, M.P.: Optimization by Simulated Annealing. Science 220(4598), 671–680 (1983)

Arithmetic in the Finite Fields Using Optimal Normal and Polynomial Bases in Combination

Sergej Gashkov[1], Alexander Frolov[2], Sergej Lukin[2], and Olga Sukhanova[2]

[1] Lomonosov Moscow State University, MSU, Faculty of Mechanics and Mathematics, Russia, 119991, Moscow, GSP-1, 1 Leninskiye Gory, Main Building
sbgashkov@gmail.com
[2] National Research University Moscow Power Engineering Institute, Krasnokazarmennaya, 14, Russian Federation, 111250, Moscow
{abfrolov,suhanovaok}@gmail.com, ieha4@mail.ru

Abstract. In this chapter the idea of using optimal normal bases (o.n.b.) of second and third types in combination with polynomial basis of field $F(q^n)$ is detailed using a new modification of o.n.b. called reduced optimal normal basis $-1, \beta_1, \dots, \beta_{n-1}$ corresponding to a permuted o.n.b. $\beta_1, \dots, \beta_{n-1}$ Operations of multiplication, rising to power q^j, rising to arbitrary power and inversion in reduced o.n.b. in combination with polynomial basis as well as converting operations between these bases in the fields of characteristic three has been described, estimated and expanded to the fields of characteristic two. This allows get efficient implementations of cryptographic protocols using operation of Tate pairing on supersingular elliptic curve.

Keywords: finite fields, polynomial basis, optimal normal basis, multiplication, rising to power, inversion, Tate pairing.

1   Introduction

Implementation of cryptographic protocols on elliptic curves over finite fields of small characteristic is based on arithmetic in these fields. It is well known that polynomial basis is most appropriate for multiplication and inverting whereas normal bases possesses essential preliminarily with respect to polynomial bases in implementation of the operation rising to power equal to field characteristic degree because this operation corresponds to cyclic shifting. Operation of raising to arbitrary power as well as other more complex operations are composed from these operations.

In this chapter, we consider implementation of arithmetic in finite fields of small characteristic using polynomial and normal bases in combination tacking into account specificity of mentioned protocols.

Let x be the root of irreducible polynomial of degree n over finite field F_p of characteristic q, Then $F_q(x)$ is an algebraic extension of the field F_q and the set $\{1, x, \dots x^{n-1}\}$ is its polynomial basis. The set $\{x, x^p, \dots x^{p^{n-1}}\}$ if possesses basis property is called normal basis [1,2]. Polynomial bases allows fast implementation of

© Springer International Publishing Switzerland 2015
W. Zamojski et al. (eds.), *Theory and Engineering of Complex Systems and Dependability*,
Advances in Intelligent Systems and Computing 365, DOI: 10.1007/978-3-319-19216-1_15

multiplication using A. Karatzuba [3], A. Toom [4] or A. Schönhage [5] multiplication methods followed by reducing operation. Normal bases possesses essential preliminarily with respect to polynomial bases in implementation of the operation rising to power q^j because this operation corresponds to cyclic shifting. In [5] there were discovered optimal normal bases (o.n.b.) with multiplication of quadratic complexity. In [6] conversion algorithms of complexity $O(n \ln n)$ between the bases polynomial and optimal normal bases of second and third types (o.n.b.-2 and o.n.b.-3) has been described. Simultaneously in [6] it was proposed using of polynomial and normal bases in combination converting the operands in polynomial basis for consequent multiplication and in normal basis for consequent rising to power q^j. In [7,8] the method of multiplication in o.n.b.-2 of fields of characteristic two using multiplication algorithm in the ring $GF(2)[X]$ followed by converting the product to normal basis has been described. Analogous multiplication algorithm with original scheme of such converting is described in [9]. Converting algorithms of complexity $O(n \ln n)$ are implemented in [7,8,9] for the same purposes to take advantage of various bases in the multiplication or exponentiation q^j. In this chapter we develop the idea on the use of optimal normal and polynomial bases in combination by applying it the so-called reduced optimal normal basis of the second or of the third type in fields of small characteristic. In details, we consider the fields of characteristic 3. Approach to the fields of characteristic two is similar. In second section we consider modification of o.n.b.-2 and o.n.b.-3 of fields of characteristic three and converting algorithms. In the third section multiplication, rising to power q^j rising to power and inversion algorithms has been described and estimated. In conclusion we summary the results expanding them to the fields of characteristic two and discus them with respect to Tate pairings algorithms with and without operation of root extraction [10].

2   Optimal Normal Bases (o.n.b.) of Second and Third Types of Fields of Characteristic Three and Their Modifications

Let $p = 2n + 1$ be prime such that p divides $3^{2n} - 1$. Let $\alpha \in GF(3^{2n})$ be such that $\alpha \neq 1$, $\alpha^p = 1$. Let the set of modulo p degrees of 3 coincides with the set of all non zeros modulo p numbers (i.e. 3 be a primitive root module p or coincides with the set of all quadratic residues modulo p, in which case -1 is quadratic non residue and $3^n = 1 \pmod{p}$).

Consider sequence $\beta_i = \alpha^i + \alpha^{-i} = \alpha^i + 1/\alpha^i \in GF(3^{2n})$ for all integers i (in case $3^n \equiv 1 \pmod{p}, \beta_i \in GF(3^n)$).

Obviously, $\beta_i = \beta_{-i}, \beta_0 = -1 \in GF(3)$,

$$\beta_i \beta_j = (\alpha^i + \alpha^{-i})(\alpha^j + \alpha^{-j}) = \left(\alpha^{i+j} + \alpha^{-(i+j)}\right) + \left(\alpha^{i-j} + \alpha^{-(i-j)}\right) = \beta_{i+j} + \beta_{i-j}.$$

Taking into account

$$\beta_{3^k} = \left(\alpha^{3^k} + \alpha^{-3^k}\right) = (\alpha + \alpha^{-1})^{3^k} = \beta_1^{3^k} = \beta^{3^k}$$

where $\beta = \beta_1$ we get

$$\beta_{3^k+i} = \beta^{3^k}\beta_i + \beta_{3^k-i},$$

$$\beta_{2\cdot 3^k+i} = \beta^{3^k}\beta_{3^k+i} - \beta_{-i} = \beta^{3^k}\left(\beta^{3^k}\beta_i - \beta_{3^k-i}\right) - \beta_i = \beta^{2\cdot 3^k}\beta_i - \beta^{3^k}\beta_{3^k-i} - \beta_i,$$

Consider basis $\{\beta_1, \ldots, \beta_n\}$ in $GF(3^n)$. It is obtained by permutation π

$$\pi(j) = \begin{cases} 3^j \bmod p & \text{if } 3^j \bmod p \le n, \\ p - 3^j \bmod p & \text{if } 3^j \bmod p > n, \end{cases}$$

from the o.n.b. $\{\xi_1, \ldots, \xi_n\}$, where $\xi_i = \beta^{2^{i-1}}, i = 1, \ldots, n$, and is called *permutated* optimal normal basis (o.n.b.) : for $j = 0, \ldots, n-1$: $b_j = a_{\pi(j)}$. The inverse conversion corresponds to inverse permutation: π^{-1}: $c_j = b_{\pi^{-1}(j)}$, $j = 1, \ldots, n$. If 3 is non quadratic residue, o.n.b. $\{\xi_1, \ldots, \xi_n\}$ is called o.n.b. of the second type, else it is called o.n.b. of the third type.

It can be verified that $\sum_{i=1}^{n}\beta_i = \sum_{i=1}^{n}\alpha^i + \sum_{i=1}^{n}\alpha^{-i} = \sum_{i=1}^{2n}\alpha^i = -1 = \beta_0$ because

$$\alpha^{-i} = \alpha^{p-i} = \alpha^{2n+1-i}, \qquad \sum_{i=0}^{2n}\alpha^i = \sum_{i=0}^{p-1}\alpha^i = (\alpha^p - 1)/(\alpha - 1) = 0.$$

As well, we will use reduced o.n.b. of 2 and 3 types $\{\beta_0, \ldots, \beta_{n-1}\}$, $\beta_0 = -1 \in GF(3)$ and redundant o.n.b. of these types $\{\beta_0, \ldots, \beta_{n-1}, \beta_n\}$.

Conversion from reduced o.n.b. $\{\beta_0, \ldots, \beta_{n-1}\}$ into permutated o.n.b. $\{\beta_1, \ldots, \beta_n\}$ is made representing β_0 as $\beta_0 = \sum_{i=1}^{n}\beta_i$: for $x \in GF(3^m)$, $x = (x_0, x_1, \ldots, x_{n-1})$ in reduced o.n.b. we get its representation in permutated o.n.b.:

$$x = x_0\beta_0 + x_1\beta_1 + \cdots + x_{n-1}\beta_{n-1} = x_0\sum_{i=1}^{n}\beta_i + x_1\beta_1 + \cdots + x_{n-1}\beta_{n-1} =$$

$$= (x_1 + x_0)\beta_1 + \cdots + (x_{n-1} + x_0)\beta_{n-1} + x_0\beta_n$$

Inverse conversion is made representing β_n as $\beta_n = \beta_0 - \sum_{i=1}^{n-1}\beta_i$:

$$x = x_1\beta_1 + \cdots + x_{n-1}\beta_{n-1} + x_n\beta_n = x_1\beta_1 + \cdots + x_{n-1}\beta_{n-1} + x_n(\beta_0 - \sum_{i=1}^{n-1}\beta_i) =$$

$$= x_n\beta_0 + (x_1 - x_n)\beta_1 + \cdots + (x_{n-1} - x_n)\beta_{n-1}$$

Conversion from redundant o.n.b. into reduced o.n.b. is made representing β_n as $\beta_n = \beta_0 - \sum_{i=1}^{n-1}\beta_i$: for $x \in GF(3^n)$ we get

$$x_0\beta_0 + x_1\beta_1 + x_2\beta_2 + \cdots + x_{n-1}\beta_{n-1} + x_n\beta_n =$$

$$= x_0\beta_0 + x_1\beta_1 + x_2\beta_2 + \cdots + x_{n-1}\beta_{n-1} + x_n\left(\beta_0 - \sum_{i=1}^{n-1}\beta_i\right) =$$

$$= (x_0 + x_n)\beta_0 + (x_1 - x_n)\beta_1 + (x_2 - x_n)\beta_2 + \cdots + (x_{n-1} - x_n)\beta_{n-1}.$$

It follows that coefficient x_n (at β_n) have to be added to coefficient at β_0 and subtracted form coefficients at β_i, $i = 1, \ldots, n$.

On the other hand, conversion from redundant o.n.b. into permutated o.n.b. one can made by adding coefficient x_0 (at β_0) to coefficients x_i (at β_i), $i=1,\ldots,n$.

Consider conversion from reduced o.n.b. into polynomial basis $\{\beta^0, \ldots, \beta^{n-1}\}$, $\beta^0 = 1$, $\beta = \beta_1 = \alpha + \frac{1}{\alpha} \in GF(3^n)$. Denote this linear transformation as $F(e)$. It can be computed recursively as follows. Let $2 * 3^k < n \le 3^{k+1}$.

Consider an arbitrary $x \in GF(3^n)$, $x = \sum_{i=0}^{n-1} x_i\beta_i$, где $x_i \in GF(3) = \{0,1,2\}$, $2 = -1$. Split sum into three parts $x = x_0 + x_1 + x_2$, where

$$x_0 = \sum_{i=0}^{3^k-1} x_i\beta_i, x_1 = \sum_{i=0}^{3^k-1} x_{3^k+i}\beta_{3^k+i}, \quad x_2 = \sum_{i=0}^{n-2*3^k-1} x_{2*3^k+i}\beta_{2*3^k+i}.$$

Using formula

$$\beta_{3^k+i} = \beta^{3^k}\beta_i - \beta_{3^k-i}, \beta_{3^k} = \beta^{3^k},$$

rewrite x_1 as

$$x_1 = \beta^{3^k}\sum_{i=1}^{3^k-1} x_{3^k+i}\beta_i - \sum_{i=1}^{3^k-1} x_{2*3^k-i}\beta_i - \beta^{3^k}x_{3^k}\beta_0.$$

Using formula

$$\beta_{2*3^k+i} = \beta^{2*3^k}\beta_i - \beta^{3^k}\beta_{3^k-i} - \beta_i, \beta_{2*3^k} = \beta^{2*3^k} + 1,$$

rewrite x_2 as

$$x_2 = \beta^{2*3^k}\sum_{i=1}^{n-2*3^k-1} x_{2*3^k+i}\beta_i - \sum_{i=1}^{n-2*3^k-1} x_{2*3^k+i}\beta_i -$$

$$- \sum_{i=3^{k+1}-n+1}^{3^k-1} x_{3^{k+1}-i}\beta_i - \beta^{2*3^k}x_{2*3^k}\beta_0 - x_{2*3^k}\beta_0.$$

Further, adding we write $x = x_0 + x_1 + x_2$ as $x = y_0 + y_1 + y_2$,

$$y_j = \beta^{j*3^k}\sum_{i=0}^{3^k-1} y_{j*3^k+i}\beta_i, j = 0,1, \qquad y_2 = \beta^{2*3^k}\sum_{i=0}^{n-2*3^k-1} y_{2*3^k+i}\beta_i,$$

where coefficients $y_i \in GF(3)$ are computed by formulae:

$$y_0 = x_0 - x_{2*3^k}, \; y_i = x_i - x_{2*3^k-i} - x_{2*3^k+i}, i = 1, \dots, n - 2 * 3^k - 1,$$

$$y_i = x_i - x_{2*3^k-i}, i = n - 2 * 3^k, \dots, 3^k - 1,$$

$$y_{3^k} = -x_{3^k}, \; y_{3^k+i} = x_{3^k+i} - x_{3^{k+1}-i}, i = 3^{k+1} - n + 1, \dots, 3^k - 1,$$

$$y_{3^k+i} = x_{3^k+i}, i = 1, \dots, 3^{k+1} - n,$$

$$y_{2*3^k} = -x_{2*3^k}, \; y_{2*3^k+i} = x_{2*3^k+i}, i = 1, \dots, n - 2 * 3^k - 1.$$

For $3^k < n \le 2 * 3^k$ converting formulae are analogous, but some coefficient are zeros.

Then each $\mathbf{y}_j = \beta^{j*3^k} \sum_{i=0}^{3^k-1} y_{j*3^k+i} \beta_i$, applying algorithm recursively transform into

$$\mathbf{y}_j = \beta^{j*3^k} \sum_{i=0}^{3^k-1} y_{j*3^k+i} \beta_i = \beta^{j*3^k} \sum_{i=0}^{3^k-1} z_{j*3^k+i} \beta^i, j = 0,1,2.$$

For $j = 1,2$, the last terms may be zeros that should be taken into account in implementation.

In the end the basis formulae will be used on the base of identity $\beta_2 = \beta^2 - \beta_0$.

It remains to collect (it is free) $\mathbf{x} = \mathbf{y}_0 + \mathbf{y}_1 + \mathbf{y}_2 = \sum_{i=0}^{n-1} z_i \beta^i$.

All this is done with the complexity $O(n \ln n)$. When $n = 3^k$ the number of actions is $L(n) = n(\log_3 n - 1) + 1$.

For arbitrary n it is defined by formula

$$L(n) = \sum_{i=0}^{k} (\left\lfloor \frac{n}{3^{i+1}} \right\rfloor (3^i - 2) + \max\{n \bmod 3^{i+1} - 3^i - 1, 0\} +$$

$$+ \max\{n \bmod 3^{i+1} - 2 * 3^i - 1, 0\}).$$

Consider conversion from polynomial basis $\{\beta^0, \dots, \beta^{n-1}\}, \beta^0 = 1, \beta = \beta_1 = \alpha + 1/\alpha \in GF(3^n)$ into reduced o.n.b. $\{\beta_0, \dots, \beta_{n-1}\}, \beta_0 = -1, \beta = \beta_1 = \alpha + 1/\alpha \in GF(3^n)$. This linera transformation denote $F^{-1}(n)$. It can be computed recursively as follows. Let us $2 * 3^k < n \le 3^{k+1}$.

It starts with the repeated application (to the sequence of triples of coefficients) of conversions on the basis of identity $\beta^2 = \beta_2 + 2$.

Then triples obtained in the form of polynomials are connected (for the first iteration, k = 1)

$$\mathbf{y} = \mathbf{y}_0 + \mathbf{y}_1 + \mathbf{y}_2 = \sum_{i=0}^{3^k-1} y_i \beta_i + \beta^{3^k} \sum_{i=0}^{3^k-1} y_{3^k+i} \beta_i + \beta^{2*3^k} \sum_{i=0}^{n-2*3^k-1} y_{2*3^k+i} \beta_i$$

Next disclose brackets and make replacement by formulae $\beta^{3^k}\beta_i = \beta_{3^k+i} + \beta_{3^k-i}$ and $\beta^{2*3^k}\beta_i = \beta_{2*3^k+i} + \beta^{3^k}\beta_{3^k-i} + \beta_i$, $\beta_{2*3^k} = \beta^{2*3^k} + 1$, i.e. replace y_1 and y_2 as follows

$$y_1 = \beta^{3^k} \sum_{i=0}^{3^k-1} y_{3^k+i}\beta_i = \beta^{3^k} \sum_{i=1}^{3^k-1} y_{3^k+i}\beta_i + \beta_{3^k}y_{3^k}\beta_0 =$$

$$= \sum_{i=1}^{3^k-1} y_{3^k+i}\beta_{3^k+i} + \sum_{i=1}^{3^k-1} y_{3^k+i}\beta_{3^k-i} + \beta_{3^k}y_{3^k}\beta_0.$$

$$y_2 = \beta^{2*3^k} \sum_{i=0}^{n-2*3^k-1} y_{2*3^k+i}\beta_i =$$

$$= \beta^{2*3^k} \sum_{i=1}^{n-2*3^k-1} y_{2*3^k+i}\beta_i + \beta^{2*3^k}y_{2*3^k}\beta_0 + y_{2*3^k}\beta_0 =$$

$$= \sum_{i=1}^{n-2*3^k-1} y_{2*3^k+i}\beta_{2*3^k+i} + \beta^{3^k} \sum_{i=3^{k+1}-n+1}^{3^k-1} y_{3^{k+1}-i}\beta_i +$$

$$+ \sum_{i=1}^{n-2*3^k-1} y_{2*3^k+i}\beta_i + \beta^{2*3^k}y_{2*3^k}\beta_0 + y_{2*3^k}\beta_0,$$

where coefficients $x_i \in GF(3)$ are computed by formulae:

$x_0 = y_0 - y_{2*3^k}$, $x_i = y_i + y_{2*3^k-i} - y_{2*3^k+i}$, $i = 1, \ldots, n - 2 * 3^k - 1$,

$x_i = y_i + y_{3^{k+1}-i}$, $i = n - 2 * 3^k, \ldots, 3^k - 1$,

$x_{3^k} = -y_{3^k}$, $x_{3^k+i} = y_{3^k+i} + y_{3^{k+1}-i}$, $i = 3^k - n + 1, \ldots, 3^k - 1$,

$x_{3^k+i} = y_{3^k+i}$, $i = 1, \ldots, 3^{k+1} - n$,

$x_{2*3^k} = -y_{2*3^k}$, $x_{2*3^k+i} = y_{2*3^k+i}$, $i = 1, \ldots, n - 2 * 3^k - 1$.

For $3^k < n \leq 2 * 3^k$ converting formulae are analogous, but some coefficients are zeros.

It remains to collect (it's free): $y = x_0 + x_1 + x_2 = \sum_{i=0}^{n-1} z_i\beta_i$.

All this is done with the complexity $O(n \cdot \ln n)$. The exact formulae are the same as for F_n.

Let us present the schemes of recursion of considered transformations.

Let $F_n(e)$ be one iteration of transformation from reduced normal to polynomial basis with known $n, 3^k < n \le 3^{k+1}$, $F_n^{-1}(e)$ be one iteration of inverse transformation. Then

a) $F_3(e_0, e_1, e_2) = ((-e_0 + e_2), e_1, e_2)$, $F_2(e_0, e_1) = (e_0, e_1)$, $F_1(e_0) = (e_0)$

(basis of recursion);

$$F_{3^k}(e) = F_{3^{k-1}}(F(e)^0) || F_{3^{k-1}}(F(e)^1) || F_{n-3^{k-1}}(F(e)^2) \text{ (step of recursion)}$$

where $||$ is concatenation, $F(e) = F(e)^0 || F(e)^0 || F(e)^2$.

b) $F_3^{-1}(e_0, e_1, e_2) = (-(e_0 - e_2), e_1, e_2)$, $F_2^{-1}(e_0, e_1) = (e_0, e_1)$, $F_1^{-1}(e_0) = (e_0)$
(basis of recursion);

$$F_{3^k}^{-1}(e) = F^{-1}(F_{3^{k-1}}^{-1}(e^0) \left|\left| F_{3^{k-1}}^{-1}(e^1) \right|\right| F_{n-3^{k-1}}^{-1}(e^2))$$

(step of recursion), where $e = e^0 || e^1 || e^2$.

Accordingly [11] these schemes confirm asymptotic estimate $O(n \ln n)$.

3   Arithmetic in the Fields of Characteristic Three

Multiplication in reduced o.n.b.-2 and o.n.b.-3 is computed through conversion into polynomial basis involving the following algorithm.

1. Convert both elements $x = (x_0, x_1, ..., x_{n-1})$, $y = (y_0, y_1, ..., y_{n-1})$ form the reduced o.n.b. $\{\beta_0, ..., \beta_{n-1}\}$ in polynomial basis $\{\beta^0, ..., \beta^{n-1}\}$:

$$x \to x' = (x'_0, x'_1, ..., x'_{n-1}), \quad y \to y' = (y'_0, y'_1, ..., y'_{n-1}).$$

2. Compute a product of degree at most $2n - 2$ in the ring $GF(3)[X]$ (or in the field $GF(3^m)$) using redundant polynomial basis $\{\beta^0, ..., \beta^{2n-2}\}$:
$z' = (z'_0, z'_1, ..., z'_{2n-2}) = x' * y'$.

3. Adding β^{2n-1} (the leading coefficient $x_{2n-1} = 0$), fulfill conversion prom extended redundant polynomial basis $\{\beta^0, ..., \beta^{2n-1}\}$ into extended reduced o.n.b. $\{\beta_0, ..., \beta_{2n-1}\}$; $z' \to z'' = (z''_0, z''_1, ..., z''_{2n-1})$.

4. Fulfill «folding» corresponding to identities $\beta_i = \beta_{2n-i+1}$, $i = 3, ..., n$, $z'''_i = z''_i + z''_{2n-i+1}$.   Initial   elements z''_0, z''_1, z''_2 ing $\beta_0, \beta_1, \beta_2)$ do not changed, the consequent element are replaced with sums $z'''_i = z''_i + z''_{2n-i+1}$, $i=3,...,n$. Result is in redundant o.n.b.:

$$z''' = (z'''_0, z'''_1, ..., z'''_n).$$

5. Convert this result into reduced o.n.b.: $z''' \to z = (z_0, z_1, ..., z_{n-1})$.

The total number of additions is

$$M(n) + 2n - 2 + \sum_{i=0}^{k} \left(\left\lceil \frac{n}{3^{i+1}} \right\rceil (3^i - 2) + \max\{n \bmod 3^{i+1} - 3^i - 1, 0\} + \right.$$

$$+ \max\{n \bmod 3^{i+1} - 2*3^i - 1, 0\}) + \sum_{i=0}^{k+1} \left(\left\lceil \frac{2n-1}{3^{i+1}} \right\rceil (3^i - 2) + \right.$$

$$+ \max\{(2n-1) \bmod 3^{i+1} - 3^i - 1, 0\} + \max\{(2n-1) \bmod 3^{i+1} - 2*3^i - 1, 0\})$$

where $M(n)$ is complexity of multiplication in the ring $GF(3)[X]$.

Asymptotic complexity of multiplication is $O(n \ln n) + O(M(n))$.

Modified operation of multiplication differs from described above in step 5 of algorithm: the result of fourth step is converted into permutated o.n.b. instead of reduced o.n.b. with the same complexity. So in modified operation multipliers are given in reduced o.n.b. and product is presented in permutated o.n.b.

Operation of multi-cubing in the reduced o.n.b. $\{\beta_0, \dots, \beta_{n-1}\}$ of $GF(3^n)$ corresponds the functionality $a^{3^j}, j \in \mathbf{N}$. The multi-cubing algorithm is the following:

1. Convert element x from reduced o.n.b. into permutated o.n.b.:

$$x' = x'_1\beta_1 + \cdots + x'_{n-1}\beta_{n-1} + x'_n\beta_n = (x_1 + x_0)\beta_1 + \cdots + (x_{n-1}+x_0)\beta_{n-1} + x_0\beta_n.$$

2. Multi cube by cyclic shift taking into account permutation:

$$y' = x'^{3^j} = \left(x'_{\pi(\pi^{-1}(1)-j)}, \dots, x'_{\pi(\pi^{-1}(i)-j)}, \dots, x'_{\pi(\pi^{-1}(n)-j)} \right).$$

3. Convert the result into reduced o.n.b.:

$$y = y_0\beta_0 + y_1\beta_1 + \cdots + y_{n-1}\beta_{n-1} =$$

$$= y'_n\beta_0 + (y'_1 - y'_n)\beta_1 + (y'_2 - y'_n)\beta_2 + \cdots + (y'_{n-1} - y'_n)\beta_{n-1}.$$

The number of additions is $2n - 2$, asymptotic complexity is $O(n)$.

Modified operation of multi-cubing does not contain first the step of described multi-cubing algorithm, i.e. input of modified operation is given in permutated o.n.b. and result is presented in reduced o.n.b. Complexity of modified algorithm is $n - 1$.

Algorithm of operation rising to power is conventional algorithm with modified operations of multiplication and rising to power 3^j in each iteration except the last one with not modified multiplication.

Inversion one can compute implementing Euclidian algorithm in $GF(p)[X]$ (generalizing binary algorithm [12]) of complexity $O(n^\sigma), 2 < \sigma < 3$ [13] . Let x be a root of $f(X)$ generates an o.n.b. -2 or o.n.b. -3. The inversion algorithm is the following:

1. Convert $\in GF(p^n)$ from reduced o.n.b. into polynomial basis.
2. $b \leftarrow 1; c \leftarrow 0; u \leftarrow a; v \leftarrow f(X);$

3. *while* deg $u > 1$:
> *while* $u_0 = 0$: $u \leftarrow u/X$;
>> *if* $b_0 = 0$: $b \leftarrow b/X$;
>> *else*: $b \leftarrow (b - f_0^{-1} b_0 f(X))/X$;
>
> *if* deg $u > 1$:
>> *if* deg $u <$ deg v: $u \leftrightarrow v, b \leftrightarrow c$;
>> $b \leftarrow (-v_0)b + cu_0, u \leftarrow (-v_0)u + vu_0$.

4. Convert (bu) into reduced o.n.b. and return.

In Table 1 there are represented algorithms of raising to power in fields $GF(2^n)$ or $GF(3^n)$ (Algorithm 1) and algorithm of scalar multiplication on supersingular elliptic curve using reduced o.n.b. (Algorithm 2 possessing the same scheme). Operations x^{-1}, x^j, xy are operations in fields, $-P, q^i P, B + P$ are in the groups $EC(GF(q^n))$, $q = 2,3$. Inversions (points 1 of algorithms) are omitted if d does not contain negative elements.

Table 1.

Algorithm 1	Algorithm 2
Input:	Input:
$a \in GF(q^n), d \in \{0,-1,1\}, q \in \{2,3\}$.	$P \in EC(GF(q^n)), d \in \{0,-1,1\}, q \in \{2,3\}$.
Output: $a^d \in GF(q^n)$.	Output: $dP \in EC(GF(q^n))$.
1. $\bar{a} = a^{-1}; b = 1; j = 1$.	1. $P' = -P; B = 0; j = 1$.
2. For $i = \overline{0,\|d\|}$:	2. For $i = \overline{0,\|d\|}$:
if $d[\|d\| - i - 1] = 0$: $j = j + 1$	if $d[\|d\| - i - 1] = 0$: $j = j + 1$
else:	else:
if $d[\|d\| - i - 1] = -1$:	if $d[\|d\| - i - 1] = -1$:
$b = b^j; b = b * \bar{a}$;	$B = q^j * B^j; B = B + P'$;
else:	else:
$b = b^j; b = b * a$;	$B = j; B = B + P'$;
$j = 1$;	$j = 1$;
if $d[\|d\| - i - 1] = 0$: $b = b^j$.	if $d[\|d\| - i - 1] = 0$: $B = q^j * B^j$;
3. Return b.	3. Return B.

Supposing $(n, 6) = 1$, we represent elements of $GF(3^{2n})$ as $a_0(x)\beta + a_1(x)\beta^3$ (shortly as pairs $(a_0(x), a_1(x))$) with coefficients $a_0(x), a_1(x)$ in reduced o.n.b. of $GF(3^n)$ using the optimal normal basis β, β^3 of the first type where β is a root of polynomial $2 + Y + Y^2$ over the field $GF(3^n)$. This allows quick multi-cubing in $GF(3^{2n})$. For multiplication we use polynomial basis $1, \beta$ and explicit formulae for reducing. Elements of $GF(3^{6m})$ we represent as $(a_0^0(x), a_1^0(x)) + (a_0^1(x), a_1^1(x))\beta' + (a_0^2(x), a_1^2(x))\beta'^3)$ with coefficients in $GF(3^{2n})$ using the reduced o.n.b. of the second type with the root β' of polynomial $(2,0) + (1,0)Z + (1,0)Z^2 + (1,0)Z^3$ over $GF(3^{2n})$. This allows quick multi-cubing in $GF(3^{6n})$. For multiplication we use polynomial basis $1, \beta', \beta'^2$ and explicit formulae for reducing.

4   Conclusion

In this chapter, the idea of using optimal normal bases (o.n.b.) in combination with polynomial bases was detailed using a new modification of o.n.b. called reduced optimal normal basis. Arithmetic operations in these bases were described and estimated for the fields of characteristic three. But the second and third sections can be rewritten in terms of fields of characteristic two. In this case fields elements are recursively split into two parts and schemes of recursion are analogous to schemes considered in [9]. Arithmetic operations in the fields of distinct small characterics are of the same asymptotic complexity. Implementing optimal normal basis of the second or the third types in combination with polynomial basis, one can show that algorithm of Tate paring with operation of root extraction can be preferable whereas it is well known that Tate paring implemented in polynomial basis and using algorithm without root extraction is much faster than algorithm using this operation.

Acknowledgements. This research was carried out with the financial support of the Russian Foundation for Basic Research, project 14-01-00671a. We thank our reviewers for support and remarks.

References

1. Lidl, R., Niderreiter, H.: Finite fields. Addison-Wesley publishing Company, London (1983)
2. Jungnickel, D.: Finite fields: Structure and arithmetics. Wissenschaftsverlag. Mannheim (1993)
3. Karatzuba, A.A., Offman, Y.P.: Umnozhenie mnogoznachnych chisel na avtomatah. DAN USSR 145(2), 293–294 (1962) (in Russian)
4. Schönhage, A.: Shnelle Multiplikation von Polynomen der Körpern der Charakteristik 2. Acta Informatica 7, 395–398 (1977)
5. Mullin, R.C., Onyszchuk, I.M., Vanstone, S.A., Wilson, R.M.: Optimal Normal Bases in GF(pn). Discrete Appl. Math. 22, 149–161 (1988/1989)
6. Bolotov, A.A., Gashkov, S.B.: On quick multiplication in normal bases of finite fields. Discrete Mathematics and Applications 11(4), 327–356 (2001)
7. Shokrollahi, J.: Efficient implementation of elliptic curve cryptography on FPGA. PhD thesis, universitet Bonn (2007)
8. von zur Gathen, J., Shokrollahi, A., Shokrollahi, J.: Efficient multiplication using type 2 optimal normal bases. In: Carlet, C., Sunar, B. (eds.) WAIFI 2007. LNCS, vol. 4547, pp. 55–68. Springer, Heidelberg (2007)
9. Bernstein, D.J., Lange, T.: Type-II Optimal Polynomial Bases. In: Hasan, M.A., Helleseth, T. (eds.) WAIFI 2010. LNCS, vol. 6087, pp. 41–61. Springer, Heidelberg (2010)
10. Kwon, S.: Efficient Tate Pairing for Supersingular Elliptic Curve over Binary Fields. Cryptology ePrint Archive. Report 2004/303 (2004)
11. Aho, A., Hopcroft, J., Ullman, J.: The design and analysis of computer algorithms. Addison-Wesley Publishing Company, Reading (1978)
12. Hankerson, D., Hernandez, J.L., Menezes, A.: Software implementation of elliptic curve cryptography over binary fields. In: Paar, C., Koç, Ç.K. (eds.) CHES 2000. LNCS, vol. 1965, pp. 1–23. Springer, Heidelberg (2000)
13. Gashkov, S.B., Frolov, A.B., Shilkin, S.O.: On some algorithms of inversion and division in finite rings and fields. MPEI Bulletin (6), 20–31 (2006) (in Russian)

Development of Domain Model Based on SUMO Ontology

Bogumiła Hnatkowska, Zbigniew Huzar, Iwona Dubielewicz, and Lech Tuzinkiewicz

Wroclaw University of Technology, Wyb. Wyspianskiego 27, 50-370 Wrocław
{Bogumila.Hnatkowska,Zbigniew.Huzar,
Iwona.Dubielewicz,Lech.Tuzinkiewicz}@pwr.edu.pl

Abstract. Domain model is the first model in the scope of interests of software developers. Its creation, especially for complex domain, can be very costly and time consuming, as it extensively involves domain experts. On the other side, domain knowledge could be included in existing ontologies, and can be extracted, with the support of domain experts, from them. That way of knowledge extraction should lead to better model quality, preventing from misunderstandings between business analysts and domain experts. The paper presents an approach to business model development on the base of SUMO ontologies. The approach is explained with the use of simple, but real example. The results are promising. Domain models, created using this approach, could be perceived as valuable input for further development.

Keywords: software development, ontology, SUMO.

1 Introduction

The quality of software strongly depends on the quality of initial stages of its development process. Assuming the Model Driven Architecture (MDA) approach to software development, a high quality of business or domain models, historically called Computation Independent Model [8], is required. The domain model provides basic knowledge for a particular domain of discourse that is necessary for problem domain understanding and its presentation. Special role plays here description of the application domain – the workspace of the future software systems. The description relates to the actual people, places, things, and laws of domain.

The domain model provides a background for problem domain representation. A high quality domain model should be fully consistent and complete with respect to the domain of discourse. Basically, there are two main sources of knowledge on a given application domain: the knowledge of domain experts or the knowledge covered by domain ontologies. In couple of last years, ontologies have become popular in several fields of information technologies like software engineering and database design, especially in building the domain models.

There two possible roles for an ontology in domain model building. In the first role, the ontology is the main source of knowledge for derivation of the domain model. In the second role, the ontology is a base for validation of a domain model, which has been elaborated in other way.

© Springer International Publishing Switzerland 2015

W. Zamojski et al. (eds.), *Theory and Engineering of Complex Systems and Dependability,*
Advances in Intelligent Systems and Computing 365, DOI: 10.1007/978-3-319-19216-1_16

163

In further, we concentrate on the first role of ontology. By definition, ontologies represent an objective and agreed viewpoint of the universal domain of discourse, which is independent of possible applications, while business models have to represent a specific problem on the background of the domain of discourse. So, in the process of domain model building we have to extract from a given ontology only this knowledge that is relevant for problem domain solution.

There are many high-level ontologies currently developed like BFO [12], Cyc [1], GFO [3], SUMO [10], etc. The last one, SUMO, seems to be one of very promising because it became the basis for the development of many specific domain ontologies.

Let us remind that informally an ontology is defined in [14] as: *a formal (machine-readable), explicit (consensual knowledge) specification (concepts, properties, relations, functions, constraints, axioms are explicitly defined) of a shared conceptualization (abstract model of some phenomenon in the world)*.

Ontologies may be grouped into high-, middle-, and low-level of abstraction. Typically, a high-level ontology is the base for some middle-level ontologies, and these in turn for low-level or domain ontologies. The number of different ontologies, developed in last three decades, exceeds 10 thousand [2].

Practical use of ontologies for research and applications in search, linguistics and reasoning, etc. needs some level of formalization. There are many approaches to ontology representation and formalization. Most commonly, first order predicate language or its sublanguages are used. There are also languages designed specifically for ontologies, e.g. RDFS, OWL, KIF, and Common Logic [13].

The Suggested Upper Merged Ontology (SUMO) seems to be particularly interesting because of its properties (http://www.ieee.org). SUMO is a formal ontology based on concepts elaborated by [10]. A particular useful feature is that the notions of SUMO have formal definitions and at the same time are mapped to the WordNet lexicon (http://wordnet.princeton.edu). SUMO and related ontologies form the largest formal public ontology in existence today. The number of notions covered by the core of SUMO exceeds 1000 and by all related ontologies more then 20 000. The ontologies that extend SUMO are available under GNU General Public License.

The SUMO ontology is formally defined in declarative language SUO-KIF (Standard Upper Ontology Knowledge Interchange Format). SUO-KIF was intended primarily a first-order language, which is a good compromise between the computational demands of reasoning and richness of representation. Notions in SUMO are defined by set of axioms and rules expressed as SUO-KIF formulas. It is possible to understand the meaning of expressions in the language without appeal to an interpreter for manipulating those expressions.

In the paper, we discuss how SUMO ontology may be applied in domain model development. A domain model is aimed to describe an institution or a company in the way which enables software developers to understand the environment in which a future software system will be deployed and operated. The domain model delivers the rationale of how the organization creates, delivers, and captures value in economic or other contexts. The model should describe organization units, business processes and business rules concerning given organization. The domain model reflects a software engineering perspective. It is a view of a system from the computation independent viewpoint; it does not show details of the structure of systems; it uses a glossary that is familiar to the practitioners of the domain in question.

There are many languages for domain modeling, e.g. UML, BPMN, SysML. The UML seems to be the most popular and the most general modeling language in the entire cycle of software development. With UML we can represent both static and dynamic aspects of reality. In further, we restrict ourselves to the static aspect, which may be defined by two elements: a class diagram and an object diagram being an instance of this class diagram. The class diagram represents all possible configurations of the modeled domain, i.e. a set of objects and links among them, while the object diagram represents exactly one configuration – an initial configuration.

The aim of the paper is the presentation of a method how to derive a domain model for a software system provided that there is a suitable SUMO-like domain ontology. The derivation of the model is driven by software requirements expressed at general abstraction level in vision document.

In the Section 2 an approach to domain model construction is outlined. The approach is illustrated in the next Section where small fragment of a case study is presented. The last Section 4 presents some final remarks.

2   Outline of the Method

The starting point while developing software system is to gather stakeholders needs. On the base of these needs system requirements are defined and specified in a software system vision document. The system vision statement can be accompanied by a system scope description which sets the rough boundaries of what the constructed system will attempt to do.

In further considerations we assume that a system vision and ontology for the domain of interest are given. Additionally, we assume that this ontology is complete, i.e. it contains specification of all interesting entities in reality as well as relationships existing among them.

Basing on these assumptions we proceed with the construction of a domain model. The construction process consists of five subsequent activities:

1. Development of a set of initial notions (IN) - a glossary.
2. Mapping of the initial notions IN into an initial subset $IE \subseteq SUMO$ of all SUMO entities.
3. Extraction of a subset of SUMO entities RE and their definitions, which are related to the subset IE.
4. Transformations of SUMO extract, i.e. the set $IE \cup RE$ into a UML class diagram.
5. Refinement and refactoring of the class diagram.

The main problem in a domain model construction is the knowledge extraction from ontology which contains plenty of possible irrelevant notions. This problem is addressed by activities 2-3.

Further in the paper, each activity is described by its goal, input and output data, and process of transformation input into output. Examples of input and output documents (at least part of it) are presented in the case study. An activity is decomposed into few steps if the input data transformation is complex. There are also discussed extraordinary situations, if any, appearing while performing an activity.

2.1   Development of Initial Glossary

Input: software vision document (i.e. a concise description of user needs).
Output: initial domain glossary – a set of initial notions.
Transformation Process:

The investigation of the vision document is performed and elicitation of all nouns appearing in it is done; found nouns are candidates for being (business) notions as well as environment elements (e.g. user roles).

Note. It is the only way for finding notions as no business process description neither business domain glossary exist.

In the literature [4], [6] it can be met different structures of the vision document.

In this paper we assumed that this document together with the scope statement includes:

- Definition of the problem which the system is supposed to solve.
- Definition of the environment (actors) of the system.
- A list of future system features (i.e. generally expressed functional require-ments); these features describe the actions of the system and the domain enti-ties which are subjects of these actions.

Each feature is well-structured element of the form:

As an <actor> I want <an activity> [<for the purpose>]

2.2   Mapping of Initial Glossary into SUMO Notions

Input: the initial glossary (result of the previous step).
Output: SUMO entities equivalences for the glossary notions.
Transformation Process:

The goal of this activity is to find the set *IE* of SUMO entities which are semanti-cally equivalent to the glossary notions.

```
For each notion n∈ IN:
if there is a class entity c∈ SUMO which has the semantics of n
then c is included to IE
else if there is a role r at the end of a relation as which has
the semantics of n
then the role r and the relation with the classes at all ends of
this relation are included to IE
else there is no semantically equivalent entity in the ontology.
```

If there is not possible to represent a given notion by a single entity then a domain expert should define a transformation of the notion into a set of entities. If the trans-formation cannot be defined it means that the ontology is not complete and should be extended.

2.3   Extraction of Related Notions in SUMO

Input: the set *IE* of SUMO notions.

Output: a set *RE* – an extract from SUMO definitions that are in the scope of interest (represent business notions and relationships among them)

Transformation Process:

To extract some additional knowledge about the set of entities *IE*, the searching for the entities related to *IE* is performed.

```
For each e∈IE:
    if e is a SUMO relationship then
        SUMO definition of e and the definitions of the classes at
        all ends of e are accepted for further considerations;
    else if e is a SUMO class then
    begin
      compute the set SR(e) of all SUMO relationships in which e
      is involved;
     for each r∈SR(e):
       if r isn't in scope of interest than continue;
       else if r is defined in SUMO between e₁..eₖ∉IE but it holds
       between instances of eⱼ..eₘ∈IE and eⱼ..eₘ are subclasses of
       e₁..eₖ then
            rewrite SUMO relation in the form (r eⱼ .. eₘ)
       else
            accept this definition for further considerations;
    end
```

We take into consideration separately each SUMO notion from the initial set of domain notions *IN*. If it is a relationship, we extract its SUMO definition, especially domains.

If the SUMO notion is a class we read all the rules for the notion trying to identify static relationships the class is a part, e.g. a class is a subclass of some other class. For interesting relationships we extract their definitions. Sometimes, an interesting relationship is used in specific ontology, e.g. Hotels, but is defined in the upper ontology, e.g. merge. In such a case, we redefine the relationship, treating classes as instances, e.g. (properPart HotelRoom HotelBuilding).

2.4   Transformation of SUMO Extract into UML Class Diagram

Input: extract from SUMO – the set $IE \cup RE$

Output: UML class diagram

Transformation Process:

The goal of this activity is to transform, manually or automatically, the extract from SUMO into UML class diagram. We have proposed some transformation rules in [5]. It should be mentioned that some elements of UML diagram, e.g. multiplicities, are not easy to obtain from automatic transformation rules.

In table 1 we have presented selected transformation rules from SUMO to UML, later used in the case study.

Table 1. Examples of SUMO to UML transformation rules

Id	SUMO element	UML element	Rule description
1	class	Class	SUMO class is translated into UML class with the same name
2	subclass relation	generali- zation	subclass relation between 2 SUMO classes is translated into UML generalization
3	proper part relation	composition	*properPart* relation instance between 2 SUMO classes is translated into UML composition relation (with 1..* multiplicity)
4	binary predicate between 2 classes	association	SUMO binary predicate (when both ends are SUMO classes) is translated into UML association with the same name; if role end is described (e.g. in documentation of predicate) it is represented as UML association role name
5	binary predicate between 2 classes from which one is a simple type, e.g. Integer	Attribute	SUMO binary predicate (when one end is a class and the other is a primitive type, e.g. Integer) is translated into UML attribute in the class

2.5   Refinement and Refactoring of Class Diagram

Input: UML class diagram
Output: UML class diagram more appropriate for software development; the semantics of the input diagram should be preserved but the syntax, especially names of selected classes, can be changed. Multiplicities for associations should be defined.
Transformation Process:

The class diagram developed in the previous activity is a subject first to refinement, and next to refactoring. First of all the associations must be analysed – if they are necessary, and should be preserved, or if they could be deleted. As in SUMO there are no exact information about multipicities of entities being in the relation, the multiplicities of associations ends should be reviewed and corrected.

Other changes are under jurisdiction of an IT expert decision. They aim in improvements of the structure of the resulting class diagram, making it more readable (we would like a diagram to become a part of ubiquitous language shared among business experts and IT developers). There are many papers where UML refactorings are defined, e.g. [7], [11]. Refactorings are organized into groups, depending on the UML element which is a subject of refactoring, e.g. inheritance refactorings, association refactorings. Typical, and easy to define are rename refactorings, when we change a name of a class, class attribute or operation.

The reason which stands for this activity is to achieve high quality of our UML model i.e. to fulfill 6C quality goals [9]. Hence, especially the model semantic correctness (with business reality), comprehensibility and confinement (e.g. removing irrelevant entities/ relations/attributes) should be guaranteed. The model characteristic completeness is ensured (as the ontology is complete) and the consistency and changebility are also guaranteed (by construction).

3 Case Study

A domain of our interest is a hotel, and hotel management process. The input of our method is a system vision. Below, we present only a small part (2 user stories) of it.

3.1 Development of Initial Glossary

Input:

1. As a *potential customer*, I want to see information about *hotel*, *hotel rooms*, *rooms' amenities* and *prices* so that to be able to decide if to become a *customer*.
2. As a *potential customer*, I want to check availability of selected room (*room availability*) in a given *period* so that to be able to decide if to make *reservation* or not.

Output: initial glossary: potential customer, hotel, hotel room, room amenity, room price, customer, room availability, period, reservation

3.2 Mapping of Initial Glossary into SUMO Notions

The result of the second activity is presented in the table 2. For selected notions SUMO definition is also presented.

Table 2. Results of activity 1

Vision term (input)	Sumo term (output)	Type	Sumo definition
Potential customer	Potential customer	Rel. end	"(potentialCustomer ?CUST ?AGENT) means that it is a possibility for ?CUST to participate in a financial transaction with ?AGENT in exchange for goods or services"
Hotel	Hotel building	Class	"A residential building which provides temporary accommodations to guests in exchange for money."
Hotel room	Hotel room	Class	"hotel room refers to a room that is part of a hotel (building) that serves as a temporary residence for travelers"
Room amenity	Room amenity	Rel. end	"(room amenity ?ROOM ?PHYS) means that traveler accommodation provides physical ?PHYS in hotel unit ?ROOM"
Room price	Price	Rel. end	"(price ?Obj ?Money ?Agent) means that ?Agent pays the amount of money ?Money for ?Obj."
Customer	Customer	Rel. end	"A very general relation that exists whenever there is a financial transaction between the two agents such that the first is the destination of the financial transaction and the second is the agent."
Room availability, Period	Reservation Start, Reservation End	Rel.	(reservationStart ?TIME ?RESERVE) means that the use of a resource or consumption of a service which is the object of ?RESERVE starts at ?TIME")
Reservation	Hotel reservation	Class	"hotel reservation refers to a reservation specifically for traveler accommodation stays"

3.3   Extraction of Related Notions in SUMO

The result of the third activity is an extract from SUMO hotel ontology. Some parts are presented below:

```
(domain potentialCustomer 1 CognitiveAgent)
(domain potentialCustomer 2 Agent)
(instance potentialCustomer BinaryPredicate)
(properPart HotelRoom HotelBuilding)
(subclass HotelRoom HotelUnit)
(subclass HotelRoomAttribute
    RelationalAttribute)
(attribute HotelUnit HotelRoomAttribute)
(instance StandardRoom HotelRoomAttribute)
(instance SuiteRoom HotelRoomAttribute)
(instance DeluxeRoom HotelRoomAttribute)
```

We selected representative examples of business knowledge extractions from SUMO ontology to present how the process is performed. In the *HotelBuilding* definition we can find following rule:

```
(=>
(instance ?HOTEL HotelBuilding)
(exists (?ROOM)
 (and
  (instance ?ROOM HotelRoom)
  (properPart ?ROOM ?HOTEL))))
```

The rule says that there exists *properPart* relation between *HotelBuilding* and *HotelRoom*. In SUMO *properPart* is defined at very general level, as a spatial relation between two *Objects*. So, for the purpose of further transformation (SUMO to UML) we "rewrite" this relation to (the instance definition):

```
(properPart HotelRoom HotelBuilding)
```

The semantics of *properPart* relation is stronger, and we know that an instance of *HotelBuilding* must have at least 1 instance of *HotelRoom*.

3.4   Transformation of SUMO Extract into UML Class Diagram

The result of the forth activity is presented in the Fig 1. The process was done manually, however, the exact transformation rules were defined by us (examples of transformation rules were presented in Chapter 2) and applied here.

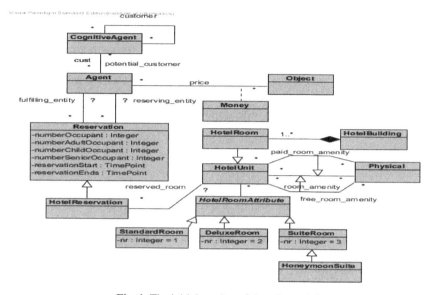

Fig. 1. The initial version of domain model

3.5   Refinement and Refactoring of Class Diagram

The result of the fifth activity is presented in the Fig 2. Within refinement part we defined missed multiplicites. We also changed the way *price* relationship is modelled.

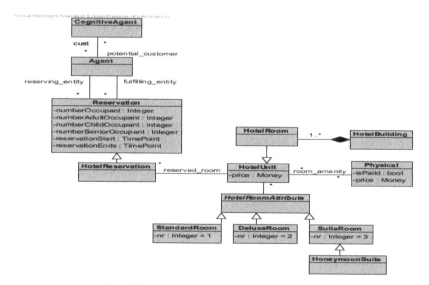

Fig. 2. The refactored version of domain model

Oryginally it was a relationship between *Agent* paying for *Object*. We had to decide what elements in our diagram have price (must be subclasses of Object in SUMO), and decided that are *HotelUnit*, and *Physical* (being a room amenity). Information about agent paying the money now is neglected. We also removed customer relationship as being out of scope of our interests. Within refactoring part we removed two associations (i.e. *paid_room_amenity, and free_room_amenity*) using appropriate attribute (*isPaid*) instead in *Physical* class.

4   Conclusions

The purpose of a business model is to give representation of a problem domain omitting implementation issues. Business modeling can benefit from the use of semantic information represented by ontology. Performing analysis at the business level in the context of ontology reduces the risk of incorrect interpretation of a problem domain by IT world.

To utilize this opportunity we have proposed the procedure which in systematic way, starting with a system vision as an input, creates a domain model (UML class diagram) as its output. The creation process consists of five activities among which the most important are two: mapping of glossary notions into SUMO notions (2nd activity) and extraction of domain-related notions in SUMO (3^d activity).

There are the following benefits while using ontology in business modeling:

- Ontology helps in identifying and understanding the relevant elements in a specific domain and the relationships between them.
- Ontology supports easier communication and shared understanding of business domain notions among stakeholders.
- When treated as a reference model of domain ontology enables to validate an existing domain model, if any exists.

On the other hand we observed that the activity of business modeling based on knowledge represented in ontology is a difficult and time-consuming. There are the following reasons for that:

- problems in finding corresponding notions in a given domain and ontology if the names are the same but their semantics are different or if the names are different but their semantics are the same;
- difficulties in identifying and retrieving properties of classes and relationships from ontology in the case when ontology is specified at different levels of abstractions;
- different types of logics used to knowledge representation in SUMO and UML;
- necessity to gain skills in retrieving and interpreting knowledge from ontology.

The use of ontologies in the design of information systems creates new opportunities for improving the quality of the artifacts, and may reduce the cost of software development. Our experience shows that the key issue is the acquisition of domain knowledge from the ontology. This is due to the size and complexity of the ontology.

The mapping problem of domain concepts into a set of terms of the ontology motivates to address this problem. Therefore, our future works will be devoted to the development of detailed rules of mapping the concepts and further, analysis of the capabilities of automation of this process.

References

1. Foxvog, D.: Computer Applications of Ontology: Computer Applications. In: Poli, et al. (eds.), pp. 259–278 (2010)
2. Garbacz, P., Trybuz, R.: Ontologies behind ontology. Wydawnictwo KUL, Lublin (2012) (in Polish)
3. Herre, H.: General Formal Ontology: A Foundational Ontology for Conceptual Modelling. In: Poli, et al. (eds.), pp. 297–345 (2010)
4. Highsmith, J.: Agile Project Management—Creating Innovative Products. Addison- Wesley, New York (2010)
5. Hnatkowska, B., Huzar, Z., Dubielewicz, I., Tuzinkiewicz, L.: Problems of SUMO-like ontology usage in domain modelling. In: Nguyen, N.T., Attachoo, B., Trawiński, B., Somboonviwat, K. (eds.) ACIIDS 2014, Part I. LNCS, vol. 8397, pp. 352–363. Springer, Heidelberg (2014)
6. Leffingwell, D., Widrig, D.: Managing Softwaer Requiremetns. Addison Wesley (1999)
7. Massoni, T., Gheyi, R., Borba, P.: Formal Refactoring for UML Class Diagrams. In: 19th Brasilian Symposium on Software Engineering, pp. 152–167 (2005)
8. Model Driven Architecture (MDA), MDA Guide rev. 2.0, Object Management Group, OMG Document ormsc/2014-06-01 (2014) (access date: January 2015)
9. Mohagheghi, P., Dehlen, V., Neple, T.: Definitions and approaches to model quality in model-based software development – A review of literature. In: Information and Software Technology, pp. 1646–1669 (2009)
10. Niles, I., Pease, A.: Origins of the IEEE Standard Upper Ontology. Working Notes of the IJCAI 2001 Workshop on the IEEE Standard Upper Ontology, pp. 4–10 (2001)
11. Pereira, C., Favre, L., Martinez, L.: Refactoring UML Class Diagram. In: Khosrow-Pour, M. (ed.) Innovations Through Information Technology. Information Resources Management Association, USA (2004), http://www.irma-international.org/viewtitle/32412/ (access date: January 2015)
12. Spear, A.D.: Ontology for the twenty first century. Technical Report 2006, Institute für Formale Ontologie und Medizinische Informationswissenschaft, http://ifomis.uni-saarland.de/bfo/documents/manual.pdf (access date: January 2015)
13. Staab, S., Studer, R. (eds.): Handook of Ontologies. Springer (2004)
14. Studer, R., Benjamins, V., Fensel, D.: Knowledge Engineering: Principles and Methods. Data and Knowledge Engineering 25, 161–197 (1998)

Verification of UML Class Diagrams against Business Rules Written in Natural Language

Bogumiła Hnatkowska and Piotr Mazurek

Wroclaw University of Technology, Wyb. Wyspiańskiego 27, 50-370 Wrocław
bogumila.hnatkowska@pwr.edu.pl,
mazipiotrek@gmail.com

Abstract. Business rules are an important part of requirement specification, and an essential input for software analysis and design. Usually, at the beginning, they are expressed in natural language, which is later translated by a business analyst to a more formal representation, e.g. UML diagrams. The translation process is error prone because business analysts can misinterpret or omit informally expressed business rules. The aim of the paper is to present an approach to automatic verification of UML class diagrams against business rules, expressed in a semi natural language, i.e. SBVRSE. The proposed approach has been implemented as a tool, and tested on representative examples. At that moment it supports structural business rules. In the future the method will be extended to cover also other types of business rules.

Keywords: business rules, SBVR, SBVRSE, UML, class diagrams.

1 Introduction

Business rules are an important part of requirement specification, especially for enterprise systems. They must be taken into account during all further stages of software development, i.e. analysis, design, implementation and tests. Usually, at the early stages of software development, business rules are defined by business analysts in natural language, which is understood by all involved parties, e.g. business experts, programmers, testers. Later, business rules can be translated (manually or automatically) into more formal representations, e.g. UML diagrams. In the case when the transformation process is done manually, there is a need of verification if the output of transformation is still consistent with its input, i.e. business rules expressed in natural language. Typically, the number of business rules is very large, what makes the verification process time consuming. It is reasonable to consider to what extend and how the verification process could be done automatically.

The aim of the paper is to present an approach to automatic verification of consistency of UML class diagrams with business rules written in natural language. In order to make this process feasible SBVRSE [1] notation was selected to represent business rules. At that moment the research covers only structural rules, which are represented by UML class diagrams. The proposed approach has been implemented in a tool, and tested with representative examples. The tool served mainly as a proof of concept of the proposed approach to UML model verification.

© Springer International Publishing Switzerland 2015   175
W. Zamojski et al. (eds.), *Theory and Engineering of Complex Systems and Dependability,*
Advances in Intelligent Systems and Computing 365, DOI: 10.1007/978-3-319-19216-1_17

The paper is organized as follows. Section 2 provides a definition of business rule, and presents a selected classification of business rules. Section 3 presents business rules in the context of MDA, and shortly presents SBVRSE notation and supporting tools. Section 4 considers how different kinds of structural business rules can be mapped to UML class diagram. Section 5 describes the proposed approach to verification of UML class diagram against business rules and the implemented tool for that purpose. Section 6 brings a simple case study. Last Section 7 concludes the paper and presents it in the wider context.

2   Classification of Business Rules

According to [2], the notion of business rule can be defined from two different perspectives: (a) from the business perspective, and (b) from the information system perspective. We are interested in the later perspective, in which "a business rule is a statement that defines or constraints some aspect of the business. It is intended to assert business structure, or to control or influence the behaviour of the business" [2].

Many different classifications of business rules exist in the literature, e.g. [2-4]. We decided to use the classification proposed by [1]. It covers all types of rules we can find in other classification, is commonly accepted and often referenced. This classification splits the business rules into 3 categories which are further divided into subcategories. The presentation below lists all kinds of business rules, and gives short explanations for those we have taken into consideration.

Category 1: Structural assertions – describe static structure of the business
- Business terms – elements of business glossary that need to be defined, e.g. invoice
- Common terms – elements of business glossary with commonly known meaning, e.g. car
- Facts – relationships between terms:
    o   Attribute – a feature of a given term, e.g. colour, name
    o   Generalization – represents "as-is" relationship, e.g. car is a specific case of vehicle
    o   Participation – represents semantic dependency between terms, e.g. student *enrols* for courses
        ▪   Aggregation – represents "a whole-part" relationship, e.g. book consists of pages
        ▪   Role – describes way in which one term may serve as an actor, e.g. customer may be *a buyer in* a contract
        ▪   Association – used when other kinds of participation are inappropriate

Category 2: Action assertions – concern some dynamic aspects of the business
- Conditions (excluded from further consideration)
- Integrity constraints – represent invariants that always must be true
- Authorizations (excluded from further consideration)

Category 3: Derivations – represent derived rules; include mathematical calculations, and inference rules (excluded from further consideration)

3   Business Rules within MDA

Business rules can be considered at different levels of abstraction which strongly influence the way they are represented. In software engineering the abstraction levels can be defined in the context of Model Driven Architecture, shortly MDA [5].

MDA is an approach to software development in which models play the key role. It introduces three kinds of models:

(a) Business or domain model (former Computational Independed Model, CIM) – "models of the actual people, places, things, and laws of a domain. The "instances" of these models are "real things", not representations of those things" [5]; business rules at that level are usually expressed in natural language.

(b) Platform Independent Model (PIM) – presents the internal content of the targeted system (from white-box perspective) in a manner that abstracts from specific information of a selected platform, libraries, tools; business rules at that level can be expressed with the use of any formal or semi-formal notations, e.g. UML.

(c) Platform Specific Model (PSM) – similarly to the former model presents the content of the targeted system but in a platform depended way; e.g. that model can show how to achieve persistency or security with the use of existing platform support; business rules at that level can be expressed in dedicated languages, e.g. programming languages (e.g. java, c#), languages used in business rule engines, relational database schema definitions.

The paper focuses on business rules expressed at domain and PIM levels. As was mentioned above, at CIM level business rules are usually written in natural language. To make the rules more readable and consistent business analysts can apply specific guidelines, e.g. RuleSpeak [6], polish translation of RuleSpeak [7]. However, RuleSpeak recommendations are not supported by any tools, so it seems, that using more restrictive standards, like SBVR is a better choice.

SBVR (Semantics of Business Vocabulary and Business Rules) is the OMG standard issued in 2008 [1]. On one side it allows to describe business rules in a way which is platform independent and can be processed by computers, on the other it separates the meaning of defined elements from their representation what enables to use different notations to express the same thing. One of such accepted representation is SBVR Structured English (SBVRSE) which enables expressing business rules in controlled natural English.

SBVR provides means for defining business glossaries and business rules. Any entry in business glossary can have many synonyms.

Business rules in SBVR are classified into four groups [1]: (a) structural business rules, (b) operative business rules, (c) permissions, and (d) possibilities.

In further we limit our interests mainly to structural business rules which are divided into:

- necessary statements (e.g. it is necessary that each student is enrolled to at least 3 courses),
- impossibility statements (e.g. it is impossible that the same person is studying at more than 2 universities at the same time), and

- restricted possibility statements (e.g. it is possible that a student retakes an exam only if he/she failed the exam on the first date).

We exclude from our consideration impossibility statements as they cannot be represented on class diagrams directly (but only with the use of OCL [8]).

Additionally, we partially consider permissions, i.e. permission statements (e.g. it is permitted that a customer requests a loan).

Defining business rules in SBVRSE is supported by many tools from which freely available are e.g. [9], UML2SBVR [10], VeTiS [11]. We found the last especially useful. This tool works as a plug-in to MagicDraw tool. It supports defining glossaries and business rules, and next their transformation to UML [12] class diagram with OCL constraints [8].

4   UML Class Diagram as Business Rules Representation

Before we are able to define verification rules we need to have an idea how different types of business rules can be represented on class diagrams. Table 1 presents proposed mappings between business rules taken into consideration and the elements of UML class diagram.

Table 1. Mapping between business rules' types and elements of UML class diagram

Business rule type	Business rule subtype		UML possible representation
Fact	Attribute		Attribute in a class
	Generalization		Generalization relationship
	Partici-pation	Association	Association
			Association class
		Aggregation	Aggregation
			Composition
		Role	Role
Term	Business term, common term		Class, Enumeration
Action assertion	Integrity constraint		Multiplicity
			Class constraints, e.g.{unique},{sorted}
			Qualifier
Permission	Permission statement		Operation in a class

Most of the mapping rules presented in Table 1 was also implemented in VeTIS tool [11]. However, the tool introduces some constraints on the form of SBVRSE which is supported. First of all business analyst can define only binary relationships. Next, terms (glossary entries) are disallowed to contain spaces. A business analyst is forced to use specific key words in a specific context, e.g. "has" or "is_property_of" to represent attributes ("name is_property_of person").

5   Proposed Approach

We would like to verify if an UML class diagram is consistent with a set of business rules expressed in SBVRSE. We limit our consideration mainly to structural business rules with exception of integrity business rules that belong to action assertions. Because we were interested in building a proof of concept, in other words we would like to verify that the method of UML model verification against business rules has the potential to be used, we decided to use VeTIS tool not only to define business rules in SBVRSE, but also to generate a class diagram being the representation of business rules. As the result, our task has become a task of comparison of two UML class diagrams, from which one serves as an object to be checked, and the second serves as a pattern (represents business rules).

We have identified and implemented several verification rules that check if any of business rules (an element of the pattern class diagram) was omitted on the original class diagram. Table 2 contains informal definition of the defined verification rule set.

Table 2. Verification rule set – informal definition

ID	Pattern diagram element	Original diagram element
R01	Class	Corresponding (corr.) class with identical or synonymous name
R02	Attribute in a class	In corr. class exists attribute with identical name
R02.1	Attribute with defined type	Corr. attribute has defined identical type
R03	Operation in a class	In corr. class exists operation with identical name
R03.1	Operation with defined returned type	Corr. operation has defined identical return type
R03.2	Operation with input parameters	Corr. operation has identical types of input parameters
R04	Binary generalization relation	Between corr. classes exists binary generalization relationship
R04.1	Binary generalization relation with generalization set	Corr. binary generalization relation has defined identical generalization set
R05	Binary composition relation	Between corr. classes exists binary composition relation
R05.1	Binary composition relation with defined name	Corr. binary composition relation has identical name
R05.2	Binary composition relation with defined roles	Corr. binary composition relation has identical roles names
R05.3	Binary composition relation with defined multiplicity	Corr. binary composition relation has identical multiplicity
R06	Binary association relation	Between corr. classes exists binary association relation
R06.1	Binary association relation with defined name	Corr. binary association relation has identical name
R06.2	Binary association relation with defined roles	Corr. binary association relation has identical roles names
R06.3	Binary association relation with defined multiplicity	Corr. binary association relation has identical multiplicity

The checks of the equality of class names on both diagrams are case insensitive. From multi-parts names white signs as well as other separators (e.g. "_", "-") are eliminated.

Below, the algorithm of R01 is presented:

1. Get a list of classes from the pattern class diagram
2. For each class C_i:
   2.1. Get a name N of C_i class
   2.2. Check if the original class diagram contains a class C_j with N name
   2.3a. If C_j exists
       2.3a.1. Remember a pair (C_i, C_j)
       2.3a.2. R01 is fulfilled for C_i and C_j
   2.3b. Otherwise
       2.3b.1. Check if the original class diagram contains a class C_j with K name where K is a synonym for N
       2.3b.1a: If C_j exits
           2.3b.1a.1. Remember a pair (C_i, C_j)
           2.3b.1a.2. R1 is fulfilled for C_i and Cj
       2.3b.1b: Otherwise
           2.3b.1b.1. Rule R1 is broken for C_i class

The verification result W is calculated according to the formula (1):

$$W = \frac{\sum_{i=1}^{N} i_i \cdot w_i}{\sum_{i=1}^{N} p_i \cdot w_i} \cdot 100\,\% \tag{1}$$

where:

N – the total number of verification rules
p_i – the total number of elements in the pattern class diagram addressed by i rule
i_i – the total number of elements in the original class diagram (being verified) addressed by i rule
w_i – the weight of rule i (can be set by a user)

At that moment checks are one-directional, i.e. we check in all business rules are presented on the original class diagram, but the original class diagram can have additional elements.

The implemented verification tool is called *Verifica*. Figure 1 presents its settings screen. A user is asked for a path to a file with verification rule set (it contains error messages as well as weights for particular rules). Additionally, the user can decide if not to use or use synonyms, and from which source (SBVR vocabulary or WorldNet).

On the main screen a user is asked for two basic inputs: a path for an XMI file with a pattern class diagram, and a path for an XMI file with an original class diagram (to be checked against business rules).

The application prepares a report with verification results in html format. An example of it (small part) is presented in Figure 5. The report consists of several parts – the introduction, general summary (where W measure is calculated), and summary part for every rule.

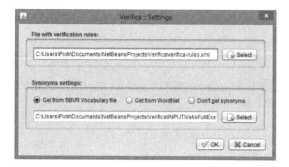

Fig. 1. *Verifica* – settings screen

6   Case Study

This chapter presents an example of application of proposed verification method. We start the presentation with a specification of business rules expressed in SBVRSE. Next, we show how these rules are transformed to UML class diagram by VeTIS tool. After that we present a class diagram acting as original class diagram, which consistency with the set of business rules we want to check. The diagram contains several inconsistencies, intentionally introduced by us. At the end we present interesting parts of generated report showing inconsistencies discovered by *Verifica* tool.

The listing below presents business vocabulary (terms, and facts), and business rules for our example in SBVRSE.

Business vocabulary contains:

- 6 terms (2 with synonyms): customer (synonym client), loan (synonym credit), instant_loan, regular_loan, name, amount
- 4 facts, and 2 generalizations: customer has name, loan has amount, customer gets loan, customer request loan, instant loan inherits from loan, regular loan inherits from loan

We have defined 2 business rules on the base on terms and facts. The first is permission statement, and the second – integrity constraint.

```
customer
        Synonym: client
name
General_concept: text
customer has name
loan
        Synonym: credit
instant_loan
        General_concept: loan
regular_loan
        General_concept: loan
amount
        General_concept: integer
loan has amount
```

```
customer gets loan
customer requests loan
It is permitted that a customer requests a loan.
It is possible that a customer gets at most 3 loan.
```

The class diagram, produced by VeTIS tool on the base of SBVRSE specification is shown on Figure 2 (a).

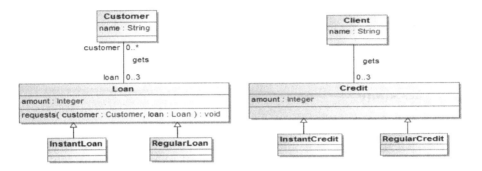

Fig. 2. (a) The pattern class diagram with business rules; (b) Original class diagram – the subject of verification

Figure 2 (b) presents a class diagram assumed to be build by business analyst which consistency with the set of business rules we would like to check.

This diagram differs from the pattern class diagram with the following elements:

- lack of multiplicity at *Client* class side
- lack of operation in *Client* class
- lack of roles at the ends of *gets* association
- wrong name of classes being descendants of *Loan* class

The report prepared by *Verifica* tool says that 51% of the original class diagram elements is consistent with defined set of business rules. Figure 3. presents a part of the generated html report.

UML Class Mapping

Pattern Class Name	Input Class Name
Customer	Client
Loan	Credit
InstantLoan	Corresponding class not found
RegularLoan	Corresponding class not found

UML Operations Mapping

Pattern Class Name	Pattern Operation Name	Pattern Operation Return Type	Pattern Operation Parameters Types	Input Class Name	Input Operation Name	Input Operation Return Type	Input Operation Parameters Types
Loan	requests	void	Customer Loan	Credit	Corresponding operation not found		

Fig. 3. Verification report – selected parts

7   Conclusions

The paper presents an approach to automatic verification of UML class diagram against mainly structural business rules. This is the first step in solving more general problem of verification of UML models against business rules expressed in semi natural language, e.g. SBVRSE.

The proof-of-concept implementation brought very promising results and showed that the solution of the general problem is feasible. At that moment a user is able to include/exclude verification rules from the rule set, and to decide about the weights of particular rules.

We did not found any research that addresses the problem of verification of UML diagrams against business rules expressed in SBVR. There are some works with similar subjects, e.g. [10] or [13], [14]. The former presents a tool called UMLtoSCP, and the verification is a side effect of UML to SCP transformation. The tool verifies UML class diagram against OCL constraints. The two later papers propose equivalence rules between two class-diagrams, and the possible transformations between equivalent diagrams. They are interesting for the approach in which business rules expressed in SBVR are first translated to a class diagram, and next used for verification purposes. At that moment, our implementation follows the same idea, but we did not consider the rules proposed by Gogolla.

In the future we plan to extend the scope of consideration to other UML diagrams and operational business rules. We are also considering either to resign from SVBRSE to UML class diagram transformation and to process directly SBVR sentences or to extend the internal representation of business rules (other UML diagrams) to express other kind of business rules.

References

1. OMG, Semantics of Business vocabulary Rules (SBVR), Version 1.2 (2013), http://www.omg.org/spec/SBVR/1.2/PDF (access date: January 2015)
2. Business Rules Group, "Defining Business Rules - What Are They Really?" (2001), http://www.businessrulesgroup.org/first_paper/br01c1.htm (access date: January 2015)
3. Von Halle, B.: Business Rules Applied: Building Better Systems Using the Business Rules Approach. John Wiley & Sons (2002)
4. Wiegers, K.: Software Requirements. Microsoft Press (2003)
5. OMG, MDA Guide rev. 2.0 (2014), http://www.omg.org/cgi-bin/doc?ormsc/14-06-01 (access date: 2014)
6. Ross, R.G.: RuleSpeak® Sentence Forms: Specifying Natural-Language Business Rules in English. Business Rules Journal 10(4) (2009), http://www.CRComunity.com/a2009/b472.html (access date: January 2015)
7. Hnatkowska, B., Walkowiak, A., Kasprzyk, A.: (2014), http://www.rulespeak.com/pl/ (access date: January 2015)
8. OMG, Object Constraint Language, Version 2.4 (2014), http://www.omg.org/spec/OCL/2.4/ (access date: January 2015)

9. Maurizio, D.T., Pierpaolo, C.: SVeaVeR business modeler editor (2006),
   `http://sbeaver.sourceforge.net/` (access date: January 2015)
10. Cabot, J., Pau, R., Raventós, R.: From UML/OCL to SBVR Specifications: a Challenging Transformation (2010), `http://www.sciencedirect.com/science/article/pii/S030643790800094X` (access date: December 2014)
11. Kaunas University of Technology (KUT), CreatingUML&OCL Models from SBVR Business Vocabularies Business Rules. VeTIS User Guide (2009),
    `http://www.magicdraw.com/files/manuals/VeTISUserGuide.pdf`
    (access date: January 2015)
12. OMG, Unified Modeling Language, Superstructure Version 2.4.1 (2011),
    `http://www.omg.org/spec/UML/2.4.1/Superstructure/PDF/`
13. Gogolla, M., Richters, M.: Equivalence Rules for UML Class Diagrams. In: UML 1998 – Beyond the Notation. First International Workshop, Mulhouse, France, pp. 87–96 (1998)
14. Gogolla, M., Richters, M.: Transformation Rules for UML Class Diagrams. In: Bézivin, J., Muller, P.-A. (eds.) UML 1998. LNCS, vol. 1618, pp. 92–106. Springer, Heidelberg (1999)

Increased Safety of Data Transmission for "Smart" Applications in the Intelligent Transport Systems

Sergey Kamenchenko and Alexander Grakovski

Transport and Telecommunication Institute, Riga, Latvia, Lomonosova 1, LV 1019
freeon@inbox.lv, avg@tsi.lv

Abstract. The problem of measuring vehicle's weight-in-motion (WIM) is one of the most important research topics in the field of transport telematics. It is important not only for development of intelligent systems used for planning and cargo fleet managing, but also for control of the legal use of transport infrastructure, for road surface protection from early destruction and for safety support on roads. Data protection plays one of the crucial role in such kind of systems as data transmitted over internet network can be not only intercepted and disclosed, but also rigged and be used as a tool for attack on your equipment or system. Traditional data protection methods are increasingly becoming an easier barrier for implementing a successful hacker attack, but time for breaking existing encryption algorithms, which recently have had a high cryptographic strength, is gradually decreasing thus opening the door for developers to create new or upgrading existing encryption algorithms, whose characteristics will be able to withstand modern hacker's threats.

Keywords: WIM, fibre-optic pressure sensors, involutory matrix, matrix ecryption.

1 Introduction

The worldwide problems and costs associated with the road vehicles overloaded axles are being tackled with the introduction of the new weigh-in-motion (WIM) technologies. WIM offers a fast and accurate measurement of the actual weights of the trucks when entering and leaving the road infrastructure facilities. Unlike the static weighbridges, WIM systems are capable of measuring vehicles travelling at a reduced or normal traffic speeds and do not require the vehicle to come to a stop. This makes the weighing process more efficient, and in the case of the commercial vehicle allows the trucks under the weight limit to bypass the enforcement.

The fibre optic weight sensor is the cable consisting of a photoconductive polymer fibres coated with a thin light-reflective layer [1]. A light conductor is created in such a way that the light cannot escape. If one directs a beam of light to one end of the cable, it will come out from the other end and in this case the cable can be twisted in any manner. To measure the force acting on the cable, the amplitude technology is more appropriated for the measurements based on measuring of the optical path intensity, which changes while pushing on the light conductor along its points.

© Springer International Publishing Switzerland 2015
W. Zamojski et al. (eds.), *Theory and Engineering of Complex Systems and Dependability,*
Advances in Intelligent Systems and Computing 365, DOI: 10.1007/978-3-319-19216-1_18

Fibre optic load-measuring cables are placed in the gap across the road, filled with resilient rubber (Fig. 1). The gap width is 30 mm. Since the sensor width is smaller than the tyre footprint on the surface, the sensor takes only part of the axle weight. The Area method [2] is used in the existing system to calculate the total weight of the axle. The following formula is used to calculate the total weight of the axis using the basic method [3]:

$$W_{ha} = \int_{t_{end}}^{t_{front}} (A_t(t) \cdot P_t(t)) dt \tag{1}$$

where W_{ha} – weight on half-axle, $A_t(t)$ – dynamic area of the tyre footprint, $P_t(t) \sim V(t)$ – air pressure inside the tyre and, according to Newton's 3rd law, it is proportional to the axle weight.

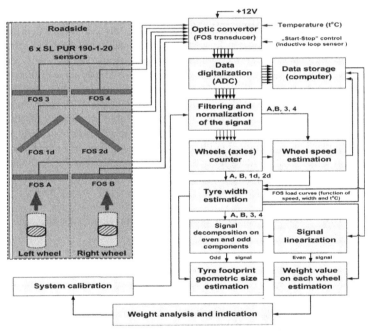

Fig. 1. Fibre optic sensors position against the wheel and tyre footprint and the algorithm of the weight in-motion measurement station

At these points the deflection of a light conductor and reflective coating occurs, that is why the conditions of light reflection inside are changed, and some of it escapes. The greater the load the less light comes from the second end of the light conductor. Therefore the sensor has the unusual characteristic for those, familiar with the strain gauges: the greater the load the lower the output is.

As we can see the exact values of the formula (1) factors are unknown. The area of the tyre footprint is calculated roughly by the length of the output voltage impulse, which, in its turn, depends on the vehicle speed. The Area Method uses the assumption that the area under the recorded impulse curve line, in other words – the integral, characterizes the load on the axle. To calculate the integral, the curve line is approximated by the trapezoid. In this case the smaller the integral – the greater the load. This method does not require knowing the tyre pressure, but it requires the time-consuming on-site calibration.

2   Tyre Footprint and Weight Estimation

There was the set of measurement experiments with the roadside FOS sensors on April, 2012 in Riga, Latvia [3-4]. Loaded truck was preliminary weighed on the weighbridge with the accuracy < 1%. The output signals from FOS sensors for truck speeds 70 km/h and 90 km/h are demonstrated on Figure 2.

It is evident that the signals for the different speeds have been changing by amplitude and the proportion of amplitudes does not fit the axle weights (Fig.2).

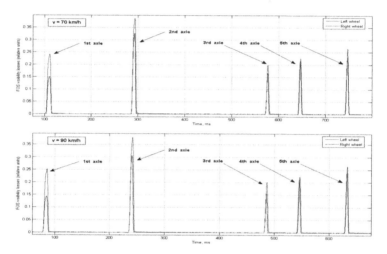

Fig. 2. Examples of FOS signals from A and B sensors (Fig.1) for vehicle's speeds 70 km/h and 90km/h respectively

The reason of this behaviour may be explained by FOS properties such as weight (pressure) distribution along the sensor length as well as sensor non-linearity and temperature dependence [4].

3   Vehicle Speed and Tire Contact Width Evaluation

Using FOS A (FOS B) and FOS 1 (FOS 2) symmetric signals, which are shown in Figure 3, it is possible to calculate the speed of each axle, also the truck speed by calculating the average of the values found before. In order to do this, it is necessary to normalize the signals, filter out the noise and obtain symmetrical signal components. Then impulse peak time value of these components will be used in the axle speed calculation. Distance between FOS A and FOS 1 (or FOS B and FOS 2) should be known in advance; in our case it is equal to 3 m (see Figure 1).

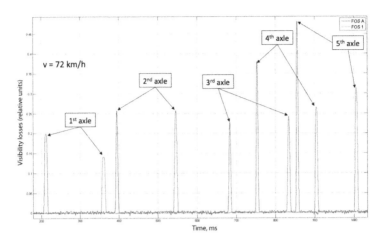

Fig. 3. FOS vertical weight component (symmetric) of s1_A, B, 3, 4_70km_27_09_2013 signal

Calculated axle and vehicle speeds, based on the FOS signal peak time of symmetric components, are shown in Table 1.

Table 1. Calculated speed values of s1_A, B, 3, 4_70km_27_09_2013 signal

Speed/axle	1staxle	2staxle	3staxle	4staxle	5staxle	Vehicle
Calculated speed [km/h]	72.34	72.00	71.63	71.56	71.93	71.89

Using FOS 1d and FOS A (or FOS 1), which are shown in Figure 4(a), as well as the symmetric FOS pair signals, it is possible to evaluate left and right tyre footprint widths. In order to do this, it is necessary to normalize the signals, filter out the noise and make linearization of the signals according to the pre-calculated axial velocity and temperature of the FOS.

Then pulse widths of perpendicular and diagonal FOS (see Figure 4(b)) are measured on experimentally chosen level of 0.4, multiplying this width subtraction by corresponding axle speed will be the evaluation of tyre footprint.

Table 2. Evaluated tyre footprint width of s1_A, B, 1d, 2d_90km_27_09_2013 signal

Parameter/axle	1staxle	2staxle*	3staxle	4staxle	5staxle
Footprint width [mm]	315	680	385	385	385
Evaluated footprint width [mm]	310.317	890.633	399.993	375.202	387.925
Error [%]	-1.487	30.975	3.894	-2.545	0.499

* - dual wheels (the distance between two neighbour dual wheels approximately is 40-50 mm and it cannot be measured exactly)

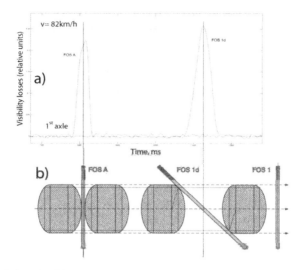

Fig. 4. (a) FOS A and FOS 1d filtered s1_A, B, 1d, 2d_90km_27_09_2013 signal; (b) Tyre footprint interaction with FOS

4 Intelligent Transport System Security Issues

Intelligent transport systems, such as WIM, smart control of traffic light and others mainly are based on the following structure (Figure 5).

Intelligent transport systems are using wireless or wired network connection to transmit data between nodes, which is the main gap in data protection because transmitted data are vulnerable for man-in-the-middle attack [6]. The problem of data security transmission through internet network is not unique, and there are technologies to transfer data over internet network in encrypted form. Virtual private network, which is based on cryptographic protocols such as IPSEC and TLS/SSL, is one of those technologies. Cryptographic protocols IPSEC and TLS/SSL are using AES algorithm for data encryption.

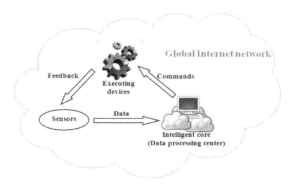

Fig. 5. Main structure of intelligent transport systems

AES is based on Rijndael algorithm, which is using 128 bits block size and variable key length of 128, 192 or 256 bits. AES was announced by NSA as a new encryption algorithm in 2000 and after this time the algorithm has become as a target for cryptanalysts from around the world (table 3).

Table 3. The history of breaking encryption algorithm DES and AES-192

Encryption algorithms	Build	Hacked	Rounds	Method	Data	Time	Memory	Reference
DES	1970	1998	16	Brute-force	-	3.5 hours	-	[7]
AES-192	1997	2000	7	Square-functional	2^{32}	2^{140}	2^{84}	[8]
		2007	7	Impossible	$2^{115.5}$	2^{119}	2^{45}	[9]
		2010	8	Square-multiset	2^{113}	2^{172}	2^{129}	[10]
		2009	full	RKABA	2^{123}	2^{176}	-	[11]
		2011	full	Bicliques	2^{80}	$2^{189.74}$	2^{8}	[12]
		2013	7	Collision attack	2^{32}	$2^{69.2}$	-	[13]
		2014	11	MITM	2^{117}	$2^{182.5}$	$2^{165.5}$	[14]

To improve cryptographic strength of the AES algorithm its developers recommend using 18 - 24 rounds that lowers the algorithm speed. The actual task of constructing new or upgrading existing encryption algorithms which are able to surpass characteristics of algorithm AES, is relevant to the present day. One of the variant for implementing such kind of algorithm is presented in this article.

4.1   Modernization of Cryptographic Hill Algorithm Based on Involutory Matrices

Lester Hill created an encryption algorithm, which is based on algorithm which replaces the sequence of plaintext with encrypted sequence of the same length in 1929 [15]. Hill Cipher is using orthogonal matrix A for data encryption/decryption:

$$A \cdot A^{-1} = I \tag{2}$$

$$Encryption; c = (A \cdot p)mod(n) \tag{3}$$

$$Decryption; p = (A^{-1} \cdot c)mod(n) \tag{4}$$

where A - encryption matrix (key), and A^{-1} - decryption matrix (key), p – plain text, c – cipher text and n – alphabet length.

If we are using involutory matrix A, for which the initial and its inverse matrix form are the same, then we do not need to spend time for calculating inverse (decryption) matrix, and the initial matrix A is a key for encryption and for decryption.

$$A \cdot A = A \cdot A^{-1} = A^{-1} \cdot A = A^{-1} \cdot A^{-1} = I \tag{5}$$

If involutory matrix consists only with elements $\pm 2^n$, where n – is minimum possible integer, then processor will use only shift and addition operations for data encryption/decryption. Involutory matrices consisting only with elements $\pm 2^n$, where n – is minimum possible integer, can construct only with size $2^k \times 2^k$, where

$$2 \leq k \in N, \tag{6}$$

then we found that the total number of involutory matrices is equal

$$m \cdot 2^{2^{k+1}}; where\ 1 < m \equiv const \tag{7}$$

As an example, from any involutory matrix with size 4×4, that contains numbers $\pm 2^n$, where n – minimum possible number, we can obtain a full set of involutory matrices, total number of which is equal to 1152, so based on expression (7) our constant m is equal to 4.5. Any involutory matrix forms a unified basis for the formation of a large number of involutory and mutually inverse pairs of matrices A and B. Total number of mutually inverse pairs of matrices that can be obtained from involutory matrices of the same size is under investigation [16].

$$A \cdot B = I => A^{-1} \cdot A \cdot B = A^{-1} => B = A^{-1}\ or\ A = B^{-1} \tag{8}$$

As an example, let's look to the new method of forming involutory matrices. If in any involutory matrix we replace elements $1 \rightarrow 2$ and $2 \rightarrow 1$, then we obtain a new involutory matrix.

$$A_1 = \begin{pmatrix} -1 & 2 & 1 & -1 \\ 2 & 1 & 1 & 1 \\ -2 & -2 & -2 & -1 \\ 2 & -2 & -1 & 2 \end{pmatrix}_{A_1 = A_1^{-1}} => A_2 = \begin{pmatrix} -2 & 1 & 2 & -2 \\ 1 & 2 & 2 & 2 \\ -1 & -1 & -1 & -2 \\ 1 & -1 & -2 & 1 \end{pmatrix}_{A_2 = A_2^{-1}} \tag{9}$$

We proposed to use several matrices (combination of involutory and mutually inverse pairs of matrices) for data encryption/decryption instead of one, as in expression (3) and (4). Each plaintext block should be encrypted with different keys (combination of matrices) which will be generated from master key (matrix) using special algorithm for forming involutory and mutually inverse pairs of matrices [16]. For secure delivery of master key we suggest to use hybrid encryption model, where asymmetric cryptography algorithms are used to encrypt/decrypt and securely key (matrix) distribution.

Figure 6 shows one example of implementing modified cryptographic Hill algorithm, where m - is a length of plaintext block which depends on matrix size, and n - is selected number of matrices used in one plaintext block encryption/decryption processes.

Let's consider data encryption/decryption algorithm based on using only involutory matrices with size 16×16, which consist only with elements ± 1 and ± 2, and then our plaintext block is equal to 128 bits. Let's look how cryptographic strength of a modified Hill algorithm changes when we are using different number of matrices to encrypt one plaintext block, and compare it's resistance for brute-force attack with AES-128/192/256. For easier calculation let's our constant m is equal to 1, then cryptographic strength for one encryption block is shown in the table 4.

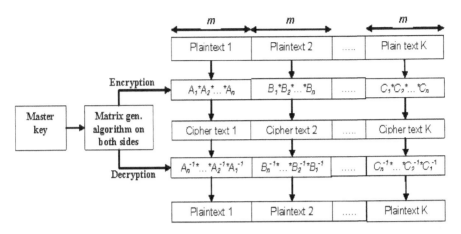

Fig. 6. Modified cryptographic Hill algorithm in electronic codebook mode

Table 4. Cryptographic strength for one encryption block used in modified Hill algorithm

No	Involutory matrix number for 1 plain text block encryption	One cipher block strength for brute-force attack	Comments
1	2	2^{64}	Not enough strength. Brute-force attack on 2^{56} operations took 2 days in 1998.
2	4	2^{128}	Complexity like an AES-128 brute-force attack
3	6	2^{192}	Complexity like an AES-192 brute-force attack
4	8	2^{256}	Complexity like an AES-256 brute-force attack
5	10	2^{320}	Full rounds on AES-128, but more stronger
6	12	2^{384}	Full rounds on AES-192, but more stronger
7	14	2^{448}	Full rounds on AES-256, but more stronger

If we used six different involutory matrices to encrypt one plaintext block then we will get cryptographic strength for one cipher text block which is equal for cryptographic strength for whole cipher text encrypted with algorithm AES-192. But if we use 12 different involutory matrices, that is equal for full encryption rounds in AES-192 algorithm, then we will obtain cryptographic strength for one cipher text block 2^{384} which is more stronger than cryptographic strength for whole cipher text in AES-192 algorithm.

5   Conclusions

Consider the WIM problem through the FOS sensor's accuracy point of view we can conclude that the installation of it into the pavement produce some probabilistic influence to the measurement results, depending not only on sensor's geometric position and properties of components, but also on weather conditions.

The experimental results [4] show that the range of the vehicles velocity from 50 to 90 km/h seems more appropriate for WIM based on fibre-optic sensors let to allow B+(7) class according to COST 323 for the high speeds and D2 according to OIML R134 for the low speeds [5].

From the analysis of the results we can assume, that most exact result can be obtained for total weight of the truck, not for each axle of it, because of longitudinal oscillations exists here (especially between 1st and 3rd axles). It can be caused by incorrect cargo distribution along the length of the truck and trailer, but it is separate task on future. Additionally, the transfer of the data from the sensor's position to measurement station or centre of data collection, is required an increased level of data security and protection.

The rapid development of information and computer technologies contributes for appearance of a large number of equipment, which are using global internet network for data transmission. Time for breaking existing strong encryption algorithms is steadily decreasing (table 3) due to the rapid development of computer technologies, which constantly challenged cryptographers to create a new or upgrade existing encryption algorithm which will be able to withstand for modern threats. Modified cryptographic Hill algorithm is based on involutory matrices, which consist only with elements $\pm 2^n$, where n - is minimal possible number. As all matrices contain the numbers of computer arithmetic's $\pm 2^n$, then processor for data encryption/decryption will used only shift and addition operations.

Cryptographic strength for one plaintext block encrypted with modified Hill algorithm ceteris paribus is much stronger than cryptographic strength for whole cipher text encrypted with AES-128/192/256 algorithm. Modified cryptographic Hill algorithm allows using not only involutory matrices but also mutually inverse pairs of matrices for data encryption/decryption processes. Combinations of involutory and mutually inverse pairs of matrices allow increasing cryptographic strength for one encrypted plaintext block with less number of encryption/decryption matrices than obtained result in table 4.

Acknowledgements. This work was supported by Latvian state research program project "The next generation of information and communication technologies (Next IT)" (2014-2017).

References

1. SENSORLINE GmbH (© Sensor Line) (2010), SPT Short Feeder Spliceless Fiber Optic Traffic Sensor: product description,
   `http://sensorline.de/home/pages/downloads.php`
2. Teral, S.: Fiber optic weigh-in-motion: looking back and ahead. Optical Engineering 3326, 129–137 (1998)
3. Grakovski, A., Pilipovecs, A., Kabashkin, I., Petersons, E.: Tyre Footprint Geometric Form Reconstruction by Fibre-Optic Sensor's Data in the Vehicle Weight-in-Motion Estimation Problem. In: Ferrier, J.-L., Gusikhin, O., Madani, K., Sasiadek, J. (eds.) Informatics in Control, Automation and Robotics. LNEE, vol. 325, pp. 123–137. Springer, Switzerland (2015)
4. Grakovski, A., Pilipovecs, A., Kabashkin, I., Petersons, E.: Weight-in-motion estimation based on reconstruction of tyre footprint's geometry by group of fibre optic sensors. Transport and Telecommunication 15(2), 97–110 (2014)
5. O'Brien, E.J., Jacob, B.: European Specification on Vehicle Weigh-in-Motion of Road Vehicles. In: Proceedings of the 2nd European Conference on Weigh-in-Motion of Road Vehicles, pp. 171–183. Office for Official Publications of the European Communities, Luxembourg (1998)
6. Ghena, B., Beyer, W., Hillaker, A., Pevarnek, J., Halderman, A.: Green Lights Forever: Analyzing the Security of Traffic Infrastructure. In: Proceedings of the 8th USENIX Workshop on Offensive Technologies (2014)
7. Wiener, J.: Efficient DES Key Search and Update. RSA Laboratories Cryptobytes 3(2), 6–8 (1996)
8. Gilbert, H., Minier, M.: A Collision Attack on 7 Rounds of Rijandel. In: AES Candidate Conference, pp. 213–230 (2000)
9. Zhang, W., Wu, W., Feng, D.: New results on impossible differential cryptanalysis of reduced AES. In: Nam, K.-H., Rhee, G. (eds.) ICISC 2007. LNCS, vol. 4817, pp. 239–250. Springer, Heidelberg (2007)
10. Dunkelman, O., Keller, N., Shamir, A.: Improved single-key attacks on 8-round AES-192 and AES-256. In: Abe, M. (ed.) ASIACRYPT 2010. LNCS, vol. 6477, pp. 158–176. Springer, Heidelberg (2010)
11. Biryukov, A., Khovratovich, D.: Related-key cryptanalysis of the full AES-192 and AES-256. In: Matsui, M. (ed.) ASIACRYPT 2009. LNCS, vol. 5912, pp. 1–18. Springer, Heidelberg (2009)
12. Bogdanov, A., Khovratovich, D., Rechberger, C.: Biclique cryptanalysis of the full AES. In: Lee, D.H., Wang, X. (eds.) ASIACRYPT 2011. LNCS, vol. 7073, pp. 344–371. Springer, Heidelberg (2011)
13. Kang, J., Jeong, K., Sung, J., Hong, S., Lee, K.: Collision Attacks on AES-192/256, Crypton – 192/256, mCrypton-96/128, and Anubis. Journal of Applied Mathematics 2013, 1–10 (2013), doi:10.1155/2013/713673
14. Li, L., Jia, K., Wang, X.: Improved Single-Key Attacks on 9-Round AES-192/256. In: Cid, C., Rechberger, C. (eds.) FSE 2014. LNCS, vol. 8540, pp. 127–146. Springer, Heidelberg (2015)
15. Eisenberg, M.: Hill Ciphers and Modular Linear Algebra. Mimeographed notes, University of Massachusetts (1998)
16. Kamenchenko, S.: GPS/GLONASS tracking data security algorithm with increased cryptographic stability. In: Proceedings of the 14th International Conference "Reliability and Statistics in Transportation and Communication" (RelStat 2014), Riga, Latvia, pp. 154–164 (2014)

Secure Hybrid Clouds: Analysis of Configurations Energy Efficiency

Vyacheslav Kharchenko[1], Yurii Ponochovnyi[2], Artem Boyarchuk[1],
and Anatoliy Gorbenko[1]

[1] National Aerospace University "Kharkiv Aviation Institute", Kharkiv, Ukraine
V.Kharchenko@khai.edu
[2] Yu. Kondratiuk Poltava National Technical University, Poltava, Ukraine
pnch1@rambler.ru

Abstract. The paper examines energy efficiency of running computational tasks in the hybrid clouds considering data privacy and security aspects. As a case study we examine CPU demanding high-tech methods of radiation therapy of cancer. We introduce mathematical models comparing energy efficiency of running our case study application in the public cloud, private data center and on a personal computer. These models employ Markovian chains and queuing theory together to estimate energy and performance attributes of different configurations of computational environment. Finally, a concept of a Hybrid Cloud which effectively distributes computational tasks among the public and secure private clouds depending on their security requirements is proposed in the paper.

Keywords: security, hybrid cloud, energy efficiency, Markovian chain, IT-infrastructure.

1 Introduction

Modern tendencies of computing globalization and increase of computational tasks complexity stipulate the development and application of large-scale distributed systems. Such systems can be built through the use of Grid technologies, cluster technologies and supercomputing, but recently the concept of Cloud Computing has become an industrial trend which is primarily used for deploying modern distributed application and delivering computing services.

Centralization of computing facilities within large data centres on the one hand requires and on the other creates opportunities for the implementation of Green technologies enhancing energy efficiency of computing [1]. Therefore, the problem of energy-aware computational tasks scheduling and distribution among available computing resources, both private and public, becomes crucial.

However, many companies are still unwilling to place their IT infrastructure in the cloud because of reasonable fears over data security and unawareness about the benefits offered by cloud computing [2]. In this concern a hybrid cloud can provide a compromise solution storing and processing private data in in-house computational

© Springer International Publishing Switzerland 2015
W. Zamojski et al. (eds.), *Theory and Engineering of Complex Systems and Dependability,*
Advances in Intelligent Systems and Computing 365, DOI: 10.1007/978-3-319-19216-1_19

resources and delegating non-critical computations to public clouds. In this paper we examine various configurations of using private and cloud computing to process sensitive security information considering the overall energy efficiency.

Measure p_{ue} is used as a factor of energy efficiency of computing in the Cloud. It is equal to the ratio of the power allocated and distributed for a data center to the power consumed by computer equipment [2]:

$$P_{UE} = P_{uptime} / P_{active} = (P_{active} + P_{downtime}) / P_{active} \tag{1}$$

where

P_{active} is the power consumed by equipment in the active operating mode,
P_{uptime} is the total power allocated and distributed for this data center,
$P_{downtime}$ is the power consumed by this data center in the idle mode.

Energy consumption of data centers with different p_{ue} values is compared in the table 1. It shows a relative ratio between p_{ue} and $P_{downtime}$ values provided the constant total power allocated and distributed for data centers ($P_{uptime} = 1$, where 1 is a conventional unit).

Table 1. Comparison of data centers energy consumption

p_{ue}	P_{uptime}	$P_{downtime}$
3	1	0.66
1.25	1	0.2
1.16	1	0.138

According to [3] and [4], a commodity rack-mounted server (taking out of considerations power overheads related to rack cooling, etc.) consumes three times less energy than a standard personal computer (39-45 W versus 0.15 kW).

Discussing the concept of green and sustainable cloud computing, the main attention is traditionally given to the technologies which aim at energy consumption reduction [5] without considering security-related power overheads. In [6] researches analyse the concept of change-over to the public Cloud without accounting security issues. At the same time, studies [2, 3] discuss security risks of deploying IT infrastructures in the public Clouds without employing additional security mechanisms. Some aspects of energy efficiency and security for different Cloud platforms are analysed in [7].

In this work we discuss main principles of secure task solving in the private and public Clouds and also introduce models evaluating its energy efficiency.

The rest of the paper is organised as follows. The second section contains the formalization of problems related to secure and public computing. A few models of the deployment of computing in the public Cloud with consideration of additional protective measures for provision of security are represented in the third part. The fourth part describes two examples of the energy efficiency computation for models of deployment of IT in the public Cloud without and with additional security countermeasures. The fifth section discusses the results and present future steps of research and development.

2   Criticality Models for Secure and Public Computing

The set of particular tasks, into which a general computing task can be divided, uses in a certain way functions of software systems of the private Cloud, which differ by the level of criticality in accordance with the international regulatory framework [8,9]. Criticality will be ultimately determined by damage that may occur as a result of the corresponding events (emergencies).

To reduce risks of security to acceptable values, input data for tasks planned to be transferred to the public Cloud must be encrypted to some extent. In this case several encryption levels for improvement of confidentiality of data being transferred. With relation to energy expenditures, each encryption/decryption operation requires some resources. Therefore, the general set of tasks is proposed to be divided into four subsets with the reduction of criticality: A, B, C, D.

With consideration of this proposal, when the general task is being solved in the hybrid Cloud, the complex of tasks MT is being formed, and these tasks can be divided by criticality into four sets:

$$MT = TA \cup TB \cup TC \cup TD. \tag{2}$$

Each set is described by the following complex of tasks:

$$TA = \{f_{Ai}, i = 1,\ldots, b_A\},$$

$$TB = \{f_{Bj}, j = 1,\ldots, b_B\},$$

$$TC = \{f_{Ck}, k = 1,\ldots, b_C\},$$

$$TD = \{f_{Dl}, l = 1,\ldots, b_D\},$$

where $\forall N,M \in \{A,B,C,D\}, N \neq M: TN \cap TM = \varnothing$,

$$Card\ MT = b_A + b_B + b_C + b_D.$$

Attributes of criticality for the tasks $f_z \in MT$ are represented in the table 2.

Each function f_z of the set MT is characterized by level of criticality (information security) u_{cr}, intended purpose a_{cr}, method m_{cr} and complexity w_{cr} of realization of $f_z \sim \{u_{cr,z}, a_{crz}, m_{crz}, w_{crz}\}$.

Accordingly, the subset TD containing tasks, which are not critical for information security and can be solved in the public Cloud, as well as the subsets of tasks TB and TC, which can be transferred to the public Cloud after the additional processing of input data (see Fig.1), may be singled out from the totality of particular tasks.

Apart from serial processing of the task flow shown in Fig.1, stage-by-stage or cyclic formation of the flow of tasks, which are critical for information security, is possible for some tasks. Options of formation and processing of the task flow in case of transfer of some tasks to the public Cloud are shown in Fig.2.

Table 2. Attributes of tasks for the hybrid Cloud depending on security criticality

Critica-lity	Attributes of tasks T			
	Level of criticality, u_{cr}	Intended purpose, a_{cr}	Method of realization, m_{cr}	Complexity of realization, w_{cr}
TA	Maximum	1. Processing of extra confidential data 2. Solving tasks for which even intermediate data are strictly confidential	Solving in the private Cloud only	The simplest realization for the reduction of en-ergy expenditures
TB	Increased	Solving tasks for which multiple encryption must be applied to input data	Solving in the private Cloud or in the public Cloud	More complex means than ones for realization of TA tasks
TC	High	Solving tasks for which single encryption must be applied to input data	Solving primar-ily in the public Cloud	Means of various complexity
TD	Non-critical tasks	Tasks may be solved in the public Cloud without any additional encryption applied to input data	Solving in the public Cloud	Means of various complexity

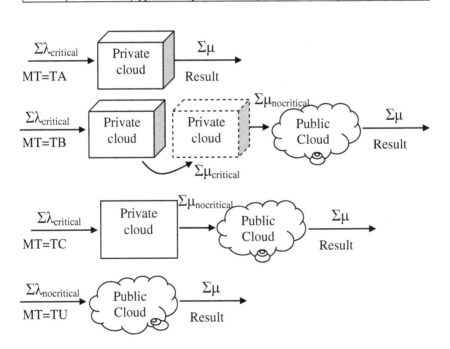

Fig. 1. Options of the task flow processing in case of transfer to the public Cloud

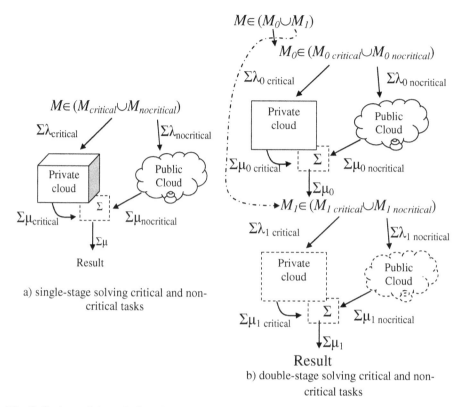

Fig. 2. Options of the task flow formation and processing in case of the transfer of their part to the public Cloud

3   Models of Task Flows Processing in Secure Green Cloud

It is necessary to develop appropriate models for the assessment of energy efficiency and other parameters of systems. They may be based on models of queuing systems and networks [10]. See below conceptual examples of such models.

Model 1 is represented in Fig.3 (a). The private Cloud is represented as a queuing network with batch arrivals, which includes a broker and queue-forming computing elements.

A query sent by the user routes to the queue of the broker of services, and if the broker's device is free, the query immediately routes for processing. After its processing by the broker, the query becomes divided into subqueries depending on the number of providers of cloud computing services K, where after each provider processes the subquery received by him.

After the processing of the subquery received, each provider sends the response to the client, and when the client receives responses to all subqueries of the query sent by him, it is considered that the user have received the response to his query.

Let the random time of the response of k-th provider of cloud computing services be t_k. In this case t (time from initiation of the query by the user till the receipt of the response to all subqueries included into the query by him) will be the time of the response of the whole system. It will be determined by the formula $t = \max t_k$, $1 \leq k \leq K$.

Let r_k be limitation for the length of the k-th queue, μ_k be intensity of processing of queries by the k-th provider, μ_0 be intensity of processing of the query by the broker of services, λ be intensity of receipt of queries from users. It is assumed that the number of elements of the group coming out of the broker's node is fixed and is equal to K, i.e. the number of providers of services.

In case of periodic use of elements of the private Cloud, a part of the incoming task flow will be formed on the basis of results of intermediate computing (see Model 1, Fig.3,c).

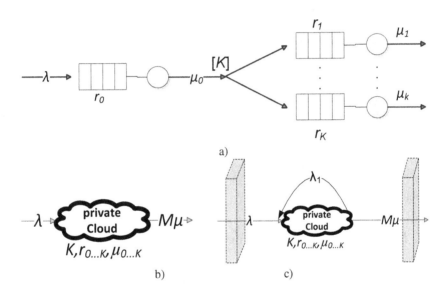

Fig. 3. Model 1 of the solving tasks in the hybrid Cloud (a – queuing network, b – simplified description, c – description with consideration of paralleling and assembly units)

Model 2 (see Fig.4,a). The difference is that the repeated use of resources of the private Cloud may be reduced by means of separation of non-critical tasks and their transfer to the public Cloud. The reduction of the task flow depends on operability of the DMS unit. Model 3 (see Fig.4,b). The initial division of the task flow into critical and non-critical tasks is typical for this model. The model of the public Cloud is also represented by a queuing network with batch arrivals.

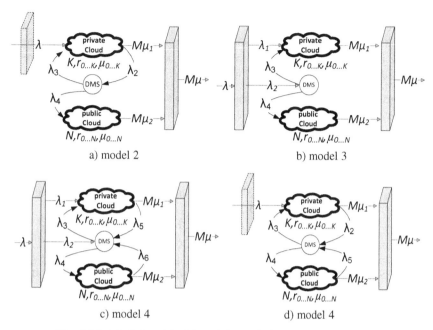

Fig. 4. Models 2, 3 and 4 of the solving tasks in the hybrid Cloud

Model 4 (see Fig.4,c,d). In contrast to previous models, resources both of the private Cloud and the public Cloud may be used repeatedly after the proper audit of incoming tasks for their compliance with criteria of security.

4   Typical Models of Energy Efficiency Analysis for Solving Tasks in Hybrid Clouds

4.1   Analysis of the Hybrid Clouds Energy Efficiency with Minimum Requirements for Security

At first let us consider the simplest model with no requirements for information security, in which case computing may be provided both in private Cloud and in the public Cloud.

Conditions of the task modelled are described in [11]. The timely formation of examination results after the tomography examination is the essence of this computing. The results may be obtained in the following ways:

a) after the examination of the patient, its results are processed on the PC. The characteristic time for realization of the algorithm for serial computing was about 20 minutes for the processor Intel Core i5 3 GHz;

b) results after paralleling and adaptation are transferred to the Grid, computing time goes down to 5-10 seconds that gives 100-200 times improvement of the performance, but in this case substantial delays due to waiting for the performance of tasks in the queue and high energy expenditures related to computing take place;

c) results after paralleling and adaptation are transferred to the Cloud, computing time is the same as in the case of the Grid, delays related to waiting are minimized, energy expenditures go down.

According to [11], the modern health center plans to serve 10 – 12 thousand patients per year. Let's solve the problem of reduction of energy expenditures for the following input data.

Incoming flow: 12000 applications per year (251 working days of 365, 1807.2 working hours per year) = 48 patients per working day or 8 patients per hour.

Let the Grid cluster is 10 servers (cluster consisting of 10 computers with licensed operating systems Windows XP and LINUX installed, with the following parameters: 2 GHz, 1024 Mb RAM, 500 Gb HDD) [11].

Energy efficiency $p_{ue} = 3$ for the Grid, and $p_{ue} = 1.25$ for the Cloud.

Let us consider the mathematical model of computing in the Grid and in the Cloud as a process of serving in a multichannel queuing system which appears as a system of the type $M/M/m/n$ (where M is the exponential distribution, m is the number of service facilities, n is the length of the queue) in Kendall's notation [10]. The state and changeover graph which describes functioning of the system is shown in Fig.5.

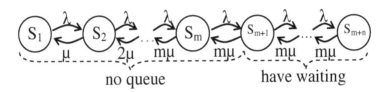

Fig. 5. Graph of functioning of the queuing system of the type M/M/m/n

In the case of consideration of the queuing system of the type M/M/m/n, two interrelated conditions pertaining to normal operation of the system are considered: $\rho_m \leq 1$ and $q > 90\%$.

Moreover, the additional limiting condition is that the results for patients must be formed in due time, i.e. all 8 applications must be served within an hour.

In the case of the use of the Cloud, let's consider that each application can be divided into 100 separate computing tasks, and in this case the time for solving one task will be 100 times less than the time for serving one application.

Moreover, additional problems of paralleling and folding ($t = 0.1 * t_{serv}$) are considered; i.e. these tasks are equal to 10 computing tasks in regard to time for their solving. Accordingly, the task flow will be routed to the input of the Cloud with the intensity $\lambda_1 = 8*120 = 960$ 1/hour, and in this case the intensity of serving of the single task will be $\mu_1 = \mu*100 = 300$ 1/hour.

The research of the system has shown that 4 or more servers are needed for timely serving of the indicated task flow.

The minimum length of the queue for the queuing system with m=4 service facilities which ensures the condition q > 90% will be n=3. If the number of service facilities in the system exceeds 4, the minimum length of the queue will be n=1.

To estimate total power used for computing it is necessary to determine the average value of total computer (server) operating time spent for computing operations (T_{serv}), for waiting solutions of tasks in a queue (T_{queue}), and the average value of total downtime of servers operated ($T_{downtime}$).

Total power used for computing within an hour will be subsequently determined in the following way:

$$P_{cloud} = P_S + P_Q + P_D = \left[t_{service} \cdot \lambda_1 \cdot P_{uptime} \right] + $$
$$+ \left[t_{queue} \cdot \lambda_1 \cdot P_{uptime} \cdot \left(1 - 1/ \left(2 \cdot p_{UE} \right) \right) \right] + $$
$$+ \left[\left(m - t_{sist} \cdot \lambda_1 \right) \cdot P_{uptime} \cdot \left(1 - 1/ p_{UE} \right) \right].$$

The table 3 contains computations of the power consumed by the Cloud at $P_{uptime} = 1$ (conventional unit) and $p_{ue}=1.25$.

Table 3. Results of energy consumption computation

m	n	q	P_S	P_O	P_D	P_{cloud}
4	3	0.919181	2.941381	0.342469	0.097568	3.381417
4	4	0.939272	3.005669	0.467419	0.04306	3.516148
5	1	0.924606	2.958739	0.045237	0.393173	3.397149
5	2	0.953969	3.0527	0.098392	0.356663	3.507755
5	3	0.971383	3.108426	0.147086	0.329286	3.584798
5	4	0.982015	3.142447	0.187606	0.308975	3.639028
5	5	0.98862	3.163585	0.21961	0.29408	3.677275
5	6	0.99277	3.176863	0.244051	0.283277	3.704191
6	1	0.967179	3.094972	0.019693	0.574441	3.689106
6	2	0.982797	3.144949	0.039998	0.557678	3.742624
6	3	0.990908	3.170906	0.056	0.547152	3.774058
7	1	0.987244	3.15918	0.007654	0.765613	3.932446
7	2	0.994202	3.181448	0.014567	0.758855	3.954869
8	1	0.995548	3.185754	0.002671	0.961959	4.150384
9	1	0.998594	3.195502	0.000843	1.160619	4.356964
10	1	0.999595	3.198705	0.000243	1.360178	4.559126

A curve that characterizes power consumption in the specific configuration of the Cloud (m and n) is shown in Fig.6.

In addition to the results shown on the three-dimensional chart in Fig.6, it is necessary to consider quality limitation and limitation related to proper regularity of maintenance. Results of the optimal cloud's configuration research with consideration of the mentioned limitations are summarized in the table 4.

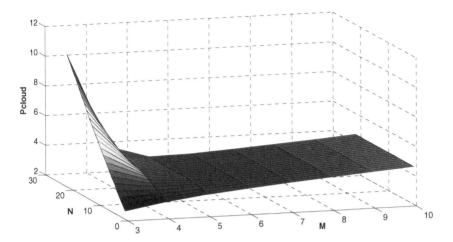

Fig. 6. Dependencies for the consumed power P_{cloud} in the specific cloud's configuration

Table 4. Comparison of energy consumption for the Cloud, Grid and standard PC

q	3 PC	10 PC Grid (p_{ue}=3)	Optimal Cloud (p_{ue}=1.25):
q > 90%	P = 9	P = 7.733	P = 3.381; m=4, n=3
q > 99%	P = 9	P = 7.733	P = 3.704; m=5, n=6
q > 99.9%	P = 9	P = 7.733	P = 3.756; m=5, n=11

Let us consider the procedure of solving the same task provided that the limitation related to time allotted for solving is removed. In this case, the queuing system with m=4 servers ensures the quality q > 99.9% at the length of the queue n=22; in this case the power consumed P_{cloud} = 4.576. This value is not optimal.

This means that it is necessary to choose an optimal structure of the public Cloud, which depends on the intensity of submission of applications and their serving, in order to minimize energy expenditures.

4.2   Analysis of the Hybrid Clouds Energy Efficiency with Advanced Requirements for Security

The problem of the estimation of compliance of the computed medical information with the actual one for finding similar images is also described in [11]. This problem is generally closely related to the recognition of images and automatic methods of establishing diagnosis.

Data repositories containing medical records and other similar data must comply with advanced requirements for data confidentiality. Therefore, their storing "as is" in repositories of the public Cloud is unacceptable. Some confidential data must be encrypted, and the reverse conversion must be realizable only in the private secure Cloud.

Let us consider the procedure of solving this problem with the use of the simplified method. Let it is necessary to process a conventional information repository with 1000000 (10^6) records.

Let the performances of all servers of the private and the public Cloud are the same, like in the previous task.

Processing of one record by the single server requires t_{serv}=10 minutes (μ_1 = 6 1/hour) on the average, but this operation enables paralleling. The pre-processing of each record made for concealing confidential data requires 100 times less computations on the average (μ_2 = 600 1/hour). It is also necessary to take into account operations related to the audit of encryption of confidential data and to the transfer of changed records to the public Cloud (μ_3 = 120 1/hour); it is assumed that performance of these operations is provided by a single serving system.

Let $m_{private}$ = 10 servers with the length of the queue $n_{private}$ = 6 functions in the private Cloud, like in the previous task; these servers have low energy efficiency (p_{ue1} = 3). Parameters of the public Cloud: m_{public} = 50 servers with the length of the queue n_{public} = 24, p_{ue2} = 1.25. And, finally, the modelling of the performance of operations of the audit of security and transfer to the public Cloud is provided by the queuing system, the characteristics of which are similar to the ones of the private Cloud (m_{at} = 10, n_{at} = 6, p_{ue3} = 3).

The task set can be solved in different ways. Let us consider some of them and estimate their efficiency in regard to energy and time consumption.

A) Solution of the whole task in the private Cloud

The private Cloud is modelled by means of the queuing system M/M/10/6. Following from these parameters, the value q>0.999 is guaranteed for the intensity of incoming applications $\lambda 0 < 33$ 1/hour. In case of such intensity of applications, total number of records (10^6) will be processed in T = 10^6/33 = 30 303.03 hours.

In this case, aggregate energy expenditures based on the consumption rate 1 conventional unit of power (CUP) per 1 hour will be as follows: P_{cloud} = 8.51 CUP.

Total number of conventional units required for solving the task will be P=30303.03*8.51=257 880.483.

B) Solution of the whole task in the public Cloud

The public Cloud is modelled by means of the queuing system M/M/50/24. Following from these parameters, the value q>0.999 is guaranteed for the intensity of incoming applications $\lambda 0 < 259$ 1/hour. In case of such intensity of applications, total number of records (10^6) will be processed in T = 10^6/259 = 3861 hours.

In this case, aggregate energy expenditures based on the consumption rate 1 conventional unit of power per 1 hour will be as follows: P_{cloud} = 45.0086 CUP.

Total number of conventional units required for solving the task will be P=3861*45=173 778.41.

C) Scenario of serial computing in a hybrid cloud

This option provides the stage-by-stage processing of the whole database; its whole scope is processed at every stage.

The private Cloud is modelled by means of the queuing system M/M/10/6. Following from these parameters, the value q>0.999 is guaranteed for the intensity of incoming applications $\lambda 0 < 3378$ 1/hour. In case of such intensity of applications, total number of records (10^6) will be processed in T = $10^6/3378$ = 296.03 hours.

In this case, aggregate energy expenditures based on the consumption rate 1 conventional unit of power per 1 hour (CUP) will be as follows: P_{cloud}=8.555 CUP.

Total number of conventional units required for task solving at first stage will be equal P=296.03*8.555=2532.586.

The audit of security and transfer of changed records to the public Cloud is also modelled by means of the queuing system M/M/10/6. Following from these parameters, the value q>0.999 is guaranteed for the intensity of incoming applications $\lambda 0 < 675$ 1/hour. In case of such intensity of applications, total number of records (10^6) will be processed in T = $10^6/675$ = 1481.48 hours.

In this case, aggregate energy expenditures based on the consumption rate 1 conventional unit of power per 1 hour will be as follows: P_{cloud}=8.553 CUP.

Total number of conventional units required for task solving at the second stage will be P=1481.48*8.553=12671.615.

3861 hours and 173778.41 conventional units of power will be required for solving the task in the public Cloud, like in the previous option.

T (total time of solving the general task in a hybrid cloud) will be equal to 5638.51 hours, and energy expenditures will P=188982.6 conventional units.

In comparison with solving the task only in a private cloud, time of solving has reduced 5.37 times, and energy expenditures have reduced 1.36 times. Accordingly, the transfer of computing to the public Cloud gives the possibility to have both the increase in the speed of decision making and the reduction of energy expenditures.

Moreover, the private Cloud will not be operated within the most part of the task solving time, and this gives additional possibilities in regard to the increase in the speed of the task solving.

D) Option of the prompt task solving with the reduction of the number of servers in the private Cloud.
A task can be solved in a hybrid cloud more promptly in case of organization of continuous processing of separate tasks at all three stages. In this case the maximum intensity of incoming applications is limited by the "slowest" queuing system [10]. Results of the computations provided for the previous option show that the public Cloud is such system, and the relevant intensity of submission of applications and their serving λ_0 is accordingly less than 259 1/hour.

In order to minimize energy expenditures at the first and second stages of the procedure of solving, let's determine the optimal number of serving devices in the private Cloud and the system of the audit of security and data transfer (provided that q>0.999).

Optimal configuration of the private Cloud is m=1, n=7; in this case, aggregate energy expenditures based on the consumption rate 1 conventional unit of power per 1 hour will be as follows: P_{cloud}=0.864 CUP. P (total number of conventional units required for solving the first stage of the task) will be equal to 3861*0.864=3337.333.

Optimal configuration of the queuing system of the audit of security and data transfer is m=4, n=8; in this case, aggregate energy expenditures based on the consumption rate 1 conventional unit of power per 1 hour will be as follows: P_{cloud}=3.426 CUP. P (total number of conventional units required for solving the first stage of the task) will be equal to 3861*3.426=13228.75.

Tasks being solved in the public Cloud will take 3861 hours and 173778.41 conventional units of power, like in the previous option.

T (total time of solving the general task in a hybrid cloud) will be equal to 3861 hours, and P (energy expenditures) will be equal to 190344.5 conventional units.

In comparison with solving the task only in a private cloud, time of solving has reduced 7.85 times, but in this case the power consumption is increased by 1361.9 conventional units (in comparison with the previous stage-by-stage option).

For illustration purposes results of computations for all options of task solving are summarized in the table 5.

Table 5. Comparison of energy consumption and speed of task solving in private and hybrid clouds

Possible solution	Stage	m	n	p_{ue}	P_S	P_Q	P_D	P_{cloud}	T (hours)	P
A) Private Cloud		10	6	3	5.50	0.06	2.96	8.51	30303.03	257880.48
B) Public Cloud		50	24	1.25	43.13	0.76	1.12	45.01	3861	173778.41
C) Hybrid Cloud, stage by stage	1	10	6	3	5.62	0.07	2.86	8.56	296.03	2532.59
	2	10	6	3	5.62	0.07	2.87	8.55	1481.48	12671.62
	3	50	24	1.25	43.13	0.76	1.12	45.01	3861	173778.41
	total								5638.51	188982.60
D) Hybrid Cloud, stage by stage, continuously	1	1	7	3	0.43	0.27	0.16	0.86		3337.33
	2	4	8	3	2.16	0.20	1.07	3.43	3861	13228.75
	3	50	24	1.25	43.13	0.76	1.12	45.01		173778.41
	total									190344.5

The partial use of total number of servers of the private Cloud provides reserves for the increase in the speed of solving the general task due to the additional load of unused servers. In this case energy expenditures may be increased, because servers of the private Cloud have low energy efficiency.

5   Conclusions

Security and energy efficiency assurance are key challenges for Cloud-based IT-infrastructures. Balancing of these characteristics depends on criticality of the solving tasks. Initial description of tasks being performed in the Cloud by levels of criticality of consequences for security in case of the performance of tasks in the hybrid Cloud has been presented by set-theoretical models.

Besides, analytical models of energy efficiency evaluation during the transfer of tasks to a public cloud can be applied to make decisions related to selection of the scenarios. These models have been added by queuing networks and Markovian chain-based models.

The results allow making the following conclusions:

- transfer of computing tasks from the set TD (without encryption) to the public Cloud gives the possibility to gain a few times reduction of energy expenditures in comparison with the use of standard PC;
- in order to minimize energy expenditures, it is necessary to select the optimal structure of the public Cloud, which depends on intensity of applications and serving;
- transfer of computing tasks from the set TC (with single encryption) to the public Cloud gives the possibility to have both the increase in the speed of decision making (for described examples in 5.37 times) and the reduction of energy expenditures (1.36 times);
- in case of organization of computing in the hybrid Cloud, time of solving the task may be reduced (7.85 times) with insignificant increase in the energy consumption due to the rational utilization of resources of a public cloud.

As a further field of work, the tools of the assessment and provision of security and resilience in a hybrid cloud infrastructure will be developed. For that more complex Markovian models can be used taking into account component vulnerability data, types of the attacks and maintenance strategies [12].

References

1. Banerjee, A., Agrawal, P., Iyengar, N.C.S.N.: Energy Efficiency Model for Cloud Computing. IJEIC 4(6), 29–42 (2013)
2. Nimje, R., Gaikwad, V.T., Datir, H.N.: Green Cloud Computing: A Virtualized Security Framework for Green Cloud Computing. IJARCSSE 2(4), 642–646 (2013)
3. Sammy, K., Shengbing, R., Wilson, C.: Energy Efficient Security Preserving VM Live Migration in Data Centers for Cloud Computing. IJCSI 9(2), 33–39 (2012)
4. Breakthrough Security Capabilities and Energy-Efficient Performance for Cloud Computing Infrastructures, https://software.intel.com/file/m/26765
5. Murugesan, S., Gangadharan, G.R.: Harnessing Green IT: Principles and Practices, 432 p. Wiley - IEEE (2012)
6. Pattinson, C., Oram, D., Ross, M.: Sustainability and Social Responsibility in Raising Awareness of Green Issues through Developing Tertiary Academic Provision: A Case Study. IJHCITP 2(4), 1–10 (2011)
7. Kharchenko, V. (ed.): Green IT-Engineering. Systems, Industry, Society, vol. 2, 628 p. National Aerospace University KhAI, Ukraine (2014)
8. EGEE-JRA3-TEC-487004-DJRA3.1-v-1-1. GSA, https://edms.cern.ch/document/487004/
9. Common Criteria for Information Technology, CCRA, http://www.commoncriteriaportal.org

10. Filipowicz, B., Kwiecień, J.: Queuing systems and networks. Models and applications. Bulletin of the Polish Academy of Sciences Technical Sciences 56(4), 379–390 (2008)
11. Medical and technical study of the project of the Northwest inter-regional center of high-tech methods of radiation therapy of cancer, `http://alice03.spbu.ru/alice/hadron/`
12. Kharchenko, V., Abdul-Hadi, A.M., Boyarchuk, A., Ponochovny, Y.: Web Systems Availability Assessment Considering Attacks on Service Configuration Vulnerabilities. In: Zamojski, W., Mazurkiewicz, J., Sugier, J., Walkowiak, T., Kacprzyk, J. (eds.) Proceedings of the Ninth International Conference on DepCoS-RELCOMEX. AISC, vol. 286, pp. 275–284. Springer, Heidelberg (2014)

Functional Readiness of the Security Control System at an Airport with Single-Report Streams

Artur Kierzkowski and Tomasz Kisiel

Wroclaw University of Technology, 27 Wybrzeże Wyspiańskiego St., 50-370 Wrocław
{artur.kierzkowski,tomasz.kisiel}@pwr.edu.pl

Abstract. The article presents a developed universal simulation model support-ing the design process of the security control area at the airport. The universali-ty of the simulation model allows for its use for the adaptation of the size of the security control area which consists of security check stations with a single flow of passenger streams to the forecast intensity of reporting passengers. The pre-sented model is mostly intended for regional airports, where the security control is conducted using a metal detector and an x-ray device. The functional readi-ness of the designed system is analysed in terms of the forecast intensity of passenger reports. The functioning of the simulation model is based on time characteristics determined on the basis of research conducted on a real system which allowed for the verification of the functioning of the model. It is also possible to introduce one's own characteristics to optimise another existing real system.

Keywords: simulation model, security control, functional readiness.

1 Introduction

Air transport of passengers and goods requires high security levels. Various legal regu-lations are introduced which determine the method of management or operation of the air transport system. The main act which is responsible for keeping the security level of air transport in Poland, is [4] which refers to the following regulations [6-8]. Therefore, the security management system is controlled by both domestic and international law.

Contemporary conviction of the priority of security principles is erroneous. Air carriers are business organisations, for which security assurance as the main objective is disadvantageous in financial terms. Thus, a management dilemma appears here which is related with business and security issues - the "2P dilemma" [12] (protection, production). The main objective of aviation organisations is effective service provision, which is in conflict with security aspects.

The dynamic development of the air transport network and air transport operations often forces the carriers to perform take-off and landing operations within specific time limits (slots) which largely depend on the readiness of the critical infrastructure, i.e. airports. The necessity of timely ground handling of planes makes it necessary for air-ports to improve the effectiveness of performed processes on a continuous basis. Thus, it is required that effectiveness is increased, while maintaining all security aspects.

W. Zamojski et al. (eds.), *Theory and Engineering of Complex Systems and Dependability*,
Advances in Intelligent Systems and Computing 365, DOI: 10.1007/978-3-319-19216-1_20

2   Functioning of Air Transport Systems - Literature Review

The passenger terminal is divided into areas with various access levels. The passenger beginning the trip at a given airport reports to the generally accessible landside area, from which he/she is obliged to move to the departure hall, which is situated in the airside area. The airside area can be entered only after showing an appropriate document entitling the passenger to stay in restricted areas. It is a boarding card for passengers which allows the passenger to board the plane at a further stage of the check-in. Pursuant to regulation [8] and the annexes [6-7], the airport operator demarcates boundaries between individual areas and marks them in an unambiguous manner using physical obstacles preventing unauthorised access. The restricted area can be accessed by access control at security control points. An example of the process execution diagram - typical of regional airports - is presented in Figure 1.

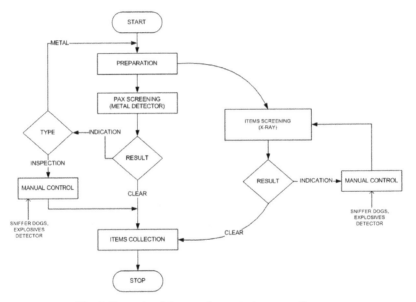

Fig. 1. Example of the security control process diagram

Both passenger and luggage control has a positive result only if the process operator can unambiguously decide that the requirements of the control process have been met. Each negative signal from control devices must be identified and eliminated. Otherwise, the passenger or the luggage goes through subsequent control or is refused admittance onto the aircraft. After the end of the security process, passengers and their cabin luggage are protected against groundless intervention until take-off unless the passenger or baggage exists (contact with a passenger or luggage that has not been screened).

Preliminary issues connected with the use of terminal areas are presented in [11], [14-15], [31], [35]. The need for solving problems connected with bottlenecks in passenger flows by the organisation of terminal operations. The possibility of minimising costs connected with the construction of the terminal and subsequent changes in the infrastructure resulting from the variable intensity of passenger flows. There are numerous publications which focus on passenger terminal design models [17] [20-21], [28]. Examination of travel comfort is also particularly important [10]. The model, apart from the need to ensure security, also indicates the appropriateness of minimisation of inconveniences which may have a negative influence on passenger service quality [9]. There exists a range of models which divide the system operation into a collection of individual queuing systems. A lot of models have been developed for these systems, which are used to estimate basic indices and parameters [3]. However, for complex processes, such as security control, where there exist a range of external factors forcing the operation of the system in a queuing system, it is necessary to develop a model which allow for taking these factors into account. There are already initial considerations that allowed for the introduction of a multi-criteria analysis for passenger flows at the passenger terminal [5]. The problem of queuing time minimisation with minimal use of resources have been brought up on numerous occasions [1], [19]. Also, several models for the selection of an appropriate number of technical resources to handle the assumed traffic flow [27], [34]. The authors of the article show the randomness in the passenger flow through the control point in a given unit of time and show this with reference to unreliability to pay special attention to the necessity of performing a broader analysis for regional airports, in particular, as they, due to low passenger traffic intensity, often use a basic analysis which can cause considerable disturbances if temporary high fluctuations in traffic intensity occur. This study presents a security control system operation model. Other parts of air transport system are described for example in [29-30], [32-33].

An analysis of a transport system using the Petri net was presented in [16]. The authors have discussed basic limitations of reliability modelling and system operation methods. At present, a lot of studies deal with the possibility of using fuzzy logic in systems [41]. The model will also take into consideration aspects connected with the reliability of the technical system [22-24], the effectiveness of its operation [2], [13], [18], [25], [26], [38], [43] and susceptibility aspects [39], [40], [42].

The studies listed in the literature review are limited to general solutions to the analysed problem or strong assumptions are adopted which are not reflected in real systems. The developed model makes it possible to analyse passenger flows, depending on the set timetable. The use of the model is of particular importance in the case of variable traffic intensity at an airport. The advantage of the presented security control model is the possibility of pre-defining the capacity of the individual areas in the security control process to determine the functional readiness index. The presented model also allows for determining the minimum capacity of the individual areas to obtain the assumed level of the functional readiness index.

3   Simulation Model of the Security Check Stations

The functional readiness index was adopted as the measure of functional readiness (1)[36], the probability of an event (the variable U_t), for which the user appearing at a given time is served within the assumed time limits t_{max} with the assumption that the system (the variable S_t) is fit (is reliable - able to perform the assumed functions under specific conditions and within the specified time limits [37]).

$$Af(t) = Pr(U_t|S_t) = Pr\ (response\ time(T) < t_{max}|T = t\ S_t) \qquad (1)$$

The simulation model was developed using the FlexSim computer simulation tool in the Flexscript dedicated programming language. The objective function is obtaining the functional readiness of the system equal to 1 for pre-defined input parameters.

For the purposes of the development of the simulation model, individual check stations are divided into three main areas: entry area $Q_{IA}(j)$, control area $Q_M(j)$ and exit area $Q_{OA}(j)$, where j means the next station index.

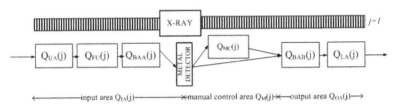

Fig. 2. The structure of a single security control station

Sub-areas were designated inside main areas (Fig 2.), where individual stages of the security control are held:

- $Q_{UA}(j)$ – a sub-area of the passenger's preliminary preparation for security control. In this area, the passenger prepares for security control on their own in accordance with their knowledge pertaining to limitations for hand baggage. The time of performing this activity is determined by the time $t_{1stunl}(i)$,

- $Q_{FU}(j)$ - a sub-area of the passenger's preliminary preparation for security control. In this area, the passenger prepares for security control with the assistance of a security control area employee. The time of performing this activity is determined by the time $t_{finunl}(i)$,

- $Q_{BAA}(j)$ - a sub-area of the passenger's waiting for security control. The waiting time depends on the availability of the metal detector.

- - the metal detector sub-area where the passenger goes through the security screening process. If the passenger is indicated by the metal detector, he/she goes to the hand search area; otherwise, he/she goes to the baggage-waiting area,

- $Q_{MC}(j)$ – a sub-area where a hand search is performed. The time of performing this activity is determined by the time $t_{mcont}(i)$,

- $Q_{BAB}(j)$ – a buffer sub-area where the passenger waits for his/her baggage which is checked separately. The time of performing this activity is determined by the time $t_{wfi}(i)$,

- $Q_{LA}(j)$ the baggage and coat collection area from conveyors after the security control. The time of performing this activity is determined by the time $t_{load}(i)$,

Each of the defined sub-areas is characterised by a capacity of the number of passengers which can stay in it at a time: $c_{UA}(j)$, $c_{FU}(j)$, $c_{BAA}(j)$, $c_{MC}(j)$, $c_{BAB}(j)$, $c_{LA}(j)$. The system user enters the capacities of the main areas $c_{IA}(j)$ i $c_{OA}(j)$ as the input data. The capacity of the $c_{FU}(j)$, $c_{MC}(j)$ sub-areas is determined at 1. The other areas change their capacities dynamically in further units of the simulation time and are equal to the number of passengers present in a given area at the time. However, relationships (2), (3) must be met which limit the possibility of exceeding the capacity of the entry and exit areas.

$$c_{IA}(j) \geq c_{UA}(j)_{(t_{simulation})} + c_{FU}(j) + c_{BAA}(j)_{simulation)} \qquad (2)$$

$$c_{OA}(j) \geq c_{BAB}(j)_{(t_{simulation})} + c_{LA}(j)_{(t_{simulation})} \qquad (3)$$

The simulation model allows for determining the functional readiness of the security control area which consists of any number of stations with variable input parameters. For this purpose, the structure is recorded in accordance with (4).

$$| NS | n | t_{max} | c_{IA}(j = 1), \dots , c_{IA}(j = n)| c_{OA}(j = 1), \dots , c_{OA}(j = n)| MC\% | \quad (4)$$

where:

— NS means the distribution of the intensity of the passenger report stream provided in the Flexscript programming language,
— n means the number of passenger service stations,
— t_{max} means the assumed maximum time spent by the passenger in the system provided in minutes,
— $c_{IA}(j = 1), \dots , c_{IA}(j = n)$ means the capacity of areas $c_{LA}(j)$ for further stations,
— $c_{OA}(j = 1), \dots , c_{OA}(j = n)$ means the capacity of areas $c_{OA}(j)$ for further stations,
— $MC\%$ means the percentage of passengers selected for the hand search.

For example, for a stream of reports defined by the exponential distribution with the parameter $\lambda = 0.08$, in the security control area consisting of 3 identical stations with capacities of c_{LA} (j)= 5 and c_{OA} (j)= 4, where 10% of passengers are selected for the hand search with a maximum time of stay in the system of less than 8 minutes, the formal notation (5) is used.

$$| exponential\ (0,12.5,0) | 3 | 8 | 5,5,5 | 4,4,4 | 10 | \qquad (5)$$

Such an approach allows for determining the functional readiness with the assumed configuration of the security control area. If the user wants to determine solutions for some assumptions, the system performs an analysis for further configurations of natural numbers until the maximum functional readiness is obtained. By modifying equation (5) and assuming that the number of stations is known, we will obtain (6).

$$| exponential\ (0,12.5,0) | n | 8 | 5,5,5 | 4,4,4 | 10 | \qquad (6)$$

The system will perform simulations for $n = 1$, $n = 2$, …, $n = k$ where for k, the functional readiness equal to 1 will be obtained.

The first stage of the passenger service process performance at the security control point is performed by entering the input data. The passenger at the time of the report is placed in the queue Q_E with an unlimited capacity (no loss system). Next, at the first moment of the entry areas at any station or in accordance with the shortest queue selection principle, the passenger is moved to one of the stations. The operation of the simulation model of the security control process implementation is shown in Figure 3, where times of performance are generated at individual stages (e.g. $t_{1stunl}(i)$) and next the passenger is directed to the next area (e.g. $c_{FU}(j)$).

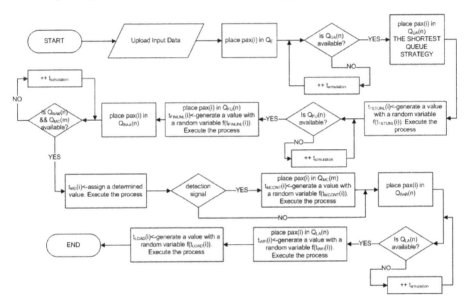

Fig. 3. Passenger service model algorithm

The model operation was verified at the Wrocław Airport using the system in operation. On the basis of the research conducted, the input data describing durations of the activities described in Chapter 2.1 are determined. These characteristics are presented by the following formulas (7-11).

$$f(t_{1stunl}) = \frac{t_{1stunl}^{3.42}}{13.74^{4.42} \cdot \Gamma(4.42)} e^{-\frac{t_{1stunl}}{13.74}} \tag{7}$$

$$f(t_{finunl}) = \frac{e^{-\frac{23.62}{t_{finunl}-0.94}}}{23.62 \cdot \Gamma(2.02) \cdot \left(\frac{t_{finunl}-0.94}{23.62}\right)^{3.02}} \tag{8}$$

$$f(t_{mcont}) = \frac{(t_{mcont}-16.95)^6}{2.13^7 \cdot \Gamma(7)} \cdot e^{-\frac{t_{mcont}-16.95}{2.13}} \tag{9}$$

$$f(t_{wfi}) = \frac{1}{B(0.67;2.56)} \cdot \frac{(t_{wfi}-2.93)^{-0.33} \cdot (139.10-t_{wfi})^{1.56}}{136.17^{2.23}} \tag{10}$$

$$f(t_{load}) = \frac{2.01}{23.62} \cdot \left(\frac{t_{load}-0.94}{23.62}\right)^{1.01} e^{-\left(\frac{t_{load}-0.94}{23.62}\right)^{2.01}} \tag{11}$$

The consistency of empirical and theoretical distribuants was verified using the Kolmogorov consistency test at a significance level of α = 0.05. In all cases, the distribuants were found to be consistent with the ones proposed above (7-11) (values lower than the limit value $\lambda_{0.05} = 1.36$). The model users can also introduce their own characteristics.

Verification was performed on the basis of average passenger stream intensities with a continuous stream of reports in an hourly unit of time for the real system and values obtained from the simulation model. Various capacities of areas c_{IA} and c_{OA}. The values observed at the operating facility and the values obtained in the simulation model are presented in Table 1, where NO denotes configurations which were not verified.

Table 1. Average passenger traffic intensity for theoretical and empirical values

| | | c_{IA} | | | |
		3	4	5	6
	3	NO	NO	108 *(106)*	NO
c_{OA}	4	NO	115 *(111)*	127 *(130)*	136 *(135)*
	5	NO	119 *(118)*	143 *(139)*	153 *(155)*
	6	NO	121 *(119)*	144 *(140)*	158 *(159)*

where:

108 – average observed passenger intensity,
(106) – average passenger intensity obtained from the simulation model.

The chi-square test at $\alpha = 0.05$ was used to verify the hypothesis on the consistency between the empirical and theoretical data obtained from the simulation model. The chi-square statistics value was lower than the limit value $\aleph^2_{0.05}=16.919$ which proves the consistency between the data.

4   The Use of the Simulation Model of the Security Check Stations

While designing the security control area, it was assumed that the forecast intensity of passengers reporting for the security check is described by the exponential distribution with the parameter λ=0.08. The Airport intends to perform the process at a maximum of 3 service stations. It was also assumed that 25% of passengers will be subjected to the hand search and the time spent by the passenger in the security control area should not exceed 10 minutes. The input data of the simulation model are defined in accordance with (12).

$$| \, exponential(0,12.5,0) \mid 3 \mid 10 \mid C_{IA} \mid C_{OA} \mid 25 \mid \qquad (12)$$

Therefore, the designation of the capacity of the entry and exit zones is of key importance so that the functional capacity equal to 1 is obtained for the pre-defined assumptions. The results obtained from the simulation model are presented in Table 2.

Table 2. Simulation results

		c_{IA}				
		1	2	3	4	5
	1	0	0	0	0	0
	2	0	0	0	0	0
c_{OA}	3	0	0	0.3	0.3	0.3
	4	0	0	0.9	1	1
	5	0	0	1	1	1

On the basis of the results obtained, it can be noticed that the smallest of the available stations meeting the pre-defined parameters create configurations (13), (14).

$$| exponential(0,12.5,0) | 3 | 10 | 3,3,3 | 5,5,5 | 25 | \tag{13}$$

$$| exponential(0,12.5,0) | 3 | 10 | 4,4,4 | 4,4,4 | 25 | \tag{14}$$

A broader analysis of functional readiness of the proposed configurations depending on changes in the passenger report intensity stream (Figure 5) shows that configuration (14) is characterised by a higher efficiency which allows for serving approx. 70 passengers more at 3 stations during an hour.

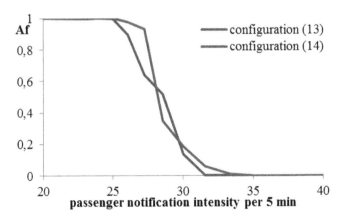

Fig. 4. Functional readiness of determined configurations depending on the report stream

5   Summary

The study presents a simulation model, the universality of which allows for its use both in the process of planning and optimisation of the security control area which consists of stations with a single stream of the passenger flow. The developed simulation model takes into account regulations connected with the security assurance and it was verified using the real system of the Wrocław Airport. The main aspect of the model functioning involves the assurance of functional readiness of the system depending on the forecast stream of reports.

Further work is being conducted to develop a logistics support mode for the Wrocław Airport operation. This model will take into account both the security control system and the check-in system, the departure hall system, the baggage-handling system, the aircraft ground-handling system. These systems will be dependent systems and the operation of each of them will have a significant influence on the functional readiness index of the whole system. Analytical models for system operation will be determined. Reliability aspects of device operation will be taken into account.

Acknowledgements. The project was co-financed by the National Research and Development Centre the Applied Research Programme. This publication presents the results of research conducted in the project: " Model of logistical support for the functioning of the Wrocław Airport" realized by the Wrocław University of Technology and Wrocław Airport consortium.

References

1. Bevilacqua, M., Ciarapica, F.E.: Analysis of check-in procedure using simulation: a case study. In: IEEE Int. Conf. Industrial Engineering and Engineering Management (IEEM), pp. 1621–1625 (2010)
2. Bujak, A., Zając, P.: Can the increasing of energy consumption of information interchange be a factor that reduces the total energy consumption of a logistic warehouse system? In: Mikulski, J. (ed.) TST 2012. CCIS, vol. 329, pp. 199–210. Springer, Heidelberg (2012)
3. Cooper, R.B.: Introduction to queueing theory, 2nd edn. Elsevier North Holland, New York (1981)
4. Ustaw, D.: The Regulation of the Minister of Transport, Construction and Maritime Economy of 31 July 2012 on the National Civil Aviation Security Program. Journal of Laws of the Republic of Poland 2012 Item 912
5. Eilon, S., Mathewson, S.: A simulation study for the design of an air terminal building. IEEE Transactions on Systems, Man and Cybernetics 3(4), 308–317 (1973)
6. EC, Commision Regulation (EC) No 300/2008 of 11 March 2008 on common rules in the field of civil aviation security and repealing Regulation (EC) No 2320/2002
7. EC, Commission Regulation (EC) No 272/2009 of 2 April 2009 supplementing the common basic standards on civil aviation security laid down in the Annex to Regulation (EC) No 300/2008 of the European Parliament and of the Council
8. EU, Commission Regulation (EU) No 185/2010 of 4 March 2010 laying down detailed measures for the implementation of the common basic standards on aviation security
9. Gkritza, K., Niemeier, D., Mannering, F.: Airport Security Screening and changing passenger satisfaction: An exploratory assessment. Journal of Air Transport Management 12(5), 213–219 (2006)
10. Greghi, M., Rossi, T., de Souza, J., Menegon, N.: Brazilian passengers' perceptions of air travel: evidences from a survey. Journal of Air Transport Management 31, 27–31 (2013)
11. Hamzawi, S.G.: Lack of airport capacity: exploration of alternative solutions. Transportation Research Part A 26(1), 47–58 (1992)

12. ICAO, Safety Management Manual (SMM), 3rd edn., International Civil Aviation Organization (2013)
13. Jodejko-Pietruczuk, A., Werbińska-Wojciechowska, S.: Analiza parametrów modeli obsługiwania systemów technicznych z opóźnieniami czasowymi. Eksploatacja i Niezawodność – Maintenence and Reliability 16(2), 288–294 (2014)
14. Kierzkowski, A., Kisiel, T.: Wyznaczanie podstawowych charakterystyk procesu kontroli bezpieczeństwa dla zimowego sezonu lotniczego z wykorzystaniem modelu symulacyjnego łączonego stanowiska obsługi. Prace Naukowe Politechniki Warszawskiej. Transport, z 103, 113–123 (2014)
15. Kierzkowski, A., Kisiel, T.: Wyznaczanie podstawowych charakterystyk dla zimowego rozkładu lotów modelu symulacyjnego funkcjonowania pojedynczego stanowiska kontroli bezpieczeństwa. Logistyka (3), 2910–2919 (2014)
16. Kowalski, M., Magott, J., Nowakowski, T., Werbinska-Wojciechowska, S.: Analysis of transportation system with the use of Petri nets. Eksploatacja i Niezawodnosc – Maintenance and Reliability (1), 48–62 (2011)
17. Lemer, A.C.: Measuring performance of airport passenger terminals. Transportation Research Part A: Policy and Practice 26(1), 37–45 (1992)
18. Magott, J., Nowakowski, T., Skrobanek, P., Werbinska, S.: Analysis of possibilities of timing dependencies modeling-Example of logistic support system. In: Martorell, S., Guedes Soares, C., Barnett, J. (eds.) Safety, Reliability and Risk Analysis: Theory, Methods and Applications, vol. 2, pp. 1055–1063. CRC Press; Taylor & Francis, Boca Raton (2009)
19. Manataki, I.E., Zografos, K.G.: Assessing airport terminal performance using a system dynamics model. Journal of Air Transport Management 16(2), 86–93 (2010)
20. McCullough, B.F., Roberts, F.L.: Decision tool for analysis of capacity of airport terminal. Transportation Research Record 732, 41–54 (1979)
21. McKelvey, F.X.: Use of an analytical queuing model for airport terminal design. Transportation Research Record 1199, 4–11 (1989)
22. Nowakowski, T., Werbińka, S.: On problems of multicomponent system maintenance modelling. International Journal of Automation and Computing 6(4), 364–378 (2009)
23. Nowakowski, T., Zając, M.: Analysis of reliability model of combined transportation system. In: Advances in Safety and Reliability - Proceedings of the European Safety and Reliability Conference, ESREL 2005, pp. 147–151 (2005)
24. Plewa, M.: Assessment of influence of products' reliability on remanufacturing processes. International Journal of Performability Engineering, 463–470 (2009)
25. Restel, F.: The Markov reliability and safety model of the railway transportation system. In: Safety and Reliability: Methodology and Applications - Proceedings of the European Safety and Reliability Conference, pp. 303–311 (2014)
26. Restel, F.: Train punctuality model for a selected part of railway transportation system. In: Safety, Reliability and Risk Analysis: Beyond the Horizon - Proceedings of the European Safety and Reliability Conference, pp. 3013–3019 (2013)
27. Roanes-Lozano, E., Laita, L.M., Roanes-Macas, E.: An accelerated-time simulation of departing passengers' flow in airport terminals. Mathematics and Computers in Simulation 67(1-2), 163–172 (2004)
28. Saffarzadeh, M., Braaksma, J.P.: Optimum design and operation of airport passenger terminal buildings. Transportation Research Record 1703, 72–82 (2000)
29. Siergiejczyk, M., Krzykowska, K.: Some issues of data quality analysis of automatic surveillance at the airport, Diagnostyka (2014)

30. Siergiejczyk, M., Krzykowska, K., Rosinski, A.: Parameters Analysis of Satellite Support System in Air Navigation. In: Selvaraj, H., Zydek, D., Chmaj, G. (eds.) Progress in Systems Engineering: Proceedings of the Twenty-Third International Conference on Systems Engineering. AISC, vol. 330, pp. 673–678. Springer, Switzerland (2015)

31. Skorupski, J., Stelmach, A.: Selected models of service processes at the airport. Systems Science 34(3), 51–59 (2008)

32. Stańczyk, P., Stelmach, A.: The use on-board flight recorder in the modeling process of air-craft landing operations. In: Nowakowski, T. (red.) Safety and Reliability: Methodology and Applications – Proceeding of the European Safety and Reliability Conference, ESREL 2014. Wydawnictwo Politechniki Warszawskiej (2015) ISBN 978-1-138-02681-0, ss. 2029, doi:10.1201/b17399-278

33. Stańczyk, P., Stelmach, A.: Artificial Neural Networks Applied to the Modeling of Aircraft Landing Phase. In: 10th European Conference of Young Research and Scientists - Proceedings, Zilina, pp. 169–173 (2013) ISBN:978-80-554-0690-9

34. Solak, S., Clarke, J.-P.B., Johnson, E.L.: Airport terminal capacity planning. Transportation Research Part B: Methodological 43(6), 659–676 (2009)

35. Stelmach, A., Malarski, M., Skorupski, J.: Model of airport terminal area capacity investigation. In: Proceedings of the European Safety and Reliability Conference 2006, ESREL 2006 - Safety and Reliability for Managing Risk, vol. 3, pp. 1863–1868 (2006)

36. Walkowiak, T.: Symulacyjna ocena gotowości usług systemów internetowych z realistycznym modelem czasu odnowy. Eksploatacja i Niezawodność – Maintenence and Reliability 16(2), 341–346 (2014)

37. Ważyńska-Fiok, K., Jaźwiński, J.: Niezawodność Systemów Technicznych. Państwowe Wydawnictwo Naukowe, Warszawa (1990)

38. Werbinska-Wojciechowska, S.: Time resource problem in logistics systems dependability modelling. Eksploatacja i Niezawodnosc – Maintenance and Reliability (4), 427–433 (2013)

39. Vališ, D., Koucký, M., Žák, L.: On approaches for non-direct determination of system deterioration. Eksploatacja i Niezawodnosc - Maintenance and Reliability 14(1), 33–41 (2012)

40. Vališ, D., Vintr, Z., Malach, J.: Selected aspects of physical structures vulnerability – state-of-the-art. Eksploatacja i Niezawodnosc – Maintenance and Reliability 14(3), 189–194 (2012)

41. Vališ, D., Pietrucha-Urbanik, K.: Utilization of diffusion processes and fuzzy logic for vulnerability assessment. Eksploatacja i Niezawodnosc - Maintenance and Reliability 16(1), 48–55 (2014)

42. Zając, M., Świeboda, J.: Analysis of the process of unloading containers at the inland container terminal. In: Safety and Reliability: Methodology and Applications - Proceedings of the European Safety and Reliability Conference, ESREL 2014, pp. 1237–1241 (2014)

43. Zając, P.: The idea of the model of evaluation of logistics warehouse systems with taking their energy consumption under consideration. Archives of Civil and Mechanical Engineering 11(2), 479–492 (2011)

Functional Readiness of the Check-In Desk System at an Airport

Artur Kierzkowski and Tomasz Kisiel

Wroclaw University of Technology, 27 Wybrzeże Wyspiańskiego St., 50-370 Wrocław
{artur.Kierzkowski,tomasz.kisiel}@pwr.edu.pl

Abstract. The article presents a developed simulation model of the check-in process at an airport. The developed algorithm of the process allows for an analysis of the entire system for any preset input parameters. The effectiveness of the functioning of check-in desks is measured by functional readiness, which can confirm both the reliability of the process in terms of the expected effectiveness and indicated the quality of the service offered (the time spent by the passenger in the system). The simulation model also allows for conducting an analysis of the possibility of controlling the input stream for security control, which is the next sub-process in the departing passenger check-in structure. Characteristics pertaining to the operation of the system were implemented in the model, such as the passenger service time and the structure of passenger reports to the system, which were determined on the basis of research conducted at a real airport.

Keywords: simulation model, check-in, functional readiness.

1 Introduction

The dynamic development of the air transport system makes it necessary to conduct continuous research in increasing the throughput capacity of the transport network. The critical infrastructure, i.e. passenger terminals, is also of key importance here. Already in 2015, an increase in the air passenger transport is expected up to the level of 7.2 billion [9], which constitutes an increase by 0.18% as compared to 2013. The current infrastructure of the system results in numerous air traffic delays, which especially affect busy airports. According to [10], the highest percentage of delayed arrivals (42.8%) in Europe is observed in London LHR. The highest percentage of delayed departures (54.7%), on the other hand, is observed in Rome.

The air transport system is an extremely complex system, in which the timely performance of tasks is determined by a lot of factors. The check-in process consists of a large number of synergistic stages in various service subsystems. These are mostly anthropotechnical subsystems, in which there exists a strong relationship between human-machine-environment relations. In the cybernetic approach, humans are an integral part of a complex system [7]. The description of such a system using mathematical equations would require the use of several hundred differential equations [15].

© Springer International Publishing Switzerland 2015
W. Zamojski et al. (eds.), *Theory and Engineering of Complex Systems and Dependability,*
Advances in Intelligent Systems and Computing 365, DOI: 10.1007/978-3-319-19216-1_21

Thus, a systemic approach should be used, which would take into account the use of computer simulation [4] allowing the representation of a real system as well as its multi-criteria assessment.

In the study, the authors describe a check-in subsystem at an airport. A process, which involves passengers checking in hold luggage or those, who have not checked in online, takes place at check-in desks. The stream of departing passengers in the landside area is divided into direct reports for security control and reports after the check-in. Due to the fact the security control subsystem is often a bottleneck at numerous airports, proper management of processes preceding it is very important as they may have a direct influence on the stream of reports for security control. This fragment of the system is presented in Figure 1, in which the scope of the developed simulation model is presented. From the point of view of the system manager, the reliability of the process and the possibility of controlling the passenger stream are important.

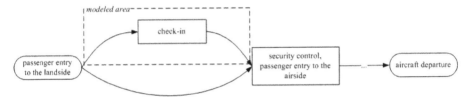

Fig. 1. Simplified diagram of departing passenger service

Preliminary issues connected with the use of terminal areas are presented in [5], [8], [12], [13], [27], [31]. There exist a range of models, which divide the system operation into a collection of individual queuing systems. A lot of models have been developed for these systems, which are used to estimate basic indices and parameters [3]. Already, initial considerations are allowed for introducing a multi-criteria analysis for passenger flows at the passenger terminal [6]. The problem of queuing time minimisation with minimal use of resources have been brought up on numerous occasions [1], [17]. Also, several models are adopted for the selection of an appropriate number of technical resources to handle the assumed traffic flow [24], [28]. An important element is also the modelling of an air transport system in a broader approach, e.g in the airport area [25], [29], [30], or along the flight route [26], which allows for obtaining a full picture of the management process for the entire air transport system.

An analysis of a transport system using the Petri net was presented in [14]. The authors have discussed basic limitations of reliability modelling and system operation methods. In many studies, the authors show that it is advisable to estimate parameters of the technical system operation [11]. The authors conducted analyses, which allowed them to determine the influence of the time delay parameter estimation level on the technical system operation. The analysis performed in the study mostly concerned the expected time delay value on the level of operation of a multi-element technical system. The problems with the modelling of multi-phase system reliability are presented in [18]. From the models obtained, the authors have estimated primary measures of system reliability, such as: the probability of proper performance of the logistic task, probability of error occurrence in the time function, etc. At present, a lot of studies focus on the possibility of using fuzzy logic in complex systems [34]. The model will also take into consideration aspects connected with reliability of the technical system [19], [20], [21], the effectiveness of its operation [2], [16], [22], [23], [36], [38] and susceptibility aspects [32], [33], [37].

The simulation model, which is presented in a further part of the study, focuses on system reliability analysis in terms of the passenger service quality – the time spent by passengers in the system. Due to the fact that the developed model constitutes a part of the work on developing a passenger service model at the entire passenger terminal, it also focuses on the analysis of possibilities of controlling the stream of passengers reporting for security control.

2   Simulation Model of the Check-In Desk System at an Airport

There are three ways of conducting the check-in. The first one involves the check-in in a flight system. Separate check-in desks are assigned to each flight and the stream of reporting passengers of a given flight is separated from the other ones. Common check-in involves assignment of joint check-in desks for one carrier, several carriers or all flights. For this check-in method, passenger streams for various flights join in a queue to check-in desks. The third way is the mixed method. Common check-in is the main method; however, there may be desks assigned to specific flights in a selected unit of time.

Depending on the predefined timetable, the check-in system structure is generated. An example of such a structure is presented in Figure 2.

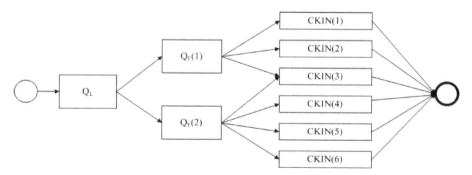

Fig. 2. An example of the check-in system structure

To describe appropriate elements of the system and relationships between them, the following assumptions were adopted. The check-in process begins two hours before the planned departure. Q_L is the landside area, to which passengers generated by the system are directed. Next, the queue $Q_F(fn)$ for the check-in is generated for each flight stream. fn is the assigned flight number. Passengers are served at $CKIN(X)$ desks, where X is the consecutive desk number. As a result, Figure 1 presents a diagram, in which 2 flights are planned in the timetable. The check-in for flight 1 is to be held at desks $CKIN=\{1,2,3\}$, while the check-in for flight 2 at desks $CKIN=\{3,4,5,6\}$. Depending on the additional input data, if the check-in time for the presented flights overlaps to any extent, desk $CKIN(3)$ will be used in the common check-in system, while the other ones will be used in the flight system. If the check-in is conducted at different times, the check-in will be held in the flight system.

The flight entered in the system is characterised by appropriate structural notation:

$$| fn | dep_{fn} | lpax_{fn} | S_{fn} | \tag{1}$$

where:
fn – flight number,
dep_{fn} – departure time expressed in seconds, where 0 means midnight on the first day of simulation.
$Lpax_{fn}$ – number of passengers reporting for the check-in
S_{fn} –set of check-in desks assigned to the flight e.g. $\{1,2,3\}$

The flow of the stream of subsequent passengers through the system is described by random variables of the cumulative distribution function) (2) and (3), which were developed on the basis of research conducted at the Wrocław Airport (EPWR). The research was conducted for traditional flights and for low-cost airlines.

$$f\left(t_{NOT(i)}\right) = 1 - exp\left(-\left(\frac{t_{NOT(i)}-572.54}{5419.34}\right)^{3.52}\right) \tag{2}$$

$$f\left(t_{H(i)}\right) = \Phi\left(\sqrt{\frac{131.37}{t_{H(i)}-19.27}}\left(\frac{t_{H(i)}-19.27}{64.10}-1\right)\right) + \tag{3}$$

$$\Phi\left(-\sqrt{\frac{131.37}{t_{H(i)}-19.27}}\left(\frac{t_{H(i)}-19.27}{64.10}+1\right)\right)exp(2\cdot131.37/64.10)$$

where:
$t_{NOT(i)}$ - means subsequent reports of passengers to the terminal,
$t_{H(i)}$ – means passenger service times at the desk.

The value of the statistics test for $f\left(t_{NOT(i)}\right)$ is 0.96 and for $f\left(t_{H(i)}\right)$, it is equal to 0.63. These values are appropriately lower than the critical value of $\lambda_{0.05}=1.36$. Therefore, on the basis of the Kolmogorov test, there are no grounds for rejecting the hypothesis. The theoretical data are consistent with the empirical data. The model user can also introduce their own characteristics.

The simulation mode developed with the use of the FlexSim simulation tool performs the algorithm of the parallel passage of individual passengers (Fig. 3) through the system. Input data are collected for the system and, next, in accordance with (2), passengers marked as $PAX_{i,fn}$, are generated, where i denotes the next ordinal number assigned to the subsequent passengers and fn the number of the flight, to which a given passenger is assigned, are moved to the landside part of the terminal when they wait for the check-in to begin. The moment the check-in starts or if the check-in is already available after being generated, the passenger is moved to an appropriate queue, which has been assigned to their flight. At that time, the beginning of the queuing time is recorded on the passenger's label, $t_{ENT(i)}$.

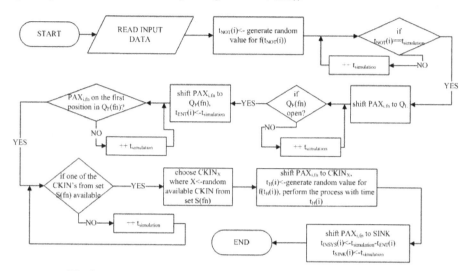

Fig. 3. The algorithm for a single passenger passage through the system

In a one-flight queue in the flight system, passengers are served according to the FIFO principle. For common check-in desks, the passenger with the longest waiting time is selected for service, considering all flights served by the common check-in desk.

When the passenger is the first in the queue and at least one check-in desk assigned to his/her flight is available, the passenger is moved to a random available desk. The service time consistent with (3) is generated and the process is conducted. After the end of the service, labels with time spent in the system (4) from joining the queue $Q_{F(fn)}$ and the time of leaving the system (5) are assigned to passengers. The data are recorded in the global table and the passenger is removed from the system.

$$t_{INSYS(i)} = t_{simulation} - t_{ENT(i)} \tag{4}$$

$$t_{SINK(i)} = t_{simulation} \tag{5}$$

where:
$t_{simulation}$ – current simulation time.

3   Functional Readiness of the Check-In Desk System

In study [35], the author defines functional readiness $(Af(t)$ as the probability of an event (variable U_t), for which a user appearing at a given time will be served within the assumed time limits t_{max} with the assumption that the system (variable S_t) is fit.

$$Af(t) = \Pr(U_t|S_t) = \Pr\left(responsetime(T) < t_{max}|T = t \; S_t\right) \tag{6}$$

This index can be used as the indicator of the reliability of the system, in which it is assumed that the critical value of the time spent by the passenger in the system may not exceed a certain value due to the assumed passenger service quality at a given airport.

It was assumed that this value should not exceed $t_{max}=600$ of simulation time units. Due to the nature of passenger reports to the passenger terminal and the time devoted to the check-in, the first aircraft departure is possible at 10800. It was assumed that at an airport, at which the system consists of 15 check-in desks, five flights departing at the same time will be served according to the input data (Table 1.). It was assumed that functional readiness would be determined for four different scenarios, in which the number of check-in desks was adopted as a variable. The first three scenarios assume the check-in in the flight system. The fourth system assumes that each flights has two dedicated check-in desks in the flight system and one additional common check-in desk for all flights served.

Table 1. Input data for 4 simulation scenarios

| | | | simulation scenario | | | |
| | | | 1 | 2 | 3 | 4 |
fn	dep	$lpax$	S	S	S	S
1	10800	100	{1,2,3}	{1,2}	{1}	{1,2,3}
2	10800	100	{4,5,6}	{4,5}	{4}	{1,4,5}
3	10800	100	{7,8,9}	{7,8}	{7}	{1,7,8}
4	10800	100	{10,11,12}	{10,11}	{10}	{1,10,11}
5	10800	100	{13,14,15}	{13,14}	{13}	{1,13,14}

Functional readiness was determined for all assumed simulation scenarios. The results of the simulation for the flight system are presented in Figure 4. The check-in for the defined flights begins at $t_{simulation}=3600$. It can be noticed that the process runs in the correct manner only for the first scenario. The reduction in the number of check-in desks assigned to each flight to 2 limits the reliability of the process for nearly the whole check-in time, which does not meet the assumed expectations of the process manager. The assignment of just one check-in desk to each flight is characterised by a complete lack of functional readiness virtually for the entire check-in time for the defined flights. The selection of the first scenario, however, is connected with the total costs incurred by the handling operator. It is necessary to provide 15 process operators.

In a real system, with uneven flight distribution, a temporary peak may make it necessary to ensure a larger number of operators over a short period of time, which generates system operation costs. Scenario 4 of the simulation shows that the opening of one common check-in desk for any flights in the system and additional flight-based check-in at two desks assigned to each flight makes it possible to achieve high functional readiness of the system for such a predefined timetable (Fig. 5). In this case, only 11 operators are needed for the process.

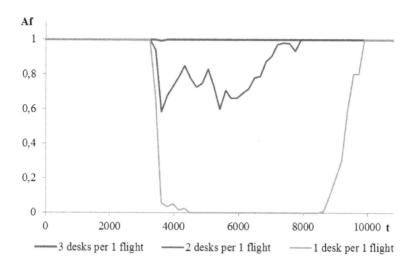

Fig. 4. Simulation results for scenarios 1-3

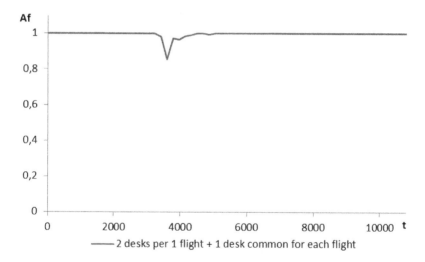

Fig. 5. Simulation results for scenario 4

Apart from determination of functional readiness in the selection of the number of check-in desks, it is also important that the timetable should be developed in a proper manner. The check-process has a direct influence on the security control process. The structure of the passenger flow through the check-in process influences significantly the structure of reports for security control. Inadequate management of the process may result in peaks in the stream of reports for security control.

The simulation model at the current phase of development makes it possible to examine the structure of the stream leaving the check-in system. Fig. 6 presents characteristics of the output stream depending on the predefined timetable.

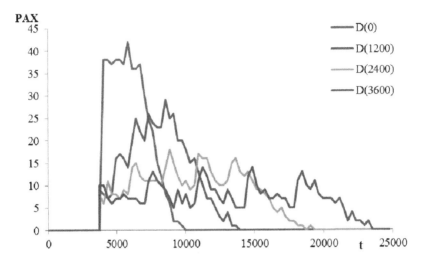

Fig. 6. The output stream structure for a predefined timetable

Scenario 4 was adopted as output point *D(0)* (Table 1.). Next, simulation was conducted shifting the departures of the next flights in time, so that relationship (7) is met.

$$D(Y) = dep_{(fn+1)} - dep_{(fn)} = Y \tag{7}$$

where:
Y – means the difference in time between the departures of two subsequent flights.

The results show that an inadequate timetable may result in considerable differences in the output stream from the check-in. At a further stage of the research, there are also characteristics pertaining to the probability of passenger reports for security control in time *t*, depending on the time of leaving the check-in using domes, which will allow a broader analysis of relationships between the two passenger service subsystems departing from a given airport.

4   Summary

The article presents a universal model of a passenger check-in system at an airport terminal. The presented simulation model allows for performing an analysis of the process functioning in terms of the functional readiness obtained. Additionally, for the preset functional parameters, it was possible to determine the output stream leaving the system, which will be used in a further part of model development as input data for the security control system.

Further work focuses on determining stream flow characteristics through the check-in system, taking into account the type of carrier (traditional, low-costs, charter) and on determining characteristics of the passenger stream flow between leaving the check-in system and entering the security control system. This will allow for coupling two system, thus creating a planning and management system for passenger service in the entire landside area of the passenger terminal.

Acknowledgement. The project is co-financed by the National Research and Development Centre under the Applied Re-search Program. This publication presents the results of research conducted in the project: "Model of lo-gistical support for the functioning of the Wrocław Airport" realized by the Wrocław University of Technology and Wrocław Airport consortium.

References

1. Bevilacqua, M., Ciarapica, F.E.: Analysis of check-in procedure using simulation: a case study. In: IEEE Int. Conf. Industrial Engineering and Engineering Management (IEEM), pp. 1621–1625 (2010)
2. Bujak, A., Zając, P.: Can the increasing of energy consumption of information interchange be a factor that reduces the total energy consumption of a logistic warehouse system? In: Mikulski, J. (ed.) TST 2012. CCIS, vol. 329, pp. 199–210. Springer, Heidelberg (2012)
3. Cooper, R.B.: Introduction to queueing theory, 2nd edn. Elsevier North Holland, New York (1981)
4. Dacko, M.: Systems Dynamics in Modeling, Sustainable Management of the Environment and Its Resources. Polish Journal of Environmental Studies 19(4) (2010)
5. De Lange, R., Samoilovich, I., Van Der Rhee, B.: Virtual queuing at airport security lanes. European Journal of Operational Research 225(1), 153–165 (2013)
6. Eilon, S., Mathewson, S.: A simulation study for the design of an air terminal building. IEEE Transactions on Systems, Man and Cybernetics 3(4), 308–317 (1973)
7. Gomółka, Z.: Cybernetyka w zarządzaniu. Placet, Warszawa (2000)
8. Hamzawi, S.G.: Lack of airport capacity: exploration of alternative solutions. Transportation Research Part A 26(1), 47–58 (1992)
9. International Air Transport Association (IATA). Simplifying the Business e Passenger Facilitation (2012), http://www.iata.org/whatwedo/stb/Documents/pf-presentation-2012.pdf (latest accessed on August 06, 2014)
10. International Air Transport Association (IATA). CODA Digest Delays to Air Transport in Europe – Annual 2013 (2014), https://www.eurocontrol.int/sites/default/files/content/documents/official-documents/facts-and-figures/coda-reports/coda-digest-annual-2013.pdf (latest accessed on March 04, 2014)
11. Jodejko-Pietruczuk, A., Werbińska-Wojciechowska, S.: Analiza parametrów modeli obsługiwania systemów technicznych z opóźnieniami czasowymi. Eksploatacja i Niezawodność – Maintenence and Reliability 16(2), 288–294 (2014)

12. Kierzkowski, A., Kisiel, T.: Wyznaczanie podstawowych charakterystyk procesu kontroli bezpieczeństwa dla zimowego sezonu lotniczego z wykorzystaniem modelu symulacyjnego łączonego stanowiska obsługi. Prace Naukowe Politechniki Warszawskiej. Transport 103, 113–123 (2014)

13. Kierzkowski, A., Kisiel, T.: Wyznaczanie podstawowych charakterystyk dla zimowego rozkładu lotów modelu symulacyjnego funkcjonowania pojedynczego stanowiska kontroli bezpieczeństwa. Logistyka (3), 2910–2919 (2014)

14. Kowalski, M., Magott, J., Nowakowski, T., Werbinska-Wojciechowska, S.: Analysis of transportation system with the use of Petri nets. Eksploatacja i Niezawodnosc – Maintenance and Reliability (1), 48–62 (2011)

15. Łukaszewicz, R.: Dynamika systemów zarządzania. PWN, Warszawa (1975)

16. Magott, J., Nowakowski, T., Skrobanek, P., Werbinska, S.: Analysis of possibilities of timing dependencies modeling-Example of logistic support system. In: Martorell, S., Guedes Soares, C., Barnett, J. (eds.) Safety, Reliability and Risk Analysis: Theory, Methods and Applications, vol. 2, pp. 1055–1063. CRC Press; Taylor & Francis, Boca Raton (2009)

17. Manataki, I.E., Zografos, K.G.: Assessing airport terminal performance using a system dynamics model. Journal of Air Transport Management 16(2), 86–93 (2010)

18. Nowakowski, T.: Problemy modelowania niezawodności systemów wielofazowych. Eksploatacja i Niezawodność – Maintenence and Reliability 52(4), 79–84 (2011)

19. Nowakowski, T., Werbińka, S.: On problems of multicomponent system maintenance modelling. International Journal of Automation and Computing 4, 364–378 (2009)

20. Nowakowski, T., Zając, M.: Analysis of reliability model of combined transportation system. In: Advances in Safety and Reliability - Proceedings of the European Safety and Reliability Conference, ESREL 2005, pp. 147–151 (2005)

21. Plewa, M.: Assessment of influence of products' reliability on remanufacturing processes, International Journal of Performability Engineering, 463–470 (2009)

22. Restel, F.: The Markov reliability and safety model of the railway transportation system. In: Safety and Reliability: Methodology and Applications - Proceedings of the European Safety and Reliability Conference, pp. 303–311 (2014)

23. Restel, F.: Train punctuality model for a selected part of railway transportation system. In: Safety, Reliability and Risk Analysis: Beyond the Horizon - Proceedings of the European Safety and Reliability Conference, pp. 3013–3019 (2013)

24. Roanes-Lozano, E., Laita, L.M., Roanes-Macas, E.: An accelerated-time simulation of departing passengers' flow in airport terminals. Mathematics and Computers in Simulation 67(1-2), 163–172 (2004)

25. Siergiejczyk, M., Krzykowska, K.: Some issues of data quality analysis of automatic surveillance at the airport. Diagnostyka 15, 25–29 (2014)

26. Siergiejczyk, M., Krzykowska, K., Rosinski, A.: Parameters Analysis of Satellite Support System in Air Navigation. In: Selvaraj, H., Zydek, D., Chmaj, G. (eds.) Progress in Systems Engineering: Proceedings of the Twenty-Third International Conference on Systems Engineering. AISC, vol. 330, pp. 673–678. Springer, Switzerland (2015)

27. Skorupski, J., Stelmach, A.: Selected models of service processes at the airport. Systems Science 34(3), 51–59 (2008)

28. Solak, S., Clarke, J.-P.B., Johnson, E.L.: Airport terminal capacity planning. Transportation Research Part B: Methodological 43(6), 659–676 (2009)

29. Stańczyk P., Stelmach, A.: The use on-board flight recorder in the modeling process of air-craft landing operations. In: Nowakowski, T. (red.) Safety and Reliability: Methodology and Applications - Proceeding of the European Safety and Reliability Conference, ESREL 2014. Wydawnictwo Politechniki Warszawskiej (2015) ISBN 978-1-138-02681-0, ss. 2029, doi:10.1201/b17399-278

30. Stańczyk, P., Stelmach, A.: Artificial Neural Networks Applied to the Modeling of Air-craft Landing Phase. In: 10th European Conference of Young Research and Scientists - Proceedings, Zilina, pp. 169–173 (2013) ISBN:978-80-554-0690-9

31. Stelmach, A., Malarski, M., Skorupski, J.: Model of airport terminal area capacity investi-gation. In: Proceedings of the European Safety and Reliability Conference 2006, ESREL 2006 - Safety and Reliability for Managing Risk, vol. 3, pp. 1863–1868 (2006)

32. Vališ, D., Kouský, M., Žák, L.: On approaches for non-direct determination of system de-terioration. Eksploatacja i Niezawodnosc - Maintenance and Reliability 14(1), 33–41 (2012)

33. Vališ, D., Vintr, Z., Malach, J.: Selected aspects of physical structures vulnerability – state-of-the-art. Eksploatacja i Niezawodnosc – Maintenance and Reliability 14(3), 189–194 (2012)

34. Vališ, D., Pietrucha-Urbanik, K.: Utilization of diffusion processes and fuzzy logic for vulnerability assessment. Eksploatacja i Niezawodnosc - Maintenance and Reliabil-ity 16(1), 48–55 (2014)

35. Walkowiak, T.: Symulacyjna ocena gotowości usług systemów internetowych z realistycznym modelem czasu odnowy. Eksploatacja i Niezawodność – Maintenence and Reliability 16(2), 341–346 (2014)

36. Werbinska-Wojciechowska, S.: Time resource problem in logistics systems dependability modelling. Eksploatacja i Niezawodnosc – Maintenance and Reliability (4), 427–433 (2013)

37. Zając, M., Świeboda, J.: Analysis of the process of unloading containers at the inland con-tainer terminal. In: Safety and Reliability: Methodology and Applications - Proceedings of the European Safety and Reliability Conference, ESREL 2014, pp. 1237–1241 (2014)

38. Zając, P.: The idea of the model of evaluation of logistics warehouse systems with taking their energy consumption under consideration. Archives of Civil and Mechanical Engi-neering 11(2), 479–492 (2011)

Performance Issues in Creating Cloud Environment

Marcin Kubacki and Janusz Sosnowski

Institute of Computer Science, Warsaw University of Technology, Nowowiejska 15/19
Warsaw 00-665, Poland
{M.Kubacki,J.Sosnowski}@ii.pw.edu.pl

Abstract. The paper deals with the problem of analyzing performability of creating cloud environment within a server taking into account various hardware and software configurations. Within the hardware we consider the impact of using SATA and SSD discs for holding data and program codes of the involved processes. Within software we checked single and multiple thread approaches. The effectiveness of the activated processes was evaluated using performance monitoring based on vSphere tool. The performed measurements were targeted at CPU, RAM and disc usage as well as execution time analysis. We present the measurement methodology and interpretation of the obtained results, which proved their practical significance.

Keywords: Performance analysis, system monitoring, cloud environment, IaaS.

1   Introduction

Developing complex information systems we face the problem of optimizing hardware and software environment to assure appropriate level of performance, which is an important attribute of system dependability. This process can be supported with monitoring various parameters using hardware or software mechanisms. The first approach may base on available in-system mechanisms or external test equipment (including emulators). Contemporary systems comprise built in performance counters [2,3,6] (used for fine-grained monitoring at machine level) or more complex monitoring interfaces (e.g. IPMI). Software approaches to performance analysis can base on sophisticated simulators (e.g. QEMU [1]) or built in performance monitors. The first approach assures high controllability and observability, however this results in big simulation time overhead and problems in dealing with real time issues. So, it is difficult to analyze complex systems (e.g. multiprocessor servers with RAID controllers and hypervisor). Moreover, the availability of simulated resources is limited. Most universal are monitoring mechanisms embedded into COTS systems, e.g. Windows performance analyzer [3,12], vSphere [14]. They provide a capability of monitoring various variables related to performance of hardware and software resources. They can be easily adapted to the considered studies, moreover their processing overhead is low and we can assume that they practically do not interfere with monitored system. We have gained some experience in this technology in previous studies [3-5,9-11].

235

W. Zamojski et al. (eds.), *Theory and Engineering of Complex Systems and Dependability,*
Advances in Intelligent Systems and Computing 365, DOI: 10.1007/978-3-319-19216-1_22

In the paper we use the monitoring technology in a powerful server with virtualization capability. Monitoring such platforms is rarely encountered in the literature [7]. We have checked this approach in analyzing the impact of hardware and software configuration on the effectiveness of creating cloud environment (designed for IBM Poland). The performed measurements were targeted at checking CPU, RAM and disc usage as well as execution time analysis of the involved processes. In section 2 we specify the monitored system and present the methodology of its monitoring for various configurations (they covered different disc resources, single and multiple threading processing). Section 3 presents and explains results obtained for various configurations. Final conclusions are comprised in section 4.

2   Cloud Environment and Test Methodology

Deployment of multiple virtual machines simultaneously in cloud environment may lead to high degree of system load, in particular the storage subsystem, due to cloning of the virtual disks and booting the cloned virtual machines. Provisioning of the virtual machine influences also CPU and RAM memory resources. Hence, looking for optimal solutions some exploratory tests are needed.

An important issue is checking system performance for various storage configurations and the number of cloning threads, as well as monitoring the hypervisor behavior during the provisioning of virtual machines. For the experiment purposes we have used SATA and SSD disc drives, which differ significantly in performance. This work is a continuation of the research related to the computer systems performance conducted in the Institute [5,10,11,]. In this study an IaaS (Infrastructure-as-a-Service [13]) model has been applied, it was dedicated to quickly provision new virtual machines used during training workshops, testing and development activities. This solution has been designed by the first author for IBM Poland. The IaaS used in this experiment is much different from the typical solutions of this kind. It allows us to use almost any VMware virtual machine image without the need to impose specific requirements on the image itself (i.e. file systems used, disk configuration, etc.).

The developed cloud environment based on the IaaS solution provides the capability of using linked-clones mechanism. As opposed to the full-clones, where each virtual machine has a set of its own virtual disks, linked-clones share a base virtual disk, while having only copy-on-write disk space, where the differences from the base image are written to. Hence, creating another linked-clone is very fast, we discuss this in section 3.2.

Another advantage of the developed IaaS is the possibility to use a cluster of virtual machines, which is a set of virtual environments, with network connectivity between them, but isolated from the other environments. It allows administrator to prepare quickly multiple environments, consisting of several VMs each, even with a complex configuration inside (services on different VMs communicate with each other on each instance of such topology) or running a DHCP server (which in some other IaaS solutions may be problematic). Moreover, the possibility to run multiple clones of the image with no modifications (even with the same IP or even MAC address) is crucial to run multiple instances of the service, which configuration is bound to the network settings on the virtual machine. In many cases the author of the VM image prepares software components configuration based on arbitrarily chosen network settings or passwords, which must be preserved on each clone of the VM.

In the performed experiments, the IaaS solution is installed on the VM running on the same x86 server (with installed hypervisor VMware ESXi 5.1), used for cloned machines. However, it uses separate storage (D1), which will not be affected by the other running VMs. Moreover, high priority for resources (CPU/disk/memory) has been set on the VM running IaaS solution, to be able to control the process of provisioning, even if the workload is high. The IaaS solution called TEC Cloud was a basis of the cloud environment. The server had two AMD Opteron 6134 2.30GHz processors (total 16 cores) and 32 GB of RAM. Virtual machine containing IaaS solution services was assigned 2 virtual CPUs and 4 GB of RAM. The IaaS VM doesn't use all of its assigned CPU resources (it was estimated, that only 1.5% of CPU processing power was used before experiment stared). The following disc data stores are used:

D1 – one 300GB SATA disk - it is used by the IaaS solution and hypervisor, it minimizes the impact on disk arrays tested during experiments;

D2 – an array of 6 SATA disks (RAID5) - attached to the Intel controller;

D3 – an array of 6 SSD disks (RAID5) - attached to the SMC controller.

Disc arrays are used for storing VM image and deployed clones. In the described experiments network activity is not considered.

The performed studies have been targeted at two sets of test scenarios:

S1 – testing provisioning (creation) time of 6 VMs and behavior of hypervisor during their deployment in the case of full cloning of the machines;

S2 – testing the effectiveness of linked clones approach, which does not need cloning of data. At the beginning, virtual disk shared by all linked-clones is cloned once, and from this moment, every clone of the virtual machine will be attached directly to it, while copy-on-write mechanism will save all of the changes written by the virtual machine in its own area.

We have analyzed various system configurations related to the number of cloning threads, location of the image repository and deployment storage space (in this case SATA or SSD drive sets). The virtual machine image used in the experiment is running CentOS 6.3 Linux (64-bit), on which a typical, small-scale J2EE environment has been installed - in this case IBM WebSphere Application Server 8.5.5 [15] as an application server and DB2 as database management system (DBMS). The VM image occupies around 11 GB of storage space. Its size is the most important factor from the cloning time perspective. On the other hand, contents of the image (OS and its services) result in certain amount of load during the boot phase. In this case, multiple VMs containing application server and database will significantly impact the storage during the OS start. After a while, only the small application polling database will be active, which means that the load should gradually stabilize on a certain level after a while. Multiple virtual machines starting in a short time one after the other, or in the worst case, simultaneously, will heavily impact on hypervisor performance.

This situation is called a boot-storm, when too many environments are trying to access resources at the same time during the boot phase (this is analyzed in scenario S2).

For each configuration of experiment scenarios S1 and S2 we monitor various performance variables which characterize the behavior of the systems and allow us to evaluate time required for performing involved processes. In particular, we measure such features as: environment deployment time, disc load and effectiveness (read/write data transfers, latency, etc.), CPU load (average CPU usage on all cores, wait time), RAM memory usage (*granted* – virtual memory allocated to virtual environment machines, *active* – estimated by hypervisor virtual memory size to be used by virtual environments, *consumed* - physical memory size reserved/consumed at the level of hypervisor, *shared* - size of the memory shared by the VMs, *balloon* – memory swapped out from VM forced by hypervisor). CPU load includes also such metrics as: *ready time* (amount of time CPU is ready to execute the instructions) and *co-stop time* (amount of time a SMP virtual machine was ready to run, but caused delay due to co-vCPU scheduling contention). A vCPU is a virtual CPU assigned to the virtual machine. If a VM has more than 1 virtual CPU it is called SMP (Symmetric Multi-Processing) virtual machine. Running more and more virtual machines will lead to additional latency due to the process of resource scheduling.

In ESXi's terminology, 100% of CPU usage time relates to: time spent on running instructions + ready time + co-stop time + wait time. While a VM is waiting for CPU scheduler to let it run on a physical CPU, the time is counted as „ready time" (the VM has instructions that need to be executed, but this cannot be done immediately). Co-stop time is an amount of time spent in the co-deschedule state, which applies to SMP VMs (VMs with more than 1 virtual CPUs). It is an additional time spent by one vCPU, if it executes more instructions than the remaining vCPUs. In other words this forces all vCPUs of the VM to spend equal time on the physical processor. Finally, wait time is the time when the virtual machine is waiting for some VMkernel (hypervisor's kernel) resource, i.e. it includes I/O wait time (when the waiting involves I/O operation) and idle time (when the VM is actually doing nothing).

The considered load metrics in the experiments are provided by vSphere Client [14]. It is a standalone client for managing and monitoring VMware infrastructure - in this case: ESXi hypervisor. All metrics were sampled every 20 seconds.

3   Experiment Results

In this section we present and comment main performance features related to test scenarios S1 and S2 specified in section 2, S1 covers configurations C1-C8 and full cloning of 6 machines, scenario S2 relates to experiments with boot storm involving linked cloning up to 39 machines in two configurations C9 and C10.

3.1   Test Scenarios S1

The first tested scenario was to use SATA drives both for image repository (the source of the cloning process) and deployment area (data store, on which virtual machines can run). The storage load is related both to the reads/writes of the cloning tasks and to boot processes of the subsequent virtual machines. Both single and multi-thread processing has been tested.

In the configuration with a single thread (C1) the time of cloning 6 machines was 1h 15 minutes. The time of cloning subsequent machines increased from 11.6 to 13.5 minutes. Disc read and write data transfers were almost constant for each cloning 12 MB/s and 22MB/s, respectively. Delays in reading (access time) were in the range of 30 ms, for writing operations they systematically increased from 1 to 6 ms, due to activation of subsequent machines by IaaS solution – each new environment generated additional load on the storage subsystem. This activation also resulted in subsequent increase of CPU usage from 3 to 8% in a stepwise way, CPU waiting time increased from 30 to 170 s in subsequent cloning subphases. In a similar way increased consumed memory (from 2 to 11 GB) and granted memory (from 1GB to 9 GB). Memory related to ballooning was at constant level (1.9 GB). The ballooning mechanism forces inactive virtual machines to swap out memory to reclaim unused part of the physical memory on hypervisor, in order to assign these pages to the virtual machines that currently need it - this mechanism heavily utilizes storage when a lot of data is being swapped.

The VM operating system uses an agent (of VMware Tools) running inside the VM to swap out some memory pages. The hypervisor is not aware of the exact number of needed pages, so it uses this agent to allocate as much memory as possible, while tagging those pages which may not reside in RAM (it knows that those pages may not reside in its physical memory). VM OS is not aware of this, and tries to swap out least used memory pages, so the agent can allocate memory. Since OS swapped out least used memory pages for this allocation, and hypervisor is aware of memory allocation made by the agent, the additional free memory can be used by the other VMs. Notice, that swapping in and out memory pages impacts storage performance.

In the configuration with 3 threads (C2) the first set of 3 machines was cloned in 39 minutes and the second set in 41 minutes, so the whole process of cloning 6 machines was 1.33 h (about 16 minutes more than in the single thread configuration). Disc data transfer rate was about 15% lower than in the first configuration C1 (due to longer access time resulting from simultaneous disc operation requests via the same controller). In the case of configuration C1 this access had a more sequential form. Moreover, the number of handled requests in the second phase was higher due to operation of three activated machines. CPU waiting time was 140 and 210s for the first and second cloning phase, respectively (consumed memory was 8 and 14 GB). Active memory grew quickly when a new set of VMs was starting (10 GB and 12 GB for each phase), and after a few minutes it gradually decreased because of VMs small activity (related to WebShere). Notice, that it was higher than consumed memory on the hypervisor, since it is just estimated based on the activity inside VM, while consumed memory is the amount of memory measured on the hypervisor level.

The subsequent configurations included also SSD drives. In this case SSD drives contained image to be cloned, while on the SATA drives the deployed virtual machines were running. Here, we also considered single and three cloning threads.

In the single thread configuration (C3) the cloning of 6 machines took 1 hour and 13 minutes, this is a little surprising that it is higher by 8 minutes as compared with C1, despite similarities in monitored performance parameters. However, the delay for SSD discs was very low (in the range of 1 ms) and for SATA discs write operations showed higher level (10 ms appeared during the first cloning due to some disturbance with ballooning process). Hypervisor noticed that there is a significant amount of free physical space available and released memory pages allocated by the agent on the IaaS virtual machine, which resulted in the OS swapping back the memory pages.

In the configuration with 3 threads (C4) the cloning process took 1 hour and 11 minutes. Consumed and granted memory spaces were in the range 8 and 14 GB in the first and second cloning phase, respectively, and the active memory around 10 and 12 GB with gradual decrease after 12 minutes to 2,5 and 4,5 GB. Notice that in the case of starting just one thread, fluctuations are smaller than in case of starting 3 VMs at the same time, which results from starting all of the services in the OS. The more VMs start at the same time, the bigger the fluctuation is.

Using SSD drives both for the image repository and deployment resulted in significant speed up of cloning. The configuration with a single thread (C5) assured cloning of 6 machines in 13 minutes (about 5 times faster than for C1). Reading and writing data transfer rates are illustrated in fig. 1, the attained values are comparable with the previous configurations but the reading and writing operation delays were much lower (2-4 ms for reads, and less than 1ms for writes). The number of handled read and write requests is shown in fig. 2. After the machine cloning we have requests (mostly writes – existing VMs stabilize, and some reads are from OS level cache) related to the running WebSphere applications. CPU and memory usage is similar as in previous configurations but with higher dynamic due to faster creation of machines. The shapes of plots on fig. 1 and 2 are different, especially after cloning phase. Deployed VMs stabilize, and the only activity is database polling, which generates I/O operation, but no significant amount of data is being transferred.

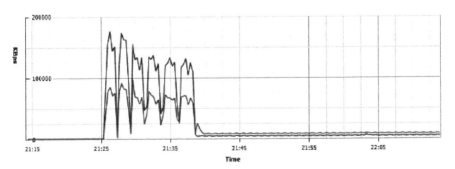

Fig. 1. SSD disc transfer rates for read (lower) and write (higher) operations: configuration C5

In the case of configuration with 3 threads (C6) the cloning process terminated after 10 minutes. This speed up resulted mostly from higher disc transfer speed for reads and writes (about 20% increase). The main performance parameters were similar to those of configuration C5.

Finally, we have checked the impact of using both SATA and SSD drives. It has to be noticed, that SATA and SSD drives were configured on 2 different controllers. In this scenario, we expected some improvement as compared with SSD only scenario, because we could utilize more effectively controllers in the machine. In this case, all of the data of the running VMs is being stored on the SSD drives (deployment area), so the randomly accessed data is stored on the faster drives, while continuous reads refer to SATA drives (image repository).

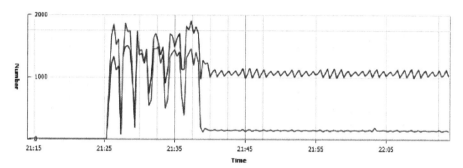

Fig. 2. The number of handled read (lower) and write (higher) requests: configuration C5

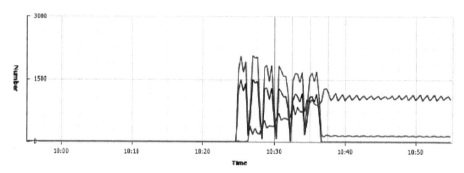

Fig. 3. The number of handled I/O requests for SSD discs: reads (stepwise increase), writes (higher pulses) and for SATA discs: reads (middle pulses) – configuration C7

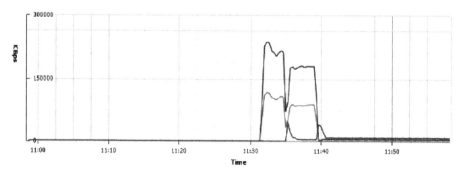

Fig. 4. Data transfer rates for SSD discs operations: read (small peaks of VM OS starts), writes (high pulses – cloning) and SATA read discs operations (middle pulses) – configuration C8

For configuration (C7) with a single thread the cloning time was similar to C5 – 12 minutes (1 minute shorter). Disc IO requests are given in fig. 3 (lower IO read requests, as compared with writes is due to some reads covered by OS cache on the existing VMs). More interesting is configuration with 3 threads (C8). It assured the lowest cloning time of 8 minutes. In particular, this resulted from higher transfer rates (fig. 4) due to two separate disc controllers (SATA and SSD) as compared with single SSD controller used in configurations C5 (compare fig. 1) and C6.

In scenarios S1 with full cloning we have observed about 2 times higher write transfer rates than for reads. This results from storage thin provisioning technology.

3.2   Test Scenarios S2

In this group of experiments the goal was checking the behavior of the storage during boot-storm in the cloud environment. Here, we used linked-clones, so the virtual machines involved already the cloned base virtual disks (read only) and the provisioning process performed only snapshotting of the virtual machine state to create a new storage area (separate for each virtual machine) for writes done by the virtual machines. Linked-cloning is similar to the copy-on-write mechanism in RAM. Having an image with a single virtual disk A, and doing full cloning will result in a set of copies: A1, A2, A3… Linked-cloning will result in a single copy A-base, and additional files storing only differences from the base image: A1-diff, A2-diff…, pointing to the parent disk A-base. Required data blocks are read from the appropriate file, either An-diff, or A-base.

Snapshotting generates some I/O operations, but since it is done before virtual machine is started, it doesn't have to save virtual machine state. As a result of using linked-clones, the IaaS doesn't have to clone all data for every clone. The load imposed on the storage in this experiment is related mainly to the process of booting operating systems and services - in this case we expected to face boot-storm issue and checked how hypervisor and storage behaved.

The base virtual disk used later by linked-clones has been cloned earlier, so no cloning was required during the benchmark, and provisioning time of each VM only depends on times of the snapshotting and powering on the VM (this operation is not instant, it also needs to allocate swap area on the deployment data store for each VM). In S2 experiments, the provisioning time of the 39 virtual machine clones have been measured, as well as other metrics (like in test scenarios S1). Here, we used three cloning threads in order to boot new instances faster and considered two configurations C9 (using only SATA discs) and C10 (using only SSD discs). The results of total cloning time T_{SATA} and T_{SSD} for these two configurations in function of the number of cloned machines (N) are given in tab. 1. For C9 the cloning time increased with N due to the activation of applications after cloning, however some decrease for N=39 appeared due to environment stabilization. After some time already existing VMs stabilized, so the storage could be better used by provisioning process. Disc read transfers decreased slightly as subsequent machines started running the application, write operation transfers increased systematically with increment 2 MB/s per clone (which implies that there were many I/O operations with short reads and writes generated by each clone). Delays in read operations ranged from 20 ms up to 500-1500 ms for high N. Some saturation effect has been observed at the moment when the used RAM memory approached to the level of available (granted) memory. At this point the ballooning mechanism involved additional load to server discs resulting in low speed of virtualized systems. CPU usage during cloning increased systematically up to 50%.

Table 1. Time (in seconds) needed to deploy N machines for configurations with SATA and SSD discs: ΔT_{SATA} and ΔT_{SSD} denote time increments related to deployment time of each subsequent set of 3 clones (columns), hence for N=39 T_{SATA} = 42 and T_{SSD} = 4.8 minutes

N	3	6	9	12	15	18	21
ΔT_{SATA}	91	125	130	145	170	244	240
ΔT_{SSD}	23.5	23.0	26.2	26.3	26.4	30.1	49.5
N	24	27	30	33	36	39	-
ΔT_{SATA}	265	315	320	405	505	475	-
ΔT_{SSD}	46.5	45.1	49.1	49.2	52.5	50.1	-

In configuration C10 the speed up is significant as compared with C9, in particular it results from faster data transfer for read and write operations up to 60 and 30 MB/s. SSD response times were in the range of 20ms and decreased to 4 ms after cloning process termination. CPU usage was close to 100% during cloning and then stabilized at the level of 80% for running application on all instantiated machines. As opposed to S1 scenarios with full cloning in S2 the impact of disc thin provisioning is low, so disc data transfer rates for reads are higher than for writes.

4   Conclusion

The presented results confirm that monitoring performance variables is a powerful tool helpful in optimizing system efficiency. In the case of complex information systems the involved measurement instrumentation overhead (CPU and memory usage) is negligible, so the obtained results are reliable. An important issue is the capability of verifying system properties for different configurations. In contemporary virtualization environment this can be assured easily by available software. Another problem relates to selection of variables for observation, this is dependent on the monitored system and the goal of the analysis. In the performed experiments we have got relatively good flexibility in allocating space for various data within SATA and SSD discs, as well as in the scope of the number of used processing threads or generated cloud clones. This resulted in a quite wide space of checked configurations and measured parameters characterizing the efficiency of the system. Exploring such wide space it is possible to detect unexpected properties. Hence, interpreting results we had to drill down into details of employed technologies in the system and the operation of the vSphere monitor.

Further research is targeted at combining performance monitoring with event logs in virtualized systems. In particular, the event logs can be addressed to specific analyzed properties. This approach can be targeted not only at performance analysis but more at problem predictions, as it is widely reported for non-virtualized environment (e.g. [8] and references therein).

We are grateful to dr P. Gawkowski for his help in preparing ZOiAK department server for the tests.

References

1. Bellard, F.: QEMU a Fast and Portable Dynamic Translator. In: Proc. of the Annual USENIX Conference, ATEC 2005, pp. 41–46 (2005)
2. Knauerhase, R., Brett, P., Hohlt, B., Li, T., Hahn, P.: Using OS Observations to Improve Performance in Multicore Systems. IEEE Micro, 54–66 (May-June 2008)
3. Król, M., Sosnowski, J.: Multidimensional Monitoring of Computer Systems. In: UIC-ATC 2009 Symposium and Workshops on Ubiquitous, Autonomic and Trusted Computing in conjunction with the UIC 2009 and ATC 2009, pp. 68–74. IEEE Computer Society (2009)
4. Kubacki, M., Sosnowski, J.: Enhanced Instrumentation of System Monitoring. In: Information Systems in Management XVI: Modern ICT for Evaluation of Business Information Systems, SGGW, pp. 29–40 (2012) ISBN 978-83-7583-469-7
5. Kubacki, M., Sosnowski, J.: Creating a knowledge database on system dependability and resilience. Control and Cybernetics 42(1), 287–307 (2013)
6. Reinders, J.: VTune Performance analyzer essentials, Measurement and tuning techniques for software developers. Intel Press (2007)
7. Salánki, Á., et al.: Qualitative Characterization of Quality of Service Interference between Virtual Machines. In: ARCS 2011: 24th International Conference on Architecture of Computing Systems (2011)
8. Salfiner, F., Lenk, M., Malek, M.: A survey of failure prediction methods. ACM Computing Surveys 42(3), 10.1–10.42 (2010)
9. Sosnowski, J., Gawkowski, P., Cabaj, K., Kubacki, M.: Analyzing Logs of the University Data Repository. In: Bembenik, R., Skonieczny, Ł., Rybiński, H., Kryszkiewicz, M., Niezgódka, M. (eds.) Intelligent Tools for Building a Scientific Information Platform: From Research to Implementation. SCI, vol. 541, pp. 141–156. Springer, Heidelberg (2014)
10. Sosnowski, J., Kubacki, M., Krawczyk, H.: Monitoring Event Logs within a Cluster System. In: Zamojski, W., Mazurkiewicz, J., Sugier, J., Walkowiak, T., Kacprzyk, J. (eds.) Complex Systems and Dependability. AISC, vol. 170, pp. 257–271. Springer, Heidelberg (2012)
11. Sosnowski, J., Poleszak, M.: Online monitoring of computer systems. In: Third IEEE International Workshop on Electronic Design, Test and Applications, DETLA 2006, pp. 327–331 (2006)
12. Yen, N.: Secure computer and network systems, Modeling analysis and design. John Wiley (2008)
13. Wang, L., et al. (eds.): Cloud computing, methodology, systems and applications. CRC Press (2012)
14. vSphere Client for ESXi 5.1, `https://my.vmware.com/web/vmware/info/slug/datacenter_cloud_infrastructure/vmware_vsphere/5_1`
15. WebSphere Application Server Network Deployment, Version 8.5.5 documentation, `http://www-01.ibm.com/support/knowledgecenter/SSAW57_8.5.5/as_ditamaps/was855_welcome_ndmp.html`

A Modified Clustering Algorithm DBSCAN Used in a Collaborative Filtering Recommender System for Music Recommendation

Urszula Kużelewska[1] and Krzysztof Wichowski[2]

[1] Bialystok University of Technology,
15-351 Bialystok, Wiejska 45a, Poland
u.kuzelewska@pb.edu.pl
[2] Graduate of the Bialystok University of Technology
vltr@o2.pl

Abstract. Searching in huge amount of information available on the internet is undoubtedly a challenging task. A lot of new web sites are created every day, containing not only text, but other types of resources: e.g. songs, movies or images. As a consequence, a simple search result list from search engines becomes insufficient. Recommender systems are the solution supporting users in finding items, which are interesting for them. These items may be information as well as products, in general. The main distinctive feature of recommender systems is taking into account personal needs and taste of users. Collaborative filtering approach is based on users' interactions with the electronic system. Its main challenge is generating on-line recommendations in reasonable time coping with large size of data. Appropriate tool to support recommender systems in increasing time efficiency are clustering algorithms, which find similarities in off-line mode. Commonly, it involves decreasing of prediction accuracy of final recommendations. This article presents an approach based on clustered data, which prevents the negative consequences, keeping high time efficiency. An input data are clusters of similar items and searching the items for recommendation is limited to the elements from one cluster.

1 Introduction

Recommender systems (RS) are electronic applications with the aim to generate for a user a limited list of items from a large items set. In case of personalised RS the list is constructed basing on the active user's and other users' past behaviour. People interact with recommender systems by visiting web sites, listening to the music, rating the items, doing shopping, reading items' description, selecting links from search results. This behaviour is registered as access log files from web servers, or values in databases: direct ratings for items, the numbers of song plays, content of shopping basket, etc. After each action users can see different, adapted to them, recommendation lists depending on their tastes [6,14,13,2].

Recommender systems are used for many purposes in various areas. The examples are internet news servers (e.g. Google News, http://news.google.com/),

W. Zamojski et al. (eds.), *Theory and Engineering of Complex Systems and Dependability*,
Advances in Intelligent Systems and Computing 365, DOI: 10.1007/978-3-319-19216-1_23

which store the type and content of articles, count the time spent on each of them and update propositions of articles to read after each reading, tourism, where personalisation helps users to plan their journeys, e-shops (e.g. Amazon, http://amazon.com) proposing products, which are the most similar to the content of customers' shopping baskets. They are particularly useful in media services, such as Netflix, LastFM or Spotify, recommending movies, songs, artists or propositions to playlists.

Collaborative filtering (CF) techniques are one of approaches to personalisation, which searches similarities among users or items [1,3]; however only archives of registered users behaviour are analysed. As an example, similar users have mostly the same products in their baskets and similar items are bought by the same customers. They can be classified into model-based and memory-based methods. The first approach builds a model on the ratings, which is then used for generating recommendations. The other approach calculates recommendations by searching neighbourhood of similar users or items in the whole archived data.

Recommender systems face many challenges and problems [2]. They particularly concern collaborative filtering, which is one of the most effective and precise approach. In case of a new visitor,without any recorded profile, an issue called cold-start problem appears. Another problem occurs when a new item is added to the offer. In case of CF methods, it has not been assigned yet to any user and cannot be recommended to anyone [10]. In arbitrary recommender system application, a number of offered items is large, whereas a user during one session visits a few to tens of them. It results in sparsity of input data and lower reliability in terms of measuring the similarity between customers [17]. Finally, however, vitally important challenge in the field of on-line recommendations is scalability [10,18]. RS deal with large amount of dynamic data, however the time of results generation should be reasonable to apply them in real-time applications. The user, while reading news, expects to see next offer for him/her in seconds, whereas millions of archived news have to be analysed.

This paper contains results of experiments on collaborative filtering recommender system, which is based on similarities among items identified a priori as clusters. The set of clusters was generated by modified DBSCAN algorithm with different values of their input parameters and evaluated with respect to their genre homogeneity. Finally, quality of prediction (RMSE) and time efficiency of examined system was calculated and compared to other recommenders: memory-based CF, a recommender system based on k-means clusters [8], SVD model-based approach [19], SlopeOne [9].

2   Description of the Clustering Algorithm Used in the System

Application of clustering algorithms can solve several problems in the field of recommender systems. Clustering is a domain of data mining which had been applied in a wide range of problems, among others, in pattern recognition, image processing, statistical data analysis and knowledge discovery [7]. The aim of cluster analysis is organising a collection of patterns (usually represented as a vector of measurements, or a point in a multi-dimensional space) into clusters based on their similarity [5]. The points within one cluster are more similar to one another than to any other points from the remaining clusters.

In hybrid recommender systems clusters can be used to increase neighbour searching efficiency. In contrast, in memory-based collaborative filtering to identify neighbours is used kNN algorithm, which requires calculating distances between an active user and the registered all ones. Clusters are generated in off-line phase, which additionally reduces time of neighbours searching. Possessing additional information about users, e.g. demographics attribute values, one can create clusters and solve a new user cold-start problem, by recommendation of items, which are popular in the most similar group.

One of the first approaches, where clustering was used to partition users' preferences is described in [16]. The authors used clusters identified in off-line mode by k-means instead of on-line neighbourhood calculated by kNN method. As similarity measure they used Pearson correlation. Finally, quality of predictions was slightly reduced, however time efficiency and scalability increased significantly.

DBSCAN is one of recent clustering algorithms [4]. It identifies clusters as highly dense groups of points. It is particularly effective in case of arbitrary shaped clusters. For recommendations was used in [12] as initial partitioning on demographic attributes of users.

Another example is a modified method proposed in [15]. Modification concerned similarity computation in a clustering procedure. The authors assumed, that points are similar if they have the same neighbours (see Equation 1).

$$neighbour(x_i) = \{x_j : sim_R(x_i, x_j) \leq Eps, x_i \in I, x_j \in I\} \tag{1}$$

It is assumed the following notation:

- I - is a set of items,
- n_I - is a size of set I,
- x_i, x_j - are items in input data,
- Eps - is a parameter of modified DBSCAN (MDBSCAN) related to minimal similarity threshold,
- U - is a set of users,
- n_U - is a size of set U,
- r - a possible value from a ratings set of range $[0, \ldots, r_{max}]$,
- $U_r(x_i)$ - is a set of the users who rated item x_i at rating r.

Basing on the neighbourhood definition, it can be determined formula of neighbour vectors $\vec{nb_k}$:

$$\vec{nb_k}(x_i) = [v_1, \ldots, v_{n_I}]^T \quad , \text{where } v_k = \begin{cases} 1, & \text{if } x_k \in neighbour(x_i) \\ 0, & \text{in the other cases} \end{cases} \tag{2}$$

Similarity between neighbour vectors is calculated by cosine similarity function (Equation 3), whereas between points - using similarity definition from information retrieval (Equation 4).

$$sim_N(x_i, x_j) = \frac{\vec{x_i} \cdot \vec{x_j}}{|\vec{x_i}| \cdot |\vec{x_j}|} \tag{3}$$

Cosine similarity prefers items, which have more common neighbours thereby reduces influence of noise. Similarity sim_R has higher values for pair of items, which are composed of more the same ratings from users.

$$sim_R(x_i, x_j) = \frac{|\cup_{r=0}^{r_{max}} U_r(x_i) \cap U_r(x_j)|}{|\cup_{r=0}^{r_{max}} U_r(x_i) \cup U_r(x_j)|} \tag{4}$$

The procedure of neighbourhood vectors forming and similarity calculation is illustrated by the following example for data from Table 1.

Table 1. Example ratings data

	Items					
Users	x_1	x_2	x_3	x_4	x_5	x_6
u_1	1	2	3	1	5	3
u_2	1	2	5	5	5	5
u_3	3	3	4	5	4	5
u_4	5	1	1	5	4	1

For the example data similarity between items are as follows:

$$sim_R(x_1, x_2) = \frac{0+0+1+0+0}{3+2+2+0+1} = 0.125 \qquad sim_R(x_1, x_4) = \frac{1+0+0+0+1}{3+0+1+0+4} = 0.25$$

$$sim_R(x_2, x_3) = \frac{1+0+0+0+0}{2+2+2+1+1} = 0.125 \qquad sim_R(x_2, x_6) = \frac{1+0+0+0+0}{2+2+2+0+2} = 0.125$$

$$sim_R(x_3, x_4) = \frac{0+0+0+0+1}{2+0+1+1+4} = 0.125 \qquad sim_R(x_3, x_5) = \frac{0+0+0+1+1}{1+0+1+3+3} = 0.25$$

$$sim_R(x_3, x_6) = \frac{1+0+1+0+1}{2+0+1+2+3} = 0.375 \qquad sim_R(x_4, x_5) = \frac{0+0+0+0+1}{1+0+0+2+5} = 0.125$$

$$sim_R(x_4, x_6) = \frac{0+0+0+0+2}{2+0+1+0+5} = 0.25 \qquad sim_R(x_5, x_6) = \frac{0+0+0+0+1}{1+0+1+2+4} = 0.125$$

The remaining values are 0. Let $Eps = 0.2$, thereby neighbour vectors are:

$$\vec{nb}(x_1) = [0\ 0\ 0\ 1\ 0\ 0]^T \quad \vec{nb}(x_2) = [0\ 0\ 0\ 0\ 0\ 0]^T$$
$$\vec{nb}(x_3) = [0\ 0\ 0\ 0\ 1\ 1]^T \quad \vec{nb}(x_4) = [1\ 0\ 0\ 0\ 0\ 1]^T$$
$$\vec{nb}(x_5) = [0\ 0\ 1\ 0\ 0\ 0]^T \quad \vec{nb}(x_6) = [0\ 0\ 1\ 1\ 0\ 0]^T$$

Taking into account cosine similarity between them, clusters on example data are composed of 5 points: $C_1 = \{x_3, x_4\}$ (the common neighbour is x_6) and $C_2 = \{x_1, x_5, x_6\}$ (the common neighbour is x_4).

3   Experiments

The recommender system used in the described experiments is composed of the following steps, which technical aspects are presented in Figure 1:

1. Filtering and preprocessing of input data.
2. Generating a set of clustering schemes of vectors of items' values.
3. Evaluation of the clustering schemes and selection the most appropriate partitioning.
4. Generating recommendation for active users.

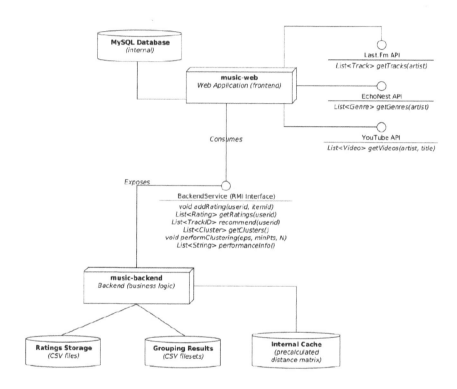

Fig. 1. Components of the recommender system used in the experiments

The main part of the system is *music − backend*, which is an item-based collaborative filtering method implemented in Apache Mahout framework (http://mahout.apache.org). It uses ratings file, clusters and similarity values stored in cache files. A part of application exposed to users is denoted as *music − web*. It gather information from users and external music services (e.g. EchoNest, YouTube, LastFM) and presents them recommendations. Additional database (relational MySQL) stores artists names, tracks titles and users' functional data.

Initial input data for the system were taken from LastFM service (repository at http://dtic.upf.edu.pl/~oelma/MusicRecommendationDataset/lastfm-1K.html) and contain users' number of song listenings. The higher value of the listening correlates with the higher level of interesting of a particular song. After filtering outdated songs, the data contained 992 users and 2 213 450 songs Then the data were supplemented by users who used the system. Users could listen the music as well as rate it. The ratings were integral values from interval [1...5] and expressed how much a user liked the particular song. To unify the numbers of listening with the ratings the following procedure was performed. If a user listened a song only once, it is assumed, that he/she didn't like it. The related rating is 1. The following values were correlated with the numbers of listening with assumption of even decomposition of the numbers in ratings (see Table 2).

Table 2. Decomposition of listening numbers in ratings

Rating value	Range of listening numbers	Numbers of listening
1	1	1 850 078
2	2	667 612
3	3-4	588 676
4	5-8	651 398
5	> 9	642 311

The first evaluation concerned homogeneity of clusters with respect to kind of music of the songs belonging to them. Table 3 contains the best results satisfying these requirements selected from data. High homogeneity is not strictly correlated with the parameters of the MDBSCAN, however there were some values sets connected with the best results: high homogeneity and appropriate number of clusters. Very small number of groups are not desirable, because the neighbourhood search space is not very limited in this case. On the other hand, it is not possible to precisely estimate preferences basing on numerous tiny clusters.

Finally, parameter Eps was limited to range [0.3,0.42]. Results on values less than 0.3 were composed of very small (the most often size was equal 1) numerous clusters. Values greater than 0.42 lead to a few large clusters. The most reasonable value of $MinPts$ was 5-8 for the reasons mentioned above. Size of neighbourhood N had to be quite large: greater than 7000 objects.

Another important issue was the ratio of input data which were located within clusters. The highest values was 20.43%, which was not very satisfying.

To increase the number of clustered input data a procedure of complement clustering was performed [15]. For every not clustered point its distance to the formed previously clusters was calculated according to Equation 5.

$$sim_N(x_i, C_j) = \max(sim_N(x_i, x_k)),, \text{ where } x_k \in C_j \tag{5}$$

Points are joined to the nearest group, if its distance is closer than parameter γ. This parameter has to be much smaller than Eps and its appropriate value is

Table 3. Clustering results of songs from LastFM dataset

Eps	MinPts	N	Ratio A number of groups	Ratio A number of clustered data [%]	Homogeneity [%]
0.3	7	8300	1911	5.75	19.14
0.42	7	8300	11912	20.43	77.67
0.39	7	8300	2484	8.18	97.20
0.39	5	8200	6533	9	97.20
0.41	7	8250	5115	13.39	97.47
0.41	7	8300	89	0.47	97.48
0.4	7	8300	5122	13.3	98.96
0.35	8	7100	1209	1.43	98.97
0.35	5	8200	6573	8.91	98.97
0.34	6	7200	1255	1.49	99.09

selected during experiments. Table 4 contains results of complementary cluster-ing with $\gamma = 0.05$. The best result contains 91.47% of input data in clusters for the parameters set: $Eps = 0.8$, $MinPts = 10$ and $N = 8300$ with homogeneity greater than 99%. Number of clusters in this case was equal 48619.

Table 4. Clustering results of items LastFM dataset with complement clustering

Eps	MinPts	N	Ratio A number of groups	Ratio A number of clustered data [%]
0.39	7	8300	11964	20.85
0.4	7	8300	11954	20.86
0.7	10	8300	8246	46.67
0.75	10	8300	7808	57.99
0.8	10	8300	48619	91.47

The second evaluation of clustering was performed in item-based collabora-tive filtering recommender system, in which the best results was used as input data. Table 5 contains RMSE values and time of recommendations calculation for MDBSCAN as well as for other recommender systems used for compari-son. It concerns item-based collaborative filtering (CFIB), user-based collabora-tive filtering (CFUB) with k-nn procedure (k=10 and k=1000) for neighbours searching, SlopeOne algorithm, SVD based recommender. This table contains also comparison to another clustering-based collaborative filtering recommender system. The clustering method was k-means with selected number of clusters: ncl=20 and ncl=1000. In all the mentioned methods as a similarity measure cosine-based was selected. The methods from Table 5 marked with * were ex-amined on smaller dataset due to time or memory problems on greater data.

The values of RMSE errors were computed for every objects from input dataset estimating for them 30% of randomly removed existing preferences and

Table 5. RMSE results on LastFM dataset

Method	Parameters	RMSE	Time of recommendation [s]
MDBSCAN	Eps=0.8, MinPts=10, N=8300	0.22	0.042
MDBSCAN	Eps=0.75, MinPts=10, N=8300	0.63	0.031
CFIB		0.58	118.77
CFUB*	k=10	1.22	0.19
CFUB*	k=1000	1.09	1.04
SlopeOne*		0.68	48.87
SVD*		0.58	69.81
CF k-means	ncl=20	0.71	0.019
CF k-means*	ncl=1000	0.64	0.02

comparing the estimated ones to the real ratings. In every system in this experiments a cosine similarity measure was applied. Time of recommendations generation was calculated for every user from input data with length of recommendation lists equal 7.

The fastest recommender systems were methods based on k-means clusterings, however their RMSE values were higher than in case of MDBSCAN for different values of its parameters. The best results (RMSE=0.22) were generated for MDBSCAN and its input parameters: $Eps = 0.8$, $MinPts = 10$ and $N = 8300$. The methods: CFIB and SVD-based generated good results, however the time of recommendations was inappropriately high (more than 60 s).

4   Conclusions

Recommender systems become an important part of internet services effectively supporting people with searching interesting products or information in huge amount of data. Collaborative filtering approach to this problem faces many challenges. One of vital issue is poor scalability.

In this article a clustering approach to CF recommendations is presented as one of solutions to scalability problem. Clusters were generated using modified DBSCAN method and given to input of collaborative item-based recommender system. As a result, in on-line stage of recommendations generation, searching similarities of an active users' favourite songs was limited to the clusters they belong to. Finally, time effectiveness of the system increased. Additionally, prediction ability of the method also increased in comparison to techniques such as: memory-based collaborative filtering user-based and item-based systems and hybrid recommenders using k-means clusters.

Acknowledgements. This work was supported by Rectors of Bialystok University of Technology Grant No. S/WI/5/13.

References

1. Adomavicius, G., Tuzhilin, A.: Toward the next generation of recommender systems: A survey of the state-of-the-art and possible extensions. IEEE Transactions on Knowledge and Data Engineering 17(6), 734–749 (2005)
2. Bobadilla, J., Ortega, F., Hernando, A., Gutiérrez, A.: Recommender systems survey. Knowledge-Based Systems 46, 109–132 (2013)
3. Candillier, L., Meyer, F., Boullé, M.: Comparing state-of-the-art collaborative filtering systems. In: Perner, P. (ed.) MLDM 2007. LNCS (LNAI), vol. 4571, pp. 548–562. Springer, Heidelberg (2007)
4. Ester, M., et al.: A density-based algorithm for discovering clusters in large spatial databases with noise. In: Proc. 2nd Int. Conf. on Knowledge Discovery and Data Mining, pp. 226–231. AAAI Press (1996)
5. Jain, A.K., Murty, M., Flynn, P.J.: Data clustering: a review. ACM Computing Surveys 31(3), 264–323 (1999)
6. Jannach, D., et al.: Recommender systems: an introduction. Cambridge University Press (2010)
7. Kuželewska, U.: Advantages of Information Granulation in Clustering Algorithms. In: Filipe, J., Fred, A. (eds.) ICAART 2011. CCIS, vol. 271, pp. 131–145. Springer, Heidelberg (2013)
8. Kuželewska, U.: Clustering Algorithms in Hybrid Recommender System on MovieLens Data. Studies in Logic, Grammar and Rhetoric 37(1), 125–139 (2014)
9. Lemire, D., Maclachlan, A.: Slope One Predictors for Online Rating-Based Collaborative Filtering. In: Proceedings of SIAM International Conference on Data Mining, vol. 5, pp. 1–5 (2005)
10. Lika, B., Kolomvatsos, K., Hadjiefthymiades, S.: Facing the cold start problem in recommender systems. Expert Systems with Applications 41(4), 2065–2073 (2014)
11. Luo, X., Xia, Y., Zhu, Q.: Incremental collaborative filtering recommender based on regularized matrix factorization. Knowledge-Based Systems 27, 271–280 (2012)
12. Moghaddam, S.G., Selamat, A.: A scalable collaborative recommender algorithm based on user density-based clustering. In: 3rd International Conference on Data Mining and Intelligent Information Technology Applications, pp. 246–249 (2011)
13. Park, D.H., et al.: A literature review and classification of recommender systems research. Expert Systems with Applications 39(11), 10059–10072 (2012)
14. Ricci, F., et al.: Recommender Systems Handbook. Springer (2010)
15. Rongfei, J., et al.: A new clustering method for collaborative filtering. In: International IEEE Conference on Networking and Information Technology, pp. 488–492 (2010)
16. Sarwar, B., et al.: Recommender Systems for Large-Scale E-Commerce: Scalable Neighborhood Formation Using Clustering. In: 5th International Conference on Computer and Information Technology (2002)

17. Sarwar, B., et al.: Item-based collaborative filtering recommendation algorithms. In: Proceedings of the 10th ACM International Conference on World Wide Web, pp. 285–295 (2001)
18. Su, X., Khoshgoftaar, T.M.: A survey of collaborative filtering techniques. Advances in Artificial Intelligence 4, 1–19 (2009)
19. Zhang, S., et al.: Using singular value decomposition approximation for collaborative filtering. In: Seventh IEEE International Conference on E-Commerce Technology, CEC, pp. 257–264 (2005)

Evaluation of the Location of the P&R Facilities Using Fuzzy Logic Rules

Michał Lower and Anna Lower

Wroclaw University of Technology, Wybrzeże Wyspiańskiego 27, 50-370 Wrocław, Poland
{michal.lower,anna.lower}@pwr.edu.pl

Abstract. The trend of limiting vehicular traffic to the benefit of public transport is developed in contemporary urban planning. One of the tasks is determining location of the collective parking places in the P&R system. Criteria for assessing the quality of the selected location are formulated generally and descriptively. However, the factors to be assessed are often ambiguous and fuzzy, difficult to be precisely determined but possible for the evaluation by an expert. Due to the large number of parameters of criteria the practice has shown that the choice of the location of these sites in a way that is intuitive without a detailed analysis of all the circumstances, often gives negative results. Then the existing facilities are not used as expected. The authors have used fuzzy inference to the evaluation of the location of the P&Rs based on fuzzy input parameters. The obtained results of the analysis allows to determine the degree of attractiveness of the selected place on the basis of a broad set of the expert input data. The proposed evaluation method has been tested on three existing facilities for which the effect is already known.

Keywords: fuzzy logic inference, park and ride location, P&R, PnR car parks.

1 Introduction

Urban indicators are an important factor to be taken into account in the spatial planning. Decisions of location of particular functions in the city area are preceded by detailed analyzes of local determinants. These analyzes are based on collected data as well as on intuitive parameters based largely on expert knowledge. The enormity and variety of parameters make that analyzing them without the help of any special algorithms seems to be quite difficult. Therefore, we have decided to analyze the problem using fuzzy logic inference. Fuzzy logic is commonly used for inference in such cases. In particular, it is often used when the parameters on the basis of which the assessment is made are difficult to determine and are based on intuitive expert knowledge. The examples of such applications can be found in publications [1-5].

Location of parking lots requires analysis of many factors, difficult to determine, especially considering the park and ride system. Park and ride (P&R) has been widely adopted in Europe in the last 20 years. The system takes different forms but in the most common form there are car parks adjacent to an intermodal transfer point where travelers can change from car to public transport – bus or rail and then reach their

© Springer International Publishing Switzerland 2015
W. Zamojski et al. (eds.), *Theory and Engineering of Complex Systems and Dependability,*
Advances in Intelligent Systems and Computing 365, DOI: 10.1007/978-3-319-19216-1_24

final destination [6]. The consequence of choosing public transport is a decrease in congestion, an expected reduction of air pollution due to decreasing number of cars in the city center. This is confirmed by scientific research [11]. Location of P&Rs should be carefully chosen although criteria for assessing the quality of the selected location are formulated generally and descriptively. To choose the best location, cities commission experts to undertake the necessary analyzes. The results are often estimated and conclusions are drawn in an intuitive way without using unequivocal algorithms. Such descriptive studies can be found in the research carried out on order of cities [13]. The kind of data that we can obtain from an expert are often ambiguous and fuzzy, difficult to be precisely determined but possible for the evaluation by an expert at the level of the linguistic variable values. The developed expert analysis usually do not exhaust all the possibilities and conclusions for decisions are drawn on the basis of similar examples examined by an expert. They are intuitively compared. Due to the large number of parameters of criteria the practice has shown that the choice of the location of these sites in a way that is intuitive without a detailed analysis of all the circumstances, usually gives negative results. Practice and research [7] show that P&R facilities are not used if they are not located conveniently from the point of view of potential users, even if a level of congestion is high.

Most literature on P&R facilities is focused on two main directions. The first group of papers investigates mathematical models that analyze the potential impact of P&R facilities. The second group considers the policy implementation and effects of P&R schemes [7] Studies on parking location algorithms are conducted in a much narrower range.

2   Park and Ride Facilities Location

In the structure of P&R system a location of individual objects plays a key role. An inconvenient location, from the users' point of view, can lead to a lack of interest and consequently to incomplete use of the object. Site selection decision is not easy, mainly due to many criteria which must be taken into account. This is a research problem undertaken by scientists, e.g. the authors of the paper [9] define locations in a very precise way on the basis of the input data. The use of such a model is very limited because obtaining such detailed and precise data requires additional effort and research means. It is also time consuming. In addition, this model was developed for the particular city – Columbus, Ohio. It is based on detailed data and analysis carried out on the existing communication system and also for the existing P&R system.

The model is based on past experiences and is intended to largely improve the existing state and to extend the system. Observations of the authors can be very helpful, but their model is not universal and it is hard to move it directly to other city, thus its implementation to other circumstances would bring further problems to solve. All these inconveniences make it that, in practice, it might be impossible to do. However, the paper shows how important is the issue of optimal distribution of P&R facilities and that there is a need to construct analytical methods for choosing location for them.

In the paper [8], the authors also make certain assumptions which largely omit the complexity of actual conditions. The authors assume that the tested city is linear, and monocentric and the main source of morning traffic is generated by housing developments located on the outskirts. Although the authors propose to consider also a model with more transport corridors, but admit that it will be particularly difficult because the catchment areas of P&R facilities would have complicated shapes such as parabolic boundaries. This direction of analyze is continued by the authors of the paper [10].

There is another group of papers where authors' studies are based on Geographic Information Systems (GIS) using statistical research [11, 12]. GIS can be very helpful but the authors do not use imprecise, estimated information, such as acoustic map which can be utilized by using fuzzy logic.

3   Fuzzy Inference Model of P&R Car Parks Location

The usefulness of the site to the location of P&R facilities results from features and objectives of P&R system. There are a number of conditions and parameters that testify to the value of the place. They can be grouped into two main fields. The first one is connected to the territorial conditionality and the second one to the public transport which is expected to take over the passenger load of individual traffic. Using this expert classification the inference has been divided into two local models of inference. The final score is the result of the fuzzy inference based on the results of inference of local models. The first local model counts the indicator of territorial conditions (IOTC). The second local model (IOPQ) counts the indicator of the public transport quality. The final result is calculated by the complete inference model (IOCM). Such an approach gives a clear assessment of the proposed location and on the basis of the results of intermediate local models indicates the components of factors of the final result.

In all three models the triangular membership functions for each value of linguistic variable are defined. The center of mass method is used for the defuzzyfication of fuzzy values. All the values of input and output parameters was decided to be determined in the range of 0-100%.

In the model, the issue of costs and profitability of the system for users has been deliberately omitted. The model is designed as a tool for cities that have P&R system or decide to introduce it. In such situations, balancing the travel costs often lies with the administrative decision. City governments seek to reduce the number of vehicles in the central zone and they solve it in different ways. One of the tools is the introduction of congestion charge fee for vehicles driving into the central zone, or very high fees for the use of temporary parking spaces with significantly reducing the number of permanent parking spaces reserved only for the residents of the city center. In parallel, they introduce financial support for the P&R users. There are free car parks for travelers who buy a public transport ticket or a substantial discount on the public transport with a relatively small parking fee. In the context of such experiences the financial analysis becomes secondary.

3.1   The First Local Inference Model – Indicator of Territorial Conditions

According to expert assessment three parameters being linguistic variables in the model are specified.

The first variable is the location in the context of the road transport system. The best is the location near the entry road to the city, connecting the residential areas located outside the city and the city center. This parameter is largely linked to the number of vehicles flowing into the city and being interested in using the car park. We have defined this parameter as *I*. The more potential users ride chosen corridor, the higher the value of *I*. The value of this variable is determined on the basis of the size of catchment area of the city, i.e. the number of housing units (settlements, towns, villages) connected to the city by the chosen corridor. Basis for estimation of this parameter can be found in the statistics data for this area. Traffic studies are sometimes carried out for some places, the results can also be the basis for the estimation of the parameter. An expert can estimate the value of the parameter building on an acoustic map of the area as well.

Another variable determining the attractiveness of a given location is the quality of access to the car park by car (*D*). This parameter includes the time and convenience of travel related to the quality of the road itself (number of lanes, width, surface quality), as well as the number of intersections and other obstacles slowing down the traffic and bandwidth of the road.

P&R facilities are often not directly supported by the road which is the main city entrance. For various organizational reasons, car parks may be spaced from the road and connected with it by a street of lower technical class which is used to direct service. This combination also affects the driver's decision to make use of the facility or not. If the object is accessible from the main road, clearly marked and with a clear indication of the place of entrance, as well as conveniently located (near and on the right side of the main road) is more likely to be used. We have denoted this parameter as *A*.

The last parameter of territorial conditions is the distance from the parking lot to the city center which is the target zone of traffic. City governments make different decisions about the minimum distance. However, there is a rule that car parks placed too close to the city center become a destination point, not a place of transfer for public transport. In this case a factor of a decrease in congestion does not exist. Medium-sized towns often hold the ring of car parks in comparable distances from the central zone. Big cities sometimes opt for several zones of distance for P&R facilities location.

The Mamdani model has been adopted for the inference model. The following parameters have been adopted as linguistic variables of the fuzzy inference model IOTC:

- *I* – traffic intensity – the number of cars aiming to the city every day (the city as a destination). Linguistic variable values: *big, medium, small*.
- *D* – quality of the road connecting traffic sources with the P&R facility (intersections, obstacles, etc.). Linguistic variable values: *high, low*.

- *A* – connection between the main road and the parking place (the level of accessibility, e.g. P&R location on the left or right side, readability of signs). Linguistic variable values: *good, bad.*
- *S* – distance from the car park to the center of the city. Linguistic variable values: *long, medium, short.*
- *TC* – territorial indicator of location, the result of inference of the IOTC model. Linguistic variable values: *very good, good, sufficient, mediocre, bad.*

In order to use the expert knowledge the authors have prepared a special table which presents the inference rules in a linguistic way. On the basis of the tables filled by an expert the inference rules of IOTC model are defined as follows:

If *I* is *big* **and** *D* is *low* **and** *A* is *good* **and** *S* is *long* **then** *TC* is *sufficient*
If *I* is *big* **and** *D* is *low* **and** *A* is *good* **and** *S* is *medium* **then** *TC* is *mediocre*
If *I* is *big* **and** *D* is *low* **and** *A* is *bad* **and** *S* is not *short* **then** *TC* is *mediocre*
If *I* is not *small* **and** *D* is *high* **and** *A* is *good* **and** *S* is not *short* **then** *TC* is *very good*
If *I* is not *small* **and** *D* is *high* **and** *A* is *bad* **and** *S* is not *short* **then** *TC* is *very good*
If *I* is *medium* **and** *D* is *low* **and** *A* is *good* **and** *S* is not *short* **then** *TC* is *very good*
If *I* is *medium* **and** *D* is *low* **and** *A* is *bad* **and** *S* is not *short* **then** *TC* is *sufficient*
If *I* is *small* **and** *D* is *low* **then** *TC* is *bad*
If *I* is *small* **and** *D* is *high* **and** *S* is *long* **then** *TC* is *mediocre*
If *I* is *small* **and** *D* is *high* **and** *S* is *medium* **then** *TC* is *sufficient*
If *S* is *short* **then** *TC* is *bad*

3.2   The Second Local Inference Model – Indicator of the Public Transport Quality

From the users' point of view the linking of the object's location place with public transport is an extremely important factor. P&R facilities should become transfer spaces located near bus or rail lines. This is the linguistic variable *K* in the model. It contains a number of conditions indicating of the attractiveness of the transfer node. First of all, it is the number of possible means of transport to choose from (bus, tram, subway, train, etc.). The quantity of various connections as part of each mode of transport is also important. It determines the flexibility of choice of the target point of the trip. Additionally, the quality and frequency of services (rolling stock and clocking) affects the complex travel time.

The last parameter, which is the value *P* in the model, is the distance that the traveler must overcome after leaving the car in order to change to public transport, which also affects the travel time. This is primarily a consequence of the way of the organization of P&R facility as a transfer hub, the adopted distance between the platforms and the conducting of pedestrian traffic.

The following parameters have been adopted as linguistic variables of the fuzzy inference model IOPQ:

- K – quantity of different means of city transport (railway, subway, bus, tram, etc.), the number of communication lines within each type, frequency of running. Linguistic variable values: *big, medium, small.*
- P – the distance from the car park to the nearest public transport stop. Linguistic variable values: *long, short.*
- S – the distance from the car park to the center of the city. Linguistic variable values: *long, medium, short.*
- PQ – public transport quality indicator, the result of inference of the IOPQ model. Linguistic variable values: *very good, good, sufficient, mediocre.*

We have decided not to use value *bad* because the location which has no possibility of using public transport was not taken into account by the expert. The expert did not choose the value *bad* not even once.

On the basis of the tables filled by an expert (the same as mentioned in 3.1) the inference rules of IOPQ model are defined as follows:

If K is *big* **and** P is *short* **then** PQ is *very good*
If S is not *short* **and** K is *medium* **and** P is *short* **then** PQ is *good*
If S is *short* **and** K is *medium* **and** P is *short* **then** PQ is *very good*
If S is not *short* **and** K is *medium* **and** P is *long* **then** PQ is *sufficient*
If S is *long* **and** K is *small* **and** P is *long* **then** PQ is *mediocre*
If S is   *medium* **and** K is *small* **and** P is *long* **then** PQ is *sufficient*
If S is not *short* **and** K is *small* **and** P is *short* **then** PQ is *sufficient*
If S is *short* **and** K is *small* **and** P is *short* **then** PQ is *good*
If S is *short* **and** K is not *small* **and** P is *long* **then** PQ is *good*
If S is *short* **and** K is *small* **and** P is *long* **then** PQ is *sufficient*

3.3   The Complete Inference Model

The inference results from both local models (IOTC, IOTQ) are used as input data for IOCM. On the basis of IOTC and IOTQ models the final indicator of location quality (*CM*) resulting from the fuzzy model IOCM is inferred. Due to the fact that the traffic intensity (parameter I) is often not constant the expert is also interested in the assessment of the location for various values of this parameter. Therefore, in order to determine the indicator, the indicator simulation depending on the parameter I is planned. At the same time the final indicator is calculated as the average value of the results of the simulation (*ACM*) according to the formula (1). The scheme of entire process of computing indicator is shown in Fig.1.

$$\text{ACM} = \frac{\int_{I_{min}}^{I_{max}} CM \, dI}{I_{max} - I_{min}} \tag{1}$$

Similarly as in section 3.1 and 3.2 the model has been constructed on the basis of the expert knowledge. Linguistic input variables for the IOCM model are *TC* and *TQ*. They are defined in the previous sections. The output linguistic variable *CM* takes linguistic variable values *very good, good, sufficient, mediocre, bad,* similarly as variables *TC* and *TQ*.

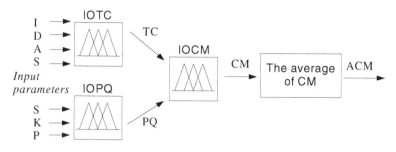

Fig. 1. The schema of the fuzzy inference model of P&R car parks location

The inference rules of IOCM model are defined as follows:

If *PQ* is *very good* **and** *TC* is *bad* **then** *CM* is *bad*
If *PQ* is *very good* **and** *TC* is *mediocre* **then** *CM* is *sufficient*
If *PQ* is *very good* **and** *TC* is *sufficient* **then** *CM* is *good*
If *PQ* is *very good* **and** *TC* is *good* **then** *CM* is *very good*
If *PQ* is *very good* **and** *TC* is *very good* **then** *CM* is *very good*
If *PQ* is *good* **and** *TC* is *bad* **then** *CM* is *bad*
If *PQ* is *good* **and** *TC* is *mediocre* **then** *CM* is *sufficient*
If *PQ* is *good* **and** *TC* is *sufficient* **then** *CM* is *sufficient*
If *PQ* is *good* **and** *TC* is *good* **then** *CM* is *good*
If *PQ* is *good* **and** *TC* is *very good* **then** *CM* is *very good*
If *PQ* is *sufficient* **and** *TC* is *bad* **then** *CM* is *bad*
If *PQ* is *sufficient* **and** *TC* is *mediocre* **then** *CM* is *mediocre*
If *PQ* is *sufficient* **and** *TC* is *sufficient* **then** *CM* is *sufficient*
If *PQ* is *sufficient* **and** *TC* is *good* **then** *CM* is *sufficient*
If *PQ* is *sufficient* **and** *TC* is *very good* **then** *CM* is *sufficient*
If *PQ* is *mediocre* **and** *TC* is *bad* **then** *CM* is *bad*
If *PQ* is *mediocre* **and** *TC* is *mediocre* **then** *CM* is *bad*
If *PQ* is *mediocre* **and** *TC* is *sufficient* **then** *CM* is *mediocre*
If *PQ* is *mediocre* **and** *TC* is *good* **then** *CM* is *sufficient*
If *PQ* is *mediocre* **and** *TC* is *very good* **then** *CM* is *sufficient*

4   The Validation of Method

The developed method has been tested on real examples. The existing P&R facilities locations, which are already known to be utilized by users, have been taken into account. For these examples the correctness of location can already be clearly evaluated. We have chosen three different examples of P&R car parks for making simulation tests.

4.1   Test I

The first example is the P&R located in Warsaw and is called Metro Mlociny. The expert has decided to determine the following values of parameters: $I_{min}= 80\%$, $I_{max}= 90\%$, $D = 70\%$, $A = 60\%$, $S = 80\%$, $K= 95\%$, $P= 70\%$.

The results of calculation of the model are:

- $TC_{Imin} = TC_{Imax} = 80\%$, it means that the territorial indicator of location is *very good* in 20% and *good* in 80%.
- $PQ= 91\%$, it means that the public transport quality indicator is *very good* in 64% and *good* in 36%.
- $CM_{Imin} = CM_{Imax} = 91\%$, it means that the final indicator of location quality is *very good* in 64% and *good* in 36%.
- $ACM=91\%$, it means that the final average indicator of location quality is *very good* in 64% and *good* in 36%.

4.2   Test II

The second example is the P&R located in Wroclaw. The place is called The Anders Hill. The expert has decided to determine the following values of parameters: $I_{min}= 80\%$, $I_{max}= 90\%$, $D = 30\%$, $A = 90\%$, $S = 20\%$, $K= 40\%$, $P= 70\%$.

The results of calculation of the model are:

- $TC_{Imin} = 32\%$, it means that the territorial indicator of location is *mediocre* in 72% and *sufficient* in 28%.
- $TC_{Imax} = 27\%$, it means that the territorial indicator of location is *mediocre* in 92% and *sufficient* in 8%.
- $PQ= 80\%$, it means that the public transport quality indicator is *very good* in 20% and *good* in 80%.
- $CM_{Imin} = 57\%$, it means that the final indicator of location quality is *sufficient* in 72% and in *good* 28%.
- $CM_{Imax} = 53\%$, it means that the final indicator of location quality is *sufficient* in 88% and *good* in 12%.
- $ACM=56\%$, it means that the final average indicator of location quality is *sufficient* in 76% and *good* in 24%.

4.3   Test III

The last example of P&R is also taken from Wroclaw and is located at the end of Grabiszynska street. The expert has decided to determine the following values of parameters: $I_{min}= 60\%$, $I_{max}= 70\%$, $D = 50\%$, $A = 80\%$, $S = 80\%$, $K= 70\%$, $P= 80\%$.

The results of calculation of the model are:

- $TC_{Imin} = 82\%$, it means that the territorial indicator of location is *good* in 72% and *very good* in 28%.

- $TC_{Imax} = 61\%$, it means that the territorial indicator of location is *good* in 44% and *good* in 56%.
- $PQ = 79\%$, it means that the public transport quality indicator is *very good* in *16%* and *good* in *84%*.
- $CM_{Imin} = 83\%$, it means that the final indicator of location quality is *very good* in 32% and in *good 68%*.
- $CM_{Imax} = 68\%$, it means that the final indicator of location quality is *sufficient* in *28%* and *good* in 72%
- $ACM = 74\%$, it means that the final average indicator of location quality is *sufficient* in *4%* and *good* in *96%*.

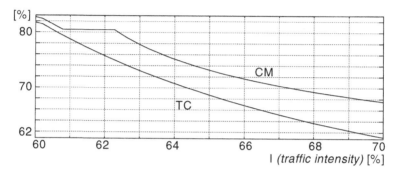

Fig. 2. The dependence of *TC* and *CM* parameters on the traffic intensity (*I*)

The relationship between the changing parameter *TC* and *CM* as a function of traffic intensity (*I*) is presented in Fig.2 and it shows the non-linear nature of the phenomenon. This graph can be very helpful for an expert. It can be seen that some operations of the system can be noticed only in the graph. In the Test 1 *I* has no influence on *CM* but in Fig.2 it can be seen that such a case occurs at a very short length.

5   Conclusions

The developed model correctly evaluates location of the P&R facilities. An important aspect of the model is the possibility of examining (jointly and separately) the chosen location in two aspects - territorial and public transport. This may be important in assessing the place attractive because of its location, but poor in means of transport. The inference based on such model allows to determine further actions that will be aimed at improving conditions. The communication aspect is more flexible - introduction of a new bus line or increasing the frequency of running is the operation relatively easy to implement. The proposed method can be used as an aid in deciding about the location of P&R car parks in cities of any size. In our opinion, this approach is extremely important in the first stage of spatial planning of the structure of communication based on the P&R system.

We intend to test the model on more actual examples to confirm the scientific credibility of the model. Additional verification of the fuzzy inference model rules with the cooperation with larger group of experts is planned.

In further research, a parameter of number of parking spaces at the P&R facility could be introduced. This is a parameter which may result from the studied attractiveness of the place and it must correspond to the size of the influx of users.

References

1. Lower, M., Magott, J., Skorupski, J.: Air Traffic Incidents Analysis with the Use of Fuzzy Sets. In: Rutkowski, L., Korytkowski, M., Scherer, R., Tadeusiewicz, R., Zadeh, L.A., Zurada, J.M. (eds.) ICAISC 2013, Part I. LNCS, vol. 7894, pp. 306–317. Springer, Heidelberg (2013)
2. Tanaka, H., Fan, L.T., Lai, F.S., Toguchi, K.: Fault-Tree Analysis by Fuzzy Probability. IEEE Transactions on Reliability R-32(5), 453–457 (1983), doi:10.1109/TR.1983.5221727
3. Lower, M.: Self-organizing fuzzy logic steering algorithm in complex air condition system. In: Håkansson, A., Nguyen, N.T., Hartung, R.L., Howlett, R.J., Jain, L.C. (eds.) KES-AMSTA 2009. LNCS, vol. 5559, pp. 440–449. Springer, Heidelberg (2009)
4. Król, D., Lower, M.: Fuzzy measurement of the number of persons on the basis of the photographic image. In: Nguyen, N.T., Borzemski, L., Grzech, A., Ali, M. (eds.) IEA/AIE 2008. LNCS (LNAI), vol. 5027, pp. 102–107. Springer, Heidelberg (2008)
5. Chen, Z., Xia, J., Irawan, B., Caulfied, C.: Development of location-based services for recommending departure stations to park and ride users. Transportation Research Part C 48, 256–268 (2014)
6. Clayton, W., Ben-Elia, E., Parkhurst, G., Ricci, M.: Where to park? A behavioral comparison of bus Park and Ride and city center car park usage in Bath. Journal of Transport Geography 36, 124–133 (2014)
7. Aros-Vera, F., Marianov, V., Mitchell, J.E.: P-Hub approach for the optimal park-and-ride facility location problem. European Journal of Operational Research 226, 277–285 (2013)
8. Wang, J.Y.T., Yang, H., Lindsey, R.: Locating and pricing park-and-ride facilities in a linear monocentric city with deterministic mode choice. Transportation Research Part B 38, 709–731 (2004)
9. Farhana, B., Murray, A.T.: Siting park-and-ride facilities using a multi-objective spatial optimization model. Computers & Operations Research 35, 445–456 (2008)
10. Holgiun-Veras, J., Yushimito, W., Aros-Vera, F., Reilly, J.: User rationality and optimal park-and-ride location under potential demand maximization. Transportation Research Part B 46, 949–970 (2012)
11. Faghri, A., Lang, A., Hamad, K., Henck, H.: Integrated Knowledge-Based Geographic Information System for Determining Optimal Location of Park-and-Ride Facilities. Journal of Urban Planning & Development 128(1), 18–41 (2002)
12. Horner, M.W., Grubesic, T.H.: A GIS-based planning approach to locating urban rail terminals. Transportation 28, 55–77 (2001)
13. Malasek, W., Seinke, J., Wagner, J.: Analiza mozliwosci lokalizacji parkingow P+R w rejonie glownych wlotow drogowych do Warszawy (Analysis of possibilities of location of P&R car parks in the area of the main inlet roads of Warsaw), Warsaw Development Planning Office J.S.C. (2009)

Quadrotor Navigation Using the PID and Neural Network Controller

Michał Lower and Wojciech Tarnawski

Wrocław University of Technology, Wybrzeże Wyspiańskiego 27, 50-370 Wrocław, Poland
{michal.lower,w.tarnawski}@pwr.edu.pl

Abstract. In this paper the neural network controller for quadrotor steering and stabilizing under the task of flight on path has been deliberated. The control system was divided into four subsystems. Each of them is responsible for setting the control values for controlling position and speed of the quadrotor and for steering rotation speed of propellers. The neural network was taught by control system with standard PID controller. This approach is used for checking how neural networks cope with stabilisation of the quadrotor under flight task. Simulation results of the neural controller and PID controller working were compared to each other. The mathematical model of quadrotor and its neural controller were simulated using Matlab Simulink software. In the paper the simulation results of the quadrotor's flight on path of are presented.

Keywords: quadrotor, control system, neural network, PID controller.

1   Introduction

A quadrotor is a flying object (drone) belonging to Unmanned Aerial Vehicle (UAV). It has been rapidly developed for a few years by scientists and commercial companies. This is the reason that this object is being analysed in the scientific literature. One of the simplest flight construction in terms of the mechanical structure is a multi-rotor. In this paper a quadrotor structure with four motors and propellers is considered. Most of quadrotors utilize the classical control theory, so they are controlled by pro-portional integral derivative-PID controllers.

PID controller theory and its application in various systems is well described in vast literature [1-4]. The procedure of tuning should be done before using PID to control the object. Changing the point of work (e.g. weight increase), the calibration procedure should be done again because the former controller settings do not guarantee the proper work of control system. Space stability of such an object is not large, rapid oscillations and disturbances can often cause loss of balance of such a system. These are disadvantages of the PID controller thus one is looking for control methods resistant to such changes.

Scientists consider more advantageous methods of control, such as artificial intelligence algorithms, e.g.fuzzy logic [5-11] or neural networks controllers [12-15]. The main advantage of fuzzy logic, in comparison to classical method, is the nonlinear

© Springer International Publishing Switzerland 2015
W. Zamojski et al. (eds.), *Theory and Engineering of Complex Systems and Dependability*,
Advances in Intelligent Systems and Computing 365, DOI: 10.1007/978-3-319-19216-1_25

character of controller and the ability to make controller using simple transformation of rules expressed in natural language. The good example of this methodology was demonstrated in our prior publications [5-9], [12]. In this paper the neural network controller for quadrotor steering and stabilizing under the task of flight on path has been presented. It allows to analyse how this type of algorithm can be cooperative with a flying object as a quadrotor. In the first stage of the study, which is described in the paper, the neural network controller was taught on the base of the PID control system. In the second stage of the study the learning process is planned to be extended to fuzzy control system.

2   Mathematical Model of Quadrotor

A quadrotor can be represented as an object with four motors with propellers in cross configuration (Fig. 1). Comparing to a classic helicopter, the division of the main rotor into two pairs of propellers in the opposite direction removes the need of a tail rotor. Usually all engines and propellers are identical, so the quadrotor is a full symmetrical flying object.

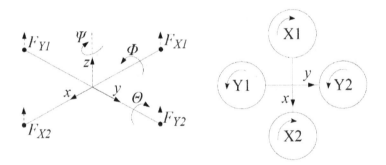

Fig. 1. The coordinate system of quadrotor

The detailed mathematical model we based on, can be found in [2]. Although, for clarity the main formulas are presented below.

The torque moment around O_y axis:

$$T_x = bl(X_1^2 - X_2^2) \tag{1}$$

where b is a thrust coefficient, l is the distance between the propeller's axis and the center of mass of the quadrotor and X_1, X_2 are rotation speeds of propellers according to the Fig. 1. As the consequence the angle Θ called pitch can be observed.

The torque moment around O_x axis:

$$T_y = bl(Y_1^2 - Y_2^2) \tag{2}$$

where Y_1, Y_2 are rotation speeds of propellers. As the consequence the angle Φ called roll can be observed.

The join torque around mass center of quadrotor:

$$T_z = d(X_1^2 + X_2^2 - Y_1^2 - Y_2^2) \tag{3}$$

where d is so called drag coefficient. As the consequence the angle Ψ called yaw can be observed. The above formulas look quite simple and so they are, thus the quadrotor position can be controlled only via propellers rotation speed changes. But these changes also have an influence on joint thrust force:

$$F_z = b(X_1^2 + X_2^2 + Y_1^2 + Y_2^2) \tag{4}$$

The main effects acting on quadrotor are presented in the system of equations (5). It is the simplified model of quadrotor's behavior according to [2], [4], [8], [9] which ignores aerodynamic drags and gyroscopic effects caused by propellers rotation, but this model is good enough to model quadrotor 's behavior in hover state and at low quadrotor values of speed.

$$\begin{cases} \ddot{x} = (cos\psi sin\Theta cos\phi + sin\psi sin\phi)F_z/m \\ \ddot{y} = (sin\psi sin\Theta cos\phi - +cos\psi sin\phi)F_z/m \\ \ddot{z} = (cos\Theta cos\phi)F_z/m - g \\ \ddot{\phi} = [\dot{\Theta}\dot{\psi}(I_y - I_z) + lT_y]/I_x \\ \ddot{\Theta} = [\dot{\phi}\dot{\psi}(I_z - I_x) + lT_x]/I_y \\ \ddot{\psi} = [\dot{\phi}\dot{\psi}(I_x - I_y) + lT_z]/I_z \end{cases} \tag{5}$$

where m is the mass of the quadrotor, g is the gravity acceleration, l is the distance between the rotor and center of the quadrotor , I_x, I_y, I_z are the inertial moment along proper axes, Φ, Θ, Ψ are roll, pitch and yaw angles, respectively.

3   Control System

In this paper we have focused on control moving by the trajectory in two axes (X,Y). It is possible by controlling velocity and rotation in all axes. Z axis depends on T_z and F_z, changing these variables have a direct influence on velocity. This dependence is very simple, so this aspect has been exluded from the paper. Changing RPM for motors triggers temporary changing of velocity and continuously changing of angular velocity (5). It has an influence on continuous changing of angle of a quadrotor, so the controller has to check this angle and correct it, otherwise the quadrotor will be rotated around its axis and an operator will not control the velocity. The study has been done for two types of controllers. One of them is based on neural network and PID controller and the second one uses only PID methodology.

In this paper the controller has been divided into three blocks (this is a cascade controller (Fig. 2)). The first block has a simple algorithm which allows to fly on path. In this block only a proportional part of classic PID controller is used. Similarly, the second block (P) uses only a proportional part of classic PID controller and it is responsible for calculating an error from linear velocity to value (*XX'*), which is transmitted to the next block.

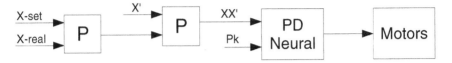

Fig. 2. The schema of the structure of the control system

The third block (PD/Neural), which has neural or classic PID algorithm, controls a RPM of motors based on its input and internal algorithm, which has a direct influence on velocity and rotation of the quadrotor.

In the first test the block (PD/Neural) was a classic PID with a proportional and derivative term and without an integrator module, schema of this block is shown in Fig. 3.

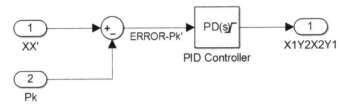

Fig. 3. Basic structure of the block (PD/Neural) which works as a PID controller

Where:

- *XX'* – output from the former block of control system,
- P_k – angle Θ,
- R_k – angle Φ.

In the second test the block (PD/Neural) consists of neural network as in Fig. 4.

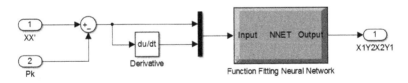

Fig. 4. Basic structure of the block (PD/Neural) which works as a NEURAL controller

The neural network used as a controller was feed-forward network with two input, sigmoid hidden neurons and one linear output neuron. Number of hidden neurons was selected experimentally and neural network was trained in off-line with Levenberg-Marquardt backpropagation algorithm. Data for training was generated from simulation of quadrotor when it was controlled by PD. This is the method called "general training" [15].

Inputs for neural block are presented in formula (6).

$$\begin{cases} Input_1 = XX' - Pk \\ Input_2 = In\dot{p}ut_1 \end{cases} \tag{6}$$

4   Simulation Tests

The mathematical model of a quadrotor is implemented for simulation in Matlab with Simulink. In the first step controllers P and PD were tuned. A quadrotor has many dynamic dependent variables so we extract potential Matlab's simulation and tune the controller experimentally. The minimization of integrated square error of velocity is an objective function. Research have been made for all axes (X, Y, Z, R). It is shown in detail in the paper [12]. The main subject of this paper is a flying on path so the authors show results of simulation with control in two axes (X, Y) in this paper. Control inputs are calculated for a given trajectory of a path.

The coefficients of simulation model of quadrotor were adopted as in the paper [2, 12]:

- propeller distance $l = 0.23[m]$,
- quadrotor's mass $m = 0.65[kg]$,
- drag coefficient $d = 7.5e^{-7}[Nms^2]$,
- thrust coefficient $b = 3.13e^{-5}[Ns^2]$,
- inertia moment $I_x = I_y = 7.5e^{-3}[kgm^2]$, $I_z = 1.3e^{-3}[kgm^2]$.

4.1   The Flight after a Square

In the first test, the trajectory of a flight path has been defined as a square with sides 3m. Both controllers, PD and neural have been tested. The simulation results are presented in the following figures. Proper execution of the task of quadrotor with neural controller is shown in Fig. 5. The results for quadrotor with PD controller were very similar. The trajectory for both controllers can be considered as satisfactory.

The difference in the results of the neural control and PID control can be observed by analyzing the graph of the changes of the P_k and R_k angles as a function of time. It may be noted that the controller based on neural network controls the system softer and more smoothly. It can be seen that the changes in the angle of P_k and R_k are less rapid. Such control can have a positive impact on the behavior of a flying object, minimizing adverse effects on the rapid changes in the value of the motors system. The experiment shows that the neural control system reduces oscillations in times of change of speed. The graph of the changes of the P_k angle as a function of time for both controllers PID and neural is depicted in Fig.6. The results of tests for R_k angle were very similar to the results of the changes of the P_k angle.

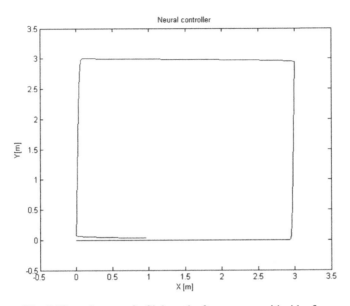

Fig. 5. The trajectory of a flight path after a square with sides 3 m

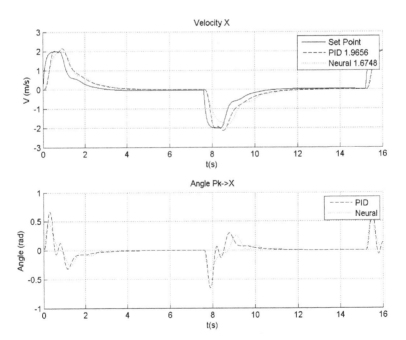

Fig. 6. Velocity V and angle P_k as a function of time during the flight after a square with 3m sides

4.2   The Flight after Path as a Parabola and Sine Function

The next research are made for the trajectory of a flight path as a parabola and sine function. The formulas of these functions are defined as follows (7),(8):

$$Y = \frac{(25-X)*X}{5} \tag{7}$$

$$Y = 2\,sin(X) + 2 \tag{8}$$

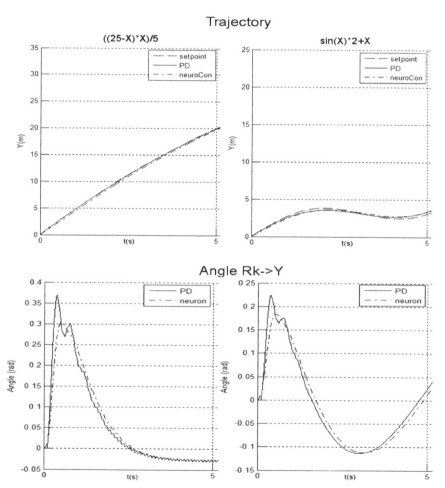

Fig. 7. The trajectory of a flight path and R_k angle as a function of time during the flight after a parabola and sine function path

The results of simulation of the flight after path as a parabola and sine function are consistent with the results obtained in the task of the flight after the square. The examples of graphs of the trajectory of a flight path and R_k angle as a function of time during the flight after a parabola and sine function path are presented in Fig.7.

4.3   The Task of Flight after the Path with the Disturbed Measuring Signals

In the studies presented in subsections 4.1, 4.2 the simulation was done for the ideal measurement signals generated by the simulation model of quadrotor. In real objects signals come from inertial sensors such as accelerometer, gyroscope, barometer, magnetometer. Measurement signals from such sensors have disturbances which complicate control.

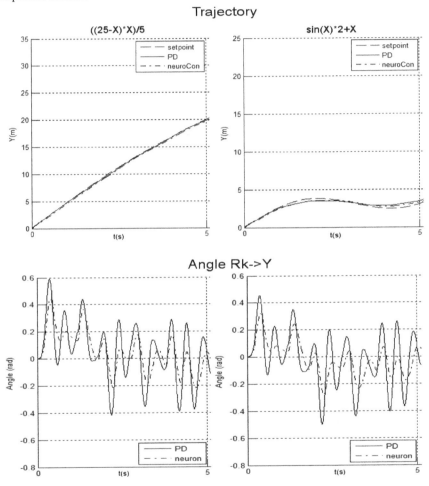

Fig. 8. The trajectory of a flight path and R_k angle as a function of time during the flight after a parabola and sine function path with disturbed the measuring signals

Many scientific publications focus on analysis of the filtering of signals using a complementary filter [16-18] or a Kalman filter [19, 20] to improve the signals input of the controller. The authors of these papers have tested the susceptibility of the proposed controller to the input signal with the noise. In order to conduct the experiment a white noise was added. The examples of graphs of the trajectory of a flight path and

R_k angle as a function of time during the flight after a parabola and sine function path for the measuring signal disturbed by a white noise are presented in Fig.8.

By analyzing the obtained data the earlier experiments [12] can be confirmed. The neural controller that was taught on the basis of PD controller can control the flying object with less overshoots and oscillations in the event of a noise signal. This feature can be seen in the graph of the trajectory of a flight path and R_k angle as a function of time under the flight after a parabola and sine function path in Fig.8.

5   Conclusions

The simulation tests presented in the paper have shown that neural controller can be used for stabilization of flight of a quadrotor. The neural network controller can be better than the PID. The presented examples show that quadrotor stabilization with the neural controller can reach the set point with lower oscillations than PID controller. Implementing such a regulator in the real flying object should allow to improve the stability of the object, as the neural controller better copes with control when measurements with noise are given into its inputs. On the basis of the research it can be expected that using the neural controller will improve the stability of a flying object equipped with inertial sensors. Previous studies have focused on improving the signal filtering and using classic PID controller for a quadrotor flight stabilization. It seems appropriate to continue studies in the subject of new control algorithms that are able to use the measurement data of inferior quality.

In the next step of research we are going to analyze another neural network configurations. It is planned to expand the space of learning data. We want to teach the neural network on the base of PID controller with different parameters which are changing according to the work point. Additionally we are going to use fuzzy controller to teach the neural network. In the last stage of research it is necessary to make tests with the real control system of a quadrotor .

References

1. Arama, B., Barissi, S., Houshangi, N.: Control of an unmanned coaxial helicopter using hybrid fuzzy-PID controllers. In: 2011 24th Canadian Conference on Electrical and Computer Engineering (CCECE), pp. 001064–001068. IEEE Press (2011)
2. Bouabdallah, S.: Design and Control of Quadrotors with Application to Autonomous Flying, Master's thesis, Swiss Federal Institute of Technology (2007)
3. Li, J., Li, Y.: Dynamic analysis and PID control for a quadrotor. In: 2011 International Conference on Mechatronics and Automation (ICMA), pp. 573–578 (2011)
4. Hoffmann, G.M., Huang, H., Wasl, S.L., Claire, E.: Quadrotor helicopter flight dynamics and control: Theory and experiment. In: Proc. of the AIAA Guidance, Navigation, and Control Conference (2007)
5. Lower, M., Król, D., Szlachetko, B.: Building the fuzzy control system based on the pilot knowledge. In: Khosla, R., Howlett, R.J., Jain, L.C. (eds.) KES 2005. LNCS (LNAI), vol. 3683, pp. 1373–1379. Springer, Heidelberg (2005)

6. Krol, D., Lower, M., Szlachetko, B.: Selection and setting of an intelligent fuzzy regulator based on nonlinear model simulations of a helicopter in hover. New Generation Computing 27(3), 215–237 (2009)
7. Król, D., Gołaszewski, J.: A Simulation Study of a Helicopter in Hover subjected to Air Blasts. In: SMC 2011, pp. 2387–2392. IEEE Computer Society (2011)
8. Szlachetko, B., Lower, M.: Stabilisation and steering of quadrocopters using fuzzy logic regulators. In: Rutkowski, L., Korytkowski, M., Scherer, R., Tadeusiewicz, R., Zadeh, L.A., Zurada, J.M. (eds.) ICAISC 2012, Part I. LNCS, vol. 7267, pp. 691–698. Springer, Heidelberg (2012)
9. Szlachetko, B., Lower, M.: On Quadrotor Navigation Using Fuzzy Logic Regulators. In: Nguyen, N.-T., Hoang, K., Jędrzejowicz, P. (eds.) ICCCI 2012, Part I. LNCS, vol. 7653, pp. 210–219. Springer, Heidelberg (2012)
10. Raza, S.A., Gueaieb, W.: Fuzzy logic based quadrotor fight controller. In: ICINCO-ICSO 2009, pp. 105–112 (2009)
11. Santos, M., Lopez, V., Morata, F.: Intelligent fuzzy controller of a quadrotor. In: 2010 International Conference on Intelligent Systems and Knowledge Engineering (ISKE), pp. 141–146 (2010)
12. Lower, M., Tarnawski, W.: Stabilisation and steering of quadrocopter using neural network. In: Świątek, J. (ed.) Information Systems Architecture and Technology: Knowledge Based Approach to the Design, Control and Decision Support, pp. 155–164. Oficyna Wydawnicza Politechniki Wrocławskiej, Wrocław (2013)
13. Kassahun, Y., de Gea, J., Schwendner, J., Kirchner, F.: On applying neuroevolutionary methods to complex robotic tasks. In: Doncieux, S., Bredèche, N., Mouret, J.-B. (eds.) New Horizons in Evolutionary Robotics. SCI, vol. 341, pp. 85–108. Springer, Heidelberg (2011)
14. Aswani, A., Bouffard, P., Tomlin, C.: Extensions of learning-based model predictive control for real-time application to a quadrotor helicopter. In: Proceedings of the American Control Conference, pp. 4661–4666 (2012)
15. Norgaard, M., Ravn, N., Poulsen, N.K., Hansen, L.K.: Neural Networks for Modelling and Control of Dynamic Systems. Springer, London (2000)
16. Euston, M., Coote, P., Mahony, R., Kim, J., Hamel, T.: A complementary filter for attitude estimation of a fixed-wing UAV. In: IEEE/RSJ International Conference on Intelligent Robots and Systems, IROS 2008, Nice, pp. 340–345 (2008)
17. Sa, I., Corke, P.: System Identification, Estimation and Control for a Cost Effective Open-Source Quadcopter. In: 2012 IEEE International Conference on Robotics and Automation (ICRA), Saint Paul, pp. 2202–2209 (2012)
18. Malatinece, T., Popelka, V., Hudacko, P.: Development of autonomous sensor based control of flying vehicles. In: 2014 23rd International Conference on Robotics in Alpe-Adria-Danube Region (RAAD), Smolenice (2014)
19. Razak, N.A., Arshad, N.H.M., Adnan, R., et al.: A Study of Kalman's Filter in Embedded Controller for Real-Time Quadrocopter Roll and Pitch Measurement. In: 2012 IEEE International Conference on Control System, Computing and Engineering, Malaysia, pp. 23–25 (2012)
20. Kenneth, D., Boizot, S., Boizot, N.: A Real-Time Adaptive High-Gain EKF, Applied to a Quadcopter Inertial Navigation System. IEEE Transactions on Industrial Electronics 61(1) (2014)

On Supporting a Reliable Performance of Monitoring Services with a Guaranteed Quality Level in a Heterogeneous Environment[*]

Piotr Łubkowski, Dariusz Laskowski, and Krzysztof Maślanka

Military University of Technology,
Faculty of Electronics, Telecommunications Institute,
ul. Gen. S. Kaliskiego 2, 00-908 Warsaw
{plubkowski,dlaskowski,kmaslanka}@wat.edu.pl

Abstract. The process of ensuring security of citizens requires access to information from sensors placed in various points of monitoring and data acquisition systems. IP video monitoring networks are nowadays the main element of a system that combines a variety of software and hardware architectures forming a heterogeneous environment. Ensuring implementation of monitoring services with a predictable quality level is an important challenge for monitoring systems. The paper includes presentation of a QoS platform that offers the possibility of support for monitoring services based on the information on video stream quality parameters. The proposed solution is part of the quality support system for monitoring and data acquisition system implemented in the INSIGMA project.

Keywords: video monitoring, reliable transfer, quality of services, heterogeneous networks.

1 Introduction

Video surveillance systems for many years are used to enhance the safety of citizens and the protected objects. With video surveillance, we meet both in public and national utility facilities, in places publicly available and in the areas with limited access. Monitoring systems are a combination of video recording (sensors), transmitting, storing and reproducing devices in one integral unit. They enable the observation of people or objects in real time as well as the event recording for later analysis. Different kind of monitoring services in the technical environment have been described in [1 – 4].

As it can be noticed, the process of detection and identification of people and events is one of the basic features of modern monitoring systems. A continuous operation mode is the characteristic feature of these systems, where a relatively short operation break may result in a significant degradation of its efficiency. Requirements for video surveillance systems are defined in the field of functional and quality limitations [5, 6].

---

[*] The work has been supported by the European Regional Development Fund within INSIGMA project no. POIG.01.01.02,00,062/09.

© Springer International Publishing Switzerland 2015   275
W. Zamojski et al. (eds.), *Theory and Engineering of Complex Systems and Dependability*,
Advances in Intelligent Systems and Computing 365, DOI: 10.1007/978-3-319-19216-1_26

Quality assurance at the network level (QoS - Quality of Service) is extremely important because it allows prediction of the size of the available bandwidth and the level of losses at the application layer. However, this information will only allow a fair distribution of bandwidth between the cameras and does not reflect the user's requirements on service severability [7]. Thus in the case of video surveillance systems it is important to consider also the quality at the application layer. The QoS at the application layer is defined in ITU-T Recommendation G.1010 and includes parameters related to the establishing, disconnecting and call-blocking of connection [8]. Within this group of parameters, service accessibility and service reliability are particularly relevant. They determine the possibility of obtaining on demand services in the specified range and under certain conditions, as well to continue it for the required period. Provision of QoS with respect to network layer and the application layer for such systems is a great challenge and requires appropriate QoS platform with traffic control mechanisms to handle video traffic transferred through such systems.

Many QoS-enabled architectures and protocols have been proposed to solve the problem of end-to-end quality of real-time video services [9 – 13]. However, the mentioned solutions do not take into account the great variability of the available data transmission rate and wireless network limitations that are currently an important element of monitoring systems. The paper presents a QoS platform, which is composed of streaming server module (STR), resources broker (BB) and client module. They are part of a mobile application characterized by the possibility of notification of QoS requirements in order to support the implementation of video monitoring services with a quality level determined by the user.

2   Concept of QoS Platform

The structure of the QoS platform for the monitoring system is based on a standard video monitoring network with quality support that is the basis for elaboration of the individual QoS modules. Solution that were proposed for the presented platform includes a signalling subsystem using the modified Real Time Streaming Protocol (RTSP), stream admission control mechanism implemented in the STR mobile streaming server, BB resources management module and application able to interact with the IP QoS network.

Monitoring network is considered as a set of end terminals, access networks and backbone networks. Communication between two end points is realized using package transmission path for which communication resources are secured.

QoS mechanisms proposed for use in the monitoring system were designed in view of ensuring adequate QoS in networks with limited bandwidth and those with oversize backbone.

In order to implement services with required quality between the monitoring network domains a Service Level Agreement is provided.

As already mentioned, the proposed QoS supporting platform consists of the QoS-Aware client application module, STR mobile streaming server and BB resources manager (Fig. 1). The QoS-Aware streaming application client is equipped with a dedicated RTSP signalling module allowing transmission of the guaranteed service request to STR. Each request contains information on the service class (represented by the expected bandwidth) and video stream priority. STR is responsible for informing the BB manager on the QoS requirements transmitted from the client application. The BB manager monitors available resources of the access router output interface of the monitoring domain and, according to the QoS requirements transmitted by STR, maintains the required resources for the selected video stream. If the resources available at the BB manager level do not allow handling the new stream, it sends an appropriate message to STR.

Following such an action, the STR server rejects the possibility of handling the new video stream in the guaranteed service mode.

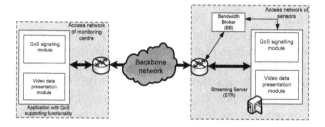

Fig. 1. Functional diagram of the QoS platform

The above described QoS, with all elements was reflected in the test environment and presented in Fig. 2.

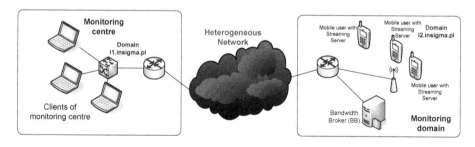

Fig. 2. The test environment with elements of the QoS platform implemented

Because of the fact that nowadays a big emphasis is put on information protection, modern monitoring network terminals are required not to pose a threat of interception of this kind of information. In this case, loss of confidentiality might occur because of using electromagnetic emissions arising during operation of IT devices to intercept classified data. In order to protect the private data the monitoring network terminals have to be protected using electromagnetic shielding [14].

3   QoS Supporting Elements

3.1   Modified RTSP

A signalling system is an essential mechanism used in the process of control assurance and exchange of QoS management information in IP networks, including those in which monitoring services are implemented. It can be also seen as an interface for implementation of advanced functions related to quality provision in QoS supporting networks. In the case of video streaming services, the RTSP is the basic protocol used in the process of initiation and implementation of the service. This protocol was used in the described platform as the main QoS requests signalling mechanism. As compared to the standard version of the protocol, modifications were introduced that are aimed at signalling QoS requirements from the streaming service client level. The mentioned modifications includes the possibility of transferring to the STR and, in consequence, to the BB as well, information relating to the declared bit rate and DSCP class in which the video streaming service is to be implemented. The proposed modifications were provided in the part used to describe parameters of the streaming session contained in the Session Description Protocol (SDP) complementary to the RTSP. The SDP describing client requirements as regards QoS was extended by two additional fields recorded in the following form: <b>=<Bandwidth> and <q>=<QoS Class>. This makes it possible to signal the requested bandwidth for the selected class of services. Resources are reserved by the BB, which then informs the STR on the possibility of video streaming initiation in a client-server relation. After completion of transmission, the reserved resources are released.

3.2   Mobile Streaming Server Module

The video streaming server was created in the form of software application running on mobile device with Android operating system. The application that acts as video server communicates with the client and resources manager by using the modified RTSP.

Upon receiving a streaming transmission request containing QoS requirements, the STR server creates new instance of MediaStream.java class that contains the reference to Packetizer abstract class. The mentioned reference is filled with data describing the selected stream data transmission mode. Further, the resources reservation process is implemented using classes H263Stream.java, BrokerConnector.java and Session.java (Fig. 3). The classes are responsible for acquisition of data on the required codec bandwidth, transfer of data on bandwidth and services class to the resources manager and control of video stream transmission. Owing to the said process, the BB receives information on the required network resources declared by the client and the quality class for the implemented service. In consequence, the BB may reserve resources or deny the request in the case of no guarantee of bandwidth appropriate for the given class of services. Block diagram of the bandwidth reservation procedure is presented in Fig. 3.

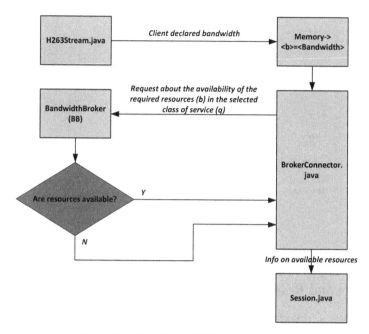

Fig. 3. The algorithm of STS application operation

3.3   Bandwidth Broker Module

General architecture of the resources manager module is presented in Fig. 4. The manager module is composed of several elements. The resources management module is the main element that is responsible for elaboration and making decision on admission of resources reservation request for implementation. This decision is made based on communication with the request control module (AC) and requested parameters for transmission of data that the management module receives using the modified RTSP signalling from the streaming server.

The quality parameters are maintained mainly through provision of adequate bandwidth for every request as part of the Service Level Agreement between the monitoring centre and sensor domains. For this purpose, the following QoS mechanisms implemented in access routers were used: Hierarchical Token Bucket (HTB) or Committed Access Rate (CAR).

The decision on resources reservation is transferred to the resources reservation module. This module cooperates with the managed network devices (technology-dependant layer) using NETCONF [15] or XML-RPC [16]. It also ensures appropriate configuration of traffic handling principles in network devices to provide support of appropriate service classes. In the case of positive decision concerning reservation of resources, this module reserves those appropriately configuring particular devices and releasing them during session closure.

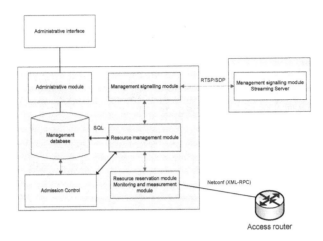

Fig. 4. General architecture of BB module

4   Testing and Results

Development tests were performed in a test environment similar to that presented in Fig. 2. In the monitoring domain, 3 streaming servers were launched (mobile user), while traffic from these servers was received in client applications in the management center domain. The domains are connected over backbone network where tunnel of predefined bandwidth is established. The BB manages the bandwidth in the tunnel and distributes it resources into defined class of services. Only under this condition we can use our mechanism for QoS support. Reasoning about the correctness of network components specification as regards data transmission reflecting information from the monitoring system is based on a statistical estimation of reliability of the software and hardware platform forming the service chain [17]. Products of renowned suppliers of hardware and software for both the systems and applications are the components of the QoS supporting platform. Therefore, it appears reasonable to conclude that the specified measuring system is a correct and highly reliable testing environment.

Access routers in both domains were built based on Linux operating system and Quagga software package. The backbone network was constructed based on Cisco routers. In a backbone network, data was transferred using priority queuing mechanisms. HTB mechanism in access routers was launched based on its implementation in the traffic control package. Communication of the resources manager and the router is provided using the XML-RPC protocol. The resources manager application was elaborated using Java language, therefore it can be launched in any node in the domain.

The streaming server runs on mobile devices equipped with a video camera and microphone, operating under Android and a dedicated application. A standard VLC programming application [18] is the client receiving the video stream.

Fig. 5 presents a signalling packets flow diagram between monitoring network nodes: streaming client (172.16.1.2), streaming server (192.168.1.101) and resources manager of the monitoring domain (192.168.1.100), received on the access router of the monitoring domain. In this case, a typical RTSP signalling data exchange is limited to exchange of information between the server and service client.

```
|Time   172.16.1.2                                  192.168.1.101                      192.168.1.100
|       |                                           |                                   |
|0.092  |         RTSP: OPTIONS rtsp://192.168.1.101:8086 RTSP/1.0                       |
|       |(49523)  ----------------------> (8086)    |                                   |
|0.209  |         RTSP: Reply: RTSP/1.0 200 OK       |                                   |
|       |(49523)  <---------------------- (8086)    |                                   |
|0.210  |         RTSP: DESCRIBE rtsp://192.168.1.101:8086 RTSP/1.0                      |
|       |(49523)  ----------------------> (8086)    |                                   |
|0.252  |                                           |RTSP/SDP: DESCRIBE rtsp://192.168.1.101 RTSP/1.0
|       |                                           |(41768) ----------------------> (6789)|
|0.691  |                                           |        RTSP: Reply: RTSP/1.0 200 OK |
|       |                                           |(41768) <---------------------- (6789)|
|0.844  |         RTSP/SDP: Reply: RTSP/1.0 200 OK   |                                   |
|       |(49523)  <---------------------- (8086)    |                                   |
|0.846  |         RTSP: SETUP 192.168.1.101:8086/trackID=0 RTSP/1.0                      |
|       |(49523)  ----------------------> (8086)    |                                   |
|2.188  |         RTSP: Reply: RTSP/1.0 200 OK       |                                   |
|       |(49523)  <---------------------- (8086)    |                                   |
|2.190  |         RTSP: PLAY 192.168.1.101:8086/ RTSP/1.0                                |
|       |(49523)  ----------------------> (8086)    |                                   |
|2.195  |         RTSP: Reply: RTSP/1.0 200 OK       |                                   |
|       |(49523)  <---------------------- (8086)    |                                   |
|12.733 |         RTSP: TEARDOWN 192.168.1.101:8086/ RTSP/1.0                            |
|       |(49523)  ----------------------> (8086)    |                                   |
|12.744 |                                           |RTSP/SDP: TEARDOWN 192.168.1.101 RTSP/1.0
|       |                                           |(41768) ----------------------> (6789)|
|12.769 |                                           |        RTSP: Reply: RTSP/1.0 200 OK |
|       |                                           |(41768) <---------------------- (6789)|
|12.812 |         RTSP: Reply: RTSP/1.0 200 OK       |                                   |
|       |(49523)  <---------------------- (8086)    |                                   |
```

Fig. 5. A data flow of signalling messages

In Fig. 5 a change in the signalling process in the form of an additional RTSP DESCRIBE package can be observed which, after being received by the STR server, transferred to the resources manager is. The structure of the package is shown in Fig. 6. The figure shows two sections of media description in SDP message, which are intended for two transmission directions and included parameters related to the required bandwidth and quality parameters.

```
⊞ Internet Protocol Version 4, Src: 192.168.1.101 (192.168.1.101), Dst: 192.168.1.100
⊞ Transmission Control Protocol, Src Port: 57538 (57538), Dst Port: smc-https (6789),
⊟ Real Time Streaming Protocol
   ⊞ Request: DESCRIBE rtsp://192.168.1.101 RTSP/1.0\n
      CSeq: 1\n
      Accept: application/sdp\n
      Date: wt., 11 mar 2014 14:51:13 GMT+00:00\n
      Content-Base: rtsp://192.168.1.101\n
      Content-type: application/sdp
      Content-length: 488
      \n
   ⊟ Session Description Protocol
      Session Description Protocol Version (v): 0
      ⊞ Owner/Creator, Session Id (o): - 3776780 3776780 IN IP4 192.168.1.101
        Session Name (s): Insigma serwer
      ⊞ Time Description, active time (t): 0 100000
      ⊞ Media Description, name and address (m): video 8086 RTP/AVP 0 2 3 97 8
      ⊞ Bandwidth Information (b): AS:300000
        Unknown: q=best 150ms 1s 3%
      ⊞ Connection Information (c): IN IP4 192.168.1.101
      ⊞ Media Attribute (a): rtpmap:0 pcmu/8000
      ⊞ Media Attribute (a): rtpmap:2 g729/8000
      ⊞ Media Attribute (a): rtpmap:3 gsm/8000
      ⊞ Media Attribute (a): rtpmap:97 speex/8000
      ⊞ Media Attribute (a): rtpmap:8 pcma/8000
      ⊞ Media Attribute (a): control:audio
      ⊞ Media Description, name and address (m): video 8086 RTP/AVP 0 2 3 97 8
      ⊞ Bandwidth Information (b): AS:300000
        Unknown: q=best 150ms 1ms 3%
      ⊞ Connection Information (c): IN IP4 172.16.1.2
      ⊞ Media Attribute (a): rtpmap:0 pcmu/8000
      ⊞ Media Attribute (a): rtpmap:2 g729/8000
      ⊞ Media Attribute (a): rtpmap:3 gsm/8000
      ⊞ Media Attribute (a): rtpmap:97 speex/8000
      ⊞ Media Attribute (a): rtpmap:8 pcma/8000
      ⊞ Media Attribute (a): control:audio
```

Fig. 6. The structure of the modified DESCRIBE package

The service will not be run without a confirmation of the possibility of its performance in the form of RTSP 200 OK package received from the manager. A similar data exchange occurs when the service is completed.

The impact of the proposed modifications on the monitoring service quality, taking into account the required quality parameters of video stream, is presented in Fig. 7. During testing, a mobile device camera operating with 320x240 resolution was used.

The client application performed the video streaming service without the guarantee of quality and, in the second case, with the guarantee of the highest quality of services (DSCP = EF). It can be noticed that in the absence of quality support, visible artefacts (fields marked with red frame) appear in relation to the video with guaranteed quality. This allows the conclusion that the provided modifications guarantee the achievement of the assumed level of video streaming.

Fig. 7. Comparison of images with quality guarantee (A) and without quality guarantees (B)

The confirmation of noticeable quality degradation in the absence of QoS supporting mechanisms are the results of qualitative tests. During the tests a single impulse event measurement method was used (in accordance with ITU-R BT 500-11 [19]).At the next figure (Fig. 8) it can be observed that perceived Mean Opinion Score (MOS) of video sequences transmitted by streaming server without QoS support decreases when a subsequent streaming servers with implemented QoS mechanisms starts transmission with increasingly higher throughput.

When subsequent video streaming servers with QoS functionality starts theirs transmission the available bandwidth decreases rapidly thus reducing the bandwidth available for servers that do not support QoS (this effect is visible in Fig. 7B). At the same time, the remaining streams are transmitted without any interruption in quality (Fig. 7A).

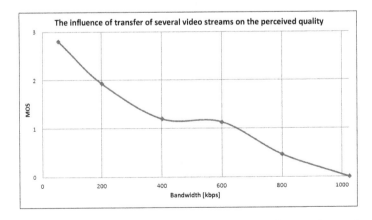

Fig. 8. The degradation of perceived quality in the absence of QoS supporting mechanisms

5   Conclusion

Results of the tests confirm that the presented QoS platform covering the proposed applications and mechanisms allows achievement of video streaming in the monitoring system with a guaranteed level of quality. This is extremely important in particular in the case of monitoring and data archiving systems in which the access to proper quality video sequences may decide on early intervention and precise identification of offenders.

References

1. Lubkowski, P., Laskowski, D.: The end-to-end rate adaptation application for real-time video monitoring. In: Zamojski, W., Mazurkiewicz, J., Sugier, J., Walkowiak, T., Kacprzyk, J. (eds.) New Results in Dependability & Comput. Syst. AISC, vol. 224, pp. 295–305. Springer, Heidelberg (2013)
2. Burdzik, R., Konieczny, Ł., Figlus, T.: Concept of On-Board Comfort Vibration Monitoring System for Vehicles. In: Mikulski, J. (ed.) TST 2013. CCIS, vol. 395, pp. 418–425. Springer, Heidelberg (2013)
3. Lubkowski, P., Krygier, J., Amanowicz, M.: Provision of QoS for multimedia services in IEEE 802.11 Wireless network. In: Information System Technology Panel Symposium on "Dynamic Communications Management" IST-062, Budapest (2006)
4. Burdzik, R., Konieczny, Ł.: Application of Vibroacoustic Methods for Monitoring and Control of Comfort and Safety of Passenger Cars. Solid State Phenomena 210, 20–25 (2014)
5. Siergiejczyk, M., Paś, J., Rosiński, A.: Application of closed circuit television for highway telematics. In: Mikulski, J. (ed.) TST 2012. CCIS, vol. 329, pp. 159–165. Springer, Heidelberg (2012)
6. Siergiejczyk, M., Paś, J., Rosiński, A.: Evaluation of safety of highway CCTV system's maintenance process. In: Mikulski, J. (ed.) TST 2014. CCIS, vol. 471, pp. 69–79. Springer, Heidelberg (2014)
7. Siergiejczyk, M., Krzykowska, K., Rosiński, A.: Reliability assessment of cooperation and replacement of surveillance systems in air traffic. In: Zamojski, W., Mazurkiewicz, J., Sugier, J., Walkowiak, T., Kacprzyk, J. (eds.) Proceedings of the Ninth International Conference on DepCoS-RELCOMEX. AISC, vol. 286, pp. 403–412. Springer, Heidelberg (2014)
8. ITU-T Recommendation G.1010, End-User Multimedia QoS Categories Series G: Transmission Systems and Media, Digital Systems and Networks Quality of Service and Performance-Study (2001)
9. AQUILA - Adaptive Resource Control for QoS Using an IP-based Layered Architecture, Project Number: IST-1999-10077, http://www-st.inf.tu-dresden.de/aquila (access date: January 2015)
10. EuQoS - End-to-end Quality of Service support over heterogeneous networks, http://www.euqos.eu (access date: January 2015)
11. Lubkowski, P., Bednarczyk, M., Maslanka, K., Amanowicz, M.: QoS-Aware end-to-end Connectivity Provision in a Heterogeneous Military Environment. In: Military Communications and Information Systems Conference, Saint Malo, France (2013)

12. Lubkowski, P., Laskowski, D., Pawlak, E.: Provision of the reliable video surveillance services in heterogeneous networks. In: Safety and Reliability: Methodology and Applications - Proceedings of the European Safety and Reliability Conference, ESREL 2014, pp. 883–888. CRC Press, Balkema (2014)

13. Natkaniec, M., Kosek-Szot, K., Szot, S., Głowacz, A., Pach, A.R.: Providing QoS Guarantees in Broadband Ad Hoc Networks. Journal of Telecommunication and Information Technology 4, 92–98 (2011)

14. Kaszuba, A., Checinski, R., Kryk, M., Łopatka, J., Nowosielski, L.: Electromagnetically Shielded Real-time MANET Testbed. In: Progress In Electromagnetics Research Symposium, PIERS 2014 Conference Proceedings, Guanzhou, China, pp. 2706–2710 (2014)

15. Enns, R.: NETCONF Configuration Protocol, RFC4741, https://tools.ietf.org/html/rfc4741 (access date: January 2015)

16. XML-RPC Homepage, http://xmlrpc.scripting.com (access date: January 2015)

17. Paś, J., Choromański, W.: Impact electric component of the electromagnetic field on electronic security and steering systems in personal rapid transit. In: Mikulski, J. (ed.) TST 2014. CCIS, vol. 471, pp. 252–262. Springer, Heidelberg (2014)

18. VLC media player, http://www.videolan.org (access date: January 2015)

19. ITU-R Rec.BT 500-11, Methodology for the subjective assessment of the quality of television pictures, https://www.itu.int/dms_pubrec/itu-r/rec/bt/R-REC-BT.500-11-200206-S!!PDF-E.pdf (access date: January 2015)

Reducing Complexity Methods of Safety-Related System

Ivan Malynyak

Stalenergo LLP, 9 Fedorenka St., Kharkiv, Ukraine
ivanmiros@gmail.com

Abstract. Traditionally, the fault tolerance is a matter of redundancy, where hardware components are replicated or additional considerable lines of program code are inserted. In spite of the widespread usage even for the basic voted-groups architectures, the problem of complexity always has to be taken into account and could lead to decrease system reliability. In this paper the combined software and hardware methods to achieve necessary system requirements without enlarged implementation price and complexity is proposed. The reducing of system complexity helps to make up to 50% savings in development life cycle stage with higher availability and higher safety.

Keywords: fault tolerant architecture, redundancy, complexity, Markov model, 1oo2D, 2oo3, safety.

1 Introduction

The challenge to provide safety requirements for systems with critical functions often brings to pure redundant design [1]. In railway applications and nuclear power plants this approach is widely used because of limited requirements for physical implementation (size, weight, power). Indeed, all controlling systems are placed on ground with other station equipment and could be easily replaced with near unlimited spare parts.

As a result, safety approach generally reduced to simple duplicate units, which are widely used because of their clear architecture and almost equal software algorithms [2, 3]. But as one could see, nature doesn't have such solutions and even the most complicated creatures have only one available decision-making organ within reliability assurance system. The Fig. 1 shows principal idea of using redundant units to achieve necessary safety and reliability levels [4].

The picture is elegant, but when it comes into practice, all hidden stones come to light. First of all, software cost of data flow synchronization within the system as a whole has to be taken into account under functional unit algorithm. Secondly, the problem of unit's coordination and synchronization leads to significant increase of data links and even basic voted-groups architectures has a surfeit of wiring. And finally, algorithm becomes huge and intricate, where one hardly knows and thinks it to be aimless trying to understand how exactly separate part of the system works [5].

© Springer International Publishing Switzerland 2015
W. Zamojski et al. (eds.), *Theory and Engineering of Complex Systems and Dependability*,
Advances in Intelligent Systems and Computing 365, DOI: 10.1007/978-3-319-19216-1_27

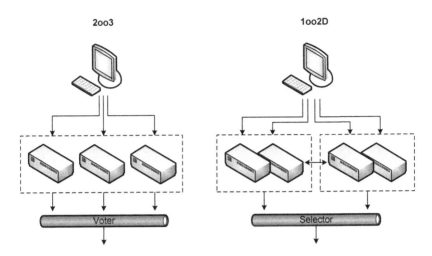

Fig. 1. Principal idea of using redundant units in safety-related systems

2   The Causes of Complexity

Railway safety-related systems very often have one of two commonly used architectures: 2oo3 or 1oo2D (shown in Fig. 1). For example, safety requirements could be achieved by quadding (4 units in 1oo2D), i.e. making two units in each channel work by pair for the purpose of failure detection. Appropriate Markov models of such architectures are presented in Fig. 2 and Fig. 3, where the simplified models of finite states of system for overcoming failures are considered. Firstly, the system assumed to be in S7 (2oo3) or S10 (1oo2D) states without failures and moves to S1 (2oo3) or S9 (1oo2D) states on an entry. In case of failure is detected the system moves to S4 (2oo3) or S8 (1oo2D) states and this channel is de-energized to perform safety response. If second failure is came before first one has been detected the system moves to states with two failures S2 (2oo3), S6 or S3 (1oo2D). For all other states the same approach is performed.

The failure rate λ is specified with the aid of standard specifications, while transition rates μ_D (test interval time) and μ_R (time to replace de-energized unit or time to repair) are constant for particular system [6]. As one could see, there are notable numbers of states in Markov models, so all transitions within architecture as a whole (in additional to core algorithm) have to be taking into account. It means that all possible connections between units have to be implemented, continually tested and verified [7, 8]. Our experience of developing 1oo2D system based on embedded FPGA Stalenergo's system "Strela-10" for signaling and control in railway domain has shown that initial thinking of making pure and clear software algorithms becomes real challenge afterwards, when unit's synchronization with all types of breakdowns and recovery has to be properly treated [9]. The requirement of necessity of all events being correctly synchronized has a big implementation price, especially in heterogeneous embedded systems with simultaneously analogue and digital parts maintenance [10].

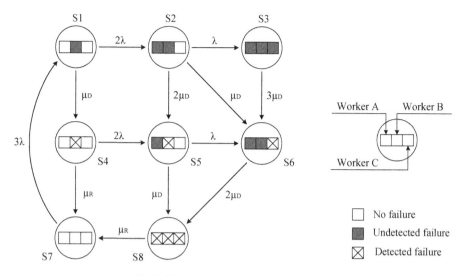

Fig. 2. Markov model of 2oo3 architecture

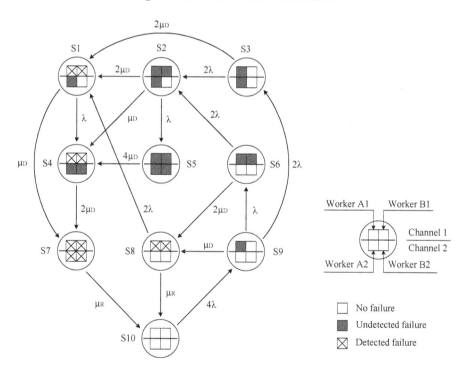

Fig. 3. Markov model of 1oo2D architecture

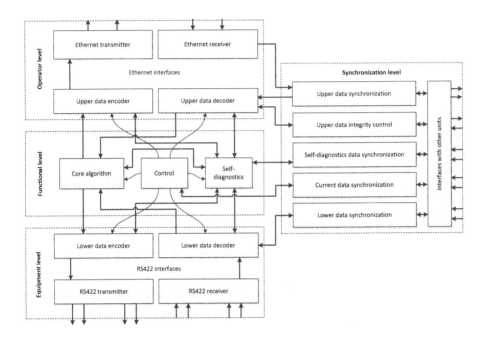

Fig. 4. "Strela-10" FPGA unit software functional scheme

From one hand it's not a problem due to present-day controllers and computers' capability. But from other hand, such project shows relatively high development expenditures to obtain necessary functional units' reliability due to complicated software where considerable amount of time has to be spent on correcting data flows. As a result, nearly 3/4 of "Strela-10" FPGA software was intended for data flow handling as shown in Fig. 4 and it's correlated with informal Gall's law about complicated systems efficiency [11]. Indeed, physically each unit of voted-groups architecture has additional 14 data lines to perform complete 1oo2D structure, which all have to be correctly synchronized and involved in core algorithm.

3   Complexity Reducing Statement

The basic principle of redundancy and fault-tolerance approach works greatly for comparably plain applications. But as far as system itself is going to be intricately big (like the voted-groups architectures) system reliability becomes a real challenge and widespread duplicating approach only makes it worse. On the basis of complexity reducing consideration the safety function circuits could be converted to sequential form as shown in Fig. 5.

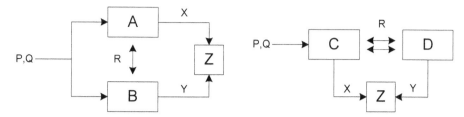

Fig. 5. The specification of parallel safety function form (left) and sequential one (right)

For the preliminary evaluation of parallel safety function it's assumed that proper $p \in P$ or fault $q \in Q$ inputs are used for both units *A* and *B* with equal functional software and data communication $r \in R$ to implement safety function *Z* which could have an intersection with unsafe output states *Z'*. For the sequential safety function form it's assumed that inputs is used only for unit *C* with both diagnostic and data communication lines *R* to unit *D* (shown in Fig. 5).

Working states where system is operated normally:

$$(\forall r, \forall p: Arp=Brp)(X=Y \Rightarrow Z)(Z \wedge Z'=\varnothing) \qquad (1)$$

Working states with fault has not been detected:

$$(\forall r, \exists q, \exists p: Arq=Brp \vee Arp=Brq)(X=Y \Rightarrow Z)(Z \wedge Z'=\varnothing) \qquad (2)$$

Nonworking safe states with fault has been detected:

$$(\forall r, \exists q, \exists p: Arq \neq Brp \vee Arp \neq Brq)(X \neq Y)(Z=\varnothing) \qquad (3)$$

Failure states where units' outputs synchronously have been changed by faults (common cause failure case):

$$(\exists r, \exists q: Arq=Brq)(X=Y \Rightarrow Z)(Z \wedge Z' \neq \varnothing) \qquad (4)$$

If software in parallel architecture is diversified as N-version programming [12], then equation 4 should not be used. Failure states where units' outputs are unsafe and equal (safety case issues):

$$(\forall r: Aq \Leftrightarrow Bq)(X=Y \Rightarrow Z)(Z \wedge Z' \neq \varnothing) \qquad (5)$$

Software for sequential form is considered to be easier and diverse with $C \wedge D=\varnothing$ equation is assumed. Number of special variables for sequential form is shown below:

- C_D – set mapped functions of *C* placed in *D*
- D_C – set mapped functions of *D* placed in *C*

Each unit in the sequential architecture has acceptable set of responses applicable to other one. For example, one unit has a set of responses of other one based on its safety diagnostic behavior, software timing, etc.

Working states where system is operated normally:

$$(\forall\, r, \forall\, p\colon Crp \subseteq C_D \wedge Drp \subseteq D_C)(X{=}Y \Rightarrow Z)(Z \wedge Z' {=} \emptyset\,) \qquad (6)$$

Working states with fault has not been detected:

$$(\forall\, r, \exists\, q, \exists\, p\colon Crq \subseteq C_D \vee Drq \subseteq D_C)(X{=}Y \Rightarrow Z)(Z \wedge Z' {=} \emptyset\,) \qquad (7)$$

Nonworking safe states with fault has detected:

$$(\forall\, r, \exists\, q, \exists\, p\colon Crq \wedge C_D{\neq}0,\ Drq \wedge D_C{\neq}0)(X{\neq}Y)(Z{=}\emptyset\,) \qquad (8)$$

Failure states where units' outputs are unsafe and equal (safety case issues) but is very rare to happen due to functional diversity in *C* and *D:*

$$(\forall\, r\colon Cq \Leftrightarrow Dq)(X{=}Y \Rightarrow Z)(Z \wedge Z' {\neq} \emptyset\,) \qquad (9)$$

Altogether, the basic framework could be made of several rules to follow: do not duplicate functions, i.e. to eliminate simple redundant approach; self-diagnostics is vital, i.e. to implement safety functions as a part of testability; move to interconnections, i.e. to push forward implementation of safety functions based on indirect factors; separate tasks, i.e. to differentiate physical functions between units.

4   Complexity Reducing Methods

In contrast to multi-chip solutions, in a single chip dual-processor architecture the memory sub-system which is shared between the processors becomes a trade-off between system integrity, memory size and performance [13]. This approach is like to put one "black box" into another one and it has a bottleneck performance problem of two cores run independently on the same memory and communication sub-systems. Such implementation reduces extra hardware wiring, although nonetheless software complexity with analogue and digital parts synchronizations remains.

Further development of safety-related systems could be similar to nature evolution with "satellite-type" design, where necessity of redundancy implements as sequential algorithm in embedded bundle of diverse units instead of parallel working of identical ones [14]. For example, parallel structure of safety-related 1oo2D architecture could be introduced as duplication of equal 1oo1D channels, where software includes: kernel algorithm part *Core*, safety-diagnostic part *SD*, input part *In*, output part *Out*, data exchange part *Data* and synchronization part *Sync* (shown in Fig. 6).

Each of unit is received equal data from the inputs and managed own part of safety output. As one could see, software is implemented with 6 software blocks where 15 possible interconnections between all of them have to be mentioned in source code.

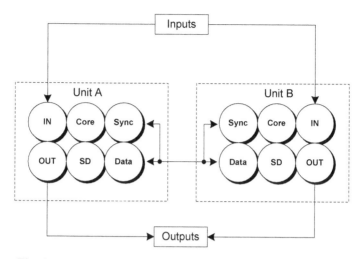

Fig. 6. Parallel software structure of safety-related architecture 1oo1D

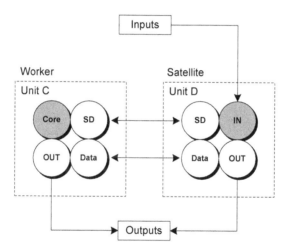

Fig. 7. Sequential software structure of safety-related architecture "1oo1S"

From "satellite" point of view, architecture without unit's functional duplication (let it be "1oo1S") has reduced software volume and eliminated strong functional synchronization between them (shown in Fig. 7). In this case software is implemented only with 3 equal (*SD, Out, Data*) and 2 different (*Core, In*) software blocks where only 9 possible interconnections between all of them are existed.

Taking into account software and hardware complexity reducing, overall saving in development and verification life cycle periods potentially could be up to 50% (by 1/3 in structure, by 1/5 in software blocks and almost by 1/2 on interconnections).

Of course, this approach has own weak points lying in additional functional requirements for data exchange between software *SD* parts which have to analyze functional integrity of units and compensate reducing wiring. The main aim of sequential

"1oo1S" architecture is to transform safety functions to additional software interconnections and avoid redundant "black box" implementation.

Markov model for architecture "1oo2S" is shown in Fig. 8. As it could be seen,
models of these architectures are equal and have ten states where transitions between
them are matter of failure entries. At the state S10 all four units (A1, B1, A2 and B2)
are working properly without failures.

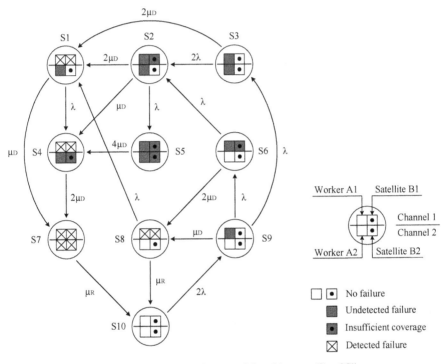

Fig. 8. Markov model of sequential architecture "1oo2S"

The system moves to state S9 when failure has occurred with transition rate 2λ (instead of 4λ as in 1oo2D) due to assumption that satellite units don't have undetected
failures, i.e. the coverage of functional testing is took 100% because of its own simplicity framework (clear algorithm, fixed number of inputs and outputs, highly testa-
ble, no software branching, minimal interrupts, etc.). Consequently, the state S9 is
represented failures in A1 or A2 worker units, which then could be detected with
transition to S8 state or would have another failure with transition to S3, or being
moved to potentially danger state S6 if satellite coverage of particular worker failure
is insufficient. The S8 is de-energized state of one of the channel which than is moved
to state S10 with faulty units are being replaced. The other states are showed system
behavior at potentially dangerous circumstances (S2, S4, S5 and S6), with safe reactions (S1, S7 and S8) or interim checking procedures (S3 or S9). The system degradation 4-2-0 is represented by S10-S8-S7 states where particular conditions for shut
down channels are described.

A numerical example is performed to illustrate the implications of the above results to see the relative performance of architectures 2oo3, 1oo2D and "1oo2S" (shown in Table 1). The calculation is done for average failure rate λ=10E-5 1/h, test interval rate μ_D= 60 1/h and time to repair μ_R=0,125 1/h. With sequential architecture "1oo2S" where satellite units are forced to simplicity, the higher availability and higher safety could be reached comparing to prevalent systems 2oo3 and 1oo2D.

Table 1. Probability of failure per hour

Probability	2oo3	1oo2D	"1oo2S"
$P_{D\,(undetected\,-\,dangerous)}$	8,3E-14	5,6E-14	2,8E-14
$P_{S\,(detected\,-\,safe)}$	3,9E-08	5,1E-08	1,3E-08
$P_{0\,(no\,failures)}$	0,99976	0,99968	0,99984

5   Conclusions

With respect to size and complexity of modern safety-critical application it could be assumed, that one way to achieve necessary system requirements without enlarged implementation price is design optimization with deletion of sophisticated synchronized function and reduction of execution rates. This paper proposes combined software and hardware applying, where up to 50% of life cycle development stage savings could be reached and overall system complexity would not exceed the uncontrollably high level with necessity of additional time and budget consuming.

Sequential safety architecture "1oo2S" combines the benefit of 1oo2D and 2oo3 systems with higher availability and higher safety levels. The price of this innovation lies in additional functional requirements to successfully analyzing integrity of units running kernel algorithm. As a result of sequential implementation with hardware and software simplified units, the higher availability and higher safety could be reached comparing to prevalent systems. The additional benefit from such diversity could be protection against vulnerability of most common cause failures when software design shortcomings or hardware faults are came to light.

References

1. Dubrova, E.: Fault-Tolerant Design. Springer (2013)
2. Johnson, B.: Fault-Tolerant Microprocessor-Based Systems. IEEE Micro 4(6), 6–21 (1984)
3. IEC 61508-6:2010: Functional safety of electrical/electronic/programmable electronic safety-related systems – Part 6: Guidelines on the application of IEC 61508-2 and IEC 61508 3
4. Nelson, V.: Fault-tolerant computing: Fundamental concepts. IEEE Computer 23(7), 19–25 (1990)
5. Brooks, F.: No Silver Bullet: Essence and Accidents of Software Engineering. IEEE Computer 20(4), 12 (1987)

6. Borcsok, J., Ugljesa, E.: Markov Models for 2004 Architecture for Safety Related Systems. In: 6th WSEAS Conference on Computational Intelligence, Man-Machine Systems and Cybernetics, Tenerife, Spain, pp. 365–370 (2007)
7. William, M., Kyle, R.: Design, development, integration: space shuttle primary flight software system. Communications of the ACM 27(9), 914–925 (1984)
8. Saposhnikov, V.L., Dmitriev, A., Goessel, M.: Self-dual parity checking – A new method for on-line testing. In: 14th IEEE VLSI Test Symposium, pp. 162–168. Princeton, New Jersey (1996)
9. Stalenergo integrated solutions for rail transport, http://en.stalenergo.com.ua/
10. Radojevic, I., Salcic, Z.: Embedded Systems Design Based on Formal Models of Computation. Springer Science + Business Media B.V. (2011)
11. Gall, J.: Systemantics: How Systems Really Work and How They Fail, 2nd edn. The General Systemantics Press, Ann Arbor (1986)
12. Avizienis, A.: The Methodology of N-Version Programming. In: Lyu, M. (ed.) Software Fault Tolerance, pp. 23–46. Wiley & Sons Ltd. (1995)
13. Ferrari, A., Garue, S., Peri, M., Pezzini, S.: Design and implementation of a dual processor platform for power-train systems. In: Proceedings of Convergence Conference (2000)
14. Mukai, Y., Tohma, Y.: A method for the realization of fail-safe asynchronous sequential circuits. IEEE Trans. Computer 23(7), 736–739 (1974)

Modeling and Reliability Analysis of Digital Networked Systems Subject to Degraded Communication Networks

Huadong Mo and Min Xie

Department of Systems Engineering & Engineering Management,
City University of Hong Kong, Hong Kong
huadongmo2-c@my.cityu.edu.hk, minxie@cityu.edu.hk

Abstract. Digital networked systems become increasingly important and perform indispensable function in safety-critical systems. However, they are exposed to various networked degradations that affect their reliability which has not been widely studied. In this paper, the reliability of such systems is estimated using event-based Montel Carlo simulation assuming a time-varying model. The degradations are described by Markov process and multi-state Markov chain subject to uncertainties. A case study is provided to illustrate the efficiency of the proposed framework.

Keywords: Reliability analysis, digital networked system, event-based Monte Carlo simulation.

1 Introduction

Digital networked systems refer to a class of spatially distributed digital systems in which the communication networks provide data exchanges among controllers, actuators, sensors and other components [1]. There have been many applications of such systems in complex and safety-critical systems, including nuclear power plant, vehicle system and chemical plants [2] and [9].

However, the communication networks have various performance degradations. Transmission delay and packet dropout are the inherent and inevitable types, and have been widely studied. The delayed or lost data packet containing output sample or control signal jeopardizes the dynamic performances of such a system by forcing all subsystems to use delayed or inaccurate information to make decisions [3] and [7].

Therefore, it is crucial to establish a model for such systems and develop methods to quantify the effects of the two degradations on the reliability. Time-varying model based on the Laplace Transform method is proposed to model and analyze the system, and it provides reasonably accurate system evolutions as a function of time [4].

For a more general description of realistic system, transmission delay and packet dropout are considered at both sensor-to-controller channel and controller-to-actuator channel simultaneously. Although the problem for such a system with discrete time delays has been well studied, there is very little literature on continuous time delays which are more of practical significance [5]. In general, the statistic information of the

© Springer International Publishing Switzerland 2015
W. Zamojski et al. (eds.), *Theory and Engineering of Complex Systems and Dependability*,
Advances in Intelligent Systems and Computing 365, DOI: 10.1007/978-3-319-19216-1_28

continuous time delay taking values is usually known and the variation range of the continuous time delay is available. The lower bound of the delay is not zero and upper bound is known as the maximum allowable bound, which ensures the stability of the system. Therefore, in this paper, a commonly use Markov process-reflected Wiener process is used to model the continuous time delays.

Multi-state Markov chains subject to uncertainties are used to describe the packet dropout. Compared with existing works, it is an improvement and assumes that the statistic information of the packet dropout is complete known, and its Markov chain only has two states with constant transition probability [6]. However, the statistic information on the characteristic of the networked degradation is always inadequate or partially unknown. Therefore, the model of packet dropout with a Markov chain subject to partially unknown transition probabilities is more practical and general.

In this paper, the partially unknown transition probabilities are modelled using the polytopic uncertainties method. By describing the quantity of packet dropouts between current period and its latest successful transmission other than the historical information of a packet is missed or not, multi-state Markov chain is defined and the relationship between adjacent periods can be presented clearly. This model releases the assumption that each period is independent of each other [4] and [6].

It is not a trivial task to derive an explicit reliability function for DNCSs, even though we obtain their time-varying model considering networked degradations. Estimating the reliability by conducting Monte Carlo Simulation (MCS) on the model becomes an alternative and efficient method. The event-based Monte Carlo simulation method is most suitable for complex system. The networked degradations are generated and then used to determine the success or the failure of the DNCSs for one given combination of operational requirements. Therefore, the reliability is then estimated as a tabulated function of the operational requirements. The result can guarantee the estimated reliability to satisfy a given precision [4] and [8].

The remains of this paper are organized as follows. Section 2 illustrates the model of a typical digital networked system considering degraded communication networks. Section 3 presents the analysis of the domain requirements and introduces the Monte Carlo simulation for reliability evaluation. Section 4 presents a case study.

2   Modeling of the System Considering Networked Degradations

2.1   A Time-Varying Model

Consider a digital networked control system (DNCS) with degraded communication networks in Fig. 1. It has two communication channels with sampling period T.

In the period k, the sensor samples the actual output $\bar{y}_{k-1}$ of the systems at the last period $k - 1$ and then sends it to the buffer through the communication networks. The controller uses the data packet based on last-in-first-out law: If not packet dropout, the buffer can receive the new data packet containing output $\bar{y}_{k-1}$, and then the controller picks it out as y_k and uses in computing the control signal u_k; otherwise, the controller has to pick up the most recent data packet $\bar{y}_{k-N_k^{sc}-1}$ as y_k in the buffer.

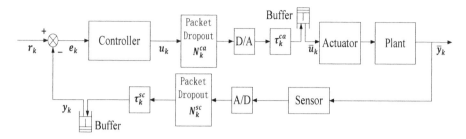

Fig. 1. DNCSs with transmission delays and packet dropouts

Thus, the system output y_k recorded by controller is:

$$y_k = \begin{cases} \bar{y}_{k-1}, & \text{if } N_k^{sc} = 0 \\ \bar{y}_{k-N_k^{sc}-1}, & \text{otherwise} \end{cases} \tag{1}$$

where N_k^{sc} is the quantity of packet dropped at the period k in the sensor-to-controller channel, which is recorded from the current period k to the last successful packet transmission at the period $k - N_k^{sc}$.

Thus, for the controller-to-actuator channel, we have the relationship between control signal u_k and the control signal $\bar{u}_k$ used by actuator which is similar to (1).

Therefore, the error e_k between system control goal and system output of the DNCSs recorded by the controller at the period k is

$$e_k = r_k - y_k \tag{2}$$

where r_k is the control goal set at period k.

In this paper, the controller uses the Proportion Integration Differentiation (PID) control strategy to compute the control signal. Therefore, the control signal u_k is

$$u_k = K_p[e_k + \frac{T}{T_i}\sum_{j=1}^{k} e_j + \frac{T_d}{T}(e_k - e_{k-1})] \tag{3}$$

where K_p is the proportional gain, T_i is the integral time constant and T_d is the derivative time constant.

Since the effects of the transmission delays on the DNCSs are more complicated, we firstly consider the case without them. As the sensor measures the system output every period T, the control signal $\bar{u}_k$ remains the same during the interval $[(k - 1)T, T]$. For ≥ 0 , the control signal $\bar{u}_k$ can be represented by a sum of steps

$$\bar{u}_k = \bar{u}_1 + (\bar{u}_2 - \bar{u}_1)I(t - T) + \cdots + (\bar{u}_k - \bar{u}_{k-1})I(t - (k - 1)T) + \cdots \tag{4}$$

where the $I(t)$ is defined as

$$I(t) = \begin{cases} 1, & \text{if } t \geq 0 \\ 0, & \text{otherwise} \end{cases}.$$

We can get the mathematic representation of the $\bar{u}_k$ in the complex frequency domain $\bar{U}(s)$ by using Laplace Transform:

$$\bar{U}(s) = \bar{u}_1 + (\bar{u}_2 - \bar{u}_1)\frac{e^{-Ts}}{s} + \cdots + (\bar{u}_k - \bar{u}_{k-1})\frac{e^{-(k-1)Ts}}{s} + \cdots \tag{5}$$

Considering transmission delays τ_k^{sc} and τ_k^{ca} in both channels, the time for the actuator acting according to the control signal $\bar{u}_k$ is $(k-1)T + \tau_k^{sc} + \tau_k^{ca}$. The effects of the transmission delays on the model are introducing a time ing $e^{-(\tau_k^{sc}+\tau_k^{ca})s}$ in each period. Thus, (6) is modified as

$$\bar{U}(s) = \bar{u}_1 e^{-\tau_1^{ca}s} + \cdots + (\bar{u}_k - \bar{u}_{k-1})\frac{e^{-((k-1)T+\tau_k^{sc}+\tau_k^{ca})s}}{s} + \cdots \tag{6}$$

where τ_1^{sc} equals to 0 as there is not communication in the sensor-to-controller channel at the first period.

Assumed the mathematic model of actuator and plant are $G_A(s)$ and $G_P(s)$, the actual system output $\bar{Y}(s)$ equals to

$$\bar{Y}(s) = \bar{U}(s)G_A(s)G_P(s) \tag{7}$$

Substituting (7) into (8) and applying the inverse Laplace transform, we deduce the system output $\bar{y}_k$ in time domain as

$$\bar{y}_k = \bar{u}_1 I(t - \tau_1^{ca})g(t - \tau_1^{ca}) + \cdots + (\bar{u}_k - \bar{u}_{k-1})$$

$$I(t - (k-1)T - \tau_k^{sc} - \tau_k^{ca})f(t - (k-1)T - \tau_k^{sc} - \tau_k^{ca}) + \cdots \tag{8}$$

where $g(t)$ and $f(t)$ are the inverse Laplace transform of $G_A(s)G_P(s)$ and $G_A(s)G_P(s)/s$.

2.2   Description of Networked Degradations

In this paper, we consider the continuous transmission delays τ_k^{sc} and τ_k^{ca} in both channels, showed in Fig. 1. Transmission delays are usually bounded [7]

$$\tau_{min}^{sc} \leq \tau_k^{sc} \leq \tau_{max}^{sc}, \text{ and } \tau_{min}^{ca} \leq \tau_k^{ca} \leq \tau_{max}^{ca} \tag{9}$$

where τ_{min}^{sc} and τ_{min}^{ca} are lower bound of the delay which is not zero. τ_{max}^{sc} and τ_{max}^{ca} are upper bound.

We model τ_k^{sc} and τ_k^{ca} as Markov process-reflected Wiener process, which take values from predefined bounds in (9). According to the reflection principle of a Wiener process based on a symmetry principle, if the path of a Wiener process $w(t)$ reaches a bound B^+ (τ_{max} or τ_{min}) at time $t = t_c$, then the subsequent path after time t_c has the same distribution as the reflection of the subsequent path about the bound.

This new process is defined in a stronger form as follows:

$$\tilde{w}(t) = \begin{cases} w(t), & \text{for } t \leq t_c \\ 2B^+ - w(t), & \text{for } t \geq t_c \end{cases} \tag{10}$$

where $w(t)$ has increments with $w(t) - w(s) \sim a\sqrt{t-s}N(0,1)$ for $0 \leq s < t$ and a is the power coefficient. In this paper, the time s and t can only take value as kT.

For packet dropouts N_k^{sc} and N_k^{ca}, they are also generally bounded

$$0 \leq N_k^{sc} \leq N_{max}^{sc}, \text{ and } 0 \leq N_k^{ca} \leq N_{max}^{ca} \tag{11}$$

where N_{max}^{sc} and N_{max}^{ca} are nonnegative integers.

In real applications, the packet dropout can be detected whether the buffer can receive the data packet before the maximal allowed transmission time.

The Markov chains take values in $N_{sc} = \{0,1,\ldots,N_{max}^{sc}\}$ and $N_{ca} = \{0,1,\ldots,N_{max}^{ca}\}$ with the transition probability matrix $P^{N_{sc}} = [\lambda_{ij}]$ and $P^{N_{ca}} = [\rho_{mn}]$, respectively. The transition probability matrix of N_k^{sc} (jumping from mode i to j) and N_k^{ca} (jumping from mode m to n) are defined as

$$\lambda_{ij} = \Pr(N_{k+1}^{sc} = i \mid N_k^{sc} = j)$$

$$\rho_{mn} = \Pr(N_{k+1}^{ca} = m \mid N_k^{ca} = n) \tag{12}$$

where $\lambda_{ij} \geq 0$, $i,j \in N_{sc}$, $\rho_{mn} \geq 0$, $m,n \in N_{ca}$, $\sum_j \lambda_{ij} = 1$ and $\sum_n \rho_{mn} = 1$. According to the definition, the transition probabilities should satisfy

$$\lambda_{ij} = 0 \text{ if } j \neq i+1 \text{ and } j \neq 0$$

$$\rho_{mn} = 0 \text{ if } n \neq m+1 \text{ and } n \neq 0.$$

We introduce the commonly used polytopic uncertainties method to describe the partially unknown transition probability [8].Thus, denote P_i be the ith row of the transition probability matrix P ($P^{N_{sc}}$ and $P^{N_{ca}}$), which is partially unknown but belongs to a convex set with known vertices $P_i^{s_i}$,

$$P_i \in \{\sum_{j=1}^{s_i} \alpha_j P_i^j, \ \sum_{j=1}^{s_i} \alpha_j = 1, \ \alpha_j \geq 0\} \tag{13}$$

3   Reliability Analysis of DNCSs

Table 1 summarizes the domain requirements used in the performance analysis and improvement of DNCSs.

Table 1. The description of domain requirements

Domain requirement		Description
Operational	Maximal rising time	The time taken by the system output to increase from a specified low value to a specified high value
	Maximal percentage overshoot	The maximal value of the system output minus the expected output divided by the expected output
	Maximal settling time	The maximal time elapsed from the application of the control goal to the time when the output has entered and remained within a specified error band
Nonfunctional	Reliability	The system ability to maintain expected performance in the presence of degraded communication networks

Conducting MCS method on the propose model, the failure of DNCSs is deter-mined by whether the real-time performance satisfies all operation requirements

$$P(x = 0) = P(PF > RT_{max} \cup PF > PO_{max} \cup PF > ST_{max}) \qquad (14)$$

where PF stands for the performance of the system output and $x = 0$ means the DNCS fails in a simulation.

We here use the event-based Montel Carlo method to assess the reliability of the DNCSs without knowing the explicit reliability function. It has two main parameters-a precision interval and a percentage of simulations belonging to this interval p. When a new simulation marked s_j is conducted, we can determine this simulation fails or not by using (14).

Thus, the reliability of the DNCSs-R_j is updated, which is determined by the num-ber of failed simulations and the total number of simulations

$$R_j = 1 - \frac{N_f}{N_t} \qquad (15)$$

If the difference between two consecutive simulations $d_j = R_j - R_{j-1}$ is within this interval, the simulation s_j is effective and the number of simulations belonging to this interval N_e increases 1. When the percentage of simulations belonging to this interval N_e/N_t exceeds a nominal threshold p, the simulation is terminated and the final reliability of the DNCSs is obtained, with required precision. A precision inter-val $\pm 2\%$ and 95% of simulations belonging to this interval means that 95% of d_j belongs to $\pm 2\%$ of R_{j-1}.

4   A Case Study

An industrial heat exchanger system is studied here to illustrate the application of the proposed framework. Fig. 2 shows a typical industrial heat exchanger system.

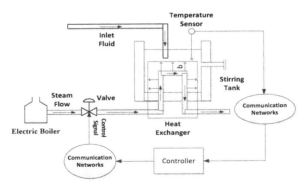

Fig. 2. The industrial heat exchanger system

4.1   System and Model Description

The top inlet pipe delivers fluid to be mixed and reacted in the stirring tank. In order to promote the chemical reactions, the controller needs to maintain the temperature of the tank liquid at a constant value by adjusting the amount of steam supplied to the heat exchanger via the control valve. We introduce the subsystems which share the information by the communication networks. Details of these subsystems are:

● Control Valve: is the actuator and implements the decisions from controller, with $G_A(s)$ is $\frac{1}{3s+1}$.

● Heat Exchanger: is the controlled plant with $\frac{5}{10s+1}$.

● Temperature Sensor: In our study, a 3-wire PT-100 RTD with a range of -200 to 600°C is adopted here and samples data every 0.5s. The feedback mechanism is unity negative feedback.

● Electric Boiler: Constant temperature steam is generated at a maximal rate-4 kg/s with a pressure which oscillates 7 and 10 bar.

● Controller: the function is to maintain operational requirements in controlling the temperature of the fluid in the stirring tank.

● Communication Networks: adopts the single-packet transmission protocol.

In this case study, the transmission delays (ms) obtained from historical data are:

$$\tau_{min}^{sc} = 10, \tau_{max}^{sc} = 50, a_{sc} = 20, \quad \tau_{min}^{ca} = 10, \tau_{max}^{ca} = 40 \text{ and } a_{ca} = 15.$$

Similarity, packet dropouts $N^{sc} = \{0,1,2\}$ and $N^{ca} = \{0,1,2,3\}$ have partially known transition probability matrix. In simulation, we have

$$P^{N_{sc}} = \begin{bmatrix} 0.9 & 0.1 & 0 \\ 0.7 & 0 & 0.3 \\ 1 & 0 & 0 \end{bmatrix} \text{ and } P^{N_{ca}} = \begin{bmatrix} 0.85 & 0.15 & 0 & 0 \\ 0.75 & 0 & 0.25 & 0 \\ 0.6 & 0 & 0 & 0.4 \\ 1 & 0 & 0 & 0 \end{bmatrix}$$

4.2   Reliability Analysis Using MCS Method

We study the reliability of DNCSs subject to different operational requirements. The initial temperature of inlet fluid is 30 °C and the optimal reaction temperature is 35 °C. The corresponding operational requirements described by Table 1 are:

● Maximal rising time RT_{max}: the time for the temperature increment rising from 10% and 90% of the expected temperature increment 5 °C is 9 s;

● Maximal percentage overshoot PO_{max}: the maximal the temperature increment should not exceed 31% of expected temperature increment;

● Maximal settling time ST_{max}: the time for the temperature increment has entered and remained within +5% of the expected temperature increment is 21 s.

We have computed the parameters of PID control strategy which are $K_p = 1.7$, $T_i = 2.5$ and $T_d = 2$ to satisfy above operational requirements when there are not networked degradations. The performances of the industrial heat exchanger system are $RT = 5.4s$, $PO = 18\%$ and $ST = 20.4s$. Thus, the pre-design PID control strategy is able to ensure the quality of the heating process.

Fig. 3 shows the real-time performances of the heat exchanger system at one simulation run. The actual performances of the heating process are RT=2.0s,   PO=59% and ST=18.5s, which do not meet all operational requirements. Based on Eq. (14), the heat exchanger system in this simulation is failed to provide a required heating process to ensure a satisfied chemical reaction. Fig. 3 (e) and (f) present the temperature increment of the perfect network and imperfect network case.

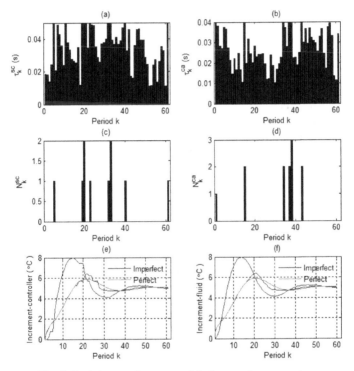

Fig. 3. Real-time performance of the heat exchanger system

Table 2 gives the reliability assessment of the heat exchanger system subject to different operational requirements with a precision interval 2% and 98% of simulations belonging to this interval. 0.8311 means that 16.89% of all simulations are defined as failures due to not satisfying the operational requirement-maximal percentage overshoot.

Table 2. Reliability assessment of the system

Operational Requirements	RT	PO	ST	Total Reliability
13s,35%,25s	1	0.8331	0.8970	0.8268
9s,31%,21s	1	0.8153	0.7604	0.7371
8s, 30%,20s	1	0.8163	0.4252	0.4068
7s,29%,19s	1	0.7800	0.1900	0.1782
6s,28%,18s	0.8317	0.8317	0.0990	0.0990

5   Conclusion

This paper proposes a time-dependent model of digital control system with degraded communication networks, which well specifies the dependence between control system and degraded communication networks. It provides an efficient method to track the state evolutions of digital control system. Applying the Laplace Transform greatly reduces the computational complexity in modeling and simulation study.

Monte Carlo simulation based on the proposed model becomes possible without knowing exact reliability function. The case study proves the accuracy of the framework and the reliability of system is determined by the networked degradations.

Future works can aim at the optimal design of the control strategy to improve the reliability of the system through the online diagnostic technique. Design of computer experiments has great potency in providing a regression model for system which can greatly reduce the simulation cost and improve efficiency.

Acknowledgement. The work described in this paper was supported by a grant from City University of Hong Kong (Project No.9380058).

References

1. Yao, J.G., Liu, X., Zhu, G.C., Sha, L.: NetSimplex: controller fault tolerance architecture in networked control systems. IEEE Transactions on Industrial Informatics 9, 346–356 (2013)
2. Hespanha, J.P., Naghshtabrizi, P., Xu, Y.G.: A survey of recent results in networked control systems. Proceedings of the IEEE 95, 138–162 (2007)
3. Wu, J., Chen, T.W.: Design of networked control systems with packet dropouts. IEEE Transactions on Automation Control 52, 1314–1319 (2007)
4. Ghostine, R., Thiriet, J.M., Aubry, J.F.: Variable delays and message losses: Influence on the reliability of a control loop. Reliability Engineering and System Safety 96, 160–171 (2011)
5. Wang, Z.D., Liu, Y.R., Liu, X.H.: Exponential stabilization of a class of stochastic system with markovian jump parameters and mode-dependent mixed time-delays. IEEE Transactions on Automation Control 55, 1656–1662 (2010)
6. Li, L., Zhong, L.: Generalised nonlinear l2 - l∞ filtering of discrete-time Markov jump descriptor systems. International Journal of Control 97, 653 – 664 (2014)
7. Al-Dabbagh, A.W., Lu, L.X.: Reliability modeling of networked control systems using dynamic flowgraph methodology. Reliability Engineering and System Safety 96, 1202–1209 (2010)
8. Yeh, W.C., Lin, Y.C., Chung, Y.Y., Chih, M.C.: A Particle Swarm Optimization approach based on Monte Carlo Simulation for solving the complex network reliability problem. IEEE Transactions on Reliability 59, 212–221 (2010)
9. Das, M., Ghosh, R., Goswami, B., Gupta, A., Tiwari, A.P., Balasubrmanian, R., Chandra, A.K.: Network control system applied to a large pressurized heavy water reactor. IEEE Transactions on Nuclear Science 53, 2948–2956 (2006)

Maintenance Decision Making Process – A Case Study of Passenger Transportation Company

Tomasz Nowakowski, Agnieszka Tubis, and Sylwia Werbińska-Wojciechowska

Wroclaw University of Technology, 27 Wybrzeze Wyspianskiego Str., Wroclaw, Poland
{tomasz.nowakowski,agnieszka.tubis,sylwia.werbinska}@pwr.edu.pl

Abstract. In the presented paper, the authors' research work is focused on the analysis of maintenance decision making process performance with taking into account necessary operational data availability. Thus, in the Introduction section, the transportation systems maintenance issues are described. Later, there is a comprehensive literature overview in the analysed research area provided. This gives the possibility to present the decision making process in the transportation systems' maintenance management area. Later, in the next Section, the case study for maintenance management processes performance in chosen passenger transportation company is investigated. Following this, the computer systems used for operational data gathering are characterised, and the data availability is investigated.

Keywords: maintenance process, decision making, transportation system.

1 Introduction

Effective performance of transportation systems needs proper operational management performance on the one hand, and adequate maintenance performance determination on the other. The decision relevance strictly depends on their accuracy and dedicated decision time. Moreover, it influences the dependability state of the system [3], [25], [55] [60]. As a result, a lot of researchers and publications in the field of maintenance decision models and techniques have been published to improve the effectiveness of maintenance process (see e.g. [39] for review).

One of the fundamental issue in the areas of technical systems operation and maintenance, both in theoretical and practical ways, is optimal decisions making problem which affects the used technical objects state and also influences other participants of the performed processes [31]. Optimal strategic decisions regard to e.g. technically, organizationally and economically reasonable deadlines for service and repair work performance, the residual lifetime of used facilities, long-term practices in the context of defined maintenance philosophy or types of performed maintenance and operational tasks [32], [26]. Natural way to support this type of enterprise activity is the use of computer tools - ranging from data management systems through to decision support systems based on Artificial Intelligence techniques implementation [4], [33].

W. Zamojski et al. (eds.), *Theory and Engineering of Complex Systems and Dependability*,
Advances in Intelligent Systems and Computing 365, DOI: 10.1007/978-3-319-19216-1_29

However, the proper implementation of maintenance management issues in complex systems (also in transportation systems) performance cannot be done without taking into account the available and reliable operational data, which e.g. give the information about the state of the system, or the possible consequences of taken decisions [56]. Following this, the presented paper is aimed at performing an analysis of the maintenance decision making process being performed in chosen passenger transportation company taking into account the operational data availability. Thus, in the next Section, a literature review in the given research area is provided. Later, the maintenance decision making process in transportation companies is investigated. This gives the possibility to present a case study of chosen transportation company, which operates in one of the biggest city in Poland. The analysis is aimed at investigation of informational flows connected with maintenance management of transportation means. The computer systems used during every day operational process performance are characterized from the point of view of gathered operational data. As a result, the critical analysis of given data from the point of their usage in maintenance decision making processes is provided.

2   Maintenance Management Issues – Literature Review

In the literature there can be found many definitions of the term of maintenance management. Following the European Standard PN-EN 13306:2010 [43] maintenance management may be defined as *all activities of the management that determines the maintenance objectives, strategies, and responsibilities and implement them by means such as maintenance planning, maintenance control and supervision, improvement of methods in the organization including economic aspects.* In [15] authors define the maintenance management as *all maintenance line supervisors, other than those supervisors that predominantly have crafts reporting to them.* Following these definitions, maintenance objectives may be classified into five groups [10]:

- ensuring technical objects functionality (availability, reliability, product quality, etc.),
- ensuring technical objects achieve their design life,
- ensuring technical objects and environmental safety,
- ensuring cost effectiveness in maintenance,
- effective use of resources, energy and raw materials.

According to Ahmad et al. [1] most of the maintenance research focuses on maintenance decision making process. Authors in their work investigated three maintenance research categories: maintenance theories, mathematical models and frameworks (management models), providing a literature review in these areas. As a result, in the presented article, authors mostly focus on the third category – maintenance management models. This research category includes tasks connected with the performance of decision-making process, by defining guidelines, procedures, or scheduling operation processes [1]. It allows managers to solve problems in a systematic way, using many known methods and statistical tools (see e.g. [1], [48]).

In response to the needs of decision-making in the area of technical objects' maintenance management many models have been developed, which comprehensive overview is provided e.g. in works [16], [48]. Moreover, in [49-50] the author presented the results of a survey of users of information computer maintenance management systems, and pointed out the main elements of systems use in practice.

The computer systems supporting maintenance management problems are developed since 1960s [30]. The literature review in the area of decision support systems designing and applications issues may be found e.g. in [2], [11-12], [34], [44], [47], and [62]. The computerized information systems used to support exploitation processes performance can be found e.g. in [22]. In this work, authors focused on Belt Conveyor Editor performance. The similar problem was investigated in [23], where authors focused on the problem of operation planning processes for machinery room. The main assumptions and structure of the system for supporting operating, repair and modernisation decisions for the steam turbines was given in [27]. The example of decision support system implementation in the area of aviation maintenance can be found in [61]. The model was based on Fuzzy Petri Nets use. The example of decision support system implementation for supporting the management of railcar operation was given in [6]. In work [37], the expert system for technical objects' reliability prediction with the use of EXSYS Professional system was developed. The similar problem for production system performance planning was also investigated in [24]. The problems of diagnostic processes supporting are analysed e.g. in works [51] and [63].

In the area of passenger transportation processes performance, the decision support systems issues being analysed in the literature regarded to e.g. timetable adjustments (e.g. [35]), scheduling means of transport maintenance activities (e.g. [17]), transportation system planning process supporting (e.g. [7], [18]), traffic control (e.g. [40-41]), transport system operation information modelling (e.g. [36]), or transportation system safety during emergencies (e.g. [59]).

Moreover, the decision making process in the analysed research area is usually a multi-criteria problem [52]. The most often used methods which support the performance of maintenance management decision systems include:

- Analytic Hierarchy Process implementation (e.g. in works [5], [57]),
- knowledge based analysis (e.g. [29]),
- neural networks implementation (e.g. in [19], [58]),
- Fuzzy Logic implementation (e.g. in [46]),
- Bayes theory use (e.g. [8]),
- Petri nets implementation (e.g. [21]).

3   Maintenance Decision Making Process in Transportation Companies

Transportation system is a system in which material objects are moved in time and space. Thus, the function of transportation is *to execute the movement of people and goods from one place to another in a safe and efficient way with minimum negative impact on the environment* [14]. Following authors of the works [14], [28], a transportation system is a very complex one with different functional characteristics depending on medium of movement, particular technology used and demand for movement in the particular medium. Aspects of these modes are e.g. vehicle, the way, control of the system, the technology of motion, intermodal transfer points, payload, drivers and pilots. As a result, the main decisions in transportation systems may be classified into three groups [14]:

- maintenance tasks which includes the definition of maintenance strategies of transportation infrastructure, system elements, or operation control systems,
- technical systems safety tasks (e.g. protection from hazard occurrence, unwanted events consequences avoidance),
- transportation tasks performance (transportation processes management), and three major areas which include [14]:
- identification of components of a system,
- identification of activities involved in putting a transportation system in place, from planning to operation and maintenance,
- identification of issues that may not be included in a transportation decision-making processes, although they may be affected by decisions.

Taking into account these considerations, and based on the Fig. 1, where the transportation system with its exploitation system elements is illustrated, the maintenance management of transportation systems may be defined as *effective performance of transportation tasks consisting in 1) the selection of means of transport in quantitative and structural way, 2) their operation and maintenance according to the intended specification, 3) continuous maintenance of operational readiness, by monitoring changes in the technical status and by conducting technically and economically reasonable replacement/repair of used vehicles, maintenance materials and spare parts.*

According to this, the model of transportation system (MST) requires four basic properties implementation, which are: the structure, characteristics of elements of the structure, transportation flows and organization [20]:

$$MST = \langle G, F, P, O \rangle \tag{1}$$

where:

G – transportation system structure graph
F – set of functions defined over structure graph elements
P – volume of transportation tasks/cargo or persons flow
O – transportation system organization

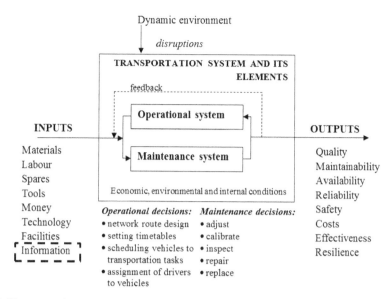

Fig. 1. Transportation system and its exploitation system's elements Source: Own contribution based on [9], [17], [20], [31], [42], [53]

The decision making process includes three steps, information gathering and analysis, available decisions definition, and optimal solution choice [45]. Exploitation decision-making process should be considered in the multi-aspect context, because decisions can regard to both, simple service and repair work for technical objects, as well as complex and multi-dimensional problems of determining the long term maintenance policy for analysed company [31]. In the area of transportation means' exploitation performance, the main decision process elements are presented in Fig. 2.

Following this, based on the literature where the maintenance decision making models are developed (see e.g. [1], [10], [31]), the exploitation decision making model (*EDM*) for transportation systems can be defined as a function of:

$$EDM = \langle X_{in}, X_{out}, Y, z \rangle \qquad (2)$$

where:

X_{in} – input parameters, which include:

$$X_{in} = \langle OS, MS, ES, EP \rangle \qquad (3)$$

where:

OS – operational strategy
MS – maintenance strategy
ES – exploitation structure (e.g. infrastructure, human resources, materials)
EP – exploitation policy (guidelines for the evaluation of the exploitation process performance)
X_{out} – output parameters:

$$X_{out} = \langle PD, EA \rangle \tag{4}$$

where:

PD – process decision (maintenance or operational decision)
EA – exploitation activity (maintenance or operational activity)
Y   – measures of decision making process quality
z   – relation: $X_{in} \rightarrow X_{out}$

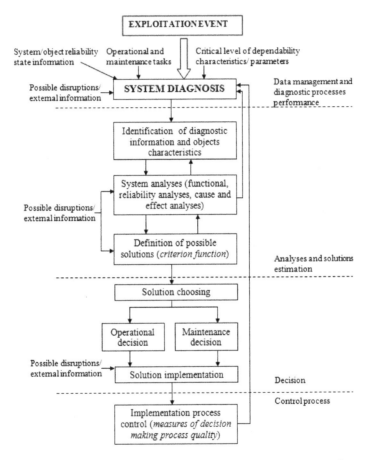

Fig. 2. Decision process in the area of transportation means' exploitation performance Own contribution based on [13], [31], [38]

Following these considerations, in the next Section there is analysed the chosen passenger transportation company's informational flows connected with maintenance management of transportation means.

4   Case Study

In the presented paper authors focus their research analyses on the municipal transport services provided by a common carrier, which operates in one of the biggest city in Poland. This company employs around 2000 workers in various positions from human resource, research department to transport and operation department. The company transports nearly 200 million passengers per year and has about 300 buses. During the year, the buses are passing about 34 million kilometres on the bus network which covers the most area of the city and is supplemented by two service depots.

To achieve daily effectiveness and continuous performance of passenger transportation tasks, maintenance and operation management plays an important role to minimize the failures and other hazard event occurrence. Following this, there can be analysed the main means of transport maintenance tasks performed in the company. There can be defined two main types of maintenance tasks:

- daily maintenance – activities performed daily by the driver to ensure technical readiness of the buses,
- periodic maintenance – specific actions to take when a bus reaches defined time between maintenance action performance and activities performed before winter and summer times.

During the daily service performance, the driver is responsible for checking in the bus the following:

- the level of exploitation fluids (including fuel and engine oil levels),
- efficiency of fire suppression system for engine compartment,
- tire pressure and their condition,
- brakes,
- exterior and interior lighting,
- efficiency of all electrical devices,
- cleanliness of windows, external cleanliness of the vehicle and passengers area,
- fire extinguishers validity,
- vehicle and operational documents completeness.

In the case of periodic maintenance performance, the type and quantity of inspected vehicle's elements depend on the type of maintenance action performance (resulting from travelled kilometres or season). The list of maintenance activities also results from the service manual which has been prepared by a producer for every bus.

4.1   Computer Systems Used in the Area of Maintenance Management and Types of Gathered Data

Recently, the company has implemented a package of IT solutions, which are aimed at improving its main and supporting processes performance. The investments concerned in particular the passenger information system implementation. Moreover, there have been developed and bought Internet service passenger system, systems for logistics warehouses and purchasing activity of the company, exploitation processes of the vehicles (mainly buses) and measurements of vehicles filling up with passengers.

The measurements area for the buses operational and maintenance process performance is supported by two main computer systems, which gather data on the basis of units being installed in vehicles. The first of the analysed systems collects data which regard to buses operational and maintenance processes. The second one records passenger movement (getting in/out at the bus stops).

In the computer system which supports maintenance management, there can be identified the exemplary software modules which collect data about: neutral gear use, engine working hours, rapid acceleration and braking, drivers login time, power supply voltage, low level of oil pressure in the engine, or fuel consumption. The exemplary window screen of this computer system is given in the Fig. 3.

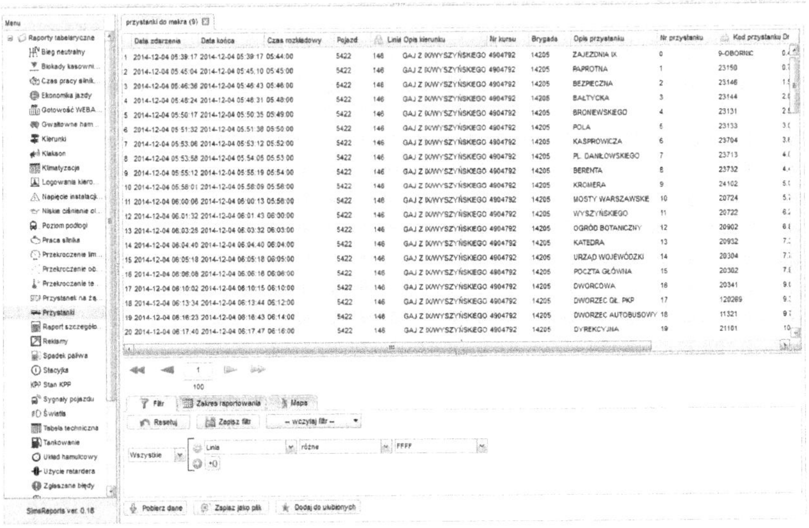

Fig. 3. The exemplary window view of Sims System

In the set of data being gathered in the computer system, one may distinguish information which are relevant to the maintenance management process performance, monitoring information of additional equipment and some information which are useless from the point of view of vehicle operation performance. The comprehensive analysis is presented in the Table 1.

Additionally, there is generated in the system a detailed report concerning:

- the number of sudden braking use (also per 100 km),
- the number of sudden acceleration (also per 100 km),
- the number of exceedances of engine rpm (also per 100 km),
- the maximum engine rpm below the acceptable level,
- maximum speed,
- fuel consumed by the engine during transportation task performance,
- working time in neutral gear,
- the percentage of engine working beyond the permissible range,
- the percentage of working time in neutral gear,
- the average fuel consumption, and the average fuel consumption per time unit [l/h].

Table 1. Analysis of gathered data taking into account their potential use in maintenance management performance

Gathered data	Maintenance data	Operational data	Monitoring data	Useless data
Working time on neutral gear		X		
Time of ticket counters locking			X	
Engine working time during stoppage		X		
Driving style of drivers analyses	X	X		
WEBASTO use analyses		X	X	
Accelerating/breaking analyses	X	X		
Travelled distance, time and fuel consumptions analyses		X		
Frequency of use of the horn			X	
Air conditioning working time			X	
Drivers login to the system and working parameters analyses	X	X		
Power supply voltage analyses		X	X	
Oil pressure analyses	X	X		
Lowering the floor of the vehicle	X		X	
Analysis of time, travelled distance and fuel consumption during engine working time		X	X	
Analysis of speed limit exceedance	X	X		
Analyses of time, travelled distance and level of engine rpm exceedance	X	X		
Analyses of time, travelled distance and level of temperature exceedance	X	X		
Switching on the "stop on demand" and number of bus stopping		X	X	
Time and distance travelled during the course, GPS status		X	X	
Advertisement availability				X
Fuel level analysis	X			
The inclusion of key switch, activation time			X	
On-board computer state			X	
Turn on and turn off the lights, the type of lighting	X		X	
Date refueling, the amount of fuel refueled		X		
Breaking use analysis	X	X		
Retarder use analysis	X		X	
Fuel consumption analysis, distance travelled	X	X		

4.2   Maintenance Analyses Performance in the Chosen Company

Taking into account the types of data being gathered by the computer systems, and following the literature in the maintenance management research area, there can be defined the types of maintenance and operation analyses in the chosen company, e.g.:

⇒ fuel consumption divided into :
   • maximal, minimal, average fuel consumption,
   • fuel consumption per bus/driver/ network route/ performed course,
⇒ driving style of drivers which takes into account the following:
   • maintenance of average speed per vehicle/network route/performed course,
   • rapid acceleration and braking,
   • road being passed in the neutral gear,
   • average fuel consumption per vehicle/ network route/ performed course,
   • exceeding the engine rpm,
⇒ punctuality of performed courses per driver/ network route/ course/vehicle,
⇒ network route/course requirements which include especially:
   • number of bus stops and number of bus stops "on demand",
   • average fuel consumptions per specified course,
   • engine working time,
   • rapid acceleration and braking per driver,
   • punctuality of specified course finishing,
⇒ vehicle load taking into account the information included in detailed report,
⇒ passenger flows intensity taking into account:
   • maximal, minimal and average passenger transportation per network route/course,
   • maximal, minimal and average passenger flow between bus stops.

Moreover, there is also possible carrying out the relevant cause and effect analyses. The exemplary ones are connected with the influence analyses of:

   • driving style of the driver on vehicle's average fuel consumption,
   • driving style of the driver on course performance punctuality,
   • driving style of the driver on vehicle's failure rate,
   • driving style of the driver on vehicle crash risk occurrence,
   • type of performed network route on the number of rapid acceleration and braking and maximal speed level.

The presented above types of possible analyses regard to the performance of single- and multi-criteria analyses. Their results should be used in operational and maintenance planning processes performance for:

   • vehicles periodic maintenance actions planning (required inspections, repairs, replacements and conservations),
   • daily maintenance actions planning (requirements connected with daily inspection of chosen elements, notifications of recorded errors),
   • transportation tasks scheduling (vehicle selection and drivers assignment for network route/courses performance).

5   Summary

Increasingly higher demands of the market for the service quality, flexibility and timeliness of transportation processes are making ever greater challenges for maintenance managers responsible for continuous operation of means of transport in road passenger transport companies. Presented in the article example clearly shows that decisions taken in the area of planning processes related to the maintenance and operation of large fleet of vehicles requires including a growing number of factors and conduction of extensive analysis on the current exploitation of vehicles. For this reason, managers need strong support in providing the necessary information from the IT systems that will provide their information needs. However, in order to fully exploit the potential of the data collected, it is necessary to determine their completeness and to define the range of needs of analysis used by managers in the decision-making processes. These activities may be supported by the maintenance management controlling system whose task is to assist managers in planning, control and information supply. For this reason, the concept of maintenance management controlling intended for road passenger transport companies is the subject of further research conducted by the authors.

The presented article is the continuation of authors' research works presented e.g. in [54], where author focused on the issues connected with maintenance management processes performed in the chosen transportation company.

References

1. Ahmad, R., Kamaruddin, S., Azid, I., Almanar, I.: Maintenance management decision model for preventive maintenance strategy on production equipment. International Journal of Industrial Engineering 7(13), 22–34 (2011)
2. Arnott, D., Pervan, G.: Eight key issues for the decision support systems discipline. Decision Support Systems 44, 657–672 (2008)
3. Babiarz, B.: An introduction to the assessment of reliability of the heat supply systems. International Journal of Pressure Vessels and Piping 83(4), 230–235 (2006)
4. Bajda, A., Wrażeń, M., Laskowski, D.: Diagnostics the quality of data transfer in the management of crisis situation. Electrical Review 87(9A), 72–78 (2011)
5. Bevilacqua, M., Braglia, M.: The analytical hierarchy process applied to maintenance strategy selection. Reliability Engineering and System Safety 70(1), 71–83 (2000)
6. Bojda, K., Młyńczak, M., Restel, F.J.: Conception of computer system for supporting the railcar maintenance management processes (in Polish). In: Proc. of TRANSCOMP – XIV International Conference Computer Systems Aided Science, Zakopane, Industry and Transport, Poland (2010)
7. Burla, M., Laniado, E., Romani, F., Tagliavini, P.: The Role of Decision Support systems (DSS) in Transportation Planning: the Experience of the Lombardy Region. In: Proc. of Seventh International Conference on Competition and Ownership in Land Passenger Transport, Molde, Norvay, June 25-28 (2001)
8. Charniak, E.: Bayesian networks without tears. AI Magazine 12(4), 50–64 (1991)

9. Cruz, P.E.: Decision making approach to a maintenance audit development (2012), `http://www.banrepcultural.org/...les/cruz_pedro_articulo.pdf` (access date: January 2015)

10. Darabnia, B., Demichela, M.: Data field for decision making in maintenance optimization: an opportunity for energy saving. Chemical Engineering Transactions 33, 367–372 (2013)

11. Despres, S., Rosenthal-Sabroux, C.: Designing Decision Support Systems and Expert Systems with a better end-use involvement: A promising approach. Eur. J. Oper. Res. 61, 145–153 (1992)

12. Eom, S.B.: Decision Support Systems. In: Warner, M. (ed.) International Encyclopaedia of Business and Management, 2nd edn. International Thomson Business Publishing Co., London (2001)

13. Florek, J., Barczak, A.: Information and decision making processes in the exploitation of technical objects (in Polish). Telekomunikacja i Techniki Informacyjne 1-2, 31–41 (2004)

14. Fricker, J.D., Whitford, R.K.: Fundamentals of Transportation Engineering. A Multimodal Systems Approach. Pearson Education, Inc., Upper Saddle River (2004)

15. Gulati, R., Kahn, J., Baldwin, R.: The Professional's Guide to Maintenance and Reliability Terminology. Reliabilityweb.com, Fort Myers (2010)

16. Garg, A., Deshmukh, S.G.: Maintenance management: literature review and directions. Journal of Quality in Maintenance Engineering 12(3), 205–238 (2006)

17. Haghani, A., Shafahi, Y.: Bus maintenance systems and maintenance scheduling: model formulations and solutions. Transportation Research Part A 36, 453–482 (2002)

18. Han, K.: Developing a GIS-based Decision Support System for Transportation System Planning. In: AASHTO GIS-T (2006), `http://www.gis-t.org/files/gnYKJ.pdf` (access date: March 2012)

19. Hurson, A.R., Palzad, S., Lin, B.: Automated knowledge acquisition in a neural network based decision support system for incomplete database systems. Microcomputers in Civil Engineering 9(2), 129–143 (1994)

20. Jacyna, M.: Modelling and assessment of transportation systems (in Polish). Publ. House of Warsaw University of Technology, Warsaw (2009)

21. Jeng, M.D.: Petri nets for modeling automated manufacturing systems with error recovery. IEEE Transactions on Robotics and Automation 13(5), 752–760 (1997)

22. Kacprzak, M., Kulinowski, P., Wędrychowicz, D.: Computerized information system used for management for mining belt conveyors operation. Eksploatacja i Niezawodnosc – Maintenance and Reliability 2, 81–93 (2011)

23. Kamiński, P., Tarełko, W.: The prototype of computer-aided system of assignment of operational tasks carried out in the engine room (in Polish). Przegląd Mechaniczny R 67, 3 (2008)

24. Kantor, J.: Computerized system for supporting maintenance decisions in production company (in Polish). Doctoral Thesis, Wroclaw University of Technology, Wrocław (2009)

25. Kierzkowski, A., Kisiel, T.: An impact of the operators and passengers behaviour on the airport's security screening reliability. In: Proc. of the European Safety and Reliability Conference, ESREL 2014, Wrocław, Poland, September 14-18, pp. 2345–2354. CRC Press/Balkema, Leiden (2015)

26. Kierzkowski, A., Kowalski, M., Magott, J., Nowakowski, T.: Maintenance process optimization for low-cost airlines. In: 11th International Probabilistic Safety Assessment and Management Conference and the Annual European Safety and Reliability Conference 2012, PSAM11 ESREL 2012, pp. 6645–6653 (2012)

27. Kosman, G., Rusin, A.: The concept of a decision support system for operation and maintenance in order to turbines durability providing (in Polish). Zeszyty Naukowe Politechniki Śląskiej, Seria: Energetyka 131(1427) (1999)

28. Kutz, M. (ed.): Handbook of transportation engineering. McGraw-Hill Companies, Inc. (2004)
29. Liberatore, M.T., Stylianou, A.C.: Using knowledge-based systems for strategic market assessment. Information and Management 27(4), 221–232 (1994)
30. Lin, L., Ambani, S., Ni, J.: Plant-level maintenance decision support system for throughput improvement. International Journal of Production Research 47(24), 7047–7061 (2009)
31. Loska, A.: Exploitation policy model for the need of supporting of decision-making process into network technical system (in Polish). In: Proc. of XVIII Międzynarodowa Szkoła Komputerowego Wspomagania Projektowania, Wytwarzania i Eksploatacji MECHANIK 7/2014, pp. 363–372 (2014)
32. Loska, A.: Remarks about modelling of maintenance processes with the use of scenario techniques. Eksploatacja i Niezawodnosc – Maintenance and Reliability 14(2), 92–98 (2012)
33. Lubkowski, P., Laskowski, D.: The end-to-end rate adaptation application for real-time video monitoring. In: Zamojski, W., Mazurkiewicz, J., Sugier, J., Walkowiak, T., Kacprzyk, J. (eds.) New Results in Dependability & Comput. Syst. AISC, vol. 224, pp. 295–305. Springer, Heidelberg (2013)
34. Martland, C.D., McNeil, S., Acharya, D., Mishalani, R.: Applications of expert systems in railroad maintenance: scheduling rail relays. Transportation Research Part A: Policy and Practice 24A(1), 39–50 (1990)
35. Mendes-Moreira, J., Duarte, E., Belo, O.: A decision support system for timetable adjustments. In: Proc. of the XIII Euro Working Group on Transportation Meeting (EWGT 2009) (September 2009)
36. Naizabayeva, L.: Information system modelling to control transport operations process. In: Proc. of International MultiConference of Engineers and Computer Scientists 2009, IMECS 2009, Hong-Kong, March 18-20, vol. 2 (2009)
37. Nowakowski, T.: Methods of predicting the reliability of mechanical objects (in Polish). Research Work of Institute of Machine Design and Operation. Wroclaw University of Technology, Wroclaw (1999)
38. Nowakowski, T., Werbińska-Wojciechowska, S.: Decision support problem in transportation means' maintenance processes performance (in Polish). In: Proc. of VI Międzynarodowa Konferencja Naukowo-Techniczna Systemy Logistyczne Teoria i Praktyka, Korytnica, Poland, September 11-14 (2012)
39. Nowakowski, T., Werbińska, S.: On problems of multicomponent system maintenance modeling. International Journal of Automation and Computing 6(4), 364–378 (2009)
40. Ossowski, S., Fernandez, A., Serrano, J.M., Perez-de-la-Cruz, J.L., Belmonte, M.V., Hernandez, J.Z., Garcia-Serramp, A.M., Maseda, J.M.: Designing Multiagent Decision Support System The Case of Transportation Management. In: Proceedings of AAMAS 2004, July 19-23, New York, USA (2004)
41. Ossowski, S., Hernandez, J.Z., Belmonte, M.-V., Fernandez, A., Garcia-Serrano, A., Perez-de-la-Cruz, J.-L., Serrano, J.-M., Triguero, F.: Decision support for traffic management based on organisational and communicative multiagent abstractions. Transportation Research Part C 13, 272–298 (2005)
42. Pinjala, S.K., Pintelon, L., Vereecke, A.: An empirical investigation on the relationship between business and maintenance strategies. Int. J. Prod. Econ. 104, 214–229 (2006)
43. PN-EN 13306:2010 Maintenance terminology
44. Power, D.J.: Understanding Data-Driven Decision Support Systems. Information Systems Management 25(2), 149–154 (2008)

45. Sala, D.: Decision support of production planning processes with the use of expert system (in Polish). Doctoral Thesis, AGH, Krakow (2007)
46. Schrunder, C.P., Gallertly, J.E., Biocheno, J.R.: A fuzzy, knowledge-based decision support tool for production operations management. Expert System 11(1), 3–11 (1994)
47. Sharda, R., Barr, S.H., McDonnell, J.C.: Decision support system effectiveness: a review and an empirical test. Management Science 34(2), 139–159 (1988)
48. Shervin, D.: A review of overall models for maintenance management. J. Qual. Maint. Eng. 6(3), 138–164 (2000)
49. Swanson, L.: An information-processing model of maintenance management. Int. J. Prod. Econ. 83, 45–64 (2003)
50. Swanson, L.: Computerized Maintenance Management Systems: a study of system design and use. Production and Inventory Management Journal 38(2), 11–15 (1997)
51. Szpytko, J., Smoczek, J., Kocerba, A.: Computer-aided supervisory system of transportation devices' exploitation process. Journal of KONES Powertrain and Transport 14(2), 449–456 (2007)
52. Triantahyllou, E., Kovalerchuk, B., Mann Jr., L., Knapp, G.M.: Determining the most important criteria in maintenance decision making. J. Qual. Maint. Eng. 3(1), 16–28 (1997)
53. Tsolakis, D., Thoresen, T.: A framework for demonstrating that road performance meets community expectations. Road and Transport Research 7(3), 79–85 (1998)
54. Tubis, A.: Improving of vehicle maintenance management process through the implementation of maintenance management controlling (in Polish). In: Proc. of XLIII Winter School of Reliability, Szczyrk, January 11-17, pp. 1–9. Publ. House of Transportation Faculty of Warsaw University of Technology, Warsaw (2015)
55. Valis, D., Vintr, Z.: Vehicle Maintenance Process Optimisation Using Life Cycle Costs Data and Reliability-Centred Maintenance. In: Proceedings of the First International Conference on Maintenance Engineering. Science Press, Beijing (2006)
56. Valis, D., Zak, L., Pokora, O.: Engine residual technical life estimation based on tribo data. Eksploatacja i Niezawodnosc – Maintenance and Reliability 16(2), 203–210 (2014)
57. Wang, L., Chu, J., Wu, J.: Selection of optimum maintenance strategies based on a fuzzy analytic hierarchy process. Int. J. Prod. Econ. 107(1), 151–163 (2007)
58. Yam, R.C.M., Tse, P.W., Li, L., Tu, P.: Intelligent predictive decision support system for condition-based maintenance. International Journal of Advanced Manufacturing Technology 17(5), 383–391 (2001)
59. Yoon, S.W., Velasquez, J.D., Partridge, B.K., Nof, S.Y.: Transportation security decision support system for emergency response: A training prototype. Decision Support Systems 46, 139–148 (2008)
60. Zając, M., Świeboda, J.: Analysis of the process of unloading containers at the inland container terminal. In: Safety and Reliability: Methodology and Applications: Proc. of the European Safety and Reliability Conference, ESREL 2014, Wrocław, Poland, September 14-18, pp. 1237–1241. CRC Press/Balkema, Leiden (2015)
61. Zhang, P., Zhao, S.-W., Tan, B., Yu, L.-M., Hua, K.-Q.: Applications of Decision Support System in Aviation Maintenance. In: Jao, C. (ed.) Efficient Decision Support Systems - Practice and Challenges in Multidisciplinary Domains, InTech (2011)
62. Zhengmeng, C., Haoxiang, J.: A Brief Review on Decision Support Systems and its Applications. In: Proc. of International Symposium on IT in Medicine and Education (ITME) 2011, December 9-11, vol. 2, pp. 401–405 (2011)
63. Zimroz, R., Błażej, R., Stefaniak, P., Wyłomańska, A., Obuchowski, J., Hardygóra, M.: Intelligent diagnostic system for conveyor belt maintenance. Mining Science – Fundamental Problems of Conveyor Transport 21(2), 99–109 (2014)

Supporting the Automated Generation
of Modular Product Line Safety Cases

André L. de Oliveira[1], Rosana T.V. Braga[1], Paulo C. Masiero[1],
Yiannis Papadopoulos[2], Ibrahim Habli[3], and Tim Kelly[3]

[1] Mathematics and Computer Science Institute, University of São Paulo, São Carlos, Brasil
{andre_luiz,rtvb,masiero}@icmc.usp.br
[2] Department of Computer Science, University of Hull, Hull, United Kingdom
y.i.papadopoulos@hull.ac.uk
[3] Department of Computer Science, University of York, York, United Kingdom
{ibrahim.habli,tim.kelly}@york.ac.uk

Abstract. The effective reuse of design assets in safety-critical Software Product Lines (SPL) would require the reuse of safety analyses of those assets in the
variant contexts of certification of products derived from the SPL. This in turn
requires the traceability of SPL variation across design, including variation in
safety analysis and safety cases. In this paper, we propose a method and tool to
support the automatic generation of modular SPL safety case architectures from
the information provided by SPL feature modeling and model-based safety
analysis. The Goal Structuring Notation (GSN) safety case modeling notation
and its modular extensions supported by the D-Case Editor were used to implement the method in an automated tool support. The tool was used to generate
a modular safety case for an automotive Hybrid Braking System SPL.

Keywords: Product lines, certification, modular safety cases, reuse.

1   Introduction

Safety cases are required by certification bodies for developing automotive, avionics,
and railway systems. Safety standards like ISO 26262 [1] and EUROCAE ED-79A
[2] require the use of safety cases. Software product lines (SPL) have been used in the
development of automotive [3] and avionics [4] systems, reducing the development
effort. Ideally, safety analysis and safety cases should be treated as reusable assets in
safety-critical SPLs. Such reuse of pre-certified elements would reduce re-
certification effort contributing to modular and incremental certification. We refer to
that concept as modular certification and we argue that it can provide means of dealing effectively with the impact of changes caused by variation of features in products
within a SPL. Note that modular certification has already been shown effective in
handling changes in the certification of a system as it is updated through its life [5].

Research in SPL provided a number of systematic approaches for managing the
impact of variations in changing the system architecture by means of feature and context models [6]. A feature model captures the main system functions offered for reuse

© Springer International Publishing Switzerland 2015   319
W. Zamojski et al. (eds.), *Theory and Engineering of Complex Systems and Dependability*,
Advances in Intelligent Systems and Computing 365, DOI: 10.1007/978-3-319-19216-1_30

in a SPL. A context model captures information about the physical, operating, support, maintenance, and regulatory environment of SPL products [7].

Clearly, each product within a safety-critical SPL would need a safety case that reflects its specifics characteristics. For such safety cases to be produced efficiently, variation must be traced transparently in feature and context models across SPL design and safety analyses [7][8]. Information like hazards, their causes, and the allocated safety requirements (derived from the hazard analysis) may change according to the selection of SPL variation in feature and context models in a particular product [7][8][9]. Such variation may also change the structure of the product safety case. SPL variability management tools and techniques like Hephaestus [10] and Pure::variants [3] can be used to manage feature and context variations in SPL design and safety analysis assets by means of definition of the configuration knowledge.

There exist numerous methodologies and tools for semi-automated model-based safety analysis that can be adapted for SPL design. One of these tools, HiP-HOPS (Hierarchically Performed Hazard Origin & Propagation Studies) [11] provides Fault Trees Analysis (FTA), Failure Modes and Effect Analysis (FMEA), and safety requirements allocation information that are mostly auto-generated and could potentially be used to inform a safety case. However, the direct translation of such information to safety arguments would very likely result in monolithic safety arguments which may lack structure and not help to identify the arguments and evidence for specific elements of the system architecture. So, while it would be useful to exploit the safety analysis assets provided by model-based safety analysis, one would need to find a 'clever' way to argue the internal stages and nuances of the argument.

In order to support the reuse of safety arguments, a modular SPL safety case architecture could be built in a way that reflects the typical system decomposition into more refined subsystems [5][12]. Thus, it would be possible to identify the properties (i.e. goals, sub-goals, and evidence) required by individual safety case modules. The establishment of a modular safety case requires the identification of boundaries (i.e. stopping points) in the system architecture to guide the rationale of decomposing the safety case into argument modules [5]. Feature and context models can provide such stopping points. These boundaries can be used to establish the structure of the argument modules and interdependences between them according to the following well-established software architecture principles: High Cohesion/Low Coupling, Divide and Conquer, Extensibility, and Isolation of Change (i.e. Maintainability) [12].

GSN [13] and CAE (Claim, Argument, Evidence) [14] are two notations largely adopted by industry and are both applicable for modular safety cases. GSN patterns [15] and modular extensions [8] support the development of reusable modular safety case architectures. Existing safety case modeling tools such as D-Case Editor [16] and ACEdit [17] support GSN and its modular extensions and provide an Eclipse-based graphical editor. Particularly, D-Case implements a formal definition of safety case modeling language based on GSN and its extensions.

Although earlier work has partly addressed SPL safety assessment [7] and safety case development [8], there is still a need to automate the traceability of SPL variation starting from architectural models to support the generation of a modular safety case architectures addressing product lines. In this paper we propose a novel approach supporting largely automated generation of modular and reusable safety arguments from system models and safety analyses. Section 2 presents the proposed method to support the generation of modular safety cases and its instantiation in a tool support, Section 3 discusses a case study and Section 4 presents the conclusion.

2   The Proposed Method

The production of a modular SPL safety case requires the traceability of SPL feature and context variation across architecture and safety analyses. Habli's SPL safety metamodel [7] can support such traceability. It includes abstractions to represent SPL variation in feature and context assets and their impact on the hazards, their causes, and the allocated requirements. The metamodel also describes the relationships between SPL variation and assurance assets (e.g. safety argument elements in GSN). The provision of automated support for this metamodel can facilitate the automatic generation of modular product line safety cases. In the following we present a method to create support for automatic generation of modular safety cases from a number of inputs which include: the SPL safety metamodel [7], the safety case modeling notation metamodel (e.g., D-Case [16] GSN metamodel) and safety case patterns [15][18].

The safety metamodel specifies core and variant system and contextual assets. It also captures the association between functional, architectural, and component-level failures with specific SPL assets specified in the product line context, feature and reference architecture models. These relationships are captured in a Functional Failure Model (FFM), an Architectural Failure Model (AFM), and a Component Failure Model (CFM). The three models support explicit traceability between development and safety assessment assets. The FFM captures the impact of contextual and functional (i.e. feature) variation on the SPL functional failure conditions, their effects, and allocated safety requirements. The AFM records the impact of contextual and design variation on the SPL architectural failure modes and safety requirements. The AFM also produces the failure mode causal chain and derived safety requirements contributing to the functional failure conditions previously captured in the FFM. Finally, the CFM records the impact of design variation on component failure modes, their effects, and means of mitigation (i.e. specific safety requirements). Details of these models can be found in [7]. In our approach, model-based development, safety analysis, and SPL variability management tools provide the information required to instantiate the SPL safety assessment metamodels. The modeling and analysis infrastructure of such tools, together with the D-Case GSN metamodel, and safety case patterns [15][18] are then used to auto-generate SPL modular safety cases.

Fig. 1 presents the structure of the proposed method in an UML activity diagram. Functional Failure Modeling requires the feature/instance model (i.e. FM or IM), and the system model (SMDL) as inputs to produce instances of FFMs. The Design of Modular Safety Case step requires the following inputs: FFMs, the SPL feature model (FM), Safety Case Modeling Notation and its metamodel to produce the structure of the safety case architecture. Architectural Failure Modeling requires the FFMs produced in the earlier step, fault trees (FTA) and requirements allocation (RAL)

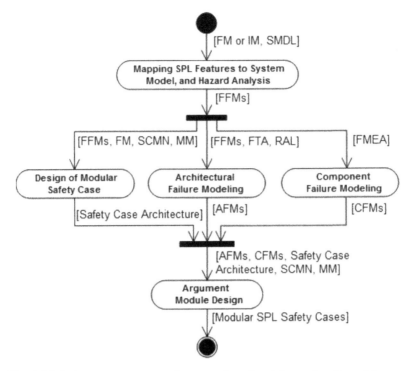

Fig. 1. Method to support automated construction of modular product line safety cases

information provided by model-based safety analysis to generate instances of AFMs. Component Failure Modeling requires FMEA information provided by model-based safety analysis as input, producing instances of CFMs as outputs. Finally, the Argument Module Design step generates the internal details of each argument module of the safety case architecture from the information provided by AFMs, CFMs, safety case architecture, Safety Case Modeling Notation and its metamodel.

A tool architecture was developed to support the method presented in Fig. 1 and it is illustrated on Fig. 2. Certain SPL assets provide inputs in this architecture instantiating FFM, AFM, and CFM models: the system model annotated with safety/hazard analysis data provided by the MATLAB/Simulink and HiP-HOPS (i.e. SMDL in Fig. 2); the feature model (FM), the instance model (IM), and configuration knowledge (CK) provided by the Hephaestus/Simulink variability management tool; and the fault trees (FTA), FMEA, and requirements allocation (i.e., RAL) generated by HiP-HOPS. The D-Case GSN metamodel and safety case patterns were used to organize the structure of the modular safety case and product-based arguments for each SPL feature and its variation. It is important to highlight that other tools can be used to develop other instances of a similar tool support. The method presented in Fig. 1 is general and can be applied manually or with the use of a set of different tools. The specific set of tools used in this paper is discussed below.

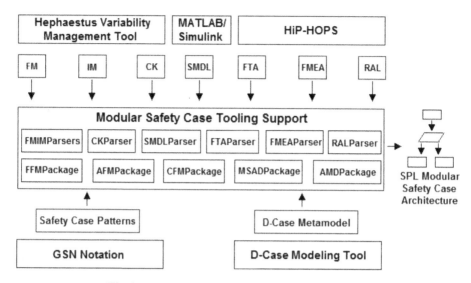

Fig. 2. Modular product line safety case tooling support

2.1 Functional, Architectural, and Component Failure Modeling

A product line variability management tool, in this case Hephaestus/Simulink [10], provides the specification of the SPL feature model (or instance model in product derivation), and the configuration knowledge, i.e., it describes the mapping between SPL features, and the design elements of the system model by associating component identifiers with feature expressions. A model-based development environment, in this case MATLAB/Simulink, provides the specification of the design elements of the system architecture. HiP-HOPS [11] was used in this work to annotate the system model with hazard/safety analysis information describing the system hazards, their causes, allocated safety requirements, and the failure logic of each design element. To obtain the necessary information for mapping SPL features, system model elements, and hazard/safety analysis data in the tool support, parsers were implemented for the SPL feature model (FMIMParsers), and configuration knowledge (CKParser) using the Java XStream[1] XML library. A Simulink parser for Java[2] was used to obtain information about the structure of the system model and its safety annotations. From the information provided by these parsers, the mapping between SPL features, design elements, and hazards is obtained by instantiating FFMs. This information is further used to provide instances for the SPL functional failure model (FFMPackage).

The HiP-HOPS tool provides fault trees, FMEA analysis, and allocation of safety requirements to architectural modules in the form of Safety Integrity Levels (in this case Automotive SILs or ASILs as our case study in an automotive system) in a XML file. This file contains the information about the failure mode causal chain of each SPL hazard, and the requirements allocated to each failure mode involved in the

---

[1] XStream: `http://xstream.codehaus.org/`
[2] Simulink Library for Java: `https://www.cqse.eu/en/products/simulink-library-for-java/overview/`

causal chain. The mapping between features, system model elements, and system hazards obtained in FFM is used together with HiP-HOPS analyses to instantiate the product line AFM. Parsers were developed to manipulate the fault trees (FTAParser) and requirements allocation (RALParser) generated by HiP-HOPS. These parsers are used together with the mapping between features and Simulink model elements to instantiate the SPL architectural failure model (AFMPackage).

CFM complements the product line AFM with a component-level view of the failure behavior of the system. CFM describes the local effects of a component failure mode, i.e. effects occurring at the boundary of the component, typically its outputs. FMEA generated by tools like HiP-HOPS provides the information required to instantiate the CFM. The developed tool reads this information (via FMEAParser) and instantiates the CFM elements (CFMPackage).

2.2 Design of Modular Safety Case Architecture

We describe how to represent the SPL variation expressed in the feature model using modular GSN [13], and how to use the D-Case tool GSN metamodel to design the architecture of a modular safety case. Czarnecki et al. [19] have provided abstractions to represent relationships between features such as inclusive-or (1-n), exclusive-or (1-1) features, and exclusive-or with sub-features (0-n). Fig. 3 shows how to specify variation expressed in the feature model into the design of a GSN modular safety case. Argument modules arguing the safety of mandatory features (Fig. 3a) is expressed by a top-level argument module connected to a module 'A' via the 'supported by' relationship (Fig. 3b). The multiplicity of such arguments capturing multiple occurrences of a mandatory feature can be represented by means of the 'supported by' relationship with a filled circle (GSN multiplicity). An argument module arguing the safety of an SPL optional feature (Fig. 3c) is represented in GSN by connecting the top-level argument to the optional module 'B' using the 'supported by' relationship with an empty circle (Fig. 3d). Argument modules arguing the safety of SPL features involved in inclusive-or (at least 1), exclusive-or (0 or n), or exclusive-or (1 of n) with grouping relationships (e.g., Fig. 3e) is encapsulated in GSN contracts (Fig 3f). The GSN optionality extension (i.e. choice) is used to connect the contract module to each argument module involved in a grouping relationship. Safety case modeling tools such as the Eclipse-based D-Case Editor [16] and ACEdit [17] provide implementations for the GSN metamodel, covering its patterns and modular extensions [13]. In the implementation of the method in the tool presented on Fig. 2, the D-Case GSN metamodel (i.e., Safety Case Modeling Notation and Metamodel) was used for developing the mechanism to automatically generate a modular safety case architecture. Initially, the tool (i.e., MSCDPackage) generates a top-level argument for the root feature, i.e., the feature representing the product line. Next, argument modules are created for each mandatory, optional, and grouping of features. The heuristics presented in Fig. 3 have also been applied to generate the structure of the modular safety case. Thus, an argument module has been created for the root feature of a group. This module is further connected to a GSN contract that encapsulates optional and alternative argument modules from the root argument module. At the end, a modular safety case reflecting the product line feature model decomposition is obtained.

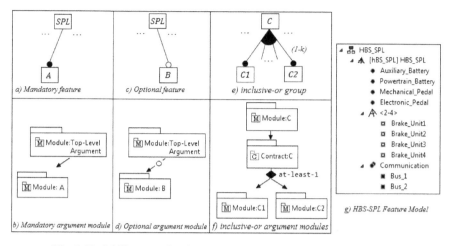

Fig. 3. Variability modeling in modular GSN and HBS-SPL feature model

2.3 Argument Module Design

This step delivers the internal structure of each argument module arguing the safety of SPL features and their variants. The mapping between SPL features, architecture, component and safety analysis data provided by functional, architectural, and component failure models is used to auto-generate the claims that compose feature-specific safety arguments. All these information together with the D-Case GSN metamodel are used as input for implementing the mechanism to auto-generate the internal structure of feature-specific argument modules in the tool support (i.e., AMDPackage). The Hazard Avoidance pattern [15] and Weaver's 'Component Contributions to System Hazards' pattern catalogue [18] have also been used to structure each feature-specific argument. Specifically, the Hazard Avoidance pattern is used to decompose the argument over the mitigation of system hazards associated to a particular feature. The Component Contributions to System Hazards pattern catalogue is used to decompose the arguments over the mitigation of the system hazards into arguments arguing the mitigation of each component failure mode contributing to the failure of a feature.

The relationships between SPL features, hazards, and allocated safety requirements provided by the functional failure model were used for arguing the mitigation of the system hazards associated with each SPL feature. The relationships between failure mode causal chains and design assets captured in the AFM, and the relationships between components, failure modes, and effects captured in the CFM have been used for arguing the mitigation of each component failure mode associated to a particular feature.

3 Case Study

The method and tool developed in this work was used for auto-generating a modular safety case architecture for a Hybrid Braking System automotive product line (HBS-SPL). The HBS-SPL is a prototype automotive braking system SPL designed in

MATLAB/Simulink. HBS-SPL is meant for electrical vehicles integration, in particular for propulsion architectures that integrate one electrical motor per wheel [20]. The term hybrid comes from the fact that braking is achieved throughout the combined action of the electrical In-Wheel Motors (IWMs) and frictional Electromechanical Brakes (EMBs). One of the key features of this system is that the integration of IWM in the braking process allows an increase in the vehicle's range: while braking, IWMs work as generators and transform the vehicles kinetic energy into electrical energy that is fed into the powertrain battery. IWMs have, however, braking torque availability limitations at high wheel speeds or when the powertrain battery is close to full state of charge. EMBs provide the torque needed to match the total braking demand. HBS-SPL components can be combined in different ways according to the constraints specified in the HBS-SPL feature model presented in Fig. 3g. The feature model was designed using the cardinality-based notation [19] and the FMP[3] modeling tool. The HBS-SPL feature model includes wheel braking alternative features: Brake_Unit1 and Brake_Unit2 front wheels braking, Brake_Unit3 and Brake_Unit4 rear wheels braking aimed to provide the braking for each wheel. The Hephaestus/Simulink variability management tool was used to manage the variation in HBS-SPL design and safety analysis assets. Thus, the mapping between SPL features and these assets has been specified in the configuration knowledge.

3.1 HBS-SPL Hazard Analysis

SPL hazard analysis was performed in HiP-HOPS from the perspective of the following product configurations: HBS four wheel braking (i.e. all product line presented in Figure 3g) (HBS-4WB); HBS front wheel braking (HBS-FWB), which includes Brake_Unit1 and Brake_Unit2 features; and HBS rear wheel braking (HBS-RWB) that includes Brake_Unit3 and Brake_Unit4 features. Table 1 presents the identified hazards for each SPL usage scenario and their allocated ASILs. Some of these hazards are common between all HBS-SPL products and some of them are product-specific (e.g., 'No braking four wheels'). The 'Value braking' hazard is common across HBS-SPL products, but its causes and allocated ASILs are different in each product. Such variation can affect the product-specific fault trees, FMEA, and ASIL allocation generated by HiP-HOPS, and the structure of the product line safety case. Thus, the impact of SPL variation in safety analysis assets is stored and managed in our tool by means of SPL functional, architectural, and component failure models.

Table 1. HBS-SPL hazard analysis

Scenario	Hazard	ASIL
HBS-4WB	No braking four wheels	D
	No braking rear	C
	Value braking	C
HBS-FWB	No braking front	D
	Value braking	D
HBS-RWB	No braking rear	D
	Value braking	D

---

[3] FMP feature modeling tool: http://sourceforge.net/projects/fmp/

3.2   HBS-SPL Safety Case Architecture

The SPL safety case architecture generated by the tool is illustrated in Fig. 4. The system boundaries specified in the feature model, and the information provided by product line FFM, AFM, and CFM were the inputs required by the tool. The figure presents the structure of the HBS-SPL modular safety case architecture in a GSN modular view organized according to the functional decomposition specified in the feature model of Fig. 3g. Mandatory, optional, and alternative feature/variation-specific argument modules are organized according to GSN variability modeling abstractions described on Fig. 3. These abstractions translate the representation of variation in feature model to variation in the safety case architecture.

The top-level argument module is supported by 'Auxiliary_Battery', 'Power-train_Battery', 'Mechanical_Pedal', 'Electronic_Pedal', and 'Communication' modules (denoted by 'supported by' relationships between the top-level module and these argument modules). 'Communication' argument module is supported by two argument modules arguing the safety of the 'Bus_1' and 'Bus_2' redundant features. 'Wheel_Braking' is a mandatory argument module supported by 'Brake_Unit1', 'Brake_Unit2', 'Brake_Unit3', and 'Brake_Unit4' alternative wheel braking argument modules encapsulated in the 'WB_Contract' module. The internal structure of the feature/variation-specific argument modules generated by the tool decomposes the argument over the safety of each feature into references to arguments modules (i.e. using GSN Away Goals) arguing the mitigation of the system hazards associated with that feature using the Hazard Avoidance pattern [15]. For example, the 'Brake_Unit1' argument module is decomposed into the 'No_Brake_4W', and 'Val_Braking', alternative hazard mitigation argument modules (Fig. 5a). Each one of these modules is further decomposed into argument modules arguing the mitigation of component faults contributing to a system hazard. Fig. 4b shows the 'Val_Braking' hazard argument. This argument is decomposed into modules arguing the mitigation

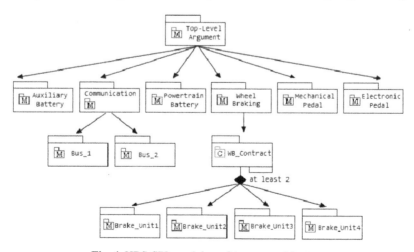

Fig. 4. HBS-SPL modular safety case architecture

of WNC and IWB component failure modes that should be addressed to minimize the hazard effects in a particular product. The selection of these argument modules is dependent upon the product definition and its operational environment. It may change the values for argument parameters (Fig. 5) changing the structure of the product argument. SPL hazard analyses provide the information required for these parameters.

By analyzing the structure of the argument modules we have identified the importance of the context in the argument structure to encapsulate the variation in product definition (i.e. feature selection) and its operational environment (i.e. context selection) at system (i.e. functional) level argumentation. So, the selection of these elements may change the structure of a product-specific modular safety case architecture. At the architectural level argumentation, context is used to encapsulate the variation in the hazards, their top-level failure conditions, and the allocated safety requirements that are subject to change according to the product definition and context specified at system level argument. Finally, at the component level, the context is used to encapsulate the variation in the component failure modes and safety integrity levels allocated to these failure modes.

The structure of this modular safety case architecture has increased the reusability of safety arguments arguing the safety of variable SPL features. The safety case architecture reflects the decomposition of the system grouping feature-specific safety arguments into cohesive and low-coupled modules. Variable argument modules such as 'Brake_Unit1' and 'Brake_Unit4' are encapsulated in a contract module. Thus, the impact of the variation on the selection of these argument modules in a product-specific safety case is isolated from reusable argument structures.

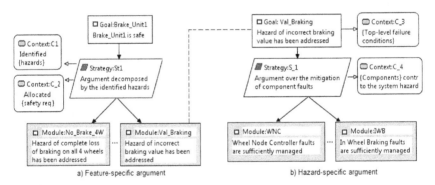

Fig. 5. Brake_Unit1 and hazard arguments

4   Conclusion

In this paper we proposed a method describing the design and implementation of a tool that supports the automated generation of a modular product line safety case architecture from outputs provided by model-based development, safety analysis, and variability management tools. The SPL safety case architecture generated in this approach is organized into modules reflecting the feature and component boundaries. This architecture has the potential to increase the reuse of safety arguments across

SPL products, which is the main novelty of our approach compared with current research on safety cases. The feasibility of the approach was tested in a small case study, but we are aware that more work is needed to assess the value and limitations of this work and the acceptability of the resultant modular safety cases. It might be the case that only partial argument structures can be constructed using this approach and such structures may need significant intervention by experts to create convincing safety arguments. Further work focuses on evaluating the quality of safety case architectures generated by the proposed method against traditional software metrics like coupling and cohesion.

Acknowledgements. Our thanks to CNPq, grant number: 152693/2011-4, and CAPES Brazilian research agencies for the financial support.

References

1. ISO. ISO 26262: Road Vehicles Functional Safety (2011)
2. EUROCAE, ARP4754A - Guidelines for Development of Civil Aircraft and Systems (2010)
3. Weiland, J.: Configuring variant-rich automotive software architecture models. In: Proc. of 2nd IEEE Conf. on Automotive Electronics, pp. 73–80 (2006)
4. Habli, I., Kelly, T., Hopkins, I.: Challenges of establishing a software product line for an aerospace engine monitoring System. In: Proc. of 11th Int'l SPL Conference, pp. 193–202. IEEE (2007)
5. Fenn, J., Hawkins, R., Williams, P., Kelly, T., Banner, M.G., Oakshott, Y.: The who, where, how, why and when of modular and incremental certification. In: IET System Safety Conference (2007)
6. Clements, P., Northrop, L.: Software Product Lines: Practices and Patterns. Addison-Wesley (2001)
7. Habli, I.: Model-Based Assurance of Safety-Critical Product Lines. Ph.D thesis, Department of Computer Science, The University of York, York, United Kingdom (2009)
8. Habli, I., Kelly, T.: A safety case approach to assuring configurable architectures of safety-critical product lines. In: Giese, H. (ed.) ISARCS 2010. LNCS, vol. 6150, pp. 142–160. Springer, Heidelberg (2010)
9. Liu, J., Dehlinger, J., Lutz, R.: Safety analysis of software product lines using stated modeling. Journal of Systems and Software 80(11), 1879–1892 (2007)
10. Steiner, E.M., Masiero, P.C.: Managing SPL variabilities in UAV Simulink models with Pure::variants and Hephaestus. CLEI Electronic Journal 16(1) (2013)
11. Papadopoulos, Y., Walker, M., Parker, D., Rüde, E., Hamann: Engineering failure analysis and design optimization with HIP-HOPS. Journal of Engineering Failure Analysis 18(2), 590–608 (2011)
12. Kelly, T.: Using software architecture techniques to support the modular certification of safety-critical systems. In: 11th Australian Workshop on Safety Critical Systems and Software, vol. 69, pp. 53–65 (2007)
13. Origin Consulting York, GSN community standard version 1 (2011), http://www.goalstructuringnotation.info/documents/GSN_Standard.pdf

14. Bloomfield, R., Bishop, P.: Safety and assurance cases: Past, present and possible future - an Adelard perspective. In: Proc. of the 18th Safety-Critical Systems Symp. Springer, London (2010)
15. Kelly, T., McDermid, J.: Safety case construction and reuse using patterns. In: Proc. of 16th Int. Conf. on Computer Safety, Reliability and Security, pp. 55–69. Springer-London (1997)
16. Matsuno: A design and implementation of an assurance case language. In: Proc. IEEE/IFIP Dependable Systems and Networks, DSN (2014)
17. ACEdit, Assurance case editor (2014), https://code.google.com/p/acedit/
18. Weaver, R.A.: The Safety of Software – Constructing and Assuring Arguments. PhD thesis, Department of Computer Science, University of York (2003)
19. Czarnecki, K., Helsen, S., Eisenecker, U.: Staged configuration using feature models. In: Nord, R.L. (ed.) SPLC 2004. LNCS, vol. 3154, pp. 266–283. Springer, Heidelberg (2004)
20. De Castro, R., Araújo, R.E., Freitas, D.: Hybrid ABS with electric motor and friction brakes. In: 22nd International Symposium on Dynamics of Vehicles on Roads and Tracks, Manchester, UK (2011)

Spatial Allocation of Bus Stops: Advanced Econometric Modelling

Dmitry Pavlyuk

Transport and Telecommunication Institute, Lomonosova 1, Riga, LV-1019, Latvia
Dmitry.Pavlyuk@tsi.lv

Abstract. This paper is devoted to discussion of econometric techniques, utilized for analysis of bus stop spatial allocation. Majority of researches are focused on bus route characteristics, averaged by route (average distance between bus stops) or by time (daily ridership on a bus stop level). Meanwhile, modern electronic ticketing systems are widely used and their detailed data can be used for analysis with advanced econometric techniques. We discussed shortcomings of existing empirical researches in this area and proposed a list of econometric models, which can be applied for deeper analysis of bus stops. We suggest possible specifications of spatial regression models, which take spatial dependence between individual bus stops and spatial heterogeneity into account. Also we discussed possible applications of the stochastic frontier model to bus stop benchmarking.

Keywords: bus stop level, econometric modelling, spatial effects, stochastic frontier.

1 Introduction

Urban public transport optimization is an integral part of many city development plans. Authorities of most cities give priority to the development of an urban public transportation system to ensure a sustainable city grow. To entice inhabitants to shift from their private cars to public transportation, city managers and public transport operators implement optimization of routing, timetables, vehicle and crew scheduling[1]. Significant efforts are also made for design and enhancing of bus stops[2]. The spacing, location, and design of bus stops significantly affect public transport attractiveness, customer satisfaction, and transit system performance.

In this paper we review econometric approaches to the problem of allocation of bus stops. A bus stop is a first point of contact between passengers and transportation services and its location is one of important factors of public transport accessibility. Bus route planners are mainly focused on stop frequency, while a location of an individual bus stop is defined on a base of inhabitant requests and traffic attraction points (i.e. shops, schools). This approach leaves a great opportunity for optimization of routes, including via mathematical techniques application. A general optimization problem is selection of bus stop locations, which ensure easy and fast access of

© Springer International Publishing Switzerland 2015
W. Zamojski et al. (eds.), *Theory and Engineering of Complex Systems and Dependability*,
Advances in Intelligent Systems and Computing 365, DOI: 10.1007/978-3-319-19216-1_31

inhabitants to demand areas, but keep minimal transportation time and operational costs. Application of modern econometric models can contribute into this problem solution.

Nowadays electronic ticketing systems, implemented in many metropolitan areas[3], provide detailed information about bus ridership. In this paper we suggest several advanced econometric models, which will allow handling of detailed ridership information on a level of individual bus stop boarding for optimization of spatial bus stop distribution and analysis of bus stop on-site enhancements.

2   Problem of Bus Stop Allocation

There are a lot of academic and empirical researches, devoted to analysis and optimization of bus routes[4, 5]. The importance of bus stop planning and empirical approaches to this task are covered by many city guidelines[6, 7].

Planning of bus stops spatial locations is a multi-criteria optimization problem for a trade-off between various passengers' and public transport operators' needs. A usual assumption says that passengers try to minimize their overall travel times[8], when operators are focused on operational costs and service quality[9]. A distance between consecutive bus stops is a core characteristic, which is frequently used for spacing. Dependence between a distance between bus stops and overall travel time can be presented by a quadratic curve (Fig. 1).

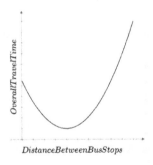

Fig. 1. Plot of an overall travel time as a function of a distance between bus stops

Shorter distances between bus stops lead to increased travel times due to a bigger number of bus acceleration/deceleration times. Conversely, longer distances reduce public transport accessibility and increase overall travel times.

Various bus stop distance patterns are used in different counties. According to Ammons[4], an average distance between two consecutive bus stops varies from 200 to 600 m in urban areas. Mainly bus stop spacing guidelines and related optimization algorithms are focused on average metrics, like an average distance between bus stops or an average time interval between buses. Average characteristics are supposed to vary for downtown, high and density residential areas, and outskirts. This frequently used approach is convenient for overall system analysis, but not informative for selection of an actual bus stop location. Frequently, a location of a bus stop is defined on the base of historical user requests and related to population activities (shops, transit points, residential areas, etc.). At the same time, a high density of bus stops allows population to choose a preferred bus stop, which better satisfy their needs. Identification of relative attractiveness of bus stops for population is an important problem, which cannot be solved by average statistics. A typical configuration of the urban bus route is presented on the Fig. 2. It can be noted that distribution of bus stops is quite irregular – areas with dense bus stops alternate with significant breaks between stops. This usual pattern is related with area specifics, which can't be taken into account with averaging approach.

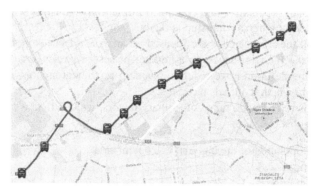

Fig. 2. A typical bus route configuration

Paying significant attention to average distance between bus stops, selection of an actual bus stop location within a particular area is weakly covered in city guidelines. Rarely guidelines[6] specify general types of bus stop locations concerning the intersections:

- Near-side (in front of the intersection)
- Far-side (after the intersection)
- Mid block (between intersections)

At the same time, if bus stops are considered as a separate economic unit, which compete for passengers, a powerful stack of theoretical and practical methods can be utilized. The theory of spatial competition is well established and there are a significant number of its applications in different economic areas. Recently models of spatial competition were applied to theatres, gas stations, retail places, hospitals, country regions and others, but, to the best of our knowledge, there are no applications to bus stop spacing. A study, frequently cited as a pioneering in the areas of spatial competition, was presented by Hotelling[10] and describes a basic case of two firms producing homogeneous goods in different locations on a line. Later this idea was developed in different ways, but Hotelling's initial model looks convenient for bus stop spacing. A bus stop, which serves a particular route ("a homogeneous good"), competes for passengers against neighbour bus stops on the same route ("on a line"). Spatial econometrics [11, 12] provides a way for identification of such spatial relationships on the base of real data.

Recently several empirical researches were published, where the problem of bus stop spacing is considered on an individual stop level. Cervero et al.[13] utilized a regression model for predicting bus rapid transit patronage in Southern California. Ryan and Frank[14] included quality of the pedestrian environment around transit stops as a factor of transit ridership and also estimated a linear regression model for San Diego metropolitan area. Pulugurtha and Agurla[15] developed and assessed bus transit ridership models at a bus stop level using two spatial modelling methods, spatial proximity and spatial weight methods, in Charlotte, North Carolina. Kerkman et al. [16] explored factors that influence transit ridership at the bus stop level for the local and regional bus transit system in the region of Arnhem-Nijmegen, Netherlands. Dill et al.[17] analysed relative and combined influence of transit service characteristics and urban form on transit ridership at the bus stop level.

Generally, all these researches are focused on analysis of bus stop catchment area characteristics and their influence on a number of served passengers. Following the human geography definition, we a define bus stop catchment area as an area around the bus stop from which it attracts passengers.

Obviously, there is a wide range of primary factors, which affect a number of passengers, served on a bus stop:

- Residential population of the bus stop catchment area
- Social conditions
- Economic conditions
- Presence of passenger traffic attractors (shopping centres, etc).
- Transit options

All mentioned studies operate on a bus stop level, but aggregate data for a time interval (usually daily). Using of daily values allows identifying of core factors, influencing bus stop ridership, but bus-specific data (a number of passengers carried by a bus from a particular bus stop) can be more informative for comparative analysis of bus stops.

Electronic ticketing systems, implemented in many large cities, provide detailed information about bus ridership. In particular, a number of passengers, who got on the bus at the bus stop, can be estimated on the base of e-ticketing data.

3   Research Methodology

3.1   Classical Regression Model

The classical regression model is widely used for analysis of a stochastic relationship between a dependent variable and a set of explanatory variables:

$$y = X\beta + v \tag{1}$$

where

— y is an ($n \times 1$) vector of a dependent variable (n is a size of the sample);
— X is an ($n \times k+1$) matrix of k explanatory variables;

— β is a $(1 \times k+1)$ vector of unknown coefficients (model parameters);
— v is an $(n \times 1)$ vector of independent identically distributed (IID) error terms.

The most frequently used dependent variable for studies at the bus stop level is a daily number of passengers, carried by buses on a bus stop. A list of selected recent studies, utilising this approach, is presented in the Table 1. Independent variables usually describe social and economic conditions of a bus stop catchment area.

Table 1. Selected existing studies at a bus stop level

Author, Year	Method	Dependent variable	Independent Variables
Kerkman et al., 2014[16]	OLS	Bus stop ridership (logarithm)	• Potential travellers (logarithm) • Income • Percent elderly • Distance to urban centre • Residential, agriculture, or socio-cultural facilities • Stop frequency (logarithm) • Number of destinations, frequency per direction • Direct connections • Competitive bus stops • Bus terminus, transfer stop or station • Dynamic Information, benches
Pulugurtha and Agurla, 2014[15]	Generalised OLS (linear, Gamma, Poisson, and Negative Binomial)	Bus stop ridership	• Population (by gender, ethnicity and age) • Households, mean income • Auto-ownership, vehicle/household • Total employment • Speed limit/functional class • One-way or two-way street lanes • Light or heavy commercial and industrial • Residential (apartments within catchment area)
Ryan and Frank, 2009[14]	OLS	Bus stop ridership	• Level of Service (number of bus routes/waiting time) • Residential density • Retail floor area ratio • Intersection density • Land Use mix • Income • No-vehicle households • Gender and ethnic structure
Cervero et al., 2009[13]	OLS	Average number daily board-ings	• Number of daily buses • Number of perpendicular daily feeder bus Lines • Number of perpendicular daily rail feeder • Distance to Nearest Stop • Park-&-Ride Lot Capacity • Population density (people within 1/2-mile buffer)

To the best of our knowledge, there are no empirical researches, where data is used in non-aggregated form (a number of passengers per bus stop per bus). This can be considered as an extensive direction for further research.

Note that though the classical regression model is widely applied, the dependent variable is limited (non-negative) by its nature. This problem will play a more important role for detailed data, because a share of bus stops without passengers for a bus will be higher. To resolve this problem, different functional specifications of the regression model can be used, i.e. binomial, Poisson. Another convenient specification is a Tobit model[18], designed for censored dependent variables. A feature of this model is a latent dependent variable y^{latent}, which is used as a base for a real output (number of passengers):

$$y = \begin{cases} y^{latent}, y^{latent} \geq 0, \\ 0, y^{latent} < 0 \end{cases} \qquad (2)$$

A comparative analysis of advantages and shortcomings of Poisson and Tobit regressions can be found in [19].

3.2   Modelling spatial effects

An important potential problem of the classical model (1) is omitting of significant explanatory variables. If a significant determinant of the dependent variable isn't included into the model, estimates of model parameters will be biased and inconsistent (well-known omitted variables bias[18]). A list of explanatory variables, utilised in the researches, mentioned above, is obviously not complete and can't include all important factors. Fortunately, factors, influencing bus ridership, are frequently clustered over space. For example, micro-district quality of pedestrian infrastructure, popular area-specific routes and destinations, ridership habits are hardly observable and measurable, but play an important role in public transport ridership. Spatial clustering of these factors (called spatial heterogeneity) allows applying of spatial econometrics for better modelling results. Consideration of another widely acknowledged spatial component, a spatial dependence, is also practicable. Spatial dependence, a relationship between numbers of passengers on neighbour bus stops, appears due to their spatial interactions. One of theoretical foundations of these interactions is spatial competition for passenger between bus stops. This factor is weakly researched, but logically can play a significant role.

A general spatial regression model[12] is expressed in linear form as:

$$Y = \rho_Y W_Y Y + X\beta + v,$$
$$v = \rho_v W_v v + \tilde{v}, \qquad (3)$$

where

— W_Y, W_v are spatial weights matrixes for output-output (spatial dependence), and error-error (spatial heterogeneity) relationships accordingly;

- ρ_Y, ρ_v are unknown parameters of spatial dependence and spatial heterogeneity accordingly;
- $\tilde{v}$ is a vector of IID symmetric disturbances.

According to the general spatial model specification (3), ridership on a bus stop y is influenced by its own explanatory factors X with parameters β and spatially weighted ridership of neighbour bus stops $W_Y Y$ with a parameter ρ_Y. Also the model includes spatial effects in random disturbances ($W_v v$ with a parameter ρ_v), which express spatial heterogeneity.

Specification of spatial weights matrixes, which describe a structure of spatial relationship, is a matter of application area. This is possible to assume that bus stop ridership is related with previous and next bus stops in the route, so the spatial weight matrix elements can be defined as binary:

$$w_{Y,ij} = \begin{cases} 1, \text{if bus stops } i \text{ and } j \text{ are consecutive in a route,} \\ 0, \text{otherwise.} \end{cases}$$

Spatial weights matrixes W_Y and W_v are generally different, but in empirical researches are frequently put to be the same subject to potential identification problems. Logically, the spatial weight matrix for spatial heterogeneity can be distance-based:

$$w_{v,ij} = \frac{1}{\text{distance}\left(bus\ stop_i, bus\ stop_j\right)},$$

where distance is defined as a geographical distance between bus stops.

3.3 Bus Stop Benchmarking

Another important public transport management problem is related with provision of the necessary amenities to a bus stop. Recently many municipalities and public transport operators make great efforts for enhancing bus stops to make them more comfortable for passengers. Developed passenger amenities include sidewalks, lighting, seating, information signs, maps, real-time data displays among others [3]. Effectiveness of these amenities for a particular bus stop is a subject of extensive analysis. Frequently effects of bus stop amendments is estimated using passenger surveys or observation of their behaviour on a bus stop[20], but comparison of bus stops is rarely utilized. Econometric benchmarking can be useful for comparative analysis of bus stops and for dynamic evaluation of changes on a bus stop.

Stochastic frontier analysis [21] is a popular econometric approach, which can be used for estimation of a relative attractiveness (or inefficiency, in terms of frontier analysis). The stochastic frontier model is formulated as:

$$Y = X\beta + v - u, \tag{3}$$

where the vector u represents individual bus stop inefficiency values.

Estimated individual inefficiency of bus stops can be a base for identification of potential problems on a bus stop, related with its location, nearby pedestrian infrastructure or on-site organization.

4   Conclusions

In this paper we discussed existing approaches to analysis of bus stop locations. A problem of optimal spatial organization of bus stops is widely acknowledged, but usually solved on an aggregated level. Majority of researches are focused on bus route characteristics, averaged by route (average distance between bus stops) or by time (daily ridership on a bus stop). Meanwhile, modern electronic ticketing systems are widely used and their detailed data can be used for advanced econometric analysis. We discussed shortcomings of existing empirical researches in this area and proposed a list of econometric models, which can be applied for deeper analysis of bus stop location and design. Our main conclusions can be summarized as:

1. Using a detailed ridership information on a level of individual bus stop boarding, available from the electronic ticketing systems, will allow utilization of advanced econometric techniques for optimal bus route planning.
2. Application of spatial econometric models will allow better estimation of factors, influencing bus stop ridership and solve problem of spatial heterogeneity and spatial dependence, typical for bus stop data.
3. Application of stochastic frontier models will allow identification of problems (inefficiencies) of individual bus stops for further bus stop enhancements or removal.

References

1. Guihaire, V., Hao, J.-K.: Transit network design and scheduling: A global review. Transportation Research Part A: Policy and Practice 42, 1251–1273 (2008)
2. Texas Transportation Institute, Texas A and M Research Foundation: Guidelines for the location and design of bus stops. National Academy Press, Washington, DC (1996)
3. Mezghani, M.: Study on electronic ticketing in public transport. European Metropolitan Transport Authorities (EMTA) 56, 38 (2008)
4. Ammons, D.N.: Municipal benchmarks: assessing local performance and establishing community standards. Sage Publications, Thousand Oaks (2001)
5. Bast, H., Delling, D., Goldberg, A., Müller-Hannemann, M., Pajor, T., Sanders, P., Wagner, D., Werneck, R.: Route Planning in Transportation Networks. Microsoft Research (2014)
6. KFH group: Guidelines for the Design and Placement of Transit Stops for the Washington Metropolitan Area Transit Authority, Bethesda, Maryland (2009)
7. Bus Network Improvement Project: Bus Stop Optimization Policy. Foursquare Integrated Transportation Planning & Jacobs Engineering (2014)
8. Murray, A.: A Coverage Model for Improving Public Transit System Accessibility and Expanding Access. Annals of Operations Research 123, 143–156 (2003)
9. Van Nes, R., Bovy, P.: Importance of Objectives in Urban Transit-Network Design. Transportation Research Record 1735, 25–34 (2000)

10. Hotelling, H.: Stability in competition. The Economic Journal 39, 41–57 (1929)
11. Anselin, L.: Spatial econometrics: methods and models. Kluwer Academic Publishing, Dordrecht (1988)
12. LeSage, J.P., Pace, R.K.: Introduction to spatial econometrics. CRC Press, Boca Raton (2009)
13. Cervero, R., Murakami, J., Miller, M.: Direct Ridership Model of Bus Rapid Transit in Los Angeles County. University of California, Berkerley (2009)
14. Ryan, S., Frank, L.F.: Pedestrian environments and transit ridership. Journal of Public Transportation 12, 39–57 (2009)
15. Pulugurtha, S., Agurla, M.: Assessment of Models to Estimate Bus-stop Level Transit Ridership using Spatial Modeling Methods (2014)
16. Kerkman, K., Martens, K., Meurs, H.: Factors influencing bus-stop level ridership in the Arnhem Nijmegen City Region (2014)
17. Dill, J., Schlossberg, M., Ma, L., Meyer, C.: Predicting transit ridership at the stop level: The role of service and urban form. Transportation Research Board 2013 Annual Meeting (2013)
18. Greene, W.H.: Econometric analysis. Pearson Addison Wesley, Harlow (2012)
19. Sigelman, L., Zeng, L.: Analyzing censored and sample-selected data with Tobit and Heckit models. Political Analysis 8, 167–182 (1999)
20. Ohmori, N., Omatsu, T., Matsumoto, S., Okamura, K., Harata, N.: Passengers' Waiting Behavior at Bus and Tram Stops. Presented at the July 17 (2008)
21. Kumbhakar, S.C., Lovell, C.A.K.: Stochastic frontier analysis. Cambridge Univ. Press, Cambridge (2003)

The Modeling Algorithm of Communication Run Operations in a Network

Henryk Piech[1], Grzegorz Grodzki[1], and Aleksandra Ptak[2]

[1] Czestochowa University of Technology, Dabrowskiego 73, 42202 Czestochowa, Poland
{henryk.piech,grzegorz.grodzki}@icis.pcz.pl
[2] Czestochowa University of Technology, Armii Krajowej 19, 42218 Czestochowa, Poland
olaptak@zim.pcz.pl

Abstract. Communication processes have to be observed because there are possibilities that a different kind of threats will occur in the processes of exchanging information in a network. These threats are connected with: the possibility of decryption, losing jurisdiction, believing in and freshness of information, message interception by intruders, etc. We also consider the run of the communication protocol operation. Security attributes have been introduced to analyze the chosen aspects of security, which are proposed by Burrows, Abadi, Needham [4] and others. They have created the system of rules that defines interrelated parts of communication operations with security aspects. In this research we continued the analysis of security in the direction of building the model of auditing and dynamic modification of chosen factors (adequate to the security aspect) with the possibility to form a prognosis.

Keywords: protocol logic, probabilistic timed automata, communication security modeling.

1 Introduction

Information is sent in the form of a message according to protocol systems which should guarantee: encryption safety, sufficient belief level, protection against intruders, and the freshness of information elements [1, 2]. Usually, we may use many mutually interleaved protocols in networks [3, 4]. Obviously, information refers to a different group of users (usually, they are grouped in a pair). Therefore, security analysis will be referred to those groups and they will be the basis of the creation of the so called main security factor [1]. Another main factor can take into account the set of messages, the public key, secret, nonce, etc. Among security attributes one may include the following: the degree of encryption, key and secret sharing, believing in sender or receiver, believing in the honesty of the user, and the freshness level of a message or nonce [1]. M. Kwiatkowska presents security attributes in the figure of probability parameters [5, 6]. This form is smart and very convenient. Therefore, the present proposition is additionally based on the transformation possibility of time attributes into probability characteristics. Apart from rule and time influences we also regard the intruder threat. The influence on security attributes is realized with the help of correction coefficients which also have a probabilistic form, according to the admitted approach. The above-mentioned rules deal on the basis of conditions that are

© Springer International Publishing Switzerland 2015
W. Zamojski et al. (eds.), *Theory and Engineering of Complex Systems and Dependability*,
Advances in Intelligent Systems and Computing 365, DOI: 10.1007/978-3-319-19216-1_32

actions which really appear in communication operations. The division of protocols into operations and operations into actions can be found in many works [1, 2]. In the proposed model we also exploit the so called tokens, which have binary character. A token directly appoints the secure or threat attribute level depending on the relation to a given security threshold. This type of approach improves the assessment reaction on security state changing and helps in the estimation of probability distribution to the next stages and thereby to one of the forms (presented in the following section) of prognosis creation. The proposed model of communication run investigation can be easily realized with probabilistic timed automata (PTA) [7 - 9] and colored Petri net [11, 12], which is especially effective in the parallel strategy variant.

2   The Procedure Concerning the Influences of Protocol Actions

At first, the main security factor(s) is (are) declared for the current action. Action usually influences one or several attributes. The first appearance of the action associated with the security factor leads to the activation of the new automata, node and the equivalent calculation node. This node contains a specific set of attributes. Let us consider the sequence of actions contained in the sample protocol. The structure of the protocol is presented in the example of ASF Handshake [1] (as one of possible main factors):

1. $A \rightarrow B : \{N_a\}_{K(a,b)}$,

2. $B \rightarrow A : \{N_a, N_b\}_{K(a,b)}$,

3. $A \rightarrow B : \{N_b\}_{K(a,b)}$,

4. $B \rightarrow A : \left\{ A \leftrightarrow^{K'(a,b)} B, N_b' \right\}_{K(a,b)}$,

where:

N_a, N_b, N_b' - are nonces.

A new session key $K'(a,b)$ for A and B is generated when the starting session key is $K(a,b)$. Initial conditions are adequate actions and may be presented as follows:

— both users agree on the starting key: A believes $A \leftrightarrow^{K(a,b)} B$, B believes $A \leftrightarrow^{K(a,b)} B$,

— A defers to $B's$ authority on session keys, A believes B controls $\forall K : A \leftrightarrow KB$,

— B generates the new session key, B believes $A \leftrightarrow^{K(a,b)} B$, - nonces are fresh: A believes N_a is fresh, B believes that N_b is fresh, B believes that N_b' is fresh.

The same parameters can play the role of the main security factor; for example: protocol, user (or a pair of users), key, message service, etc. The type of influences is practically regarded in two algorithm forms concerning attribute corrections:

— $mc = \{0,1\}$ - correction by multiplication by a given updating coefficient MCC in the case of influence pertaining to logic and heuristic rules, $mc = 1$ - the activation of this form regarding attribute correction, $mc = 0$ - the rejection of this form of correction.

— $ec = \{0,1\}$ - correction by exchanging to the current level (represented by the current coefficient value of ECC) in the case of influences relating to the lifetime or user (intruder). Therefore, it is possible to simultaneously use two forms of correction for a single attribute. So, when $ec = 1$ then the attribute value does not have to increase:

$$at_{t=k+1}(i) \xrightarrow{mc=0,ec=0} at_{t=k}(i),$$
$$at_{t=k+1}(i) \xrightarrow{mc=1,ec=0} at_{t=k}(i)*MCC,$$
$$at_{t=k+1}(i) \xrightarrow{mc=0,ec=1} ECC,$$
$$at_{t=k+1}(i)\min\{at_{t=k}(i)*MCC,ECC\}.$$

The experiments have proved that influences of heuristic rules in specific cases (for example, in the multi-usage of the same nonce) are more effective when correction is realized in the following way:

$$at_{t=k+1}(i) \xrightarrow{mhc=1,ehc=0} at_{t=k}(i)*(1-MCC),$$

or

$$at_{t=k+1}(i) \xrightarrow{mhc=1,ehc=0} at_{t=k}(i)*(1-at_{t=k}(i)),$$

where:

$mhc(i) = \{0,1\}$ - heuristic rule influence activation,

$ehc(i) = \{0,1\}$ - dishonest user influence activation.

The actual value of ECC, in the case of the lifetime type of influence, will be counted by the formula:

$$ECC = 1 - e^{t(j)-lt(i)} \tag{1}$$

where:

$t(j)$ - the time of attribute activation,

$lt(i)$ - the attribute lifetime.

In reality, the time activity is transformed into a probability attribute value according to a given attribute lifetime. The actual value of ECC, in the case of an additional user (intruder) type of influence, will be counted by the formula:

$$ECC = if\,(nus < nht)\,then\,ECC = 1,$$
$$else\,ECC = enht - nus \tag{2}$$

where:

nus - the number of users (in the environment of the main security factor),

nht - the number of honest users (in the environment of the main security factor).

In reality, the time activity is transformed into the probability attribute value according to a given number of honest users. Let us introduce the set of describing input variables:

$$c(j,k), \ cc(j,i), \ mc(i), \ mhc(i), \ ec(i), \ ehc(i) \text{ and index limits:}$$

nf	- the number of main factors: $k = 1, 2, ..., nf$;
nna	- the length of the multi-run; the number of all actions in the network: $j = 1, 2, ..., nna$;
nat	- the number of attributes (it is assumed that structures of security nodes for all main factors are the same: $i = 1, 2, ..., nat$. Obviously, it is not necessary; in such case we will use $i(k) = 1, ..., nat(k)$,
$fat(k,i)$	- the matrix of the attribute structure of main factors,
$mhc(i)$	- heuristic rule influence activation,
$ehc(i)$	- dishonest user influence activation,
mcc	- the correcting coefficient value,
mhc	- the correcting coefficient value,
$t(i)$	- the time of the i-th attribute activation,
$lt(i)$	- the lifetime of the i-th attribute $ec(i) = 1$,
nus	- the number of users,
nht	- the number of honest users,
$w(i)$	- the weight of the attribute according to communication security.

After the realization of the procedure of describing variable input, we will execute the security assessment algorithm. The stages of this algorithm are as follows:

1. action input,
2. the recognition of the attribute corrected by the action,
3. the recognition of the type of correction,
4. correction realization,
5. go to the 3-rd point until the last attribute,
6. the recognition of the main factor activated by the action,
7. token structure creation for the main factor,
8. security state estimation for the main factor,
9. go to the 6-th point until the last factor,
10. auxiliary analysis (the creation of the prognosis with respect to the threaten state the distribution of probabilities of transitions to the next stages),
11. go to the 1-st point until the last action.

The general presentation of the algorithm in the form of a pseudo-code is very convenient:

```
Procedure SECURITY ASSESSMENT
for j ← 1 to nna do
{j - the number of the action}
begin
  for i ← 1 to nat do
  {i - the number of the attribute}
  if cc(j,i) = 1 then
  begin
  if mc(i) = 1 then atm(i) ← at(i)*mcc;
  if ec(i) = 1 then ate(i) ← (1-exp(t(i)-lt(i)));
  if mhc(i) = 1 then ahm(i) ← at(i)*(1-mhc);
  if ehc(i) = 1 then if (nus > nht) then
  ahu(i) ← exp(nht-nus)else ahu(i) ← at(i);
  at(i) ← min(atm(i), ate(i), ahm(i), ahu(i));
  if at(i) < th(i) then tk(i) = 0 else tk(i) = 1;
  end;
  for k ← 1 to nnf do
  {k - the number of the main security factor}
  begin
  GFS(k) ← 0; St(j) ← 1;
  for i ← 1 to nat do
  if fat(k,i)=1 then
  begin
  GFS(k) ← GFS(k) + w(i) * at(i);
  {GFS(k)-the security level of k-th factor}
  St(j) ← St(j) + 2(i-1) * tk(i);
  {St(j) - the code of the new state}
  end
  end
end.
```

The procedure concerning the distribution of the creation probability to the next states and the prognosis procedure will be described in the following sections.

3   The Procedure of Security Distribution Analysis

The fundamental axiom is based on the impossibilities of increasing the level of communication security during the realization of a run. Therefore, the security states, which are possible to achieve can be constrained in the following way:

$$\left\{ St(j) : at(i,j) \le at(i,j-1), i = 1,2,...,\text{nat} \right\}$$

or

$$\left\{ St(j) : tk(i,j) \le tk(i,j-1), i = 1,2,...,nat \right\} \tag{3}$$

where
  j - the number of the current action

In general, the probability of achieving the threaten state $tk(i,j)=0$ by a given attribute is defined as follows:

$$prob\big(tk(i,j)=0\big) = \begin{cases} th(i)/at(i,j-1) \text{ if } at(i,j-1) < th(i). \\ \qquad 1 \qquad\qquad otherwise. \end{cases} \tag{4}$$

Let us define the description of the threaten zone.

Definition 1. A tuple (At', Tk'', Th) is a threaten zone description, where At' - the regarded attributed set (their names), Tk'' - given token boundary values (in the threaten zone all tokens are equal to or less than Tk'', Th - the set of given thresholds for all regarded attributes.

Definition 2. The scale of coming closer to the threaten zone CTZ is measured with respect to the average of distances between given (Th is used as a characteristic of the threaten zone) and current attributes. In practice, it is expressed in the following way:

$$CTaZ(k) = 1/nat(k) \sum_{\substack{i=1, \\ at(i)>th(i), \\ tk(i)\ge\sim tk''(i), \\ fat(k,i)=1}}^{nat} \big(at(i)-th(i)\big) \text{ - the attribute closeness}$$

for the k-th factor;
  or

$$CTZt(k) = 1/nat(k) \sum_{\substack{i=1, \\ tk(i)>tk''(i), \\ fat(k,i)=1}}^{nat} \big(tk(i)-tk''(i)\big) \text{ - the token closeness}$$

for the k-th factor;

where
  $\{*\}$ - added conditions referring to the index.

Axiom 1. The number of new possible states achieved from the state $ST(j)$ is equal to:

$$nnst = 2^v, \text{ where } v = \sum_{i=1;tk(i)\ge tk''(k,i)=1}^{nat} \big(tk(i)-tk''(i)\big),$$

where $tk_j(i)$ - token values in the investigated j-th state (it infers from the impossibility to achieve the security level better than the current situation).

Obviously, only attributes with the token equal to 1 will be taken into account in the analysis of the distribution probabilities concerning the transition to the next state. Attribute transition probability is estimated on the basis of the current attribute value and a given security threshold (threaten zone is placed under it) (fig.1).

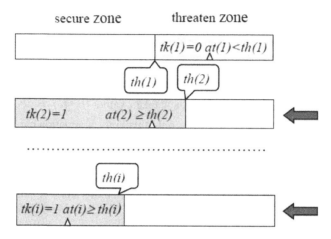

Fig. 1. The selection of relevant attributes according with: {tk(i)=1}

Collar 1. The probability of the achieved concrete possible state is defined as follows:

$$prob\big(St(j)\,St(j+1)=St(g)\big)=$$

$$\frac{prcg(k,j,g)\,prtg(k,j,g)}{\sum_{tk^{j+1}(1)=0}^{1}\sum_{tk^{j+1}(2)=0}^{1}\cdots\sum_{tk^{j+1}(nat)=0}^{1}prc(k,j)\,prt(k,j)},$$

where g - a given state from the v feasible state,

ingredient formula elements are defined as follows:

$$prcg(k,j,g)=\prod_{\substack{i=1;\\ tk^{j}(i)=1;\\ fat(k,i)=1;\\ tk^{j+1}(i)=0;\\ tk^{j+1}(i)\in St^{(g)}}}^{nat} th(i)/at(i),$$

$$prtg(k,j,g)=\prod_{\substack{i=1;\\ tk^{j}(i)=1;\\ fat(k,i)=1;\\ tk^{j+1}(i)=1;\\ tk^{j+1}(i)\in St^{(g)}}}^{nat} \big(at(i)/th(i)\big)/at(i),$$

$$prc(k,j,g)=\prod_{\substack{i=1;\\ tk^{j}(i)=1;\\ fat(k,i)=1;\\ tk^{j+1}(i)=0;}}^{nat} th(i)/at(i),$$

$$prt(k,j,g)=\prod_{\substack{i=1;\\ tk^{j}(i)=1;\\ fat(k,i)=1;\\ tk^{j+1}(i)=1;}}^{nat} \big(at(i)/th(i)\big)/at(i).$$

The result presented in the collar is based on attribute changing probability $th(i) = at(i)$ for the token (adequate for the attribute) transition from 1 to 0, and $(at(i) - th(i))/at(i)$ for staying on level 1 (transition from 1 to 1). The consideration of the transition probability refers to a given main factor k. Thus, only attributes fulfilling the condition: $\{fat(k,i) = 1\}$ would be regarded. The given next state g, for which the probability transition is defined, has to be described by the set of tokens fulfilling the condition: $\{tk^{j+1}(i) \in St^{(g)}\}$. The denominator represents the full probability space for all feasible states. It is a permutation that refers only to relevant attributes, i.e. those which can change their token value $\{tk^j(i) = 1\}$ because only tokens defined the state code. Therefore, only this kind of situation can influence the changing state: $\{tk^{j+1}(i) = 0\}$ or $\{tk^{j+1}(i) = 1\}$. Let us pay attention to two theoretical cases: token transition from 0 to 1 and from 0 to 0. The first case is impossible (axiom 1), the second is realized with probability equal to 1 (multiplication by 1 does not contribute any changes). By calculating the dominator value, we may estimate the distribution of probabilities. This task is realized by the following procedure.

```
Procedure TRANSITION PROBABILITY ANALYSIS (j, k)
u := 1; s := 0;
for tk(j+1, 1) := 0 to 1 do
{tk(j+1, i) - token value in (j+1) - th state (next
state)}
  for tk(j+1, 2) := 0 to 1 do
  ................ . .
  for tk(j+1, nat) := 0 to 1 do
  begin
    m(u) := 1;
    for i := 1 to nat do
    begin
    if (tk(j, i) = 1) and (tk(j+1, i) = 0) and
    (fat(k, j) = 1)
    then m(u) := m(u) * th(i) / at(i);
    if (tk(j, i) = 1) and (tk(j+1, i)=1) and
    (fat(k, j) = 1)
    then m(u) := m(u) * (at(i) - th(i)) / at(i)
{m(u)- denominator (7) component of probability of tran-
sition to u - th feasible state, u - state code}
    end;
    s := s + m(u); {value of denominator (7 )}
    p(j, u) = m(u) / s;
    {p(j, u) = transition probability from state j-th
    to u-th}
    u := u + 1;
  end.
```

The algorithm return values of the probability concerning the transition to all feasible next states:

$$prob\big(St(j) \rightarrow St(j+1)\big) = m(u)/s, \ u = 1, 2, ..., v. \tag{6}$$

The short time prognosis is defined by the maximum transition probability and the long term prognosis can be determined on the basis of trends or the distribution of the different types of operations in the run (this problem will not be explored here).

4   Conclusions

The simple form of procedure algorithms does not guarantee a low level of complexity but simultaneously the limited number of attributes (usually less than 10) permits us to audit long sequences of operations in a run (in the online investigation). On the other hand, the experiences pertaining to known inter- leaved protocols (Kerberos, Andrew RPC, Needham-Shredder, CCITT X.509 etc.) show and approve a short (<30 state changing) process of achieving a threaten zone in main security factors. Usually, these factors consist of several (5 - 8) security attributes. It gives the possibility to create the prognosis and give warning about different kinds of threats.

References

1. Burrows, M., Abadi, M., Needham, R.: A Logic of Authentication. In: Harper, R. (ed.) Logics and Languages for Security, pp. 15–819 (2007)
2. Zhang, F., Bu, L., Wang, L., Zhao, J., Li, X.: Numerical Analysis of WSN Protocol Using Probabilistic Timed Automata. In: 2012 IEEE/ACM Third International Conference on Cyber-Physical Systems (ICCPS), p. 237 (2012)
3. Lynch, N.: Timed and Probabilistic I/O Automata. In: 28th Annual IEEE/ACM Symposium on Logic in Computer Science (LICS), p. 12 (2013)
4. Chen, T., Han, T., Katoen, J.: Time-Abstracting Bisimulation for Probabilistic Timed Automata. In: 2nd IFIP/IEEE International Symposium on Theoretical Aspects of Software Engineering, TASE 2008, pp. 177–184 (2008)
5. Kwiatkowska, M., Norman, G., Segala, R., Sproston, J.: Automatic Verification of Real-time Systems with Discrete Probability Distribution. Theoretical Computer Science 282, 101–150 (2002)
6. Kwiatkowska, M., Norman, R., Sproston, J.: Symbolic Model Checking of Probabilistic Timed Automata Using Backwards Reachability. Tech. rep. CSR-03-10, University of Birmingham (2003)
7. Huang, Y.-S., Chiang, H.-S., Jeng, M.D.: Fault measure of discrete event systems using probabilistic timed automata. In: 2011 IEEE International Conference on Systems, Man, and Cybernetics (SMC), pp. 1218–1223 (2011)
8. Wu, J., Wang, J., Rong, M., Zhang, G., Zhu, J.: Counterexample generation and representation in model checking for probabilistic timed automata. In: 2011 6th International Conference on Computer Science & Education (ICCSE), pp. 1136–1141 (2011)

9. Thorat, S.S., Markande, S.D.: Reinvented Fuzzy logic Secure Media Access Control Protocol (FSMAC) to improve lifespan of Wireless Sensor Networks. In: 2014 International Conference on Issues and Challenges in Intelligent Computing Techniques (ICICT), pp. 344–349 (2014)

10. Liu, K., Ye, J.Y., Wang, Y.: The Security Analysis on Otway-Rees Protocol Based on BAN Logic. In: 2012 Fourth International Conference on Computational and Information Sciences (ICCIS), pp. 341–344 (2012)

11. Zhang, H., Liu, F., Yang, M., Li, W.: Simulation of colored time Petri nets. In: 2013 IEEE International Conference on Information and Automation (ICIA), pp. 637–642 (2013)

12. Li-Li, W., Xiao-jing, M., Yang, N.: Modeling and verification of Colored Petri Net in stop and wait protocol. In: 2010 International Conference on Computer Design and Applications (ICCDA), vol. 5, pp. V5-24–V5-28 (2010)

Failure Prediction in Water Supply System – Current Issues

Katarzyna Pietrucha-Urbanik

Rzeszow University of Technology Rzeszow University of Technology, Faculty of Civil and Environmental Engineering, 35-959 Rzeszow, Al. Powstańców Warszawy 6, Poland
kpiet@prz.edu.pl

Abstract. The presented method for the assessment of water supply system functioning in face of undesirable events occurrence takes into account the failure analysis. The developed analysis will allow decision support in the economic efficiency assessment of the water supply infrastructure functioning. The proportional hazard method for the failure of water pipes assessment enables to assess the reliability of water pipes and to obtain the possibility of prediction of undesirable events. The use of the proposed method can significantly support management analysis of water supply system. The establishment of the assessment parameters is very important issue that requires the use of waterworks experience. Moreover, it is based on the real data from the water supply system functioning.

Keywords: water network functioning, failure, prediction, water supply.

1   Introduction

Due to the character of the water supply system (WSS), in case of failure occurrence, it can generate high costs. The failure of such a system often causes the necessity of exclusion the part of it, which is associated with a decrease in the amount of supplied water or with a lack in its delivery. Interruptions in water supply or supplying water with inadequate parameters (eg. at too low pressure) may result in financial claims of the customers. For this reason, water network requires to have plans of periodic maintenance and repairs for a longer period of time. Inspections allow for early detection of damage and to plan possible repair [3]. Planning repair in advance will help to inform the customers and such organizing water supply that the negative consequences associated with the lack of water supply will be minimized [11]. However, it must be remembered that the periodic inspections and repairs will not eliminate sudden failures completely.

Recently, a number of water pipes failures, whose the main cause is ageing, is growing. If the number of failures associated with ageing of water supply material increases, the replacement of the entire water supply system in a particular area should be considered, because the increasing number of repairs will generate increasing costs and the number of failures will not be decreased [15].

© Springer International Publishing Switzerland 2015
W. Zamojski et al. (eds.), *Theory and Engineering of Complex Systems and Dependability*,
Advances in Intelligent Systems and Computing 365, DOI: 10.1007/978-3-319-19216-1_33

Quite often a good practice to reduce the number of emergency failures is the so called preventive renewal [16]. Within this type of renewal of water supply network the elements most vulnerable to failure are prophylactically replaced after a specified time of operation, thereby reducing the probability of failure due to their wear.

Actions taken to reduce the number of failures of water supply network are, among others:

- technical renewal of pipelines (renovations, maintenance and diagnostics),
- replacement of pipes and fittings,
- improvement in detecting places of leaks,
- network pressure limitation to the lowest permissible value,
- proper operation, design and execution of water supply system.

The increase in the reliability of water supply is affected by rational modernization and operation of water supply systems, which is related to, among others, the stabilization of pressure in the network and efficient emergency service. Undoubtedly, an important aspect influencing the reliability of drinking water supply is the use of active methods of water supply network management [2], [13], which allow to reduce the duration of unplanned interruptions in water supply.

In contrast to passive management, in which the duration of existing leaks can be up to several hundred days, the establishment of the telemetry system and the application of active management can shorten this time to a few (or dozen) days. Monitoring of the water supply network influences minimization of disruptions in water supply and reduces harmful effects on public health [8]. The economic aspect of risk management directly with taking or omitting the procedures aimed at reducing risk is of strategic importance [14], [17].

Analysis of the costs incurred to maintain a certain level of reliability is very important when planning new water supply networks or repairing the existing ones. While designing new or modernizing old water supply systems one should develop such systems that will allow to keep reliability at a fixed level, but lowering the costs of maintaining reliability, or to raise the level of reliability at the same costs [9], [10].

In addition, the detailed analysis of the costs to maintain the reliability will allow:

- to find those investment projects that have the greatest impact on improving the reliability of water supply systems,
- to adjust the level of reliability to the economic conditions,
- to ensure the level of reliability that will satisfy the water recipient,
- to plan new investments and modernize used water supply systems.

Generally, these tools are aimed at valuing reliability by:

- assessing the costs caused by losses resulting from break of water supply or supplying bad quality water,
- identification and ranking the projects aiming at maintaining the reliability on a certain level according to costs,
- estimating the capital resources to be spent on activities related to maintaining the reliability,
- calculating the reliability indices for particular water company systems.

The objective of this work is to characterize the unreliability of the water-pipe network in one of the biggest cities in the south Poland. Detailed analysis of the water network failure should constitute the main element of the managing system of the urban water networks, particularly in the strategic modernization plans [4].

The aim of the paper will be achieved through the failure analysis in order to understand and predict failure occurrence in water supply infrastructure and will be used in developing a strategy for activities in the face of a pipe renovation.

2   Materials and Methods

2.1   Materials

The calculations were made on the basis of the operational data on the water-pipe network in one of the biggest cities in the south Poland, as well as on the prepared failure protocols based on the water supply exploitation in Municipal Enterprise for Communal Economy and Municipal Water and Sewerage Company. The analysis was performed using the application contained in a Statsoft computer program Statistica.

The exemplary protocol of the water supply system failure should include:

- general information about the object (the place and the time of construction of the object),
- technical data (information about the function performed by the object, operating parameters, year of pipe build, pipe type, diameter, material (cast iron, cast iron spheroid, steel, PVC, PE, AC), connection type in case of pipe: rigid, flexible, no data, etc., failure location: pipe, connector, fittings, gate valve, damper, hydrant, water meter, others),
- failure description (address of failure, time of occurrence, the description of symptoms, type and cause of failure: transverse crack, pitting, extensive corrosion, material defect, defective installation, worn material, land displacement, heavy car traffic, leaking connector, tram traffic, pushed seal, damage during construction works, cracked cup, leaking fittings, frost, no backfill sand; ground conditions: wetland ground, rocky ground, unstable ground, normal ground; location of water pipe: in the lanes of heavy traffic; in the lanes; off the main road in the pavement, parking lot; off the main road in green areas, on the side of the road and the asphalt, concrete, flagstone, cobblestone and other pavement; method of failure removal: caulker , split sleeve, repair band, pipe replacement (specify length in (m)), valve replacement, choke seal, parts replacement, and others), persons removing failure,
- description of the consequences of failure (difficulties, threats and damages: including area occupied by repair in (m^2), water used for pipe washing in (m^3), interruption in the water supply (date), etc.),
- additional characteristic information (used material and equipment: excavator, compressor, car emptier, drainage pump, welder, generator, compactor, others).

The aforementioned information will allow for making periodic analysis of the causes and consequences of failure, which in the future will allow to reduce the probability of failure and when the failure occurs to reduce its consequences.

The length of the examined water supply network is for the mains 45,5 km, for the distributional 98,7 km and for the water connections 122,5 kilometers. The materials of the network is cast iron, AC (asbestos-cement), steel, PVC and PE. The average age of water mains is 62 years. The population served from water supply system is 39 thousands recipients. Water is provided to consumers via a water pipe network having radial-ring arrangement, which is beneficial to the reliability of water supply system. Water pipeline is supplied with water from two water treatment plants (WTP), taking raw water from two independent surface water intakes

2.2   Methods

Survival analysis describes a set of statistical methods, in which a very important issue is the time of occurrence of specific event occurrence. Using this procedure we can estimate for example patients life expectancy of or devices functioning period. Initially the survival analysis was applied in the field of biology and medicine, currently it is widespread in economics, sociology and social sciences, as well as in reliability engineering.

Types of survival analysis methods include, for example, life tables, which are one of the oldest methods, as well as the estimation of the survival function of the continuous survival using the Kaplan-Meier method and regression models, among other things: Cox proportional hazard model, exponential regression model or normal linear regression model [5].

In case of incomplete and imprecise data we can distinguish different types of censoring such as: left censored, right censored and sectional censored. The left censoring occurs when the failure occurred before the period of starting the observations, and we do not know the exact date of this fact [1]. On the contrary, if within the prescribed period of time the event is not observed, then the right censoring occurs. Such situations can be caused by loss or lack of data [6]. For the probability analysis of water pipes without failure occurrence Cox model was used, which is applied in the analysis of censored data. The probability of undesirable event occurrence determines the hazard function in the form [12]:

$$h(t) = lim_{\Delta x \to 0} \frac{P[t \leq T < t + \Delta t | X \geq t]}{\Delta t} \tag{1}$$

Because the hazard function does not accept negative values, we can write it in exponential form:

$$h(t, X) = h_0(t) \exp(X\beta) \tag{2}$$

where $h_0(t)$ is the baseline hazard level, β is a vector of regression coefficients and X is the vector of explanatory variables .

The assumption of proportional hazards model assumes a constant hazard function quotient of two cases in the concerned study, which can be written as:

$$\frac{h(t,X_1)}{h(t,X_2)} = \frac{h_0(t)\exp(\beta X_1)}{h_0(t)\exp(\beta X_2)} = \frac{\exp(\beta X_1)}{\exp(\beta X_2)} \tag{3}$$

Hazard ratio for determining relative in increase of the covariate value x_k by one unit can be determined as:

$$\frac{h(t,X_1)}{h(t,X_2)} = \frac{h_0(t)\exp(\beta X_1 + \beta_k(x_k+1) + \ldots + \beta_n x_n)}{h_0(t)\exp(\beta X_1 + \beta_k x_k + \ldots + \beta_n x_n)} = \exp(\beta_k) \tag{4}$$

Survival function, where $\hat{S}_0(t)$ is the baseline survival function, is estimated from the following formula:

$$\hat{S}(t,X) = \hat{S}_0(t)\exp(X\beta) \tag{5}$$

The main criterion for assessing the state of water pipes is the failure rate index - λ. Failure rate index estimator per year for particular type of water pipes (mains, distributional and water supply connections), was determined from the formula [7]:

$$\lambda = k/(L \cdot \Delta t) \tag{6}$$

where k is the total number of failures on the type of network [-], L is the length of the given type of network [km] and Δt is unit of time equal to one year.

The basic quality parameter of service is its availability, including the duration of interruptions in water supply. This nuisance is proportional to the size of failure, the number of people affected by failure and the duration of interruptions in water supply, what can be described in the following way:

$$IR = I_u/R_i \tag{7}$$

where IR is the customer interruption, I_u is the sum of unplanned interruptions of water supply and R_i is the total number of recipients.
and

$$T_{avg} = D_T/I_T \tag{8}$$

where T_{avg} is the average time necessary to restore the water supply in case of unplanned interruptions in water delivery, D_T is the sum of the duration of all interruptions in the water supply and I_T is the total number of water interruptions.

The service availability determined quotient of the time of the continuous water supply throughout the year and the time when there was the demand for water can be reflected through following expression:

$$SA = (R_i \cdot 1\ year - R_i \cdot D_T)/(R_i \cdot 1\ year) \tag{9}$$

where SA is the availability service indicator.

3   Results and Discussion

The lowest failure rate have distribution pipelines (λ_{Davg} = 0.24 km^{-1}a^{-1}) and the highest failure rate have main pipelines (λ_{Mavg} = 0.54 km^{-1}a^{-1}). The European criteria say that the pipeline needs repairing when the failure rate index exceeds 0.5 km^{-1}a^{-1}.

The calculations show that the distribution pipelines and the water supply connections are in good condition, but one should focus on improving the technical state of the main water pipelines.

According to the percent of failures in the main water pipes, the distribution pipes and the water supply connections depending on the material from which they were made, 53% of the failures occurring in the main water pipes happened in the cast iron pipes, which results from a significant share of this material in the construction of these pipelines and their significant age. The distribution pipelines are characterized by high failure rate of iron pipes (56% of all failures), and PVC pipes (38%). Most failures in the water supply connections, as many as 59%, occur in steel pipes, this is due to their poor technical condition. The lower number of failures in pipelines made of PVC and PE is caused by the fact that they are part of the younger sections of water network and that they are resistant to corrosion.

In the case of the diameter and material of water pipes the most often failures occur in the water connections with a diameter of 32 mm, made of steel (13% of all failures). The next frequent are the failures in pipes made of PVC and steel, with a diameter of 40 mm, respectively 15% and 8.8% of all failures.

Taking into account the failure rate in water pipes it can be concluded that pipes made of PE are characterized by the lowest failure rate. In the water supply connections the failure rate index does not exceed 0.15 km^{-1}a^{-1} and in the distribution pipelines 0.09 km^{-1}a^{-1}.

Also it was stated that iron pipelines in the main pipelines show the highest failure rate. The average failure rate is 1.02 km^{-1}a^{-1}. In the distribution pipelines the average failure rate is 0.43 km^{-1}a^{-1}. The highest failure rate in PVC pipes is seen in the water connections, from 0.65 to 1.13 km^{-1}a^{-1}, and the lowest in the distribution pipelines, from 0.26 to maximum of 0.37 km^{-1}a^{-1}. In the main water pipes the average failure rate was 0.69 km^{-1}a^{-1}. In the steel pipelines the highest failure rate is in the water supply connection, it ranges from 0.66 to over 1.34 km^{-1}a^{-1}.

Thermoplastic materials as PVC and PE should replace traditional materials. As shown the analysis the distinguished materials have a low failure rate.

The next step of the analysis was to research the influence of parameters on failure occurrence of water pipe. The examined significance level assured that there is no basis to reject the hypothesis of the lack of influence of the distinguished parameters on pipe reliability. Relative risk of pipes located in the lanes of heavy traffic is twenty times greater than located off the main road in the pavement. Also considerations about the pipe material confirmed previous calculations associated with the failure rate. Risk associated with pipes made of steel are sixteen times greater than pipes made of PE. It means that pipes made of cast iron and steel should be under special supervision, including the continuous monitoring and the replacement program.

Based on the performed analysis of the failure protocols the recovery time and the total time of repair of damaged pipelines were examined, the range of the recovery time changes for the distribution pipelines was from 1 to 15.5 hours and for the main pipelines from 1 to 17.5 and the total time of damaged water connections repair of about 8 h is longer than the repair time of distribution pipelines (6.5 h). The average values of presented indicators from the distinguished period of time, on the example of the examined water supply system, were as follow: IR = 1.7 no of failures per recipient and year, T_{avg} = 4.5 h per failure and SA = 0.9417.

4   Conclusions

The water company should have developed a scenario of conduct in case of typical failure occurrence. The development of such a scenario should be preceded by the appropriate analyses and consultation with water consumers, since the improper decisions taken at the time of failure may lead to additional costs incurred by the water supply company.

An important element influencing the lifespan of the water supply system is to monitor failures and detect these elements which cause the failures most often. While selecting the elements that cause the failure most often it should be remembered that the "lifetime" of some elements may be much shorter or longer than assumed. The operating time of the water supply network should be determined not only on the basis of theoretical data set by the designers but also on the operating data taken from the network monitoring. Such a procedure is justified because it is not common that the actual lifespan of a technical element overlaps with a theoretical working time. It should also be noted that there are cases where the real operation time of the object is much longer than the theoretical one. For the entire water supply system, as the end of service life should be considered time when the failure rate is so high that it is uneconomical to repair the failures.

The result of the further research will be to achieve renovation assessment procedures, what will constitutes a very important role in the risk analysis procedures and will lead to investment projects and strategic modernization plans.

References

1. Cox, D.R.: Regression models and life-tables. J. R. Stat. Soc. B. 34, 187–220 (1972)
2. Estokova, A., Ondrejka Harbulakova, V., Luptakova, A., Stevulova, N.: Performance of fiber-cement boards in biogenic sulphate environment. Advanced Materials Research 897, 41–44 (2014)
3. Glowacz, A.: Diagnostics of induction motor based on analysis of acoustic signals with the application of eigenvector method and k-nearest neighbor classifier. Arch. Metall. Mater. 2, 403–407 (2012)
4. Herz, R.K.: Exploring rehabilitation needs and strategies for water distribution networks. J. Water. Supply. Res. T. 47, 275–283 (1998)
5. Kaplan, E.L., Meier, P.: Nonparametric estimation from incomplete observations. J. Am. Stat. Assoc. 53, 457–481 (1958)

6. Klein, J.P., Moeschberger, M.L.: Survival analysis techniques for censored and truncated data. Springer, New York (1997)
7. Kwietniewski, M., Roman, M., Trębaczkiewicz-Kłoss, H.: Water and sewage systems reliability (in Polish). Arkady Publisher, Warszawa (1993)
8. Le Gat, Y., Eisenbeis, P.: Using maintenance records to forecast failures in water networks. Urban. Water. J. 3, 173–181 (2000)
9. Pietrucha-Urbanik, K.: Multidimensional comparative analysis of water infrastructures differentiation. In: Pawłowski, A., Dudzińska, M.R., Pawłowski, L. (eds.) Environmental Engineering IV, pp. 29–34. Taylor & Francis Group, London (2013)
10. Pietrucha-Urbanik, K.: Assessment model application of water supply system management in crisis situations. Global. Nest. J. 16, 893–900 (2014)
11. Rak, J.: Safety of water supply system (in Polish). PAN, Warsaw (2009)
12. Sokołowski, A.: Statistics in medicine - advanced methods. Course Materials (2011)
13. Tabesh, M., Soltani, J., Farmani, R., Savic, D.: Assessing pipe failure rate and mechanical reliability of water distribution networks using data-driven modeling. J. Hydroinform. 11, 1–17 (2009)
14. Tchórzewska-Cieślak, B., Rak, J., Pietrucha-Urbanik, K.: Safety analysis and assesment in the water supply sector. Reliability: Theory & Applications 1, 142–153 (2012)
15. Valis, D., Koucky, M., Zak, L.: On approaches for non-direct determination of system deterioration. Eksploat. Niezawodn. 41, 33–41 (2012)
16. Valis, D., Vintr, Z., Malach, J.: Selected aspects of physical structures vulnerability – state-of-the-art. Eksploat. Niezawodn. 14, 189–194 (2012)
17. Wieczysty, A.: Methods of assessing and improving the reliability of municipal water supply systems (in Polish), Committee of Environmental Engineering Sciences, Cracow, Poland (2001)

Freshness Constraints Semantics of RT Framework Credentials

Wojciech Pikulski and Krzysztof Sacha

Institute of Control and Computation Engineering,
Warsaw University of Technology, Warsaw, Poland
w.pikulski@elka.pw.edu.pl, k.sacha@ia.pw.edu.pl

Abstract. The paper focuses on problem of credential revocation in distributed environment where credentials are dispersed among users and there is no simple way to cancel already issued credential. To overcome presented problem, access requests acceptors must contact each credential issuer in order to check its validity status. This process is not always acceptable as it imposes more system load. The work focuses on RT Framework and RT credentials. The proposed solution associates with each credential a freshness constraint which defines how long the credential can be regarded as valid after last validity check. Constraints are propagated along a credential chain that makes the user be granted access to a shared resource. Article presents model and its semantics. The semantics associates each RT credential with a freshness constraint that should be used during credential validity check. The solution provides mechanism for cancelling RT credentials that allows controlling access to shared resources in a fine manner and limit system load.

Keywords: Credential revocation, distributed authorisation, trust management, freshness constraints, formal model.

1   Introduction

Competitive market makes many companies to cooperate in order to reach advantage over competitors. There is a need for technology that facilitates integration of different computer systems in a way that allows exchanging data, accessing shared resources and providing collaboration services. To meet these requirements, trust management models have been proposed. They provide access control mechanisms for distributed environment. Decentralised design allows users from different administrative domains to access resources located in other domains. This is achieved by issuing credentials that can be transferred to user. In order to use resource, user presents credentials and based on them, system can deduct whether access should be granted or not. Inherent characteristic of distributed design is possibility to delegate authority over roles between security domains.

The distributed design is very appealing. It does not impose any requirements for systems integration. In order to mutually exchange data, two systems only need to use the same trust management model. Then, they define a security policy and issue credentials to users. Then, resources can be accessed only based on issued credentials. Nevertheless, there are some drawbacks. As users own credentials, there is no possibility that issuer can revoke particular credential. This feature would be very helpful in many real life scenarios. For example, when employee gets dismissed the employer would like to cancel all his credentials to minimise risk of corporate data leakage. This paper focuses on a RT Framework, which is a trust management model. The article presents an enhancement that provides possibility to revoke credentials. In order to provide such feature, system accepting access request must have possibility to verify credential validity, e.g. by contacting its issuer. Administrators can define how often such verification should be performed. This allows keeping balance between system load and amount of risk of accepting faulty credential.

The paper is organised as follows. Next section briefly presents related work. Then a business scenario that illustrates the problem of revoking credentials is presented. Section 4 provides an overview of the RT Framework. Section 5 presents a proposed solution. Finally, the paper is summed up and conclusions are drawn in section 6.

2   Related Work

A problem of designing access control mechanisms suited for distributed environment has been studied by many researches. Few examples are SPKI/SDSI[1], KeyNote[2], PolicyMaker[3] and RT Framework[4]. Seamons et. al. in [5] and Chapin et. al. in [6] present requirements that trust management model should fulfil and perform survey of various solutions against them. The commonly agreed requisite is monotonicity of a model, which is regarded to be one of the factors allowing a practical implementation of the system. Credential revoking makes model non-monotonic. Nevertheless, Li et. al. in [7] proved that revocation can be implemented in a monotonic manner.

Skalka et. al. [8] created RT^R language which associates each credential with a risk of being invalid. The threat is aggregated along credential chains and when exceeds predefined threshold all subsequent credentials are being ignored. The main advantage of it is possibility to control the authority delegation depth. Thus, user can delegate authority only predefined number of times. The notion is conceptually similar to the idea of freshness constraints propagation presented in this work; however, this solution associates each credential with a maximum age of the credential that can be used in the authorisation process without being verified. If the age is exceeded, system needs to check whether credential is still valid. The result of this operation determines if credential can be used or should be ignored.

Changyu et. al. in [9] introduce a non-monotonic trust management model called Shinren. The concept is different from RT framework analysed in the paper in the sense that Shinren utilises multi-value logic with negative assertions in contrast to RT, which uses classic logic with only positive statements. Furthermore, presented freshness constraints propagation focuses on allowing for credential revocation instead of expressing a negative information in the security policy. Another concept, similar to

Shinren is the RT. language described by Czenko et. al. [10]. It also provides possibility to define negative information and also utilises multi-value logic to perform authorisation queries.

3   Business Scenario

We present a business scenario that describes the discussed problem of revoking credentials in distributed environment. The example presents a policy limiting access to medical data only to licensed doctors who are also defined by patients as their personal doctor. As the research is focused on the RT Framework, the scenario utilises RT credentials, which are described in next chapter. However, we decided to use them here as their meaning is straightforward and will facilitate understanding of later sections.

The patient named Bob gets medical check. His doctor named Alice has ordered some blood tests on the previous visit and now wants to check the results on the laboratory website. Laboratory security policy specifies that access to specific patient's data have only licensed doctors who are defined by patient as his personal doctor. Laboratory administrator delegated control over this role to the NHS (National Health System). This can be expressed as the following credential:

$$Lab.results(?p) \leftarrow NHS.license \cap NHS.hc.doctor(?p) \tag{1}$$

The doctor licenses are not issued directly by NHS, but by its branches. Alice has been licensed by Warsaw branch, denoted by WAW:

$$NHS.license \leftarrow NHS.branch.license \tag{2}$$

$$NHS.branch \leftarrow WAW \tag{3}$$

$$WAW.license \leftarrow Alice \tag{4}$$

In order to define his personal doctor, a patient needs to go to his local health centre and fill appropriate form. Bob is being served by a Promed clinic that has a contract with NHS:

$$NHS.hc \leftarrow Promed \tag{5}$$

$$Promed.doctor(Bob) \leftarrow Alice \tag{6}$$

During the medical check, Alice can present all above credentials to the laboratory and get access to the blood tests results. A ?p in above credentials is a variable denoting any patient.

The problem arises when an issuer decides to revoke some credentials. For example NHS can decide to suspend Alice's doctor license and revoke the credential (4). Similarly Bob may change his personal doctor, which will revoke credential (6). In either event, Alice still has copy of cancelled credential and can use it to access Bob's medical data.

The proposed solution addresses this problem. With each role, administrator can associate a freshness constraint, which defines how long a credential can be used in authorization process after last validation. When this time passes, the credential validity must be verified.

In presented example, laboratory administrator has decided that doctor licenses have to be validated every 10 days, and personal doctor credentials should be checked once a month.

4   The RT Framework Overview

The RT Framework [4] is a trust management model that provides authorisation mechanism for distributed environments. Users and systems are represented as entities. They can issue authorisation requests in order to access shared resource or define roles that can be identified with some privilege in the system. All members of a role representing some resource can access it. Another term is a role name. It can be used by the entity to construct a role. A role is denoted by the entity name followed by dot and then a role name. For example, NHS.doctor is a role defined by national health system by using role name doctor. The role is local to the entity who defines it, so NHS.doctor can have different meaning than Promed.doctor.

In the RT Framework, a security policy is expressed as a set of credentials. Based on them, resource owner can deduct whether user access request should be accepted or denied. The RT Framework is a set of RT languages that provides different capabilities to express security policies. This paper utilises RT_1 language, which allows using parametrised roles [11]. There are four types of credentials:

1. Simple membership: A.r ← D. With this statement A asserts that D is a member of role A.r.
2. Simple inclusion: A.r ← B.s. Issuer A asserts that all members of B.s are also members of A.r. This is a simple role delegation, since B can add new entities to A.r role by adding them to B.s role.
3. Linking inclusion: A.r ← A.s.t. Issuer A asserts that A.r includes all members of B.t role for each B that is member of A.s. This type of credential allows for authority delegation over A.r to members of A.s.
4. Intersection: A.r ← C.s ∩ D.t. This credential allows A to assert that a member of A.r is any entity that simultaneously is member of C.s and D.t. This is partial delegation to C and D.

In the RT_1 language, each role name have a set of parameters $(h_1, \ldots, h_n)$. The parameter can have fixed value (6) or be a variable (1). Based on the presented definitions, four sets can be defined:

$$\text{Roles} = \{ \text{A.r}: \text{A} \in \text{Entities}, \text{r} \in \text{RoleNames} \} \tag{7}$$

$$\text{LinkedRoles} = \{ \text{A.r.s}: \text{A} \in \text{Entities}, \text{r, s} \in \text{RoleNames} \} \tag{8}$$

$$\text{Intersections} = \{ f_1 \cap \ldots \cap f_i: f_i \in \text{Entities} \cup \text{Roles} \cup \text{LinkedRoles} \} \tag{9}$$

$$\text{RoleExpressions} = \text{Entities} \cup \text{Roles} \cup \text{LinkedRoles} \cup \text{Intersections} \tag{10}$$

Authorisation procedure is based on the credential graph, which is built using a set of credentials. Each vertex relates to role expression. Simple edges represent credentials, and derived edges are used to handle linked roles and intersections.

Credential graph is defined below. Notation $e_1 \Leftarrow e_2$ denotes an edge, while $e_1 \ll e_2$ represents a path in a graph. While constructing a graph, a relation credentialEdge is defined. It links credential with corresponding edge in a credential graph.

$$credentialEdge \subseteq C \times E_C \qquad (11)$$

Let C be a set of RT_1 credentials. The basic credential graph G_C relative to C is defined as follows: the set of nodes N_C=RoleExpressions and the set of edges E_C is the least set of edges over N_C that satisfies the following three closure properties:

1. If $A.r \leftarrow e \in C$ then $A.r \Leftarrow e \in E_C$, and it is called a credential edge. The pair $(A.r \leftarrow e, A.r \Leftarrow e) \in$ credentialEdge relation.
2. If there exists a path $A.r \ll B \in G_C$, then $A.r_1.r_2 \Leftarrow B.r_2 \in E_C$, and it is called a derived link edge, and the path $A.r \ll B$ is a support set for this edge.
3. If $D \in N_C$, and $B_1.r_1 \cap B_2.r_2 \in N_C$, and there exist paths $B_1.r_1 \ll D$, and $B_2.r_2 \ll D$ in G_C, then $B_1.r_1 \cap B_2.r_2 \Leftarrow D \in E_C$. This is called a derived intersection edge, and $\{B_1.r_1 \ll D, B_2.r_2 \ll D\}$ is a support set for this edge.

If user is a member of a certain role, there is a path in the credential graph linking vertexes representing user and this role. Such path with corresponding support sets is called a credential chain. It can be regarded as a proof that user is a member of specific role. If role represents some privilege in a system, user can perform certain action, e.g. access shared resource.

To illustrate the authorisation process Figure 1 presents a credential graph corresponding to analysed business scenario. There exists path linking Alice with Lab.results(Bob) role. Thus, Alice has access to results of Bob's blood test. As described business scenario is simple, one can observe that the graph presented on Figure 1 is also the credential chain. It is also interesting that Lab.results role has patient parameter set to Bob in the graph. This is a consequence of the credential (6) which associates the variable ?p with specific value.

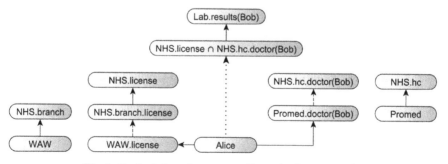

Fig. 1. Credential graph corresponding to business scenario

5   Freshness Constraints Semantics of RT Framework Credentials

5.1   Freshness Constraints

The proposed model introduces an enhancement to the RT Framework that allows revoking credentials. Generally, to provide such feature system accepting authorisation requests needs to verify if all credentials forming credential chain are valid. This can be very difficult or unwanted procedure. Very often, some credentials are being revoked on a regular basis and there is no necessity to verify them on every access request, e.g. universities revoke student cards twice a year. In other situations, users can accept some amount of risk of accepting revoked credential in return to lowering system load or network traffic.

To provide flexible solution, the proposed model utilises idea of freshness constraints. Each credential has associated fresh time f_n and a freshness requirement Δt. The fresh time denotes the moment when credential was successfully validated last time. Afterwards, the credential can be regarded as valid until $f_n + \Delta t$ passes. In other words, at the time of access decision evaluation, when the condition $t_{now} < f_n + \Delta t$ is fulfilled then credential can be assumed to be valid. Otherwise, system needs to verify credential status before using it in the authorisation process. The Δt value can be seen as a risk of accepting invalid credential that application can accept in place of less frequent credentials verification.

We think that freshness constraints should be propagated along the credentials forming credential chain. The propagation should start with the role representing resource and proceed in direction of user. This design ensures that when administrator defines a constraint for a specific role (e.g. Lab.results) then all credentials used to calculate all members of this role fulfil this constraint.

The freshness constraint can be regarded as maximum age of the credential without verification. The value can be any positive number. The set T denotes all possible values for freshness constraints:

$$T = <0, \infty) \tag{12}$$

5.2   Defining Security Policy and Freshness Constraints Propagation

In the RT Framework a security policy is defined by issuing credentials. In a business scenario, laboratory administrator creates credential (1) which specifies who has access to medical data.

In the proposed solution, there is need to define requirements for specific roles. Therefore, we introduce the f_c function:

$$f_c: \text{Roles} \rightarrow T \tag{13}$$

Table 1 presents policy defined by laboratory administrator from business case.

Table 1. Policy defined by Lab administrator – f_c values

Role	Δt [days]
NHS.license	10
NHS.hc.doctor	30

The freshness requirements should propagate along the edges of the credential chain and associate each edge with a freshness constraint. When credential validity is performed, a value associated with edge relative to the credential is being used as a freshness requirement.

As can be observed on Figure 1, the chain may be disconnected. Each derived link edge has a support set that forms a separate component of the graph. Moreover, support sets determine existence of derived edges. Therefore, freshness constraints should be propagated also through support sets.

To carry out propagation, a freshness graph FG_C is being built based on a credential graph G_C. The propagation is performed along its edges from vertex representing resource to vertex relative to the user. The function Ψ transforms a credential graph into freshness graph.

$$\Psi: G_C(N_C, E_C) \rightarrow FG_C(FN_C, FE_C) \tag{14}$$

Two graphs have the same sets of vertexes:

$$FN_C = N_C \tag{15}$$

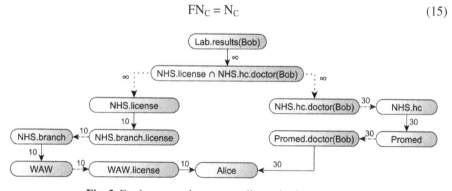

Fig. 2. Freshness graph corresponding to business scenario

As credentialEdge relation links a credential with an edge in a credential graph, the freshnessCredentialEdge relation links the same credential with an edge in a freshness graph:

$$\text{freshnessCredentialEdge} \subseteq C \times E_C \tag{16}$$

The FE_C set of edges of freshness graph is constructed as follows:

1. If $A.r \Leftarrow e \in E_C$, then $A.r \Rightarrow e \in FE_C$. Let c be a credential from credentialEdge relation, corresponding to edge $A.r \Leftarrow e$. Then a pair $(c, A.r \Rightarrow e) \in$ freshnessCredentialEdge relation.

2. If $A.r_1.r_2 \Leftarrow B.r_2 \in E_C$, then $A.r_1.r_2 \Rightarrow A.r_1 \in FE_C$ and $B \Rightarrow B.r_2 \in FE_C$

3. If $B_1.r_1 \cap B_2.r_2 \in N_C$, then $B_1.r_1 \cap B_2.r_2 \Rightarrow B_1.r_1 \in FE_C$ and $B_1.r_1 \cap B_2.r_2 \Rightarrow B_2.r_2 \in FE_C$.

Figure 2 presents a freshness graph corresponding to the credential graph from Figure 1. The propagation is performed along its edges. The process associates each edge with a freshness constraint value that is a result of combining freshness constraint defined in the policy for the role in vertex which is a beginning of the edge and values associated for each edge that ends in this vertex. The combination is performed by propagation operator ∇ defined as follows:

$$x \nabla y = \min(x, y) \tag{17}$$

The policy is defined by f_c function. However, it only specifies freshness constraints for Roles. On the other hand, the vertexes in the freshness graph may represent not only Roles but also Entities, LinkedRoles and Intersections. In order to generalize the f_c function, a calc function is defined. Based on the policy, it allows calculating a freshness constraint for any role expression. The f_c is a partial function. If it is not defined for specific role, then calc returns infinity for it.

$$\text{calc: RoleExpression} \rightarrow T \tag{18}$$

$$\text{calc}(A) = \infty \tag{19}$$

$$\text{calc}(A.r) = f_c(A.r) \tag{20}$$

$$\text{calc}(A.r.s) = f_c(A.r) \tag{21}$$

$$\text{calc}(A.r \cap B.s) = \infty \tag{22}$$

The calc function associates infinity with entities. This is for clear mathematical calculus as this ensures calc is applicable for any vertex. Nevertheless, vertexes relative to entities are always leafs of the graph and therefore doesn't take part in freshness constraints propagation.

Intersections are associated with infinity to ensure that freshness constraints defined for one element of intersection does not propagate into branch relative to another element. This is safe, as in the freshness graph the vertex representing intersection is always connected with nodes representing intersection's elements. Those observations can be seen in Figure 2.

5.3 Propagation Process

The propagation is performed in steps. At each step some freshness graph edges are associated with freshness constraints. The relation P_i represents such association at the ith step of propagation. The relation is defined later in this chapter. From this point we will assume that edges are represented as a set of ordered set of pairs of nodes.

$$FE_C \subseteq FN_C \times FN_C \tag{23}$$

The propagation starts from the edges that go out from the freshness graph root. The set of such edges is named roots.

$$roots \subseteq FE_C \tag{24}$$

$$roots = \left((a, b) \in FE_C: \not\exists_{(x,y)\in FE_C}(y = a) \right) \tag{25}$$

A freshness constraint value can be associated with edge going from node n_1 to node n_2, when all edges ending in n_1 have been processed. In order to decide whether edge is ready to be processed a predicate named complete is introduced.

$$complete: FE_C \times P_i \rightarrow \{false, true\} \tag{26}$$

$$complete\left((a, b), P_i \right) = \left(\forall_{(x,y)\in FE_C:\, y=a} \exists_{(c,d,\Delta t)\in P_i}(c = x \cap d = y) \right) \tag{27}$$

When an edge is being process by the propagation process, the prop function is used to calculate a freshness constraint value that should be associated with it.

$$prop: FE_C \times P_i \rightarrow T \tag{28}$$

$$prop\left((a, b), P_i \right) = \begin{cases} calc(a) \nabla \left(\nabla_{(x,y,\Delta t)\in P_i:\, y=a}(\Delta t) \right) & (a, b) \notin roots \\ calc(a) & (a, b) \in roots \end{cases} \tag{29}$$

The P relation represents a freshness constraints associated with freshness graph edges. It is constructed incrementally. The relation P is the least set P_i satisfying the following conditions:

$$P \subseteq FE_C \times T \tag{30}$$

1. $P_0 = \emptyset$
2. $P_{i+1} = \bigcup_{(a,b)\in FE_C} f((a, b), P_i)$
3. $f((a, b), P_i) = \{((a,b),\Delta t): (complete((a,b),P_i) \cup (a,b)\in roots) \cap \Delta t = prop((a,b),P_i)\}$

5.4   Semantics Definition

The freshness constraints semantics S_C of the set C of RT credentials associates each credential with a freshness constraint value that should be used to verify credential validity during authorization process.

To define the semantics, a freshnessCredentialEdge relation between credential and a corresponding edge in a freshness graph is used. The semantics can be defined as composition of two relations:

$$S_C = freshnessCredentialEdge \circ P \tag{31}$$

Figure 2 presents the values of P relation. The values that are associated with edges corresponding to credentials are elements of semantics S_C. Such edges are drawn with solid line. The propagation has been performed with policy presented in Table 1.

6   Summary and Conclusions

The proposed solution provides a formally defined enhancement to the RT Framework that allows for credential revocation. It is a desirable feature that helps employment of trust management in real life situations. Proposed enhancement takes into consideration a credential chain that allows user access a shared resource and therefore allows controlling a security policy in a fine manner. Freshness constraints also limit the system load by lowering frequency of credential validity checks.

The solution supports RT_0 and RT_1 languages but there should be no much effort to make it work with also other languages like RT_2 and RT^T[11].

References

1. Clarke, D., Elien, J.E., Ellison, C., Fredette, M., Morcos, A., Rivest, R.: Certificate Chain Discovery in SPKI/SDSI (1999)
2. Blaze, M., Feigenbaum, J., Ioannidis, J., Keromytis, A.D.: The KeyNote trust-management system, version 2. IETF RFC 2704 (September 1999)
3. Blaze, M., Feigenbaum, J., Strauss, M.: Compliance Checking in the PolicyMaker Trust Management System. In: Hirschfeld, R. (ed.) FC 1998. LNCS, vol. 1465, pp. 254–274. Springer, Heidelberg (1998)
4. Li, N., Winsborough, W., Mitchell, J.: Distributed Credential Chain Discovery in Trust Management. J. Computer Security 1, 35–86 (2003)
5. Seamons, K.E., Winslett, M., Yu, T., Smith, B., Child, E., Jacobson, J., Mills, H., Yu, L.: Requirements for policy languages for trust negotiation. In: Proc. Third Int. Policies for Distributed Systems and Networks Workshop, pp. 68–79 (2002)
6. Chapin, P.C., Skalka, C., Wang, X.S.: Authorization in trust management: Features and foundations. ACM Comput. Surv. 40 (2008)
7. Li, N., Feigenbaum, J.: Nonmonotonicity, User Interfaces, and Risk Assessment in Certificate Revocation (Position Paper). In: Syverson, P.F. (ed.) FC 2001. LNCS, vol. 2339, pp. 157–168. Springer, Heidelberg (2002)
8. Skalka, C., Wang, X.S., Chapin, P.C.: Risk management for distributed authorization. Journal of Computer Security 15, 447–489 (2007)
9. Dong, C., Dulay, N.: Shinren: Non-monotonic trust management for distributed systems. In: Nishigaki, M., Jøsang, A., Murayama, Y., Marsh, S. (eds.) IFIPTM 2010. IFIP AICT, vol. 321, pp. 125–140. Springer, Heidelberg (2010)
10. Czenko, M., Ha, T., Jeroen, D., Sandro, E., Pieter, H., Jerry, den H.: Nonmonotonic Trust Management for P2P Applications. Electronic Notes in Theoretical Computer Science (ENTCS) archive 157(3), 113–130 (2006)
11. Li, N., Mitchell, J., Winsborough, W.: Design of a Role-Based Trust-Management Framework. In: IEEE Symposium on Security and Privacy, pp. 114–130. IEEE Computer Society Press (2002)

Logistic Support Model for the Sorting Process of Selectively Collected Municipal Waste

Marcin Plewa, Robert Giel, and Marek Młyńczak

Wrocław University of Technology, 27 Wybrzeże Wyspiańskiego St, 50-370 Wrocław
{marcin.plewa,robert.giel,marek.mlynczak}@pwr.edu.pl

Abstract. An intensive growth of population, progressing processes of social and economic development and increasing urbanisation of fast-developing countries considerably contribute to the increase in the amount of the waste produced. This necessitates reasonable decisions to be made within waste man-agement in urban areas. Therefore, a clear increase in the interest in matters as-sociated with the selection of relevant strategies for planning the waste man-agement process has been noticeable in recent years. One of the global aims is to develop appropriate waste recycling strategies. Solving problems associated with the selection of an appropriate strategy for waste management, planning collection processes, sorting and recovery, designing infrastructures for waste treatment and monitoring of such processes often requires that a model-based approach should be applied. The main objective of this paper is to present a simulation model that aids the decision-making process as regards the planning of sorting processes for the selectively collected municipal waste at the Recy-cling Centre in WPO ALBA S.A.

Keywords: waste management, reverse logistics, simulation.

1 Introduction

An intensive growth of population, progressing processes of social and economic development and increasing urbanisation of fast-developing countries considerably contribute to the increase in the amount of the waste produced. Therefore, a clear increase in the interest in matters associated with the selection of relevant strategies for planning the waste management process has been noticeable in recent years. One of the global aims is to develop appropriate waste recycling strategies, which would result in an increased effectiveness of such recycling due to improved production at the source, collection and reprocessing of priority waste flows. Solving problems associated with the selection of an appropriate strategy for waste management, plan-ning collection processes, sorting and recovery, designing infrastructures for waste treatment and monitoring of such processes often requires that a model-based approach should be applied [2,12,18,20]. The main objective of this paper is to present a simulation model that aids the decision-making process as regards the planning of sorting processes for the selectively collected municipal waste at the Recycling Centre in WPO ALBA S.A. This is an element of a comprehensive solution devised to aid in the decision-making as

W. Zamojski et al. (eds.), *Theory and Engineering of Complex Systems and Dependability,*
Advances in Intelligent Systems and Computing 365, DOI: 10.1007/978-3-319-19216-1_35

regards municipal waste management. In the first section of this paper, the current state of knowledge, including the classification of waste man-agement models in the light of literature-provided research, is presented together with a discussion on basic model groups, which allows to assign the model in question to one of the groups and establish a direction for further research.

2   Classification of Models – Literature Overview

The discussion of waste management system/process modelling requires that basic definitions and waste classification should be presented in the first place.

Waste may refer to unnecessary products created as a result of a specific activity being completed (e.g. production, household or commercial activity) and refused by the society [14].

The simplest waste classification includes a division into municipal and industrial waste. The rules of waste classification are extensively discussed in the literature and are presented in [4, 6, 15, 27]. Much attention in the literature is also paid to the defi-nition of waste management, which includes effective planning, completion and con-trol of waste collection, transport and reprocessing (sorting, recovery and treatment) [6].

The problem of waste management process modelling has been analysed in the lit-erature since the late 1960s [7]. One of the first studies aimed at the overview of the literature in the field of waste management is [27]. Its author focused on a wide over-view of the literature discussing mathematical modelling of the planning of the waste management process. The models were classified into three main groups (static mod-els, dynamic models with periodic or continuous optimisation of parameters relating to the waste management strategy).

Other studies, which present a wide overview of the literature devoted to the field of modelling municipal waste management processes are [2, 3, 9, 10, 12, 13, 17, 19, 24, 25, 30].

The literature-based study has allowed suggesting three main groups of waste management models based on the criterion of their applicability: They are: models of locations and selection of waste collection sites/ designing waste collection networks; models of planning waste collection routes and models of support for planning waste management systems, including the use of decision support systems/ expert systems, assessment tools for system functioning and case studies.

In its later section, this paper presents studies included in the last group, as the model to be presented in the paper is to be found among decision support systems.

Classified into the first two groups and described in the literature, the models of process optimisation also provide assistance to the planners in decision making as regards designing waste collection networks (e.g. location, routes, number and type of vehicles). These, however, are not system solutions which would take into account the relationships between individual waste management sub-systems.

The last group of models includes studies suggesting comprehensive solutions in the form of Decision Support Systems (DDS), Spatial Decision Support Systems (SDSS) and

Planning Support Systems (PSS), based on the use of information tech-nologies [14]. The overview of the literature in this scope, among others, is provided in [14], the author suggested a SDSS system to support the planning of solid waste in Philadelphia, PA.

Another DSS system are presenting in papes [1,5,7,23,26].

Further, the literature provides several papers, which cover concepts of waste man-agement system planning along with the use of the multi-criterion method. An exam-ple here is [8, 11, 16, 22].

With such an insight into the literature, the scope of which covers modelling of lo-gistic processes as regards waste management, it is possible to summarise the existing state of knowledge. The subject literature still does not provide any solution, which, in a system approach, would cover factors, i.e. the fact is that the collection point can generate flows of various loads (various municipal waste fractions: plastic materials, paper, glass, etc.), requiring loadability of transport means to various extent. As to the waste collection, time of collection is important as well. Delays may cause containers to be overfilled. For municipal waste collection, it may be associated with penalties to be imposed on the company. The main decision-related problem discussed in the lit-erature is the manner of resource allocation, determination of collection areas and assignment of transport means so that all areas can be handled within the required time with the lowest total cost possible. Insufficient attention has been paid to models including many independent sources of waste and their impact on the completion of waste management processes within recovery systems. There are no models that would include changes in the morphology of waste on the completion of management of selectively collected municipal waste. There are no models that would include the product quality of the recovery process and repeatability of parameters characterizing a given product. There are no models that would include any regulation of waste flows to the recovery installations and full integration between the transport, sorting, storing, handling and recovery sub-systems. There are no models that would improve the operation of the recovery installation, including the dependency between individ-ual elements of the installation. Finally, there is also no study that would provide the information on the impact of unreliability of individual process on the efficiency of the recovery system operation.

The fact that there are no optimum solutions to be found in the literature is caused by constantly changing organization of this system type and a large number of ran-dom factors affecting the efficiency of collection process completion.

3   Model Description

Based on the literature overview provided, main objectives of this paper can be identi-fied. This paper presents a model being an element of a comprehensive solution designed to support decisions as regards municipal waste management in WPO ALBA S.A. The tool developed is a subject of a research project, which is to incorporate the development of a logistic system model, including elements of the waste management process (collection, transport to the handling site, handling, storage, sorting and transport of mixed waste to Regional Municipal Reprocessing Systems. The aim of the IT tool will be to ensure the integration of individual processes, which constitute municipal waste management. Further step will be to incorporate elements of the system reliability as well as in [21, 28, 29].

This paper presents a simulation model of the sorting process of selectively collected municipal waste at the Recycling Centre in WPO ALBA S.A.

4   Assumptions

The municipal sanitation company receives waste from households in the city of Wrocław and nearby communes. Operations of the company consist in the organiza-tion of waste collection, transport to the sorting plant and distribution of the output material.

The waste-sorting plant is designed to clean flows of selectively collected waste from contamination caused by improper waste sorting at the source. By using the system, it is possible to separate undesired waste fractions. The company employs two waste-sorting methods, i.e. positive and negative selections.

The selectively collected waste form a part of Wrocław area and surrounding communes is transported to the receiving station at the sorting plant. During the week, supply quantities are different, which results from the waste collection schedule, as waste is collected from various locations on various days. Most schedules apply weekly, and, therefore, it is assumed that waste is collected from the same locations on the same weekdays for simplification. This assumption will be removed in the course of further work on the model.

Once the waste is unloaded, the employees carry out pre-selection, during which they sort out hard plastic and large-size waste. Then the waste is fed into a system of connected belt conveyors and passes through subsequent elements of the system, in which waste flows are separated.

The flow commences at the hopper chamber and the adjacent conveyor 1_01. Then, as shown by the diagram below, the waste is transported along the conveyor 1_01 to the conveyor 1_02 to the manual sorting compartment (compartment no. 1) on the conveyor 1_05. The compartment is the final place, where the waste is trans-ported before flows of various rates are separated. Once transported by the conveyor 1_05, the waste is routed to the conveyor 1_08, which feeds it into a drum sieve, di-viding the waste into two flows with fractions smaller and larger than 200 mm. Once separated into two flows, they are conveyed to various sorting compartments (com-partments 2 and 3) to be sorted into raw materials and ballast. The latter is milled up and converted into an alternative fuel. Apart from manual sorting at the sorting plant, there are mechanical classifiers to separate, among others, ferrous and non-ferrous metals out of other fractions.

5   Key

- m_1, m_2, m_3, m_4, m_5– random variable describing the waste supply quantity on each day of of the week [Mg];
- T_z – normal working time of one shift [s];
- T_{ez}– actual working time of one shift (with downtime and shift breaks) [s];
- t_a – random variable describing downtime of the line due to a random event [s];
- T_p – break time between shifts [s];

- Z – number of shifts, on which the plastic material line is being operated [items];
- T_c – total operation time of the plastic material sorting line [s];
- s_m – random variable describing the waste mass flow [kg/s];
- c_m – random variable describing the volume of the waste mass flow [m3/s];
- v_1, v_2, v_3, v_4– speed of the conveyer 1_01, 1_02, 1_05, 1_08 [m/s];
- k – drum sieve efficiency [%];
- F_a – waste flow with a fraction up to 200 mm [kg/s];
- F_b – waste flow with a fraction over 200 mm [kg/s];
- M_{Tc} – total waste mass processed during the operation time of the plastic material sorting line [Mg];
- C_{Tc} – total volume of the waste mass flow processed during the operation time of the plastic material sorting line [m3];

Table 1 presents density functions of probability distribution for random variables describing supply quantities of selectively collected waste delivered to the Recycling Centre in WPO ALBA S.A.

Table 1. Density functions of probability distribution for random variables describing supply quantities of selectively collected waste on individual weekdays

Variable description	Variable	Density function of probability distribution	Statistical value
Quantity supplied on Monday [Mg]	m_1	$f(m_1) = \dfrac{5.04}{14.63} \cdot \left(\dfrac{m_1 - 29.83}{14.63}\right)^{4.04} e^{-\left(\frac{m_1-29.83}{14.63}\right)^{5.04}}$	0.39092
Quantity supplied on Tuesday [Mg]	m_2	$f(m_2) = \dfrac{2.66}{14.38} \cdot \left(\dfrac{m_2 - 18.2}{14.38}\right)^{1.66} e^{-\left(\frac{m_2-18.2}{14.38}\right)^{2.66}}$	0.60629
Quantity supplied on Wednesday [Mg]	m_3	$f(m_3) = \dfrac{1.06}{7.83} \cdot \left(\dfrac{m_3 - 6.6}{7.83}\right)^{0.06} e^{-\left(\frac{m_3-6.6}{7.83}\right)^{1.06}}$	0.63440
Quantity supplied on Thursday [Mg]	m_4	$f(m_4) = \dfrac{3.37}{9.67} \cdot \left(\dfrac{m_4 - 22.3}{9.67}\right)^{2.37} e^{-\left(\frac{m_4-22.3}{9.67}\right)^{3.37}}$	0.40255
Quantity supplied on Friday [Mg]	m_5	$f(m_5) = \dfrac{2.01}{11.02} \cdot \left(\dfrac{m_5 - 17.46}{11.02}\right)^{1.01} e^{-\left(\frac{m_5-17.46}{11.02}\right)^{2.01}}$	0.67293

The random variable cm describing the volume of the waste flow is given in the function of the mass flow amount. The tests completed allowed to take into account the variable amount of specific gravity of the flow of the selectively collected waste. The volume of mass flow can be represented using the following equation:

$$c_m(s_m, \tau_o) = 0.0137 \cdot s_m + \tau_o \quad [m3/s], \tag{1}$$

where:

τ_o – random variable describing the deviation of waste volume from the linear function describing the dependency of volume from mass [m3].

Table 2 presents density functions of probability distribution for random variables used to describe the operation of the waste sorting plant.

Table 2. Density functions of probability distribution for random variables describing the ope-ration of the plant

Variable description	Variable	Density function of probability distribution	Statistical value
Waste mass flow [kg/s]	s_m	$f(s_m) = \dfrac{1.64}{0.69} \cdot \left(\dfrac{s_m - 0.2}{0.69}\right)^{0.64} e^{-\left(\frac{s_m - 0.2}{0.69}\right)^{1.64}}$	0.52177
Deviation from the linear function describing the relationship between the mass flow and volume [m³]	τ_o	$f(\tau_o) = \dfrac{18.12}{5.1} \cdot \left(\dfrac{\tau_o + 4.88}{5.1}\right)^{17.12} e^{-\left(\frac{\tau_o + 4.88}{5.1}\right)^{18.12}}$	0.62055
Downtime of the sorting line caused by a random event [s]	t_a	$f(t_a) = \dfrac{2.1}{1\,941.89} \cdot \left(\dfrac{t_a + 517.89}{1\,941.89}\right)^{1.1} e^{-\left(\frac{t_a + 517.89}{1\,941.89}\right)^{2.1}}$	0.69735

The Kolmogorov at the level of significance $\alpha = 0.05$ was applied to verify the goodness-of-fit of empirical and theoretical distribution functions. All cases resulted in compliance of the distribution functions (values lower than the limit value of $\lambda_{0.05}$).

6   Simulation Algorithm

In the model developed, the influence of time is given by constant increments. The assumed unit of time is second. Each simulation experiment has a predefined random time horizon given by the number of shifts, the difference in working time on an eight-hour shift and planned downtimes and downtimes caused by random events. Completion of individual process in subsequent periods depends on the actual values of the input parameters and the decisions made during previous periods. Based on the assumptions employed while developing the model, the following simulation algo-rithm has been devised:

1. Step one: determine the length of the period to be analysed.
2. Determine input parameters for the equipment.
3. Determine the number of repeated simulations for each of the tested input parame-ters.
4. Generate the value of the random variable describing the downtime of the line dur-ing the shift.
5. Assign a unit step for the generation time of the waste flow and number of interac-tions for the flow generator.

6. Start flow transfer through the simulated equipment at the sorting plant.
7. Divide the flow into two sub-flows, depending on the parameters of waste fraction percentages.
8. Record the input parameters for the flow.
9. Last step: save the collected values of the output parameters for the simulation models to a file.

The described simulation algorithm is presented by the block diagram in Figures 1-2.

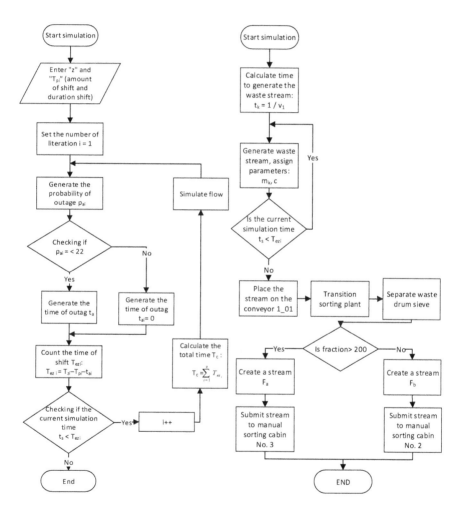

Fig. 1. Block diagram of the simulation process, time calculate (left side) and block diagram of the simulation process, stream flow (right side)

7   Model Sensitivity Analysis

The range of variability of the input parameters in the model sensitivity analysis has been selected based on the initial analysis of how the parameters affect the model results. Table 3 presents the assumed values and the variability of the model input parameters.

Table 3. Variability of the assumed input parameters of the simulation model

Item	Symbol	Output quantity	Variability (min:unit step:max)	Explanation
1	v_1	0.0676	0.0176:0.01:0.0976	speed of the conveyer 1_01
2	v_2	0.41	none	speed of the conveyer 1_02
3	v_3	0.1	none	speed of the conveyer 1_05
4	v_4	0.4	none	speed of the conveyer 1_08
5	k	100	10:10:100	drum sieve efficiency
6	z	3	1:1:3	amount of shifts

Table 4 presents the output parameters of the simulation model, collected in the course of the sensitivity analysis.

Table 4. Output parameters of the simulation model, collected in the course of the sensitivity analysis for the logistic support model

Item	Symbol	Determination method	Explanation
1	F_a^{PET}	% of F_a	PET waste flow
2	C_{Tc}	$\sum_{j=1}^{Tc} c_m^j$	Sum of volume of the waste mass flow
3	M_{Tc}	$\sum_{j=1}^{Tc} s_m^j$	Sum of waste mass flow

7.1   Analysis of the Impact of Changing Speed of the Conveyor 1_01 on the Rate of the Mass Flow

The graph 2 present an analysis of how a changing speed of the conveyor 1_01 af-fects the rate of the mass flow of the waste processed within one day, depending on the number of plastic material sorting shifts. At a higher speed of 0.0967 m/s within one shift, up to approx. 28 Mg of waste can be sorted. The same amount of waste can be processed at a speed of 0.0467 m/s on two shifts and from 0.0276 m/s and 0.0376 m/s on three shifts. It must be noted, however, that maintaining a high speed of the conveyor 1_01 for a longer time does not prove to be an optimum solution. For proper functioning of the sorting plant, batching waste mass and volume is most important. As the mass increases, the volume of the waste flow transported along the belt conveyors increases as well. An excessive increase in the amount of waste results in a decreased efficiency of manual sorting. If the volume is too large, sorters are not able to spot proper waste fractions.

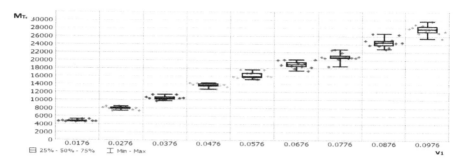

Fig. 2. An analysis of how a changing speed of the conveyor 1_01 affects the total waste mass processed during the operation time of the plastic material sorting line, depending on the num-ber of plastic material sorting shifts (results for Z=1).

The following graphs present the rate of a 600-second mass flow and the waste volume flowing through the sorting compartment, depending on the speed of the con-veyor 1_01. It can be noted that there is a volume flow of more than 1.5 m3 passing through the sorting compartment within 100 seconds, if the speed is 0.09 m/s.

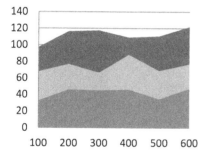
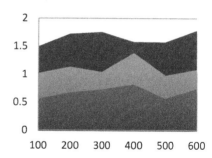

Fig. 3. Mass of waste flowing through the sorting compartment, depending on the speed of the conveyor 1_01 [kg] and waste volume flowing through the sorting compartment, depending on the speed of the conveyor 1_01 [m3]

Based on initial analyses of the tests completed, it has been determined that 0.0676 m/s is the speed, at which sorters' efficiency is satisfactory. Model sensitivity analysis showed that approx. 2.35 Mg of waste is flowing at the given speed within an hour, which corresponds to approx. 41 m₃. While increasing the speed as mentioned earlier, the sorting efficiency is considerably compromised. For the optimisation assumptions based on the maximum flow rate, without sorters' efficiency being affected at a speed of 0.0676 m/s, the sorting plant is capable of processing up to 19 Mg on one, 37 Mg on two and 54 Mg on three shifts. Having such results, it is possible to ensure an op-timum scenario for a specific week day of sorting plant operation and create a sched-ule based on the analysis of the mass of the waste supplied to the sorting plant on that day.

7.2   Analysis of How Changing the Speed of the Conveyor 1_01 Affects the Rate of the PET Fraction Mass at Varied Efficiency of the Drum Sieve.

One of the most important raw materials obtained in the process of the selectively collected municipal waste is blue PET. In the course of the model, the sensitivity analysis on the impact of speed variation, as applied to the conveyor 1_01 on the rate of the mass fraction with PET type material, has been checked. In the course of the analysis, the impact of a changing efficiency of the drum sieve on the results obtained has been investigated as well. It is justified to have such an analysis performed, as during the tests, it was noted that efficiency of the sieve decreased in relation to oper-ation time. This is caused due to clogging sieve apertures. A decreased sieve efficien-cy necessitates arranging an additional person to pick up plastic material in another sorting compartment. The analysis of how the sieve operation time affects efficiency is subject to further research.

The graph 4 presents the PET waste flow mass per hour of plant operation and how speed variation and efficiency of the drum sieve affect flow changes. The highest waste flow per hour has been obtained for a speed of 0.09 m/s and 100% efficiency of the drum sieve. If such parameters apply, the manual sorting compartment will re-ceive approx. 0.18 Mg of the plastic material per hour of sorting plant operation. At a speed of 0.06 and 0.03, the value drops to 0.1 Mg and approx. 0.05 Mg, respectively.

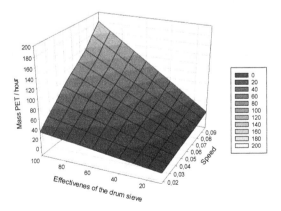

Fig. 4. PET waste flow mass per hour of plant operation and how speed variation and efficiency of the drum sieve affect flow changes

8   Summary

The main objective of this paper was to present a simple simulation model that aids the decision-making process as regards the planning of sorting processes for the selec-tively collected municipal waste at the Recycling Centre in WPO ALBA S.A. During the sensitivity analysis the impact of changing speed of the conveyor 1_01 on the rate of the mass flow and on the rate of the PET fraction mass at varied efficiency of the drum sieve, was investigated. The tests completed allowed to take into account the variable amount of specific gravity of the flow of the selectively collected waste. Which is very important in practice because as the mass increases, the volume of the waste flow transported along the belt conveyors increases as well. An excessive in-crease in the amount of waste results in a decreased efficiency of manual sorting. If the volume is too large, sorters are not able to spot proper waste fractions.

The aim of future research is to develop a model of inventory management and waste sorting planning based on the assumptions presented in this paper.

References

1. Antmann, E.D., Shi, X., Celik, N., Dai, Y.: Continuous-discrete simulation-based decision making framework for solid waste management and recycling programs. Computers and Industrial Engineering 65, 438–454 (2013)
2. Beigl, P., Lebersorger, S., Salhofer, S.: Modelling municipal solid waste generation: A review. Waste Management 28, 200–214 (2008)
3. Chang, N.-B., Pires, A., Martinho, G.: Empowering systems analysis for solid waste management: Challenges, trends and perspectives. Critical Reviews in Environmental Science and Technology 41, 1499–1530 (2011)
4. Chowdhury, M.: Searching quality data for municipal solid waste planning. Waste Management 29, 2240–2247 (2009)
5. Costi, P., Minciardi, R., Robba, M., Rovatti, M., Sacile, R.: An environmentally sustainable decision model for urban solid waste management. Waste Management 24, 277–295 (2004)
6. Demirbas, A.: Waste management, waste resource facilities and waste conversion processes. Energy Conversion and Management 52, 1280–1287 (2011)
7. Eriksson, O., Bisaillon, M.: Multiple system modelling of waste management. Waste Management 31, 2620–2630 (2011)
8. Generowicz, A., Kowalski, Z., Kulczycka, J.: Planning of waste management systems in urban area using multi-criteria analysis. Journal of Environmental Protection 2, 736–743 (2011)
9. Gentil, E.C., Damgaard, A., Hauschild, M., Finnveden, G., Eriksson, O., Thorneloe, S., Kaplan, P.O., Barlaz, M., Muller, O., Matsui, Y., Ii, R., Christensen, T.H.: Models for waste life cycle assessment: Review of technical assumptions. Waste Management 30, 2636–2648 (2010)
10. Ghiani, G., Lagana, D., Manni, E., Musmanno, R., Vigo, D.: Operations research in solid waste management: A survey of strategic and tactical issues. Computers and Operations Research 44, 22–32 (2014)
11. Huang, Y.-T., Pan, T.-C., Kao, J.-J.: Performance assessment for municipal solid waste collection in Taiwan. Journal of Environmental Management 92, 1277–1283 (2011)

12. Laurent, A., Bakas, I., Clavreul, J., Bernstad, A., Niero, M., Gentil, E., Hauschild, M.Z., Christensen, T.H.: Review of LCA studies of solid waste management systems: Part I: Lessons learned and perspectives. Waste Management 34, 573–588 (2014)
13. Lisa, D., Anders, L.: Methods for household waste composition studies. Waste Management 28, 1100–1112 (2008)
14. MacDonald, M.L.: A multi-attribute spatial decision support system for solid waste planning. Computers, Environment and Urban Systems 20(1), 1–17 (1996)
15. Mbido, K.: Optimization of municipal solid waste transportation management with composting plant, M.Sc. Dissertation, University of Dar es Salaam (2013)
16. Milutovic, B., Stefanovic, G., Dassisti, M., Markovic, D.: Multi-criteria analysis as a tool for sustainability assessment of a waste management model. Energy 74, 190–201 (2014)
17. Morrissey, A.J., Browne, J.: Waste management models and their application to sustainable waste management. Waste Management 24, 297–308 (2004)
18. National Waste Management Plan (Polish Monitor, Official Journal of the Republic of Poland, No. 101, Resolution No. 217 of the Council of Ministers of 24 December 2010). On the national waste management plan in 2014.
19. Pires, A., Martinho, G., Chang, N.-B.: Solid waste management in European countries: A review of systems analysis techniques. Journal of Environmental Management 92, 1033–1050 (2011)
20. Poulsen, O.M., Breum, N.O., Ebbehoj, N., Hansen, A.M., Ivens, U.I., van Lelieveld, D., Malmros, P., Matthiasen, L., Nielsen, B.H., Nielsen, E.M., Schibye, B., Skov, T., Stenbaek, E.I., Wilkins, C.K.: Collection of domestic waste. Review of occupational health problems and their possible causes. The Science of the Total Environment 170, 1–19 (1995)
21. Restel, F.: The Markov reliability and safety model of the railway transportation system. In: Proceedings of the European Safety and Reliability Conference, ESREL 2014, pp. 303–311 (2014)
22. Rogge, N., De Jaeger, S.: Measuring and explaining the cost efficiency of municipal solid waste collection and processing services. Omega 41, 653–664 (2013)
23. Simonetto, E. de O., Borenstein, D.: A decision support system for the operational planning of solid waste collection. Waste Management 27, 1286–1297 (2007)
24. Singh, A.: An overview of the optimization modelling applications. Journal of Hydrology 466-467, 167–182 (2012)
25. Stypka, T.: Critical review of municipal solid waste management models, http://www.wis.pk.edu.pl/media/file/.../Stypka.pdf (dostęp: October 20, 2014r)
26. Tanskanen, J.-H.: Strategic planning of municipal solid waste management. Resources, Conservation and Recycling 30, 11–133 (2000)
27. Wilson, D.C.: Strategy evaluation in planning of waste management to land – a critical review of the literature. Applied Mathematical Modelling 1, 205–217 (1997)
28. Zajac, M., Kierzkowski, A.: Uncertainty assessment in semi Markov methods for Weibull functions distributions. In: Advances in Safety, Reliability and Risk Management - Proceedings of the European Safety and Reliability Conference, ESREL 2011, pp. 1161–1166 (2011)
29. Zajac, M., Kierzkowski, A.: Analysis of the reliability discrepancy in container transshipment. In: 11th International Probabilistic Safety Assessment and Management Conference and the Annual European Safety and Reliability Conference, pp. 826–831 (2012)
30. Zhang, Y.M., Huang, G.H., He, L.: A multi-echelon supply chain model for municipal solid waste management system. Waste Management 34, 553–561 (2014)

Access Control Approach in Public Software as a Service Cloud

Aneta Poniszewska-Maranda and Roksana Rutkowska

Institute of Information Technology, Lodz University of Technology, Poland
aneta.poniszewska-maranda@p.lodz.pl

Abstract. Newest technologies like cloud computing are readily explored by both private users and organizations. Cloud brings with it ease of use and convenience for the end user. However often at the cost of convenience we may pay in terms of data security.

The paper presents a new access control security model that can strengthen data protection in public software as a service cloud. The new model gives the new opportunities and possibilities for tighter protection of data. Its constant checking of the permissions, dynamics, manageability and elasticity show that it can be introduced to the cloud environment and ensures tight security.

Keywords: Cloud computing, security of cloud, access control, software as a service.

1   Introduction

Newest technologies like cloud computing are readily explored by both private users and organizations. Cloud brings with it ease of use and convenience for the end user. However often at the cost of convenience we may pay in terms of data security. As the data in the cloud is stored outside of private computers or organization's internal networks, it is even more vulnerable to the threat of unwanted access and use. Cloud provides a new way of delivering services through a network connection, where users can usually access services through their browsers. Because data travels through many public networks it may fall into the wrong hands. Therefore with the creation of cloud technology the approach to security must change and become even tighter as the data and information is not only stored internally. Security's main goal is to ensure only legitimate users can access their data and that their resources cannot be used in any unwanted way.

With the advance of cloud technology the organizations want to achieve more innovation and improve their efficiency with less, getting high availability and computation resources tailored to their current needs at a competitive price. Cloud creates new opportunities around shared resources. It already has found many applications, environments to test and develop new software, storage and backup services or collaboration tools. It is predicted that cloud will grow significantly in the future with the new services created and more exibility given to end user. Surveys show that the current factor which stops the organizations from cloud adoption is actually security of

© Springer International Publishing Switzerland 2015
W. Zamojski et al. (eds.), *Theory and Engineering of Complex Systems and Dependability,*
Advances in Intelligent Systems and Computing 365, DOI: 10.1007/978-3-319-19216-1_36

the services and privacy. As the cloud develops and grows so will the issue of its security and proper data protection.

The paper presents a new access control security model that can strengthen data protection in public software as a service cloud. The new model gives the new opportunities and possibilities for tighter protection of data. The paper is composed as follows: section 2 presents the security concepts in cloud computing, section 3 gives the outline of access control for cloud services based on URBAC approach, while section 4 deals with cooperation of URBAC with SaaS cloud at access control level and section 5 show some technical aspects of URBAC implementation.

2   Security in Cloud Computing

National Institute of Standards and Technology (NIST) provides the following definition of cloud computing [1]:

Cloud computing is a model for enabling ubiquitous, convenient, on-demand network access to a shared pool of configurable computing resources (e.g. networks, servers, storage, applications, and services) that can be rapidly provisioned and released with minimal management effort or service provider interaction.

Cloud provides various resources through a network so that the customers do not have to install specific software or infrastructure on their devices, but instead can rent them from cloud service vendors. In cloud computing the data is stored in centralized servers and cached temporarily on clients' devices including desktop computers, notebooks or mobile phones [8]. Cloud has a multitenant architecture, where multiple independent customers also known as tenants are serviced with a single set of resources [6]. The first type of cloud services differentiation can be presented according to the provided service model. Each model represents a different level of abstraction and capabilities, providing the resources as a Service [2, 3].

Infrastructure as a Service (IaaS) provides the infrastructure resources such as storage, network and computing services. In this infrastructure the user can deploy and run any software he wishes via remote access usually applications, database or operating systems [1, 8].

Platform as a Service (PaaS) allows the user to deploy to cloud infrastructure applications which are consumer-created or purchased. Those applications must be created using the programming languages, libraries, services and tools supported by the provider [1].

Software as a Service (SaaS) allows the customer to use the provider applications running on a cloud infrastructure as shown in figure 1. User can access those applications from client devices through the web browser (e.g. web based e-mail) or program interfaces [1, 2, 5].

Once the cloud services are created they can be deployed in various ways depending on the organizational structure, security requirements and cloud provider. Those models determine who has an access to the cloud's services. We can distinguish four main types, i.e. public, private, community and hybrid, recommended by NIST.

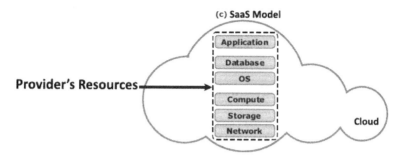

Fig. 1. Software as a Service model [6]

Private cloud provides its infrastructure for exclusive use of a single organization with many consumers. This type of cloud can be owned, managed and operated by the organization, a third party or a combination of them [1, 4]. *Public cloud* provides its infrastructure to anyone from the general public [1]. This type of cloud can be owned, managed and operated by academic, business or government organizations or a combination of them. Provider of such cloud gives public access to its services, which can be either free of charge or billed based on the user of resources.

Community cloud provides its infrastructure for exclusive use of specific community of consumers from some organizations that share for instance a mission, security requirements or policy [1]. *Hybrid cloud*'s infrastructure consists of two or more individual cloud types (private, community or public). They create a mix of internally and externally hosted services [1, 8].

With data stored and processed in many locations dispersed around the world the issue of security is very pressing. As the user of the cloud is entrusting his sensitive data over to the service provider, the risk of losing a customer's trust is very high in case of any security breach. Not all security measures used in dedicated hardware can be mapped well to the cloud environment.

Public Software as a Service cloud model is most commonly offered on the market. It enables the cloud providers to offer their software applications to the customers over the Internet. Such applications are usually globally accessible to everyone, sometimes requiring a subscription fee. This model allows for easy maintenance on the part of the provider, with software stored and managed in central locations. Users in turn can access them anywhere they are, ensuring every user has the same version of the software and enabling easy collaboration with others.

Both the provider and the customers are always responsible (in different proportions) for the security of particular services. Provider has the least control over security in Infrastructure as a Service type of cloud as the customer sets up his own system, middleware and deploys his software there and takes care of its security. In contrast, in the Software as a Service cloud nearly whole responsibility for proper security lies on the cloud provider as the end user deals with a ready-to-use application.

There are various techniques employed by the cloud providers to secure the cloud. As the cloud environment is broad and dynamic there are many aspects that have to be taken into account while designing a proper security. Common practical approaches met in cloud security measures are: data fragmentation [7], distributed file systems

[14], trusted third party authentication [13, 14], secure socket layer [16], encryption of data [4, 15], key management [16], site-to-site VPN [63]. However, the access control is also important, especially from the point of view of data in public SaaS cloud.

3   Access Control in Cloud Services

Access control is a vital part of the cloud computing security. In order to ensure tight security of the cloud services a proper access control policy plays an important role. Its goal is restricting an access to the resources only to the authorized users. It also deals with how and when the resources can be used [4, 8]. Many access control models developed because this area of knowledge is constantly expanding and adapting to the new requirements of changing technologies and new developments. There exist classical models of access control and new solutions created to suit the needs of new applications.

One of such new solution is Usage Role-Based Access Control (URBAC) [10] model developed for security of dynamic applications/information systems. URBAC offers a new approach which creates a tight fit with the cloud services. Due to the URBAC's dynamic nature it is able to address the issue of constant control over the user's permissions, continuously evaluating various aspects to check if the access predicates are still valid.

Usage Role-based Access Control (URBAC) [10] is a fairly new model in access control. The approach is designed for dynamic and complex information systems. Its main concept is based on the best features taken from two already existing models, Role-based Access Control (RBAC) [11] and Usage Control [12], combining them and further extending the model. RBAC is widely known and used due to its concept of Roles, which are an easy and natural way to define the user's permissions and obligations in the system. On the other hand, UCON model expands the concept of the usage permissions, basing them not only on authorizations, but also taking into account user's obligations and conditions of his environment. These concepts form the base of Usage Role-based Access Control. It allows to clearly define the user roles with the functions he is allowed to perform, while also dynamically evaluate the permissions whether he can at a specific moment access a resource (Fig. 2) [10].

Subject can represent users and groups of users that share the same rights as well as obligations and responsibilities. Session is the interval of time during which a user is actively logged into the system and may execute the actions in it that require the appropriate rights. User is logged in to the system in a single session, during which the roles can be activated. A *Role* can be regarded as a reflection of position or job title in an organization that holds with it the authority as well as responsibilities. It allows to accomplish certain tasks connected with processes in an organization. Users are assigned to them based on their competencies and qualifications. Therefore, role is associated with subjects, where user or group of users can take on different roles, but one role can also be shared among users. This association also contains *Subject Attributes*, like identity or credits, which are additional subject properties that can be considered in usage decision.

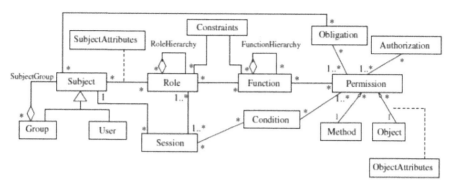

Fig. 2. Elements of Usage Role-based Access Control Model [10]

As each role specifies a possibility to perform specific tasks, it consists of many *functions*, which users may apply. Like with roles, function hierarchy can be defined with inheritance relations between specific functions. Function in turn, can be split to more atomic elements which are operations that are performed on objects. Those are granted by *permissions*. We therefore can view the functions as sets or sequences of permissions, that grant them right to perform the specified methods on a specified objects.

4   URBAC Model for Public Software as a Service Cloud

Software as a Service clouds put the entire issue of proper security on the cloud provider. It is therefore in his best interest to implement access control in a way that tightly controls what the users can access and when, but also a mechanism that can respond to any dynamic changes in the environment. The provider should be able to tailor the access control's mechanism to his own needs and the existing security policies.

Cloud brings with it distribution of data in various locations, with multiple users requesting access to huge amount of information. As the users can access data objects at any time they may modify its contents and attributes constantly. This in turn can cause a change in security policies, which must be re-evaluated dynamically on the spot, since other users may request access to the particular data. Therefore the main aspect of Usage Role-based Access Control model which makes it a good fit for the cloud services is its dynamic aspect.

This gives URBAC a significant advantage over other access control models (like RBAC or DAC) which are purely static and evaluate user's access to the resource only before access. URBAC is responsive to dynamic security policy changes which can occur when subject or object attributes change (can be modified before, during or at the end of access). The model evaluates user's permissions, including obligations, conditions and authorizations both before and during user's access to the service. Thanks to the permissions being based not only on authorizations, but also on obligations reflecting user's responsibilities and conditions checking the environment the user is working in, the model can ensure access control's continuity. Checking the permissions before and during access gives sensitivity to the dynamic changes in subject and object attributes and the system's status.

In the URBAC model a user is assigned a subset of roles:

$$User_i \mapsto \bigcup_i \left(Roles_{u_i}\right) \subseteq Roles \qquad (1)$$

Each role in turn has a subset of functions assigned to it:

$$Roles_j \mapsto \bigcup_j \left(Functions_{r_j}\right) \subseteq Functions \qquad (2)$$

Finally, each function has assigned a subset of permissions:

$$Function_k \mapsto \bigcup_k \left(Permission_{f_k}\right) \subseteq Permissions \qquad (3)$$

We can denote a permission (*Perm*) as consent for the user to perform a method (*m*) on an object (*o*). Permission is limited by a set of constraints, consisting of authorizations (*Auth*), conditions (*Cond*) and obligations (*Oblig*):

$$Perm_i = (m,o) + C_{st} \text{ and } C_{st} = \left(Auth_{Perm_i} \cup Cond_{Perm_i} \cup Oblig_{Perm_i}\right) \quad (4)$$

In turn, authorizations, obligations and conditions can be applied in two ways, before access (pre) or during access (ongoing):

$$Auth_j = \left(preAuth_{Auth_j}\right) \cup \left(ongoingAuth_{Auth_j}\right) \qquad (5)$$

$$Oblig_k = \left(preOblig_{Oblig_k}\right) \cup \left(ongoingOblig_{Oblig_k}\right) \qquad (6)$$

$$Cond_l = \left(preCond_{Cond_l}\right) \cup \left(ongoingCond_{Cond_l}\right) \qquad (7)$$

An example of URBAC authorization can be denoted as:

$$\{user \ is \ owner \ of \ file\} \in preAuth \subset Auth \qquad (8)$$

If a user of cloud's storage service decides that he wants to remove a file, one of the predicates that have to be checked for access decision is an authorization whether the user is the owner of the file. If he is the file owner, it can be removed, otherwise access will be denied. Authorizations are predicates that base on both subject and object attributes. Other examples of authorizations include:

- credits present in user's account to access an object,
- checking user's role,
- does user have any access time left?

Pre-authorizations, checked before an access, are common in all the traditional access control models. As URBAC also contains ongoing-authorizations, predicates such as access time left for a user can be checked constantly during user's access to dynamically deny usage when the time is up.

An example of the URBAC obligation concept can be expressed by:

$$\{agreement \ of \ service \ terms\} \in preOblig \subset Oblig \qquad (9)$$

Most of the cloud services require users to agree to their terms of service so that the provider can inform the user about both parties' responsibilities and obligations towards each other. Therefore, before the service access request can be granted the cloud's access control mechanism can check whether the user has agreed to the service terms. If he has agreed the access will be granted, otherwise it will be denied and the user will be informed that he has to agree to the terms first. Often there are actions required to be performed by the user in the cloud service before he can be granted access, for instance:

- waiting for set amount of time before access or object download,
- agreeing to service or copyright terms,
- watching advertisements before access,
- confirmations through e-mail,
- providing personal information.

Obligations allow to realize all those cases and directly incorporate them into the access control.

An example of URBAC condition is denoted as:

$$\{user\ location\} \in preCond \subset Cond \tag{10}$$

Services such as *Google Apps* or *Office 365* have offers tailored to the needs of the business. For the protection of data that can be accessed in such cloud the organization which uses the service may have a security policy stating that the employees may only access the service from the office building and no other location. URBAC allows to check for such condition and evaluate if an employee is currently located in the office building where he will be granted access or is for instance at his home where access should be denied. The concept of the ongoing conditions allows to continuously check whether the user has not actually moved and changed his location during his access to the service. Other examples of conditions include:

- set access time (e.g. business hours),
- system status in stable state,
- system load.

Conditions are not mutable and they do not depend on the subject or object attributes. They allow to dynamically evaluate user's environment and adjust access decisions to constant changes in the system.

The three aspects of URBAC access control, i.e. authorizations, obligations and conditions being checked both before and during access allows for an immediate response in case permission suddenly changes. If a server is under attack the permissions can change so that no users are given access to any resources. In this example the access can be immediately revoked for all users due to continuous access control.

Another very important aspect of Usage Role-based Access Control is *elasticity*. URBAC model can be tailored to the exact needs of the cloud provider. Because the model is universal it may be adapted to different types of systems and services. Furthermore not all elements that the model defines have to be implemented in the application at once. If at the moment of service creation it does not require any checks of condition elements, they can be omitted in the implementation. An example is a free public service which can be accessed from anywhere at any time and which doesn't depend on any environment factors. Elements that are not deemed necessary at the time of implementation can always be added in later stages as required. Once the model is operating in a service, security policies can be added incrementally. As needs of the service for tighter security may grow, additional policies can be added with time (it is possible to do it in dynamic fashion without service downtime). URBAC may be implemented as a standalone service or embedded into the application's code.

URBAC model takes into account the important aspect of mutability of attributes. Mutability indicates that attributes assigned to subjects (e.g. amount of credits in the account, ownership of an object or role) and objects (e.g. security level) may change constantly, before, during or after access. This allows to implement dynamic permission checking before and during user's access to reflect ongoing changes in a cloud service.

Manageability is also a significant advantage of the URBAC model. Usually access rights have to be pre-defined by developers and granted to subjects. URBAC provides the possibility for an administrator of the service to dynamically add new access permissions (defined them in a CRUD-like administrator panel) during the application's runtime. This way the rules can be instantly used in the system without any necessary downtime or code restructuring. Such dynamically defined policies can be kept in persistent storage like database or XML file. Apart from loading them from storage dynamically into the system, they can also be applied to multiple systems or used as backup. What is more, aggregation of service functionalities in a role allows to change assignments and their respective functions on the fly. Roles also allow to manage the groups of users at once, instead of dealing with function assignment for individuals.

5   Technical Aspects of URBAC Implementation in SaaS Cloud

SaaS cloud can have more than one service offer, for which access control can be implemented per service, resulting in multiple physical URBAC implementations, or one implementation for all services. From more technical point of view, URBAC model can be either implemented as a separate system or be incorporated into the application itself.

Implementing the model as a separate service or system allows to define global policies separating access control from the actual code of the application. Such service can serve as a central point for storing all access control policies and enforcing the rules. The rules may be applied to one or more cloud services. The service may be developed separately from the application or its creation can be entrusted to a third party (e.g. as a Security as a Service cloud). However such solution may prove to be too general and not able to serve some security cases that are specific to an application. As seen in figure 3 application would be able to communicate with the service through an interface allowing to send requests and get response if access can be granted or not.

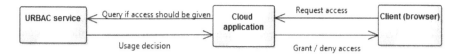

Fig. 3. Usage Role-based Access Control model as a separate service

The second solution is to embed the URBAC model into the service application. It allows tailoring the implementation to the specific needs of an application. Implementing access control into the application allows enforcing policies both on the server and client side. This can prove to be efficient in implementing dynamic aspect of rules enforcement like time limit (can be just checked on the client's side instead of sending constant requests to a separate system). The server side would store the policies and rules, which would be enforced within the application. The main advantage of this approach is ability to serve cases specific to an application. However if not designed properly, it can lead to access control being tangled up with the application's code and a change in implemented policies may be complex and require additional modifications in the code.

6   Conclusions

Data security in public Software as a Service clouds is a significant issue. Since cloud computing is a new way of delivering services and public services store vast amounts of data they can prove to be a popular target for the attackers.

As far as access control is concerned, this paper proposes applying Usage Role-based Access Control (URBAC) model in the context of security of public Software as a Service cloud. Traditional access control models are not well aligned with the needs of cloud's dynamic aspects because they focus mostly on static permissions checked before access is granted. Due to cloud's constantly changing environment and constant access of large amounts of users to huge amounts of data, often requested by multiple people at the same time, the access security has to be more complex. The main advantages that URBAC model brings public SaaS cloud security include:

1. Evaluating permissions at the time of access request but also during usage fits in well with the concept of subscription-based cloud services where user is billed per unit of time he is using a service.
2. Permissions based not only on authorizations but also incorporate concepts of obligations reflecting user's responsibilities and conditions of the environment a user is working in.
3. Dynamic evaluation of access policies takes into account mutability of attributes (attributes can change in time), e.g. user may modify data and at the same moment other user may request access to it (access rights may change due to actions of one user).
4. Utilizes common concept of Role assignment, most natural way to express the user's responsibilities and permissions that can be assigned. Roles aggregate functionalities allowing to manage permissions for groups of users at once.
5. Model ensures continuous control of customer's usage of the cloud service.

URBAC can be implemented in two ways: as a service, acting independently from particular cloud services (as central point for storing global access policies which can be queried by the cloud service) or embedded into the cloud service application's code (tailored to requirements of particular cloud service, controlling access on server and client side).

References

1. Mell, P., Grance, T.: The NIST definition of cloud computing. Special Publication 800-145, National Institute of Standards and Technology (2011)
2. Krutz, R.L., Vines, R.D.: Cloud Security: A Comprehensive Guide to Secure Cloud Computing. Wiley Publishing, Inc. (2010)
3. Hausman, K., Cook, S.L., Sampaio, T.: Cloud Essentials, Sybex by John Wiley & Sons, Inc. (2013)
4. Winkler, J.R.V.: Securing the Cloud. Cloud computer security techniques and tactics. Syngress (2011)
5. Ahson, S., Ilyas, M.: Cloud Computing and Software Services. Theory and Techniques. CRC Press (2011)
6. Gnanasundaram, S., Shrivastava, A.: Information Storage and Management, 2nd edn. EMC Corporation (2012)
7. Yang, K., Jia, X.: Security for Cloud Storage Systems. Springer (2014)
8. Poniszewska-Maranda, A.: Selected aspects of security mechanisms for cloud computing – current solutions and development perspectives. Journal of Theoretical and Applied Computer Science 8(1), 35–49 (2014)
9. Shackleford, D.: Cloud Security and Compliance: A Primer, A SANS Whitepaper (2010)
10. Poniszewska-Maranda, A.: Modeling and design of role engineering in development of access control for dynamic information systems. Bulletin of the Polish Academy of Sciences, Technical Science 61(3), 569–580 (2013)
11. Ferraiolo, D., Sandhu, R.S., Gavrila, S., Kuhn, D.R., Chandramouli, R.: Proposed NIST Role-Based Access control. ACM TISSEC (2001)
12. Zhang, X., Parisi-Presicce, F., Sandhu, R., Park, J.: Formal Model and Policy Specification of Usage Control. ACM TISSEC 8(4), 351–387 (2005)
13. Gharehchopogh, F.S., Bahari, M.: Evaluation of the data security methods in cloud computing environments. International Journal in Foundations of Computer Science & Technology (IJFCST) 3(2), 41–51 (2013)
14. Borthakur, D.: The Hadoop Distributed File System: Architecture and Design. The Apache Software Foundation, pp. 3–4
15. Kaur, S.: Cryptography and Encryption. Cloud Computing. VSRD International Journal of CS & IT 2(3), 242–249 (2012)
16. Anthony, T., Velte, A.T., Velte, T.J., Elsenpeter, R.: Cloud Computing: A Practical Approach. McGraw-Hill (2010)

Advanced Security Assurance Case
Based on ISO/IEC 15408

Oleksandr Potii[1], Oleg Illiashenko[2], and Dmitry Komin[3]

[1] JSC Institute of Information Technology, Chief of Department of Information Security,
12 Bakulina Street, Kharkov, 61023, Ukraine
potav@ua.fm
[2] National Aerospace University n. a. N. E. Zhukovsky "KhAI",
Department of Computer Systems and Networks (503),
17 Chkalova Street, Kharkov, 61070, Ukraine
o.illiashenko@csn.khai.edu
[3] Kharkov Kozhedub Air Force University, Air Force Scientific Centre,
77/79 Sumska Street, Kharkov, 61023, Ukraine
dimakomin@mail.ru

Abstract. Assessment and assurance of conformity with regulation documents assumes significant cost in modern economies. Demonstration of compliance with security standards involves providing evidence that the standards' security criteria are met in full substantiating appropriate decision. Nevertheless despite its importance such type of activity haven't been addressed adequately by the available solutions and the tool support given to conformity assessment and assurance processes is rather poor. International standards do not contain any formal technique for security evaluation, what makes performing evaluation process complicated and one-sided. In the article the approach to the security assurance evaluation Advanced Security Assurance Case (ASAC) is proposed based on refined definition of existed assurance case structure.

Keywords: information security, security assurance, advance security assurance case, ASAC, DRAKON.

1 Introduction

In attempt to define and regulate IT-products evaluation the International standard ISO/IEC 15408 [1,2] has been developed with general model and security criteria. It supposes wide application and is resulted in necessity of mutual recognition results of security evaluation. Requirements of objectivity, repeatability, reproducibility, impartiality and comparability are put forward for the evaluation results of the modern IT products. They can be met in case of supporting scope, depth and rigor of evaluation process only. Another international standard ISO/IEC 18045 [3] contains description of the methodology of security evaluation, but it doesn't contain any formal technique for evaluation, what makes performing of evaluation process complicated and one-sided [4].

© Springer International Publishing Switzerland 2015
W. Zamojski et al. (eds.), *Theory and Engineering of Complex Systems and Dependability,*
Advances in Intelligent Systems and Computing 365, DOI: 10.1007/978-3-319-19216-1_37

Both standards [1-3] are related to IT security techniques. In ISO/IEC 15408 the definition of assurance is: *"Assurance grounds for confidence that an IT product meets its security objectives"*, and in ISO/IEC TR 15443 [5]: *"Assurance provides confidence that the deliverable enforces its security objectives without examining whether the security objectives appropriately address risk and threats"*. In a broader sense assurance reduces the uncertainty associated with vulnerabilities of the IT-product, and thus the potential vulnerability is reduced leading to a reduction in the overall risk.

About to 74% of organizations base their compliance mechanism mostly on manual methods (e.g. text editors, spreadsheets, etc.), indicating lack of existence of satisfactory solutions in 37% of cases [7]. As a starting point the following statements are used [8]:

- Assurance is a measure of confidence in the accuracy of a risk or security measurement;
- Assurance is a major factor in security decisions;
- Assurance arguments are a powerful tool to reduce the uncertainty in risk or security assessments.

The goal of current paper is to present that the case as a concept should not be limited as a means of a tool for formalization of requirements, but its definition should be enhanced by decision making procedure of conformity with requirements. It'll tend to reducing of uncertainty and increasing of objectivity, repeatability, reproducibility, impartiality and comparability of assessment in order to satisfy needs of all groups of ISO/IEC 15408 in evaluation of the security properties of TOEs: consumers, developers, evaluators, others [1,2]. We propose enhancement of the assurance case concept.

2   Problem Statement. Why Do We Need Changes?

2.1   Actuality

International standard ISO/IEC 15408 [1,2] contains two concepts – "confidence" and "assurance", which are closely interrelated. Measure of confidence is assurance level. Assurance level is a set of assurance requirements. Their implementation is characterized by correctness of the functional requirements realization, IT-product abilities to resist the security threats and provide achievement of the required dependability level in the system. Assurance is a ground for confidence that the IT-product meets its security objectives. Assurance requirements are put forward to the target of evaluation (TOE). TOE is the set of software, firmware and/or hardware possibly accompanied by guidance.

To define the degree of assurance requirements the implementation of the evaluation process is carried out (see figure 1). They are based on the evaluation program and evaluation methodology within the scope of the TOE certification and according to the evaluation criteria.

Evaluation program is a documental set of assurance requirements, which are checked during the TOE evaluation process. Evaluation methodology means established methods of the assurance requirements evaluation. Certification is a procedure by which the assurance level is approved, which may be performed by independent experts. In the evaluation process the next parties are involved: evaluator, validator, owner and developer. Evaluator is an individual person with an appropriate competency to carry out the security evaluation. There are several stages within a TOE lifecycle; there should be several evaluators for each stage. Validator is an organization which prepares a validation report.

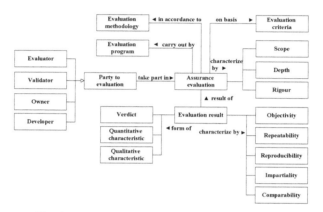

Fig. 1. Ontological model of the assurance evaluation

Evaluation criteria are formal or informal rules for making decision in relation to assurance requirements implementation. The output of the evaluation process is an assurance result. It is a documented quantitative or qualitative characteristic of the TOE. The advanced requirements to evaluation results foresee *objectivity, repeatability, reproducibility, impartiality* and *comparability*.

To assess properly security of IT product it is needed to have a unify methodology for both parties. This methodology should be based on international standards [1-3,5] and should contain requirements how to prove that the decision of conformity to the standard is solely correct. Such kind of methodology is building of cases for justification of correctness and implemented requirements. A significant reliance on people during security assurance activities and further application of assurance procedure (by the target audience of ISO/IEC 15408) is a danger that particular people, the company "experts", become a bottleneck on any project or solution. Without strict and rigorous algorithmic process of each conformity decision that they made, people become a critical resource in an organization. The loss of knowledge, lack of consistency, proficiency, maturity, inappropriate use of artefacts, lack of traceability can lead to problems in maintaining the case [6]. There is a great need for providing clear and comprehensible argumentation process which addresses the importance of demonstrating and justifying how evidence fulfills the safety requirements [7,8].

2.2   Existed Types of Cases

Most of known cases contain claim- and evidence-based justification that a system under assessment meet or should meet the specified objectives (e.g. safety, security, dependability etc.) in a particular field. The body of the case contains lots of internal interdependencies, so it's difficult to present the case as a simple document. Nowadays there exist several evaluation approaches which based on case assessment. The types of cases and their structure, notation which is used for building of cases of different types, differ, but logic of construction is comparable between each other. Among them are: *safety case, assurance case, trust case, security-informed safety case*, etc. In this section a brief description of known types of cases is provided. The goal of this article is not in rigorous analysis of existed cases; the information on them is given for reference mainly. One of the key stages of cases construction is obtaining of evidences for requirements. An exhaustive example of systematic review on existing techniques for structuring and assessment of evidences is given in [9].

Safety Case. The definition of a safety case was initially proposed by [10]. Today this definition changed to the following: "document body of evidence that provides a demonstrable and valid argument that a system is adequately safe for a given application and environment over its lifetime". Nowadays this definition is used by Adelard LLP in ASCE (Assurance and Safety Case Environment) tool. The idea of safety case (behind the development of argument structures) is to make expert judgment explicit to redirect the dependence on judgment to issues on which the judgment would be trusted [11]. In current safety case practice one of the most used approaches is proposed by Toulmin [12]. To demonstrate graphically claim structure the following notations are common: Goal Structuring Notation (GSN) [13] or Claim-Arguments-Evidence (CAE) [14,15]. Safety cases are commonly used in different critical domains: nuclear [16,17], aerospace [18,19].

Security-Informed Safety Case. This type of case is inherited from the previous one, but taking into account security considerations which can have a significant impact on a safety case, when already built safety case is reviewed from the point out of security influence to the particular claim. Initial safety case is further enhanced detailed by additional clarifying claims [20]. For safety case and safety-informed security cases the ASCE tool is provided by Adelard LLP [20, 21], but its capabilities are not enough in the sense that it doesn't allow to work with ISO/IEC 15408 in a full manner [22]. Another approach of security-informed safety assessment for safety-critical systems is described in [23].

Trust case, Assurance Case. This is a documented body of evidence that provides a convincing and valid argument that a specified set of critical claims regarding a system's properties are adequately justified for a given application in a given environment [24]. The structures of arguments here are used to demonstrate properties different from safety. The problems with such type of cases lies in: building, reviewing, maintaining, reusing due to explosion of number of the documentation presented to an expert, little structuring support, manual or based on private experience rules of evidence construction (weak guidance on review of arguments and evidence) [22]. Such type of cases is also used in critical domains, e.g. medicine industry [25].

No one type of cases listed do not contain a technique for justification that decision of an expert is solely correct and couldn't be treated in another way by any other person. Uncertainty of such kind could lead to loss of money and time. Both industry and academia need a holistic methodology for building and using the security assurance cases in a universal way. The evidence of conformity should correspond to the constructed claim in order to tie the solution with the judgment. In modern cases construction of evidence for claims (during mapping the evidence to claims) is mostly rely upon an experience of a particular person, but the logic of this construction is absent. This fact significantly increases influence of a final result, case, from the person, who is involved in process of evaluation. The decision, which is based on empirical experience of evaluator or expert, potentially increases an uncertainty (characteristics, history, observations, measurements, evaluative results, analyses, inferences) of evaluation process and the final result (which could be potentially inadequate), that may lead to hazardous consequences in the end of the day.

3   Development of Advanced Security Assurance Case

The initial assurance case concept from [8] tends to be universal enough to embrace proposed security case assurance technique. According to [15] *Assurance Argument* is *a set of structured assurance claims, supported by evidence and reasoning, that demonstrate clearly how assurance needs have been satisfied*. The process of building the security assurance cases in new way should is as follows:

1. Development of hierarchical tree of properties / attributes of a particular requirement and constriction / determination of claim;
2. To be in compliance with ISO/IEC 15408 standard, evaluation evidence should include:
   (a) the target of evaluation (TOE), which include applying of the security assurance requirements (SARs, from security target)
   (b) security target (ST)
   (c) input from the development environment (such as design documents or developer test results) – usually.

We propose to enhance the concept of assurance case, proposed in [8], which says that case consists of four elements (see table 1) with the fifth one, *a decision-making technique*, which allows decreasing uncertainty of assessment.

Table 1. Assurance case components with added decision-making technique

Argument Element	Description
Claims	Statements that something has a particular property
Evidence	Empirical data on which a judgment can be based
Reasoning	Statements which tie evidence together to establish claim
Assumption Zone	Limit of an argument where claims are accepted without evidence
Decision making technique	**The basis for decision of conformity**

We propose to name this type of assurance case – *Advanced Security Assurance Case* (ASAC). Figure 2 depicts four main stages of ASAC building. As an example of ASAC building we provide requirements for assurance and evaluation of vulnerabilities AVA_VAN.1 from vulnerability class AVA from ISO/IEC 15408.

During the *Stage 1* the ontological analysis and modeling of the domain evaluation are carried out. Analysis includes the research of the assurance requirements set ($R=\{r_1, r_2,...,r_i\}$, $i = \overline{1, N}$) advanced to the TOE, and detection the assurance properties set ($P=\{p_1, p_2,...,p_j\}$, $j = \overline{1, L}$) the TOE must possess. The assurance *properties* set P defines dependences and relations among properties (see table 2).

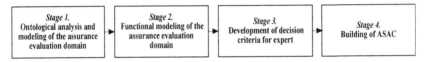

Fig. 2. Stages of Advanced Security Assurance Case building

Table 2. Assurance properties for top-level requirement

$P_i^{(j)}$	Property
P_0	Resistant of the TOE to attacks performed by an attacker possessing Basic attack potential
P_1^0	Readiness of the TOE for testing
$P_1^{0,1}$	Consistent of the TOE with ST
$P_1^{0,1,1}$	Conformity of the TOE reference with the CM capabilities (ALC_CMC) sub-activities and ST introduction
$P_2^{0,1,1}$	Consistent of the all TOE configurations with ST
$P_3^{0,1,1}$	Conformity of the testing environment to the security objectives for the operational environment described in the ST
$P_2^{0,1}$	Accuracy of the TOE installing
$P_1^{0,1,2}$	Successfulness completion of the AGD_PRE.1
$P_2^{0,1,2}$	Successfulness of the TOE install and start up, using the supplied guidance only

Analysis results are shown in form of ontological graphs $G^R = \langle R, Q_R \rangle$ and $G^P = \langle P, Q_P \rangle$, that exactly and unambiguously (in accepted notation) describe the domain (notably the main concepts and relations among them). Complete coverage of domain modeling is ensured by ontological graphs of two types: object-oriented and process oriented. To the next the hierarchical graph of the evidences set ($G^E = \langle E, Q_E \rangle$) is constructed. Evidences are getting from the TOE decomposition. For each elementary property $p_i \in P$ the set of *evidences* $Ep_i=\{e_1, e_2, ... , e_i\}$, $i = \overline{1, N}$ is defined (see table 3).

Table 3. Evidences for defined properties of AVA_VAN.1

$E_i^{(j)}$	Evidence
E_1^0	TOE is suitable for testing
E_2^0	Security Target
E_3^0	Guidance documentation
E_4^0	Information is publicly available to support the identification of potential vulnerabilities
E_5^0	Current information regarding potential vulnerabilities (e.g. from an evaluation authority)

We considered lower abstraction levels (more specific evidences) to be more useful, that is confirmed by [9,26]. Dependences between graphs G^P and G^E are shown in the form of relations kind of "property - evidence" $D[P \leftrightarrow E]$. Evaluation assurance actions ontological graph ($G^A = <A, Q_A>,$) is constructed. Dependences set ($D[A \leftrightarrow P]$) between ontological graphs of evaluation *actions* (G^A) and properties (G^P) are defined (see table 4). The resulting ontological model of *Stage 1* is shown on Figure 4.

Table 4. Actions for assessment of identified assurance properties for AVA_VAN.1

$A_i^{(j)}$	Action
A_1^{0}	Obtain the evidences
A_2^{0}	Check the conformity of the TOE reference with the CM capabilities (ALC_CMC) sub-activities and ST introduction
A_3^{0}	Check the consistent of the all TOE configurations with ST
A_4^{0}	Check the conformity of the testing environment to the security objectives for the operational environment described in the ST

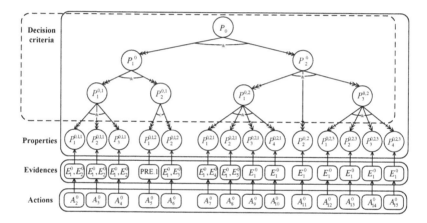

Fig. 3. Ontological model in graphical form

In the *Stage 2* the functional modeling of the assurance evaluation process is implemented. Functional modeling goal is the formalized presentation of the evaluation process.

During *Stage 3* the rules for conformity of assurances complex properties are developed. For the development of decision criteria for the assessment results of simple properties it is encouraged to use the mathematical apparatus of subjective logic [27], which allows moving away from traditional binary logic with its "Yes" or "No" and more fully express judgments of an expert on the assessment of the assurance properties. Operators of subjective logic allow formulating a final judgment on the assessment of the assurance requirements by sequentially combining judgments about individual assurance properties. This mathematical apparatus allows combining of several judgments of experts on the evaluation of the same properties, which allows comparison of the results of several independent examinations. In general, subjective logic is suitable for modeling and analyzing situations involving uncertainty, incomplete knowledge and different world views. In order to standardize and automate the creation of a report on the assessment of assurance requirements of each judgment for verification of assurance properties templates optional judgments are designed.

Stage *4* finalizes building of ASAC. During this stage the algorithm of validator for expertise is developed. We propose to use DRAKON, which advantages are clarity and simplicity of algorithms representation and fully formal representation of visual rules of structured programming. DRAKON is an algorithmic visual programming language developed within the Buran space project [28,29]. Firstly, there were attempts to use IDEF3 standard to represent such kind of activities, but the tool proved to be unsuitable in terms of scaling. The fragment of such algorithm of ASAC case in DRAKON for AVA_VAN.1 requirement is represented on figure 4. The corresponding actions of expert are also indicated.

The main peculiarities of ASAC are:

- Analysis of verbal requirements in ISO/IEC 15408 written in natural language form;
- Identification of properties that the object should possess. Verbal natural language requirements should be decomposed to the level of elementary properties which forms a tree of properties;
- Linking of properties with corresponding evidences (both quantitative and qualitative) from the set of evidences (in other types of case-based techniques there is no such technique);
- After the linking of properties with evidences the corresponding actions for the proof of evidences are defined;
- If the particular evidence is satisfied, the tree is aggregated and it is concluded that the corresponding property is satisfied;
- To prove that evidences of related properties correspond to particular requirement the actions are defined, it means that there is an algorithm of conformity decision making, and so the new concept of case is proposed – *an algorithmic concept, which is the core of Advanced Security Assurance Case.*

So, the final ASAC in DRAKON language representation will contain requirements for assessment (what should be assessed), algorithm of their validation (how and in which sequence) and result of assurance assessment. This fact will greatly reduce the workload of the expert, allowing him to start work immediately, without firstly developing a methodology for assessment.

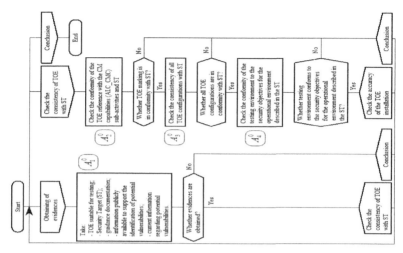

Fig. 4. Advanced security assurance case ASAC in DRAKON notation

In this article we propose a new concept of assurance as a notion. It is shown on figure 5. If the decision maker in future decides to reduce this uncertainty, we propose a structure for assurance arguments as a logical way to communicate the information used in making security decisions.

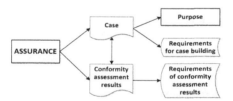

Fig. 5. Refinement of assurance definition concept

An assurance argument starts with claims about risks and then gathers all the evidence and supporting arguments into a logical hierarchical structure. The goal is that these arguments are capable of reuse in a wide variety of applications, easing the burden of security evaluations.

4   Conclusions and Future Work

The enhanced structure of security assurance case ASAC was proposed. It is characterized by introduction of technique of decision making, which is easy to scale, modify, and in compliance with ISO/IEC 15408 and with requirements to the assessment results. The future work is concerned to improve of proposed security assurance case formalization technique ASAC. We also plan to develop tool support for our methodology. In particular, we wish to explore how: to build ASACs more effectively and provide rigorous decision-making procedure for building ASACs using subjective logic. We have starting developing of ASACs in DRAKON language for each class from ISO/IEC 15408 (protection profile evaluation, security target evaluation, development, guidance documents, life-cycle support, tests, vulnerability assessment, and composition) to obtain full coverage and reduce time and cost of expertise.

References

1. ISO/IEC 15408-1:2009, Informational technology – Security techniques – Evaluation criteria for IT security, Part 1: Introduction and general model (2009)
2. ISO/IEC 15408-3:2008, Informational technology – Security techniques – Evaluation criteria for IT security, Part 3: Security assurance requirement (2008)
3. ISO/IEC 18045:2008, Informational technology – Security techniques – Methodology for IT security evaluation (2008)
4. Potii, O., Komin, D., Rebriy, I.: Method of Assurance Requirements Evaluation. In: Kharchenko, V., Tagarev, T. (eds.) Kharkiv, National Aerospace University n. a. N. E. Zhukovsky "KhAI", vol. 1, pp. 123–132 (2011)
5. ISO/IEC TR 15443-1:2012, Information technology – Security techniques – Security assurance framework – Part 1: Introduction and concepts (2012)
6. Kelly, T., McDermid, T.: Safety Case Construction and Reuse Using Patterns. In: Daniel, T. (ed.) Proceedings of the 16th International Conference on Computer Safety, Reliability and Security (SAFECOMP 1997), pp. 55–69. Springer, London (1997)
7. Cyra, L., Gorski, J.: SCF - A Framework Supporting Achieving and Assessing Conformity with Standards. Special Issue: Secure Semantic Web 33(1), 80–95 (2011)
8. Williams, J.R., George, F.J.: A Framework for Reasoning about Assurance, Document Number ATR 97043. Arca Systems, Inc. (April 23, 1998)
9. Nair, S., de la Vara, J.L., Sabetzadeh, M., Briand, L.: An Extended Systematic Literature Review on Provision of Evidence for Safety Certification. Information and Software Technology 56, 689–717 (2014)
10. Bishop, P., Bloomfield, R.: The SHIP Safety Case. In: Rabe, G. (ed.) The proceedings of the 14th Conference on Computer Safety, Reliability and Security, SafeComp 1995, Belgirate, Italy, pp. 437–451. Springer (1995)
11. Strigini, L.: Formalism and Judgement in Assurance Cases. In: DSN 2004 Workshop on Assurance Cases: Best Practices, Possible Obstacles, and Future Opportunities, Florence, Italy (2004)
12. Bloomfield, R.E., Wetherilt, A.: Computer Trading and Systemic Risk: a Nuclear Perspective. Foresight study, The Future of Computer Trading in Financial Markets, Driver Review DR26. Government Office for Science (2012)
13. Kelly, T., Weaver, R.: The Goal Structuring Notation – A Safety Argument Notation. In: Workshop on Assurance Cases, 2004 International Conference on Dependable Systems and Networks, Florence (2004)
14. Bishop, P.G., Bloomfield, R.E.: A Methodology for Safety Case Development. In: Redmill, F., Anderson, T. (eds.) Industrial Perspectives of Safety-critical Systems: Proceedings of the Sixth Safety-Critical Systems Symposium, Birmingham, pp. 194–203. Springer, London (1998)
15. ISO/IEC 15026-2:2011. Systems and software engineering — Systems and software assurance, Part 2: Assurance case (2011)

16. NPP Safety Automation Systems Analysis. State of the Art, VTT, http://www.vtt.fi/files/projects/mallintarkastus/npp_safety_au tomation_systems_analysis_state_of_the_art.pdf (access date: January 2015)

17. The Purpose, Scope, and Content of Safety Cases, ONR Nuclear Safety Technical Assessment Guide, http://www.onr.org.uk/operational/tech_asst_guides/ns-tast-gd-051.pdf (access date: January 2015)

18. Safety Case Development Manual, European Organization For The Safety of Air Navigation, http://www.eurocontrol.int/sites/default/files/article/content /documents/nm/link2000/safety-case-development-manual-v2.2-ri-13nov06.pdf (access date: January 2015)

19. Building a Preliminary Safety Case: An Example from Aerospace, http://www-users.cs.york.ac.uk/tpk/preliminary.pdf (access date: January 2015)

20. Netkachova, K., Bloomfield, R.E., Stroud, R.J.: Security-informed safety cases. In: Specification and Safety and Security Analysis and Assessment Techniques. D3.1, SESAMO project, http://sesamo-project.eu (access date: January 2015)

21. Scott, A.T., Krombolz, A.H.: Structured Assurance Cases: Three Common Standards. In: 9th IEEE International Symposium on High-Assurance Systems Engineering, http://www.acq.osd.mil/se/webinars/2010-01-19-SECIE-Structured-Assurance-Ankrum-Kromholz-brief.pdf (access date: January 2015)

22. Adelard Safety Case Development Manual, http://www.adelard.com/resources/ascad/ascad_download.html (access date: January 2015)

23. Kharchenko, V., Illiashenko, O., Kovalenko, A., Sklyar, V., Boyarchuk, A.: Security Informed Safety Assessment of NPP I&C Systems: GAP-IMECA Technique. In: 22nd International Conference on Nuclear Engineering, ICONE 22, Prague, Czech Republic. Next Generation Reactors and Advanced Reactors; Nuclear Safety and Security, vol. 3, p. V003T06A054 (2014)

24. A Method of Trust Case Templates to Support Standards Conformity Achievement and Assessment, http://citeseerx.ist.psu.edu/viewdoc/download?doi= 10.1.1.163.906&rep=rep1&type=pdf (access date: January 2015)

25. Towards an Assurance Case Practice for Medical Devices, Carnegie Mellon University, http://www.sei.cmu.edu/reports/09tn018.pdf (access date: January 2015)

26. Linling, S., Kelly, T.: Safety arguments in aircraft certification. In: 4th IET International Conference on Systems Safety 2009. Incorporating the SaRS Annual Conference, London, pp. 1–6 (2009)

27. Jøsang, A.: Subjective logic. University of Oslo, http://folk.uio.no/josang/papers/subjective_logic.pdf (access date: January 2015)

28. Parondzhanov, V.: How to improve the work of your mind. Algorithms without programmers – it's very simple! Delo. Moscow (2001)

29. DRAKON official website, http://drakon-editor.sourceforge.net/ (access date: January 2015)

Application of Data Encryption for Building Modern Virtual Private Networks

Marcin Pólkowski[1], Dariusz Laskowski[2], and Piotr Łubkowski[2]

[1] Transbit Sp. z o.o., ul. Łukasza Drewny 80, 02-968 Warszawa
polkowski.marcin@gmail.com
[2] Institute of Telecommunications, Faculty of Electronics, Military University of Technology
Gen. S. Kaliskiego 2 Street, 00-908 Warsaw
{dlaskowski,plubkowski}@wat.edu.pl

Abstract. The aim of this article is to present drawbacks of the most popular version/ application of virtual private network service based on Internet Protocol Security (IPsec) as well as to describe the most interesting alternatives used to develop modern business services. Firstly, the article presents history of virtual private network (VPN) and focuses particularly on Secure VPN, where data are encrypted. Secondly, it discusses various aspects of using IPsec VPN while requirements of enterprises are constantly rising. Thirdly, it indicates examples of DMVPN (Dynamic Multipoint VPN) and GETVPN (Group Encrypted Transport VPN) used for implementing private services in hub-and-spoke and full-mesh architecture. DMVPN is generally recommended for usage over public networks, where it creates a VPN and secures it. DMVPN is a very good security improvement for MPLS VPN. GETVPN is favorable to secure existing VPN over private networks. Contrary to DMVPN, it uses distinct polices and multiple overlays, which give limited interoperability.

Keywords: VPN, encryption, security, network.

1 Introduction

Since inception of Internet network one started to consider a concept of private network for business use. First VPN services in the form of virtual circuits, delivered by using protocols such as X.25 and Frame Relays, were created in response to requirements and needs of enterprises. The sudden development of ATM (Asynchronous Transfer Mode) technology accelerated growth of this type of service.

The real breakthrough was possible thanks to using set of protocols encrypting IPsec data to build virtual private networks. Applying it in a tunnel mode allowed a compilation of secure tunnels, in which both data and addresses of a sender and a recipient are protected against the unauthorized access (Fig. 1). The modes of secure data transfer have been tested in several scenarios reflecting the transfer of sensitive data in the industrial sector [1÷4].

© Springer International Publishing Switzerland 2015
W. Zamojski et al. (eds.), *Theory and Engineering of Complex Systems and Dependability,*
Advances in Intelligent Systems and Computing 365, DOI: 10.1007/978-3-319-19216-1_38

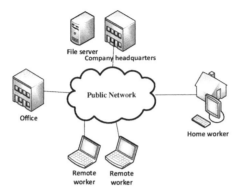

Fig. 1. General architecture of virtual private network

2   IPsec VPN

Set of IPsec protocols had been developed by various institutions over the course of several years. Its main task was to protect data transmitted via IP protocol. For this purpose various cryptographic protocols are used, such as 3DES (Triple Data Encryption Standard) for ciphering data or SHA (Secure Hash Algorithm) for hashing. IPsec contains of three mains protocols:

- AH (Authentication Header) provides authentication and integrity of data transmitted in IP packets.
- ESP (Encapsulating Security Payloads) offers AH features with additional assuring confidentiality of transmitted data and anti - replay service.
- SA (Security Associations) contains algorithms and rules for establishing secure connections between users. It uses protocol ISAKMP (Internet Security Associations and Key Management Protocol) for authentication and key exchange.

IPsec has two modes of operation: transport and tunnel (Fig. 2). As it was mentioned before, for implementing virtual private networks the tunnel mode is used. It consists in connecting two gateways by an encrypted tunnel. In this mode, the whole packet is protected, including the headline.

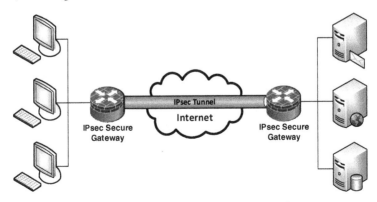

Fig. 2. General architecture of IPsec VPN

The first virtual private networks consisted in connecting business units with its main seat, which all data were stored in. The new difficulty, concerning dynamic exchange between these network locations, has arisen as a result of globalization and lease of resources from other companies. It turned out that static IPsec tunnels were not able to meet the growing requirements of users. Despite the continuously growing throughput of available links, the issue of data excess, linked to additional headlines, has remained substantial. The increasing number of tunnels reduces network's efficiency.

3   DMVPN and GET VPN

In full meshed VPN tunnels are established directly between all devices. This is a good solution for a small number of gateways. Contrary, hub and spoke is character-ized by a central device, which brokers in establishing tunnels between other sites. It is not efficient to configure and establish a lot of tunnels between all devices.

The fundamental problem in virtual private networks services using IPsec concerns lack of possibility to apply dynamic routing protocols. It is driven by the fact that IPsec is not able to serve broadcast and multicast traffic. To solve this problem one can use GRE (Generic Routing Encapsulation) combined with IPsec tunnels. Usage of GRE protocol to build tunnels requires static IP addressing for end points and routing a protocol between them. IPsec employs access list (ACL) for indicating a source of packet addresses that have been chosen to be encrypted. This situation may result in frequently changing configuration of ACL. In case of large VPN of hub and spoke type, the size of central hub's configuration file grows significantly. For example, if one VPN serves 300 branches of spoke type, the hub type router has to include about 3900 lines of configuration. DMVPN technology bases on hub and spoke architecture, where central points are of hub type (e.g. seat of the company) and the other are spokes (e.g. remote branches). In basic configuration of DMVPN (Fig. 3), communi-cation between spokes is done indirectly in a central hub, whereas in extended case, spoke to spoke, branches of company are able to connect themselves mutually through direct and dynamic tunnels. [5]

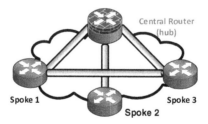

Fig. 3. DMVPN architecture

DMVPN enables to build dynamic and scalable VPN networks by combining ad-vantages of Multipoint-GRE IPsec and NHRP (Next Hop Resolution Protocol). This technology allows connecting remote branches of one company over public networks by encrypted tunnels. In case of connecting remote hosts, establishing constant links is not required contrary to full mesh topology. The NHRP protocol provides possibility to dynamically source information for hub device regarding real addresses of interfaces of

VPN's spokes. It means that all hub routers are enabled to get required information in order to connect two branches by an IPsec tunnel. This is possible due to interoperability between dynamic routing protocols and IP addresses translations. To provide high level of reliability, typical hub and spoke topology can be extended by additional hub routers or DMVPN networks, which are based on other providers' infrastructure.

Current network applications, based on the transmission of voice and video, accelerate the need to design wide WAN networks, which enable direct communication between branches of a company and ensure relevant Quality of Service (QoS). Nowadays, the companies expect from Wide Area Networks (WAN) a compromise between granting adequate quality of data transfer and ensuring security of transmitted information. The security requirements are sometimes even imposed by top-down regulations e.g. in banks and public institutions. As the requirements are growing, Cisco GET (Group Encrypted Transport) VPN eliminates necessity to set aforementioned compromise. [7,8]

With the introduction of GET technology, Cisco offers a new category of VPN network – tunnel-less VPN, where establishing point-to-points tunnels between each branch of the enterprise is no longer required. GET VPN proposes a new standard of a security model, based on concept of 'trusted' members of a defined group. By using the method of trusted groups, we get the opportunity to design scalable and secure networks, while simultaneously maintaining its intelligent properties (i.e. QoS, routing and multicast), which are essential for quality of voice and video calls.

GET VPN enables secure data transfer in various WAN environments (IP, MPLS) without the necessity to create point-to-point (P2P) tunnels. As a result, it minimizes delays in voice and video transmission. GET facilitates securing large L2 and MPLS (Multiprotocol Label Switching) networks, which require partial or full-mesh network of connections between edge devices of particular branches of the company. The MPLS VNP networks using this type of encryption are highly scalable, easily managed and meet the imposed by the government requirements on encryption.

The flexible nature of GET technology allows security conscious companies both to manage security policies of WAN network, as well as to pass the encryption services to their Service Providers (SP). Key properties of GET VPN (Fig. 4) include:

- GDOI (Group Domain of Interpretation) – a protocol managing keys, responsible for the establishment of a common security policy (IPsec SA) between routers being the members of the same 'trusted' group.
- Central Key Services – a router responsible for: sending security keys and their cyclical restoring, as well as distributing security policies among the routers which are members of the same trusted group.
- IP Header Preservation – preserve the original IP headline outside IPsec packet.
- Redundancy of Cooperative Key Server – ability to implement up to eight key servers within a single domain. The main router's database, containing keys and security policies and synchronized with the other spare key servers.
- Support for 'anty-replay' type of properties – the functionality protects the network again attacks of 'man-in-the-middle' type.
- Support for encryption - the possibility of using encryption algorithms such as DES (Data Encryption Standard), 3DES (Triple DES) and AES (Advanced Encryption Standard).

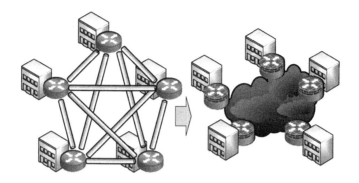

Fig. 4. Transition from full meshed VPN to GETVPN

Application of this type of solution gives the opportunity to build a scalable architecture of a full mesh type with private addressing maintained. Basing GETVPN on core MPLS networks which use Traffic Engineering (TE) and offers a classification of services (for QoS performance), we get links with low delays and jitter.

4   Test Bed Implementation and Results

The developed methodology of the research consisted of two parts: analytical (checking configuration options with accuracy and simulation applications) and validation focusing on local studies on physical devices existing hardware and software test bed platform. DMVPN and GET VPN service with MPLS has been implemented on the test network consisting of five routers forming the backbone and four client routers (Fig. 5). Desirable properties of a test bed are achieved by providing:

- efficient data routing with configuration of routing protocol metrics OSPF (Open Shortest Path First), with the function of switching MPLS label on routers PX (P1, P2, P3, Provider) and PEY (PE1, PE2, Provider Edge);
- appropriate confidentiality of encrypted IPsec tunnels established between routers CE1A-CE1B and CE2A-CE2B (CE Customer Edge).

Identification of the test environment has been presented in detail in publications. Such a configuration allowed the separation of traffic packets between CE routers attached to a PE router and additional st-digit data sent over the backbone network. PE routers were in reality virtual routers configured with two routers using classical protocols and VRRP (Virtual Router Redundancy Protocol) or HSRP (Hot Standby Router Protocol) – thanks to this widening range of options with regard to the implementation of tests was achieved.

Effective mechanisms VRRP and HSRP are redundancy protocols of transport resources and support reliability functionalities of networks. They allow you to share the same IP version 4 or 6 address by several common routers identified as the default gateway (DG) virtual network. Then that one of the routers is the primary (master), and the rest act as a backup (standby). Using one of these protocols ensures the continuity of the provision of services in the event of failure of one of the devices.

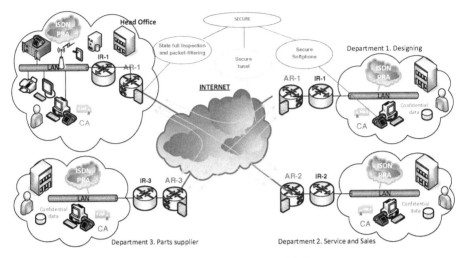

Fig. 5. Diagram of test stand [1]

The configuration process MPLS VPN technology included basic stages, i.e.:

- selection of devices and wired configuration Gigabit-Ethernet interfaces and protocols VRRP or HSSR (depending on the scenario) and OSPF,
- configuration of MPLS technology and label distribution protocol LDP (Label Distribution Protocol) with authentication using MD5 (Message Digest 5),
- create a VRF tables PE routers, the establishment of MP-BGP session between PE routers and the statement IPsec CE routers,
- launch traffic generators, analyzers and monitors the event of non-suitability.

From the point of view of security (confidentiality) and reliability (undamaged) the critical aspects of the configuration is accurate implementation of all specific conditions specific to MPLS technology and IPSEC VPN and backup routers. The many aspects of research targeted to validate the functionality of select characteristics of DMVPN and GET VPN service via the multiplicity and diversity of the tests within defined scenarios. Group of scenarios were to examine the confidentiality and undamaged backbone network using IP packet labeling variants of networking relationships: router having authorized the release of data for each class and unauthorized router "pretending to be" one of the backbone routers. Determination of resistance to unauthorized backbone network consisted of placing the connection between the two routers eligible of a valid router (unknown *id* and *password* for authentication protocol routers adjacent LDP). The study looked for reliable data as a basis for building information about the characteristics, i.e.: reaction times up routers configured with protocols HSRP and VRRP for packet generating traffic with different bit rates and different sizes depending on the bit rate data that is sent to depend on the stateness of individual resources. For example - the study of router reaction time of the main router failure after-divided into four scenarios: two HSRP protocols with the values provided for 1...1000Mbps and the corresponding two VRRP protocols. After-measure reaction time was held with the help of the traffic analyzer Wireshark.

Counted from the time it disconnects the network reconfiguration and obtain data flow. For each scenario, 30 replicates were performed for the average results. To calculate the average value of implementing a median because of its resistance to the disorders associated with significantly projecting the results of the test. As a result of the measurements were as follows average values (Table 1, Fig. 6., Fig. 7.)

Table 1. The resulting response times of substitute routers

Test	Bandwidth	DMVPN		GET VPN	
		HSRP	VRRP	HSRP	VRRP
	[Mbps]	[s]	[s]	[s]	[s]
1	1	2,81	1,785	2,66	1,635
2	10	3,783	2,597	3,493	2,147
3	100	4,473	3,261	4,203	2,971
4	1000	5,289	4,095	4,719	3,435

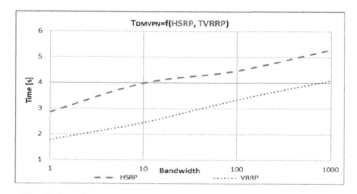

Fig. 6. Graphical representation of the situation results for DMVPN

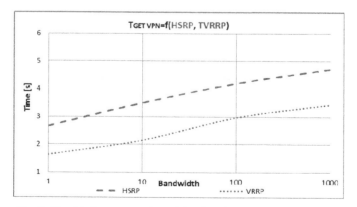

Fig. 7. Graphical representation of the situation results for GET VPN

GET VPN applications of technology resulting in a gain in the form of shorter time than for DMVPN (Table 2).

Table 2. The resulting response times of substitute routers

Test	Bandwidth	Profit GET VPN	
		HSRP	VRRP
	[Mbps]	[s]	[s]
1	1	0,2	0,15
2	10	0,49	0,3
3	100	0,27	0,38
4	1000	0,57	0,66

A recent study was to check out dependence on changes in the value of the data-flow, not assigned to any policy of qualitative QoS, from the intensity of IPsec-encrypted data priority stream. Unencrypted data were generated at a constant rate of 40% of the capacity of a single link. Traffic prepared to encrypt grew until getting 95% of the bandwidth. As you can see in the table (Table 3), shows the same situation in the network using engineering movement TE (Traffic Engineering).

Table 3. The resulting response bit rate streams for Fast Ethernet

Test	Time	Traffic without TE		Traffic with TE	
		Open	Secret	Open	Secret
	[s]	[%]			
1	1	40	35	52	49
2	20	87	83	91	89
3	40	89	86	94	91
4	60	86	81	93	90
5	80	87	86	94	92
6	100	91	87	95	92

The use of TE allowed to "fast" apparent motion switch to a different path in the network with lower occupancy of the line.

5   Summary

The continuous growth of IT industry has imposed researches about more and more modern solutions for implementation of virtual private network services. Migration from a traditional telephony to Voice over Internet Protocol (VoIP) and atomization of companies' structures has led to the situation in which standard virtual private networks based on IPsec cannot fully satisfy requirements of companies.

In response to that problem the new technologies have been developed. The aim of this article was to analyze and compare the two major technologies of VPN: DMVPN and GET VPN with MPLS. Both, DMVPN and GETVPN, have specific advantages. Backbone network with MPLS service implemented through a set of enhanced IPsec allows you to transfer confidential important data from each other. This technique helps to maintain the confidentiality of non-confidential and sensitive network resources, both local and wide area networks. It also does not affect the reliability of degrading. However, sometimes you may need to increase the level of safety, e.g. for sensitive data - there is a possibility of introduction of additional items such as protocols from FHRP group (First Hop Redundancy Protocol), which include HSRP and VRRP. Implementation of one of them, allows redundancy node / s (router / s) backbone network, which makes quicker replacement of unsuitable items by fit router waiting in the standby mode. The yield of such solutions depends on the network load and the used solution. HSRP protocol received a longer reaction time of VRRP, which was probably due to the longer time intervals between the transmitted messages about the status of routers. Despite this, resulting response time is incomparable with the time needed on a drive to the node. Thanks to this service after only a few seconds is able to work. Considering the above, it appears advisable to say that the present theoretical properties of MPLS technology have been confirmed in local researches.

DMVPN is generally recommended for usage over public networks, where it creates a VPN and secures it. DMVPN is a very good security improvement for MPLS VPN. We can reduce deployment complexity and costs in integrating voice, video with VPN, simplifies branch communications by direct branch-to-branch connectivity for business applications and improves business resiliency. DMVPN prevents disruption of business-critical services by incorporating routing with standards-based SSL, IPsec or SCIP technology. But GETVPN is favorable to secure existing VPN over private networks, because simplifies branch-to-branch instantaneous communications. This technique ensures low values jitter and latency by enabling data communications without central router. Moreover, GET VPN offered a high level security for MPLS while maintaining QoS and existing routing path in full mesh architecture. Contrary to DMVPN, it uses distinct polices and multiple overlays, which give limited interoperability. The biggest advantage is to provide a flexible traffic and peer-to-peer key management, reduce IPv4/6 pool without central router (hub).

Furthermore using IT devices with DMVPN in LAN nodes poses a threat of electromagnetic leakage of private data. In order to protect the private data the IT devices have to be protected using electromagnetic shielding [8]. That will greatly influence the ability to create dispersed organizational structures, remote workstations and will facilitate the operation of the intercontinental companies. [9] Some issues can be affected by vibration occurring in close environment of the network [10]. It can be new area of the further research based on state of art in different fields of science [11-16].

References

1. Laskowski, D., Łubkowski, P.: Confidential Transportation of Data on the Technical State of Facilities. In: Zamojski, W., Mazurkiewicz, J., Sugier, J., Walkowiak, T., Kacprzyk, J. (eds.) Proceedings of the Ninth International Conference on DepCoS-RELCOMEX. AISC, vol. 286, pp. 313–324. Springer, Heidelberg (2014)
2. Łubkowski, P., Laskowski, D.: Test of the Multimedia Services Implementation in Information and Communication Networks. In: Zamojski, W., Mazurkiewicz, J., Sugier, J., Walkowiak, T., Kacprzyk, J. (eds.) Proceedings of the Ninth International Conference on DepCoS-RELCOMEX. AISC, vol. 286, pp. 325–332. Springer, Heidelberg (2014)
3. Jankuniene, R., Jankunaite, I.: Route Creation Influence on DMVPN QoS. In: 31st International Conference on Information Technology Interfaces, pp. 609–614. SRCE Univ. Computing Centre, Croatia (2009)
4. Malinowski, T., Arciuch, A.: The procedure for monitoring and maintaining a network of distributed resources. In: Federated Conference on Computer Science and Information Systems (FedCSIS), pp. 947–954. IEEE (2014)
5. Cisco Systems, Inc.: Dynamic Multipoint VPN (DMVPN) Design Guide (2006)
6. Cisco Systems, Inc.: Group Encrypted Transport VPN (GETVPN) Design and Implementation Guide (2012)
7. Conlan, P.J.: Cisco Network Professional's Advanced Interworking Guide (2009)
8. Nowosielski, L., Łopatka, J.: Measurement of Shielding Effectiveness with the Method Using High Power Electromagnetic Pulse Generator. In: Progress In Electromagnetics Research Symposium, PIERS 2014 Conference Proccedings, China, pp. 2687–2691 (2014)
9. Sławińska, M., Butlewski, M.: Efficient Control Tool of Work System Resources in the Macro-Ergonomic Context. In: Applied Human Factors and Ergonomics, AHFE Conference, pp. 3780–3788 (2014)
10. Konieczny, Ł., Burdzik, R., Figlus, T.: Possibility to control and adjust the suspensions of vehicles. In: Mikulski, J. (ed.) TST 2013. CCIS, vol. 395, pp. 378–383. Springer, Heidelberg (2013)
11. Krzykowska, K., Siergiejczyk, M.: The impact of new technologies on the safety level of air traffic. In: Proceedings of the European Safety and Reliability Conference, Safety and Reliability: Methodology and Applications, p. 18. CRC Press/Balkema (2015)
12. Nowakowski, T., Werbińska-Wojciechowska, S.: Data gathering problem in decision support system for means of transport maintenance processes performance development, pp. 899–907. CRC Press/Balkema, Amsterdam, The Netherlands (2014)
13. Jasiulewicz-Kaczmarek, M.: Sustainability: Orientation in Maintenance Management - Theoretical Background. In: Golinska, P., et al. (eds.) Eco-Production and Logistics. Emerging Trends and Business Practices, pp. 117–134. Springer
14. Siergiejczyk, M., Krzykowska, K., Rosiński, A.: Parameters analysis of satellite support system in air navigation. In: Selvaraj, H., Zydek, D., Chmaj, G. (eds.) Proceedings of the Twenty-Third International Conference on Systems Engineering. AISC, vol. 330, pp. 673–678. Springer, Heidelberg (2015)
15. Jin, W., et al.: Scalable and Reconfigurable All-Optical VPN for OFDM-Based Metro-Access Integrated Network. Journal of Lightwave Technology 32, 318–325 (2014)
16. Simion, D., et al.: Efficiency Consideration for Data Packets Encryption within Wireless VPN Tunneling for Video Streaming. International Journal of Computers Communications & Control 8, 136–145 (2013)

Defining States in Reliability and Safety Modelling

Franciszek J. Restel

Wroclaw University of Technology, Wybrzeze Wyspianskiego 27, 50-370 Wroclaw
franciszek.restel@pwr.edu.pl

Abstract. In this paper, the author's research work is focused on state defining method for modeling of complex transportation systems reliability and safety, especially the railway transportation system. The paper begins with an introduction related to a literature review on railway transportation system functionality, reliability and safety modeling. The set of states can be divided into two subsets: availability set and failure set of states. Defining states in terms of railway transportation system is a complex issue, therefore the state classification into availability an failure is not sufficient. In addition to the technical effects associated with incorrect operation of system components, traffic consequences (traffic disruptions) of events are important, especially in case of railway transportation system. The paper ends with conclusions of the analysis, dealing with applicability of potential models to solve problems for the real system and a summary with prospects for further research.

Keywords: reliability, safety, transportation systems.

1   Introduction

The task of the transportation systems involves moving cargo and passengers from one place to another. The importance of transport systems increases with increasing mobility of society. Economic development makes it necessary to improve transport services. Passengers expect well-connected solutions with minimum travel and waiting times. On the other hand, goods recipients wanting to compete on the economic market take optimization actions, which lead to the application of technologies, such as Just-in-Time, Time Window [1, 21], etc. It can be seen that a well-functioning transport system should meet the requirements for logistics systems in the 7R formula (Right product, Right quantity, Right quality, Right place, Right time, Right customer, Right price) [14]. Thus, a reliable transport system is characterized by:

— availability of appropriate (planned) products in the transport offer,
— appropriate quantity of performed transport tasks,
— appropriate quality of performed transport tasks (passenger and cargo security)
— appropriate place of arrival, in accordance with the timetable (including transport routes),
— appropriate time of performance (punctuality),
— appropriate recipients,
— appropriate assessment (in accordance with the rates assumed in the carriers' transport plan).

© Springer International Publishing Switzerland 2015   413
W. Zamojski et al. (eds.), *Theory and Engineering of Complex Systems and Dependability,*
Advances in Intelligent Systems and Computing 365, DOI: 10.1007/978-3-319-19216-1_39

Using of the railway transportation system (performance of transportation tasks) is an organized process, in which the train presence at a given time and place is regulated by the timetable. Also, technical servicing must be scheduled and included in the timetable. As far as this aspect is concerned, the routing and circulation of rail vehicles are particularly complex [17].

The conditions for a railway transport system as regards the infrastructure result in adverse events [20] interfering with transport processes and in the propagation of such interferences.

The complexity of the rail transport system allows for compensating some phenomena, which influence the safety and reliability of the system operations in the short run. The compensation of the effects of adverse events allows for keeping the availability of the system, however, under atypical conditions. A state of a safety threat can be stated. Detailed analyses of rail accidents indicate that (conscious or unconscious) stays in these threat states becomes the cause of serious events with catastrophic outcomes.

2   The "Super Cube" Model for Defining the Space of States

The model of the functioning of a rail transport system can be presented as a process described by a set of states S ([3, 5]), which is divided in a specific number of subsets. The subsets of fitness and unfitness are distinguished most often [6, 24]. Literature studies and studies of operation data show that this approach is not sufficient for complex systems, also for rail transport systems. As a result, the so-called super-cube model (Fig. 1. and 2.) was developed, which was used successfully in the description of the parameters of a rail transport system.

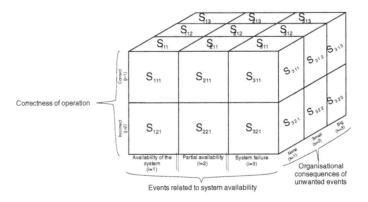

Fig. 1. Three of five dimensions of the super cube used for state defining

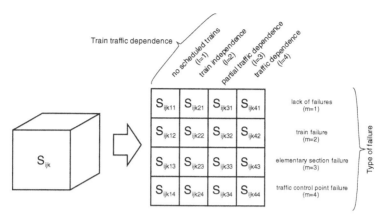

Fig. 2. Remaining two dimensions of the state defining super cube

Taking the first three features into account, the set of states is divided into subsets described in three dimensions:

$$S = \{S_{zpo}\}; \ z = 1, 2, 3; \ p = 1, 2; \ o = 1, 2, 3 \tag{1}$$

where:

- z - the index of event connected with availability of the system,
- p - the index of the correctness of the system use,
- o - the index of disruptions in the transport process.

The general description of states also includes the motion relationship and the type of adverse event, thus, the description is obtained in five dimensions:

$$S = \{S_{zporb}\} \tag{2}$$

$$z = 1, 2, 3; \ p = 1, 2; \ o = 1, 2, 3; \ r = 1, 2, 3, 4; \ b = 1,2,3,4$$

where:

- r - the index of the train traffic dependence,
- b - the index of the unwanted event type.

2.1   Events Related to System Availability

In the aspect of the rail transport system availability, the set of states was divided into three subsets:

— availability,
— partial availability,
— unavailability.

In the states of the first subset, all clusters [18] of the system are available. In the states of the partial unfitness subset, one of the clusters is unfit due to impermanent damage [27]. Partial unfitness does not take into account events connected with gradual degradation of the system and the resulting deteriorating transport capability. For single-track lines, partial unfitness states are assigned to the group of unfitness states.

The system is defined as unavailable if adverse events cause unfitness of clusters included in one of the minimum unfitness profiles. The transition from the system to the reliability structure is shown in Fig. 3. The assumption was adopted that the borders are junction stations at the end of a section. In the example below, these stations were not included in the system.

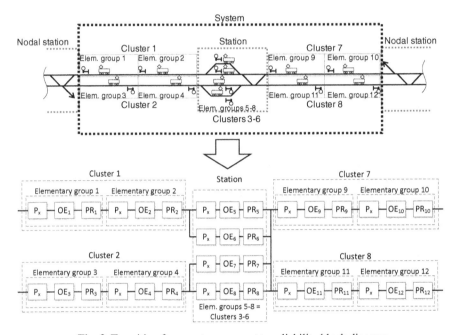

Fig. 3. Transition from system concept to reliability block diagram

In accordance with the previously adopted division of the system, elementary systems were designated, each of which consists of a train, route and the traffic control stations allowing the train to move from a given elementary section to the next one. Trains are marked with a dashed line, as the occupation of the elementary sections is variable in time (the elementary section can be free or occupied). To build elementary systems, it was necessary to assume that each of the sections was occupied (there was a train on it). For the reliability structure, the occupancy variability is taken into account by the probability of the occupancy of the elementary section. In this aspect, two separate cases are considered:

— a train with the reliability $R_{PO}(t)$ occupies an elementary section with the probability $p_{PO}(t)$,
— the train does not occupy the elementary section; therefore the reliability of the train is equal to one with the probability of $1-p_{PO}(t)$.

As a result, it can be assumed that the reliability of the train element in a given system has the following form:

$$R_{Pi}(t) = R_P(t) \cdot p_{Pi}(t) + 1 \cdot (1 - p_{Pi}(t)) \tag{3}$$

where:

- $R_{Pi}(t)$ - the reliability of a train belonging to the *i-th* elementary section,
- $R_P(t)$ - the reliability of a train,
- $p_{Pi}(t)$ - probability of the occupancy of the elementary section by a train.

There should be a traffic control station for each elementary system, which would connect the end of a given section with the next one. It was adopted that the direction of the train movement is not significant due to the need for traffic isolation of distances in both directions (which is sometimes performed in practice by various technical solutions). Each of the elementary section belonging to a non-junction station is a separate cluster.

2.2   Correctness of Use

The occurrence of damage to a transport system does not always result in unfitness for train traffic. In [13], attention was paid to the fact that damage to the elements of rail transport can reduce the readiness to perform transport tasks without causing complete unfitness of the system. Other authors also draw attention to the possibility of the implementation of transport tasks, despite of the damage [7]. Similarly, a human error does not always interrupt the traffic of trains. For this reason, the correctness-of-use feature takes into account all of these situations, in which traffic continuation occurs, despite of the damage to elements of the system or against the regulations [8, 9]. The correctness of use divides the set of system states into two subsets:

— correct use,
— incorrect use.

Therefore, incorrect use is traffic in a situation, in which procedural or device barriers do not function properly, which would protect against an accident. Thus, the incorrect use subset models situations of a direct threat to security. Fail-safe damage which is not removed [11]) and which is planned to be removed during the next servicing of a given maintenance level constitutes correct use,

2.3   Disruption of the Transport Process

In [27], permanent and non-permanent unfitness was specified. It was found that permanent damage, i.e. damage requiring technical renovation occurred less and less frequently and non-permanent damage predominates, which is called traffic disruption. In tests of the rail transport system reliability, two basic areas of disruptions were identified:

— the time of traffic tasks performance,
— the route used for the performance.

Werbińska in [25], also mentions exclusion of vehicles from traffic. However, it was adopted that at the system level, the excluded vehicle could be replaced. Thus, the consequences of events with vehicle exclusion are assigned to the time disruption category if a replacement rail vehicle performs transport tasks. An alternative is rail replacement transport, the consequences of which will be time- and route-related. It was adopted for the model that each transport task will be performed. Accidents, in which passengers die and the cargo is destroyed, are an exception here.

In the area of rail vehicles, an attempt at relating damage to consequences in the form of delays was made by Magiera in [12], although the ranges of delays used were not made dependent on the passengers' or cargo transport clients' perception. Disruptions connected with time are deviations from the planned transport schedule (the timetable), which, due to infrastructural limitations, is very important in rail transport. Deviations may be positive (delays) or negative [16]. However, it is adopted that negative time disruptions are compensated on an ongoing basis (e.g. by a longer stopping time at a station).

Route-related disruptions pertain to situations, in which the route assumed in the timetable is changed [10]. An example is a train moving along the other (incorrect) track of a double-track railway line. In this case, the passage time and the stations served may remain the same, but it constitutes a disruption in the system operation. Basiewicz [2] specifies the following route-related disruptions:

— passage along the wrong track,
— passage along the wrong track with a two-directional lock,
— combined single- and double-track passage (with additional movement between tracks).

Disruptions of the transport process were divided into three groups:

— none,
— small,
— considerable.

Small disruptions mean the occurrence of primary disruptions only. Primary disruptions are disruptions, which concerns the train directly affected by an adverse event not caused by a disruption of another train [23]. If the event occurred, e.g. in a subsystem of the infrastructure before the train enters it, the first train reaching the place of the event occurrence while it lasts is treated as disrupted in a primary manner (on condition that its movement is disrupted). The reasons for primary disruptions may include damage to elements of the system or traffic difficulties caused, for example, by collision traffic, i.e. conflicting train passages [16]. In accordance with the adopted assessment criteria as far as time is concerned, delays equal to or longer than 5 minutes will be taken into account [23].

Considerable disruptions mean primary disruptions and secondary disruptions caused by them. The transfer of primary disruptions onto other trains is called propagation of disruptions, which are called secondary disruptions [16]. Vromans [23] pays attention to the necessity of distinguishing cases of secondary disruptions. The possibility of secondary disruptions requires a timetable, in which trains depend on each other as far as traffic is concerned.

2.4   Dependence of Train Traffic

The issues of dependence between trains travelling on one railway line can be presented by means of a graphic timetable (graphs of train traffic) [4]. Trains in the system move in the network at certain distances between each other or at other technical time intervals resulting from traffic procedures. Technical intervals are minimum times between two trains. They were described by, e.g. [16].

It is possible to design the timetable based on minimum values of time intervals; in this way, the maximum throughput capacity of the infrastructure would be obtained and each primary interruption would be transferred to the other trains.

For available data [15], it was shown that the third quartile of delays has the value of 34 minutes. This value was used as the borderline between the partial dependence and independence of train traffic.

The issue of delays and time reserves was shown on the basis of [22]. In this way, four cases of dependence of train traffic were distinguished:

— no scheduled trains,
— independence,
— partial dependence,
— dependence.

Independence occurs with time intervals between trains larger than or equal to 34 minutes. For intervals longer than the minimum ones and, at the same time, smaller than 34 minutes, partial independence was adopted. If the trains are routed at minimum technical intervals, they will be classified as traffic-dependent.

It is assumed that trains in the analyzed time interval are independent if each of the trains under analysis is independent. Train dependence is assumed in the same way. Partial dependence will be established if at least one train is independent and two are dependent.

The number of dependent trains in the system increases together with an increase in traffic intensity. Traffic intensity combined with train categories and their numbers constitutes the intensity of use of the rail transport system. As a result of this dependence, the intensity of use was not taken into consideration as a separate feature describing the states of the system.

2.5   Type of Unwanted Events

The intensity of the stream of rail transport damage $\Lambda_{STS}(t)$ is a function of many parameters. The starting point is the intensity of the unfitness stream presented for a discreet transport system in [27].

The intensity can be variable in time and estimated on the basis of the intensity of damage to elements of the system (technical devices), the intensity of communication disruptions and human errors and the intensity of disruptions [26] caused by the impact of the environment of the system.

After identification of the failure stream in the system, they were grouped as regards the concept of the elementary section. The notation of an failure stream of a rail transport system was obtained:

$$\Lambda_{STS}(t) \cong \Lambda_{PO}(t) + \Lambda_{OE}(t) + \Lambda_{PR}(t) \tag{4}$$

where:

- $\Lambda_{PO}(t)$- intensity of the unfitness of a train,
- $\Lambda_{OE}(t)$- intensity of damage to the elementary section,
- $\Lambda_{PR}(t)$- intensity of unfitness of traffic control stations.

A train unfitness stream consists of intensity of damage to the rolling stock, intensity of errors committed by humans (drivers, ticket inspectors, passengers), intensity of damage to the rolling stock caused by the environmental factors. The stream of damage to the elementary section consists of the intensity of damage to the track superstructure, intensity of damage to energy devices, intensity of damage to protection devices for contact points with the surrounding area and intensity of damage to the infrastructure as a result of environmental factors. The stream of unfitness of traffic control stations includes the intensity of damage to train traffic control devices, intensity train dispatchers' errors and intensity of damage to devices as a result of environmental factors [20].

3   Application Example

The developed way of state defining has made it possible to take into account the identified, significant functional, reliability and safety qualities. Using the proposed method, states can be defined for a complex technical system. The presented hypercube model is related to the railway case, therefore an adequate application example was shown below. Off course, the hypercube model for state defining can be used also in other situations, for other technical systems. However, there is the requirement of parameterization by new system qualities.

For the presented example (Fig. 1. and 2.) the set of possible states consists of 288 states. From that should be selected states corresponding to a given type of system. In case of the regional railway transportation, 44 states corresponds to the system operation process. In addition, transitions between states must be identified and implemented. The resulting state-transition graph allows to analyze operational consequences of changes in safety, reliability or functional parameters. The described example of a model is shown in Fig. 4.

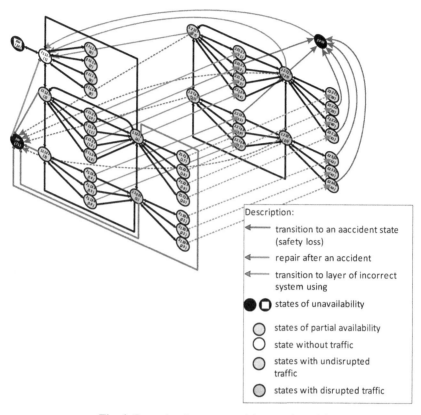

Fig. 4. Example of a state-transition graph model

Finally, using of mathematical methods, such as Markov or semi-Markov processes, gives the possibility to estimate probabilities staying in states of the system. Such models allow to identify common cause failures leading to accidents and where should be installed safety barriers. They allow also to evaluate timetable quality in relation to process reliability before its execution.

4   Summary

To take into account the registered significant features of a rail transport system for the modelling of reliability and security, a description of states in five dimensions was adopted:

— events connected with the fitness of the system,
— correctness of use,
— disruptions of the transport process,
— dependence of train traffic,
— type of adverse event.

The aforementioned features are partly dependent, which makes it necessary to assess states in the aspect of compliance of features. The type of adverse event features depends on the event connected with the fitness feature. The correctness-of-use feature and the disruptions of the transport process feature depend on the dependence of train traffic feature (which also provides information about the intensity of use). The dependence of features describing states does not influence the separateness of the defined states.

Taking into account events connected with the fitness of the system in combination with the type of adverse events allows to include in the model failures of vehicles, failures of the infrastructure and human errors. The correctness of use allows for reproducing situations, in which the system is used, despite train traffic protection devices being out of order.

Considering the correctness of use and disruptions of the transport process allow for reproducing the threat situation. Together with the traffic dependence of trains, it is possible to model the process leading to the security failure. Moreover, traffic dependence contains information about the intensity of use, which increases together with an increase in the number of traffic-dependent trains.

The developed method of defining states made it possible to take into account the identified significant functional, reliability and safety features in the context of modelling reliability and safety [19].

References

1. Akbalik, A., Penz, B.: Comparison of just-in-time and time window delivery policies for a single-item capacitated lot sizing problem. International Journal of Production Research 49(9) (2011)
2. Basiewicz, T., Rudziński, L., Jacyna, M.: Linie kolejowe. Oficyna Wydawnicza Politechniki Warszawskiej, Warszawa (2009)
3. Birolini, A.: Reliability Engineering – Theory and Practice. Springer (2010)
4. Chwieduk, A., Dyr, T.: Projektowanie ruchu pociągów. Politechnika Radomska, Radom (1997)
5. Gołąbek, A. (ed.): Niezawodność autobusów. Oficyna Wydawnicza Politechniki Wrocławskiej, Wrocław (1993)
6. Grabski, F., Jaźwiński, J.: Funkcje o losowych argumentach – w zagadnieniach niezawodności, bezpieczeństwa i logistyki. Wydawnictwa Komunikacji i Łączności, Warszawa (2009)
7. Jaźwiński, J., Smalko, Z., Żurek, J.: Związki między nieuszkadzalnością i skutecznością systemów transportowych. Materiały Konferencji Zimowa Szkoła Niezawodności, Szczyrk (2005)
8. Jodejko-Pietruczuk, A., Plewa, M.: Reliability based model of the cost effective product reusing policy. In: Safety and Reliability: Methodology and Applications - Proceedings of the European Safety and Reliability Conference, ESREL 2014, pp. 1243–1248. Taylor & Francis (2015)

9. Kierzkowski, A., Kisiel, T.: An impact of the operators and passengers behavior on the airport's security screening reliability. In: Source of the Document Safety and Reliability: Methodology and Applications - Proceedings of the European Safety and Reliability Conference, ESREL 2014, pp. 2345–2354. Taylor & Francis (2015)

10. Kwasniowski, S., Zajac, M., Zajac, P.: Telematic problems of unmanned vehicles positioning at container terminals and warehouses. In: Mikulski, J. (ed.) TST 2010. CCIS, vol. 104, pp. 391–399. Springer, Heidelberg (2010)

11. Kritzinger, D.: Aircraft system safety. Woodhead Publishing Limited, Cambridge (2006)

12. Magiera, J.: Uszkodzenia lokomotyw elektrycznych a opóźnienia pociągów pasażerskich. Materiały Konferencji Zimowa Szkoła Niezawodności, Szczyrk (2005)

13. Mazzeo, A., et al.: An Integrated Approach for Availability and QoS Evaluation in Railway Systems. In: Flammini, F., Bologna, S., Vittorini, V. (eds.) SAFECOMP 2011. LNCS, vol. 6894, pp. 171–184. Springer, Heidelberg (2011)

14. Nowakowski, T.: Niezawodność systemów logistycznych. Oficyna Wydawnicza Politechniki Wrocławskiej, Wrocław (2011)

15. Polish Railway Lines: Dane na temat zdarzeń niepożądanych w systemie transportu szynowego dla wybranych linii kolejowych w Polsce w latach 2009-2011. PKP Polskie Linie Kolejowe S.A., Warszawa (2012)

16. Potthoff, G.: Verkehrsströmungslehre (Band 1) – Die Zugfolge auf Strecken und in Bahnhöfen. Transpress, Berlin (1970)

17. Restel, F.J.: Obiegi taboru w aspekcie systemu wspomagania zarządzaniem eksploatacją. Logistyka (6) (2011)

18. Restel, F.J.: Koncepcja modelu systemu transportu szynowego w aspekcie niezawodności i bezpieczeństwa. In: Problemy Utrzymania Systemów Technicznych, pp. 183–201. Oficyna Wydawnicza Politechniki Warszawskiej, Warszawa (2014)

19. Restel, F.J.: The Markov reliability and safety model of the railway transportation system. In: Safety and Reliability: Methodology and Applications - Proceedings of the European Safety and Reliability Conference, ESREL 2014, pp. 303–311. Taylor & Francis (2015)

20. Świeboda, J., Zając, M.: Initial FMEA analysis of the container transport chain. In: Safety and Reliability: Methodology and Applications - Proceedings of the European Safety and Reliability Conference, ESREL 2014, pp. 2433–2438. Taylor & Francis (2015)

21. T'kindt, V.: Multicriteria models for just-in-time scheduling. International Journal of Production Research 49(11) (2011)

22. Vansteenwegen, P., Van Oudheusden, D.: Decreasing the passenger waiting time for an intercity rail network. Transportation Research Part B (41) (2007)

23. Vromans, M., Dekker, R., Kroon, L.: Reliability and heterogeneity of railway services. European Journal of Operational Research (172) (2006)

24. Ważyńska-Fiok, K., Jaźwiński, J.: Niezawodność systemów technicznych. Państwowe Wydawnictwo Naukowe, Warszawa (1990)

25. Werbińska-Wojciechowska, S.: Analiza niepewności danych eksploatacyjnych na przykładzie wrocławskiego systemu komunikacji tramwajowej. Materiały Konferencji Zimowa Szkoła Niezawodności, Szczyrk (2009)

26. Zając, M., Kierzkowski, A.: Uncertainty assessment in semi Markov methods for Weibull functions distributions. In: Advances in Safety, Reliability and Risk Management - Proceedings of the European Safety and Reliability Conference, ESREL 2011, pp. 1161–1166 (2012)

27. Zamojski, W. (ed.): Systemy transportu dyskretnego – modele, niezawodność. Wydawnictwa Komunikacji i Łączności, Warszawa (2007)

Macroscopic Transport Model as a Part of Traffic Management Center: Technical Feasibility Study

Mihails Savrasovs

1st Lomonosov Street, Riga, Latvia, LV 1019
savrasovs.m@tsi.lv

Abstract. This paper is result of technical feasibility study done in Transport and Telecommunication Institute, located in Riga, Latvia. As many world cities, Riga as capital of Latvia is suffering from transport congestions. There are many objective reasons of this situation, like outdated transport infrastructure, star-shaped roads schema, limited number of possibilities to cross the river etc. Riga City Traffic Department has a number of solutions, but most promising is the idea of implementation of ITS. The ITS provides a number of solutions, but the target of completed technical feasibility study is related with implementation of on-line traffic forecasting tool, which is based on macroscopic transport model and could be treated as subsystem of the traffic management system. The results of analysis give a number of "hot spots" which should be taken into account before technical implementation of the system. But the analysis results lead to the conclusion that the implementation of the on-line traffic forecasting tool is feasible for Riga city.

Keywords: feasibility study, on-line forecasting, macroscopic transport model.

1   Introduction

Modern information and communication technologies (ICT) gives a wide range of possibilities to organize intelligent system, which is able to simplify and make faster the decision making process. One of the active directions in this area is related with ITS initiatives implementation. These initiatives are supported also by EU directives (at example The European ITS directive (2010/40/EU)).The ITS is general term which describes a set of information and communication technologies used to manage all issues of the transport system. But here must be underlined that a number of subsystems could be described with reference to transport system: traffic management tools, parking management tools, accident management tools etc.

Riga, the capital of Latvia is suffering from congestions. There are a number of reasons, which course this situation; most vivid could be enumerated here:

- *Outdated transport infrastructure* – this point is a reference to the low investment level to the transport infrastructure, which results to the bad quality of the roads, outdated traffic management center, limited number of modern solutions in transport;

W. Zamojski et al. (eds.), *Theory and Engineering of Complex Systems and Dependability,*
Advances in Intelligent Systems and Computing 365, DOI: 10.1007/978-3-319-19216-1_40

- *Star-shaped schema of transport system* – the current shape of transport system is a classical star-shaped schema, which leads to concentration of traffic in city center.
- *Limited number of bridges across Daugava River* – Riga is divided on two parts by Daugava River, in same time only 4 bridges are operating and 3 of them are located in Riga city center. This is a bottleneck of the system, which leads to congestions during morning, evening, and special events times.
- *Lack of ICT in transport* – not looking to the fact, that Riga for a long time is having traffic management center, the lack of ICT is a vivid problem. The traffic management center is outdated and does only functions related with monitoring of the situation, but not management of the traffic.

Of course there are a number of activities planned by Riga City Development Department and Riga City Traffic Department, but most of these solutions are related with implementation of new transport infrastructure objects, like North Corridor, West Corridor etc. The final concept is to make rings around the parts of the city and to remove the traffic from city center. The figure 1 demonstrates the final concept of the transport system after implementation of all infrastructure projects.

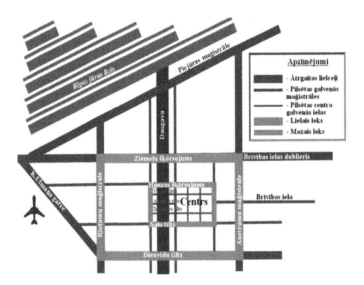

Fig. 1. Riga city development plan [1]

If all of these infrastructure projects will be implemented, it will partly solve existing problems in transport infrastructure. But there are no any plans about active use of ICT on transport in planning documents of Riga City Council.

Here is proposed the approach related with implementation of ITS in Riga, and as a very first step the implementation of on-line traffic forecasting tool is concerned.

Current paper is devoted to results of technical feasibility study done for on-line traffic forecasting tool, which could be treated as significant part of the traffic management system.

2   The Concept of On-line Traffic Forecasting Tool

It is planned that a core of future traffic management center will be on-line traffic forecasting tool, which will alert about possible future collisions in the transport network. Must be noted, that it should be a tool, which is able to provide the results of the modelling on-line. Here on-line means not now, but a feasible time up to 10 minutes after lunching of the forecasting process. The model should provide a short term forecast (forecast horizon – 1h) of situation in transport network for the whole city and districts located near the city. The figure 2 demonstrates concept of the forecasting tool, which is based on macroscopic transport model.

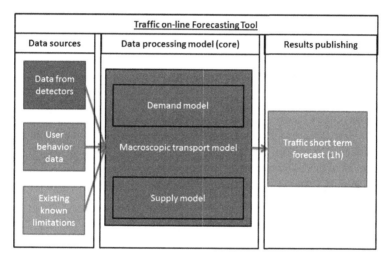

Fig. 2. The general concept of the traffic on-line forecasting tool

As could be seen from figure 2 the traffic on-line simulation tool has 3 subsystems: data sources subsystem, which is responsible for extracting, transforming, loading data, storing and providing different kind of data necessary for the forecasting; data processing model (actually the macroscopic transport model), which is feed by the data from different sources; and finally results publishing system, which is targeted to store the forecasting results in the format, which is suitable for traffic management system and for providing interfaces to the data.

Next all components of the presented traffic on-line forecasting tool will be described in detail, in order to get the view on future system technical requirement.

2.1 The Component "Data from Detectors"

This component is a vivid for the whole system as it is a source of data which is used to calibrate the macroscopic model for the current state of the transport network. The primary role of detectors is to collect the information about the traffic flow and fill the database continuously. The data required by transport model in this case are the following one: traffic flow intensity; average speed of the traffic flow and finally classification of the vehicles. It is proposed that EU level supported standard, titled DATEX II, will be used. It will allow not only to organize the local system by itself, but also to provide the data to the EU users. DATEX II has been developed to provide a standardized way of communicating and exchanging traffic information between traffic centers, service providers, traffic operators and media partners [2].

Currently, Riga has the limited number of detectors, which are able to provide the required data, it will be necessary to implement the network of the detectors with support the DATEX II standard.

2.2 The Component "User Behavior Data"

The component provides the information to the model about typical behavior of the citizens in the city. Usually the data is presented in form of origin-destination (OD) matrices, which define number of trips from one transport zone to another with specific mode of transport. In order to fill the requirements of the system it is necessary to have the OD matrices by time periods, which should be 1h long. So, each hour will have its own set of matrices. Here must be underlined that matrices also should be divided by the type of days:

- typical days: Monday, Tuesday, Wednesday, Thursday;
- untypical workdays: Monday, Friday;
- weekends: Saturday, Sunday.

Also it should be noted, that it is important to have different matrices for different seasons (summer, spring, winter and autumn). These requirements are not strict, but the quality of the forecast depends much from the quality of the data in OD matrices. Currently, Riga does have only one OD matrix, which describes morning peak hour, for typical day and for typical seasons (autumn and spring).

2.3 The Component "Existing Known Limitations"

This part of the system should be a database, which will include the information about planned restrictions in the network, because of planned reconstruction of the roads,

roads closing, because of the events etc. The temporal database approach is feasible here as a possible solution. Currently the data about limitations in transport network are available and are published in website for citizens of Riga. The information stored in database will allow to obtain realistic forecast of the situation.

2.4   The Component "Data Processing Model"

The data processing model is a core of the forecasting system. In general the core is represented by macroscopic transport model, which has two parts: demand model and supply model (see figure 3). The supply part of the model has the information about the transport network, public transport stops etc. The demand side is presented by already mentioned here OD matrices. Both components are vivid for the modeling and in order to get the results the supply model must be updated before each forecasting step.

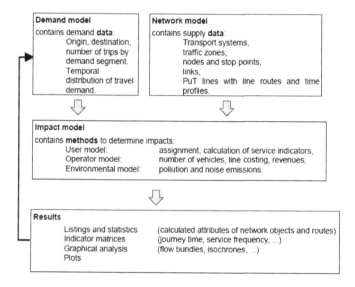

Fig. 3. Transport model [3]

The components mentioned above are sources of data which should feed the demand and supply model by the updated data. The base (current) model by itself does have the data for supply model – the network, public transport stops etc. It will be feed by information from the database of existing known limitation in the network. The demand side will be loaded for each forecasting step (OD matrices for current hour), and here the user behavior data subsystem will be used. And finally the information from detectors will be used to calibrate the model to get the current state of the network.

2.5   The Component "Results Publishing"

The last component of the proposed system is related to the representation task of the modelling and forecasting. This is vivid problem from number of sides, first of all the results must be presented graphically in order to be controlled by operators of the traffic managing center, the second – the data should be provided in format, which allows to publish the data in public places (at example in city official webpage), the third is related with interfaces which will be used to access the forecasting results by the rest components of traffic management system, and the last is targeted on providing the API (application program interface) for third-party applications and services. All mentioned issues should be taken into account during detailed specification of the system.

3   Riga Transport Model as the Base of Traffic Forecasting Tool

As was mentioned earlier the core of the traffic forecasting tool is a macroscopic transport model. Here the issues related with use of transport model are discussed in detail.

In present time there are 3 macroscopic transport models used in decision making process in Riga:

- official transport model used by Riga City Development Department, developed using EMME software;
- official model implemented in frame of Riga and Pieriga mobility plan development project, which was developed using CUBE software;
- unofficial model developed by Transport and Telecommunication Institute using VISUM software.

There no relationship between models and all of them are not covering issues required by the traffic forecasting tool. Most useful is treated a model hold by Riga City Development Department as it used for strategic planning, but also the examples of the application of VISUM model are known [4]. The CUBE model was used once for development of the mobility plan and not currently used. So there are two options for creating a core of the system: EMME and VISUM models. Both models have a number of disadvantages which should be taken into account:

- outdated OD matrices for private vehicles;
- no information for public transport;
- lack of information for cargo transport;
- OD matrices are defined only for morning peak hour, no data for rest of hours.

All points are significant problems for the implementation of the traffic forecasting tool and during implementation should be solved. The table below summarise the activities which should be taken in order to overcome the issues.

Table 1. Activities table

Issue	Impact on system implementation	Activity
Lack of information for cargo transport	LOW	• Strong cooperation and data exchange with ICT solutions providers for transport companies. Usually such solutions includes a GPS tracking, which allows to analyse the mobility of the cargo vehicles; • Implementation of the network of detectors, which has possibility to classify vehicles by types (video observations, weight-in-motion systems, speed cameras etc). This will be a source of information for calibration of the cargo transport OD matrices. • Planning of city logistics for cargo vehicles based on cooperation with industry representatives.
No information for public transport	AVERAGE	• The Riga city uses e-ticketing system for public transport; this could be a source of information for evaluation of the OD matrices for the public transport. But it must be taken into account, that system requires to do validation of the ticket only once, so the destination point of the trip could not be evaluated directly. Here is proposed to apply trip-chain approach. This will allow to estimate the OD matrices for peak hours [5]. • Passenger counting detectors installation will provide the information necessary for calibration. The linkage of the information to the GPS data will give vivid information about level of service during trips.
Outdated OD matrices for private vehicles	HIGH	• The issues could be overcome only by implementation of national survey system. Only in this case the mobility data could be treated as actual. • The use of ICT technologies should be wider. At example the local and international companies could provide useful data, like GPS tracking, mobile phone tracking, etc.
OD matrices are defined only for morning peak hour, no data for rest hours	HIGH	• The issue could be overcome in by using ICT providers in local area: GPS tracking, mobile phone tracking, e-ticketing data etc.

As could be seen from the table above it is possible to overcome all disadvantages of existing models, by application of modern ICT or by using administrative procedures, like national household and mobility surveys. Currently, Latvia do not have any strategy related to the transport surveys and data collection issues, that is why, as initial point of implementation it is proposed to use the scheme presented below in figure 4.

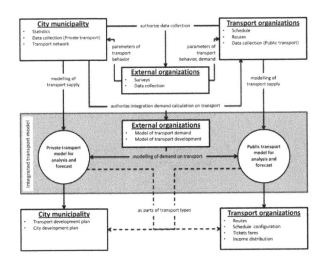

Fig. 4. The schema of transport model use on national level

The two main players according to this schema could be mentioned: the national government/local municipalities and transport organizations. Transport organizations do the main analysis based on the usage of traffic analysis tools (transport plans and programs in this case) provide output data to national government/local municipalities. National government/local municipalities use the output data when working out the future area development strategy. Furthermore, it should be mentioned that input data (country/city development plan, transport system development plan, statistical data, etc.) for the analysis are provided by national government/local municipalities.

Next table (see table 2) aggregates the data about the companies which are able to provide transport related data based on currently implemented ICT solutions. This list should be taken into account during implementation of the traffic forecasting tool.

Table 2. Review of potential data providers

Issue	Possible data providers
Mobile phone tracking data	• TELE2 Ltd. • LATVIJAS MOBILAIS TELEFONS Ltd. • Bite Latvija Ltd.
Public transport passenger data	• Rigas Karte Ltd. • Road Transport Administration Ltd. State • Riga International Coach Terminal JSC • Pasažieru vilciens JSC
Commercial cargo transportation data	• Jana Seta Ltd. • EcoTelematics Ltd. • Mapon Ltd. • Mappost, Ltd.
Traffic intensities	• Riga City Traffic Department • Latvian State Roads Ltd. • Latvian State Police

4   Feasibility Study Results

The feasibility study results could be divided on five thematic groups: technical, economic, legal, operational, and scheduling. The primary goal of the paper was to conduct the technical feasibility study. The technical feasibility study assessment is focused on gaining an understanding of the present technical resources of the organization and their applicability to the expected needs of the proposed system. It is an evaluation of the hardware and software and how it meets the need of the proposed system [6]. The analysis of the system general concept gave a number potential "hot spots" for traffic forecasting tool. Mainly these issues are related with limitations of the existing solutions and require during implementation significantly improve current systems or even to build them from the zero. Main point which could be characterized as critical for the traffic forecasting tool implementation was related with need to update existing macroscopic transport simulation model. From technical point of view it is possible to overcome the disadvantages by using the data from external sources (like private companies), but here legal issues should be analyzed in details and taken into account during project realization. All other components of the traffic forecasting tool from technical point of view are feasible and number of existing solution could be located in the market. In general it means that a realization of the traffic forecasting tool as a part of traffic management system is a technically feasible solution, with number of "hot spots", which could be overcome.

Acknowledgment. This work was supported by Latvian state research program 2014-2017 project "The next generation of information and communication technologies (NexIT)".

References

1. Riga City Development Department (RDPAD), Longterm Development Strategy of Riga City till 2025, http://www.rdpad.lv/uploads/Longterm-strategy_EN.pdf (access date: May 2014)
2. EasyWay project webpage, http://www.datex2.eu
3. PTV VISION VISUM Official Manual (2008)
4. Transport and Telecommunication Institute, Project report "Freight traffic flow research and rerouting from Riga city center", 68 p. (2014)
5. Barry, J., Freimer, R., Slavin, H.: Using entry-only automatic fare collection data to estimate linked transit trips in New York City. Transportation Research Board 2008 Annual Meeting CD-ROM, Washington, D.C. (2008)
6. O'Brien, J.A., Marakas, G.M.: Developing Business/IT Solutions. In: Management Information Systems, pp. 488–489. McGraw-Hill/Irwin, New York (2011)

Reliability Assessment of Integrated Airport Surface Surveillance System

Mirosław Siergiejczyk, Karolina Krzykowska, and Adam Rosiński

Warsaw University of Technology
{msi,kkrzykowska,adro}@wt.pw.edu.pl

Abstract. In Poland, a problem of lack of radar coverage at certain flight levels over certain regions of the country is being observed. Therefore, the Polish Air Navigation Services Agency is looking for some better surveillance solutions for airspace and airport surface.The paper presents an analysis of surveillance systems with particular emphasis on the of the possibility of integrating radar systems with multilateration and automatic dependent ones. Presented article turned out to be basis for discussion on sense of implementation the integrated system of surface surveillance at the airport where such a surveillance is carried out only visually and there is a bigger risk of causing a dangerous situation.

Keywords: surveillance, air traffic, reliability.

1   Introduction

In the case of transport, the largest technological breakthroughs in the twentieth century were in air industry. Such progress would not have been possible without parallel achievements in safety managementand risk reduction in air traffic. Still, the criteria for the construction of the technical facilities and measures to enhance the level of security is tightened after the accident or serious incident. It is worth paying special attention to the movement area of airports for example in Poland, where surveillance is done visually. Here, safety plays a very important role. Movement area is defined as part of an airport designedfor take-off, landing and taxiing of aircraft, consisting of the maneuvering area and the disc. This field is the meeting place of the air traffic system components, including aircraft, ground control agents, air navigation, traffic control, flight crew and ground personnel. If the set of attributes of the elements of this system ensures prevention of emergency situations, provide rescue in case of hardware failures, errors, crew or ground staff - this can be named the safety of the aviation system. So far, in the literature there were numerous reliability systems analysis conducted, which indicated the validity of the implementation of complex objects, and thus – integrated ones, what in this case is particularly important [4-6], [10], [11].

2   Integrated Airport Surface Surveillance System

With regard to the surveillance of airport surface traffic - it is worth paying attention to the factors that may affect the occurrence of dangerous situations. The most common

© Springer International Publishing Switzerland 2015   435
W. Zamojski et al. (eds.), *Theory and Engineering of Complex Systems and Dependability*,
Advances in Intelligent Systems and Computing 365, DOI: 10.1007/978-3-319-19216-1_41

factors are associated with the work of air traffic controllers and directing vehicles operating on the airport surface, but which are not aircraft. There can be distinguished[2-3]:

- in relation to the work of air traffic controllers:
  o a temporary forgetfulness about the aircraft,
  o mistakes determining separation between aircrafts located on the airport surface,
  o failure in identifying the aircraft or its location,
  o errors in communication of air traffic controllers,
  o distraction at work,
  o overwork,
  o lack of controllersexperience,
- in relation to a vehicle which is moving along the surface of the airport, but non-aircraft:
  o not following the instructions of air traffic control,
  o errors in communication with air traffic control,
  o lack of radio communication equipment in the vehicle.

Therefore, the ICAO (International Civil Aviation Administration) has proposed the creation of so-called hot spots - critical points noticed at themap of the airport. They are defined as locations on the movement area of an aerodrome, which pose a potential risk of collision or runway incursionsby unauthorized vehicles [1].

It is difficult to control such a critical points while surveillance of the airport takes place only in the visual form. There are often areas that are beyond the reach of visual air traffic controllers (eg. hidden behind the passenger terminal). In such situations integrated surveillance system could be used. It consists of several components, including hyperbolic systems (MLAT), airport surface surveillance radar (SMR) and automatic dependent surveillance GNSS (ADS - B)[3]. The fusion of the radar, multilateration and ADS-B data enables much better estimation of the target location using multiple data sources. The system could be used as a security tool on the runway, as it allows air traffic controllers to detect potential collision streams by providing information about the traffic on the runway and taxiways [7]. With the collected data from various sources it may, in some way, track aircrafts on the airport surface.

In conjunction with the air traffic control some information are determined related to accurate motion parameters of the aircraft. It helps to improve the quality of supervision at the airport, especially during adverse weather conditions. In addition, the data obtained through an integrated system allowstaking look at aspects such as [2]:

- correlation between information about the flight plan and aircraft position indicated on the display controller,
- elimination of dead zone,
- constant observation of the situation on the movement area including the detection and resolution of conflicts.

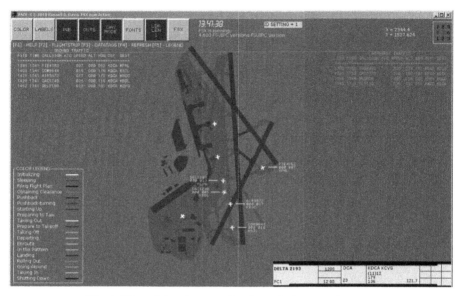

Fig. 1. View of the airport surface from the integrated system

Aviation systems implement operational programs through actions of their subsystems. These, to carry out such a program must conduct their functionalities at the same time [8]. Performing the operation of the system and its operating program is described in terms of reliability, suitability in different states. The opposite is the unsuitability condition that can be caused by extreme events, for aviation systems -eg. weather. However, in the case of the above-mentioned system, this should not matter, unless you talk about phenomena such as hurricane or snowstorm. You can also meet with the state of partial suitability of the system.

Each system has its own structure. It is possible to assess system from the point of reliability structure, or some representation, in which the reliability of the components determines the reliability of the system at the same time.

3   Reliability Assessment of Integrated Airport Surface Surveillance System

Integrated airport surface surveillance system consists of three subsystems, each of which individually damaged does not transfer whole system to the state of no operational capability [9]. This condition will be achieved only when all three subsystems are damaged. It can be therefore concluded that the subsystems operate completely independently and airport surface surveillance can function when at least one subsystem work correctly. In conducting the analysis, the relationship can be illustrated in the present system in terms of safety [11].

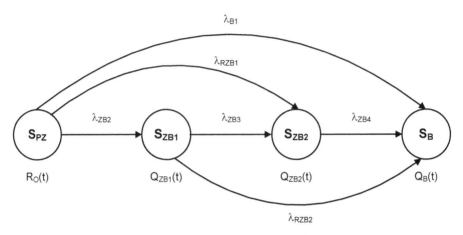

Fig. 2. .Relations in the system

Denotations in figures:

$R_0(t)$ – the function of probability of system staying in state of full operational capability,

$Q_{ZB}(t)$ – the function of probability of system staying in state of partial operational capability,

$Q_B(t)$ – the function of probability of system staying in state of no operational capability,

λ_{B1} – transition rate of the system,

$\lambda_{ZB2}, \lambda_{ZB3}, \lambda_{ZB4}$ – transition rate of subsystems

$\lambda_{RZB1}, \lambda_{RZB2}$ – transition rate of two subsystems

The system illustrated in fig. 2 may be described by the following Chapman – Kolmogorov equations:

$$R_0'(t) = -\lambda_{B1} \cdot R_0(t) - \lambda_{ZB2} \cdot R_0(t) - \lambda_{RZB1} \cdot R_0(t)$$

$$Q_{ZB1}'(t) = \lambda_{ZB2} \cdot R_0(t) - \lambda_{ZB3} \cdot Q_{ZB1}(t) - \lambda_{RZB2} \cdot Q_{ZB1}(t)$$

$$Q_{ZB2}'(t) = \lambda_{ZB3} \cdot Q_{ZB1}(t) - \lambda_{ZB4} \cdot Q_{ZB2}(t) + \lambda_{RZB1} \cdot R_0(t) \qquad (1)$$

$$Q_B'(t) = \lambda_{B1} \cdot R_0(t) + \lambda_{ZB4} \cdot Q_{ZB2}(t) + \lambda_{RZB2} \cdot Q_{ZB1}(t)$$

Given the initial conditions:

$$R_0(0) = 1 \qquad (2)$$

$$Q_{ZB1}(0) = Q_{ZB2}(0) = Q_B(0) = 0$$

Using the Laplace transform we obtain the following system of linear equations:

$$s \cdot R_0^*(s) - 1 = -\lambda_{B1} \cdot R_0^*(s) - \lambda_{ZB2} \cdot R_0^*(s) - \lambda_{RZB1} \cdot R_0^*(s)$$

$$s \cdot Q_{ZB1}^*(s) = \lambda_{ZB2} \cdot R_0^*(s) - \lambda_{ZB3} \cdot Q_{ZB1}^*(s) - \lambda_{RZB2} \cdot Q_{ZB1}^*(t)$$

$$s \cdot Q_{ZB2}^*(s) = \lambda_{ZB3} \cdot Q_{ZB1}^*(s) - \lambda_{ZB4} \cdot Q_{ZB2}^*(s) + \lambda_{RZB1} \cdot R_0^*(s) \qquad (3)$$

$$s \cdot Q_B^*(s) = \lambda_{B1} \cdot R_0^*(s) + \lambda_{ZB4} \cdot Q_{ZB2}^*(s) + \lambda_{RZB2} \cdot Q_{ZB1}^*(t)$$

Turning it gives:

$$s \cdot R_0^*(s) + \lambda_{B1} \cdot R_0^*(s) + \lambda_{ZB2} \cdot R_0^*(s) + \lambda_{RZB1} \cdot R_0^*(s) = 1$$

$$s \cdot Q_{ZB1}^*(s) - \lambda_{ZB2} \cdot R_0^*(s) + \lambda_{ZB3} \cdot Q_{ZB1}^*(s) + \lambda_{RZB2} \cdot Q_{ZB1}^*(t) = 0 \tag{4}$$

$$s \cdot Q_{ZB2}^*(s) - \lambda_{ZB3} \cdot Q_{ZB1}^*(s) + \lambda_{ZB4} \cdot Q_{ZB2}^*(s) - \lambda_{RZB1} \cdot R_0^*(s) = 0$$

$$s \cdot Q_B^*(s) - \lambda_{B1} \cdot R_0^*(s) - \lambda_{ZB4} \cdot Q_{ZB2}^*(s) - \lambda_{RZB2} \cdot Q_{ZB1}^*(t) = 0$$

Transforming further we get:

$$\left(s + \lambda_{B1} + \lambda_{ZB2} + \lambda_{RZB1}\right) \cdot R_0^*(s) = 1$$

$$\left(s + \lambda_{ZB3} + \lambda_{RZB2}\right) \cdot Q_{ZB1}^*(s) - \lambda_{ZB2} \cdot R_0^*(s) = 0 \tag{5}$$

$$\left(s + \lambda_{ZB4}\right) \cdot Q_{ZB2}^*(s) - \lambda_{ZB3} \cdot Q_{ZB1}^*(s) - \lambda_{RZB1} \cdot R_0^*(s) = 0$$

$$s \cdot Q_B^*(s) - \lambda_{B1} \cdot R_0^*(s) - \lambda_{ZB4} \cdot Q_{ZB2}^*(s) - \lambda_{RZB2} \cdot Q_{ZB1}^*(t) = 0$$

Then we get it as a schematic:

$$(R_0 \cdot Q_{ZB1} \cdot Q_{ZB2}, Q_B) \rightarrow \begin{pmatrix} \dfrac{1}{a + \lambda_{RZB1}} \\[2mm] \dfrac{\lambda_{ZB2}}{a \cdot b + a \cdot \lambda_{RZB2} + b \cdot \lambda_{RZB1} + \lambda_{RZB1} \cdot \lambda_{RZB2}} \\[2mm] \dfrac{b \cdot \lambda_{RZB1} + \lambda_{ZB2} \cdot \lambda_{ZB3} + \lambda_{RZB1} \cdot \lambda_{RZB2}}{a \cdot b \cdot c + a \cdot c \cdot \lambda_{RZB2} + b \cdot c \cdot \lambda_{RZB1} + c \cdot \lambda_{RZB1} \cdot \lambda_{RZB2}} \\[2mm] \dfrac{b \cdot c \cdot \lambda_{B1} + b \cdot \lambda_{ZB4} \cdot \lambda_{RZB1} + c \cdot \lambda_{B1} \cdot \lambda_{RZB2} + c \cdot \lambda_{ZB2} \cdot \lambda_{RZB2} + \lambda_{ZB2} \cdot \lambda_{ZB3} \cdot \lambda_{ZB4} + \lambda_{ZB4} \cdot \lambda_{RZB1} \cdot \lambda_{RZB2}}{a \cdot b \cdot c \cdot s + a \cdot c \cdot s \cdot \lambda_{RZB2} + b \cdot c \cdot s \cdot \lambda_{RZB1} + c \cdot s \cdot \lambda_{RZB1} \cdot \lambda_{RZB2}} \end{pmatrix} \tag{6}$$

where:

$$a = s + \lambda_{B1} + \lambda_{ZB2}$$

$$b = s + \lambda_{ZB3} \tag{7}$$

$$c = s + \lambda_{ZB4}$$

In the above result the symbols "*" and "s" are omitted at the residence probabilities of the system in the states highlighted R_0, Q_{ZB1}, Q_{ZB2}, Q_B.

Using the inverse transform, we get:

$$R_0(t) = e^{-(\lambda_{B1} + \lambda_{ZB2} + \lambda_{RZB1}) \cdot t} \tag{8}$$

$$Q_{ZB1}(t) = \lambda_{ZB2} \cdot \left[\frac{-e^{-(\lambda_{B1} + \lambda_{ZB2} + \lambda_{RZB1}) \cdot t} + e^{-\lambda_{ZB3} \cdot t}}{\lambda_{B1} + \lambda_{ZB2} + \lambda_{RZB1} - \lambda_{ZB3}} \right] \tag{9}$$

$$Q_{ZB2}(t) = \lambda_{ZB2} \cdot \lambda_{ZB3} \cdot \left[\frac{e^{-(\lambda_{B1}+\lambda_{ZB2}+\lambda_{RZB1}) \cdot t}}{\left(\lambda_{B1}+\lambda_{ZB2}+\lambda_{RZB1}-\lambda_{ZB3}\right) \cdot \left(\lambda_{B1}+\lambda_{ZB2}+\lambda_{RZB1}-\lambda_{ZB4}\right)} + \frac{e^{-\lambda_{ZB3} \cdot t}}{\left(\lambda_{ZB3}-\lambda_{B1}-\lambda_{ZB2}-\lambda_{RZB1}\right) \cdot \left(\lambda_{ZB3}-\lambda_{ZB4}\right)} + \frac{e^{-\lambda_{ZB4} \cdot t}}{\left(\lambda_{ZB4}-\lambda_{B1}-\lambda_{ZB2}-\lambda_{RZB1}\right) \cdot \left(\lambda_{ZB4}-\lambda_{ZB3}\right)} \right] +$$

$$+ \lambda_{RZB1} \cdot \left[\frac{-e^{-(\lambda_{B1}+\lambda_{ZB2}+\lambda_{RZB1}) \cdot t} + e^{-\lambda_{ZB4} \cdot t}}{\lambda_{B1}+\lambda_{ZB2}+\lambda_{RZB1}-\lambda_{ZB4}} \right] \tag{10}$$

$$Q_B(t) = \frac{\lambda_{B1}}{\lambda_{B1}+\lambda_{ZB2}+\lambda_{RZB1}} \cdot \left[1 - e^{-(\lambda_{B1}+\lambda_{ZB2}+\lambda_{RZB1})t}\right] + \lambda_{ZB2} \cdot \lambda_{ZB3} \cdot \lambda_{ZB4} \cdot$$

$$\cdot \left[\frac{-e^{-(\lambda_{B1}+\lambda_{ZB2}+\lambda_{RZB1})t}}{\left(\lambda_{B1}+\lambda_{ZB2}+\lambda_{RZB1}\right) \cdot \left(\lambda_{B1}+\lambda_{ZB2}+\lambda_{RZB1}-\lambda_{ZB3}\right) \cdot \left(\lambda_{B1}+\lambda_{ZB2}+\lambda_{RZB1}-\lambda_{ZB4}\right)} - \frac{e^{-\lambda_{ZB3} \cdot t}}{\left(\lambda_{ZB3}-\lambda_{B1}-\lambda_{ZB2}-\lambda_{RZB1}\right) \cdot \lambda_{ZB3} \cdot \left(\lambda_{ZB3}-\lambda_{ZB4}\right)} - \frac{e^{-\lambda_{ZB4} \cdot t}}{\left(\lambda_{ZB4}-\lambda_{B1}-\lambda_{ZB2}-\lambda_{RZB1}\right) \cdot \left(\lambda_{ZB4}-\lambda_{ZB3}\right) \cdot \lambda_{ZB4}} + \frac{1}{\left(\lambda_{B1}+\lambda_{ZB2}+\lambda_{RZB1}\right) \cdot \lambda_{ZB3} \cdot \lambda_{ZB4}} \right] +$$

$$+ \lambda_{RZB1} \cdot \lambda_{ZB4} \cdot \left[\frac{e^{-(\lambda_{B1}+\lambda_{ZB2}+\lambda_{RZB1}) \cdot t}}{\left(\lambda_{B1}+\lambda_{ZB2}+\lambda_{RZB1}\right) \cdot \left(\lambda_{B1}+\lambda_{ZB2}+\lambda_{RZB1}-\lambda_{ZB4}\right)} - \frac{e^{-\lambda_{ZB4} \cdot t}}{\left(\lambda_{B1}+\lambda_{ZB2}+\lambda_{RZB1}-\lambda_{ZB4}\right) \cdot \lambda_{ZB4}} + \frac{1}{\left(\lambda_{B1}+\lambda_{ZB2}+\lambda_{RZB1}\right) \cdot \lambda_{ZB4}} \right] \tag{11}$$

<u>Example</u>

The following quantities were defined for the system:

- test duration - 1 year (values of this parameter is given in [h]):

$$t = 8760 \ [h] \tag{12}$$

- reliability of airport surface surveillance radar:

$$R_{ZB1}(t) = 0{,}99995 \tag{13}$$

- reliability of multilateration system:

$$R_{ZB2}(t) = 0{,}99995 \tag{14}$$

- reliability of automatic dependent surveillance system:

$$R_{ZB3}(t) = 0{,}99995 \tag{15}$$

- transition rate from the state of full operational capability S_{PZ} into the state of partial operational capability S_{ZB2}:

$$\lambda_{RZB1} = 1{,}1415 \cdot 10^{-9} \left[\frac{1}{h}\right] \tag{16}$$

- transition rate from the state of partial operational capability S_{ZB1} into the state of no operational capability S_B:

$$\lambda_{RZB2} = 1{,}1415 \cdot 10^{-9} \left[\frac{1}{h}\right] \tag{17}$$

- transition rate from the state of full operational capability S_{PZ} into the state of no operational capability S_B:

$$\lambda_{B1} = 5{,}7077 \cdot 10^{-10} \left[\frac{1}{h}\right] \tag{18}$$

Knowing the value of reliability $R_{ZB1}(t)$, transition rate from the state of full ability into the state of the impendency over safety S_{ZB1} may be estimated. Provided the up time is described by exponential distribution, the following relationship can be used:

$$R_{ZB1}(t) = e^{-\lambda_{ZB2}t} \qquad \text{for} \qquad t \geq 0 \tag{19}$$

thus

$$\lambda_{ZB2} = -\frac{\ln R_{ZB1}(t)}{t} \tag{20}$$

For $t = 8760 \ [h]$ and $R_{ZB1}(t) = 0{,}99995$ we obtain:

$$\lambda_{ZB2} = -\frac{\ln R_{ZB1}(t)}{t} = -\frac{\ln 0{,}99995}{8760} = 5{,}7079 \cdot 10^{-9} \left[\frac{1}{h}\right] \tag{21}$$

Similarly, we determine the value of λ_{ZB3} and λ_{ZB4}.

For above initial values, by use of (8-11) equations, following results are obtained:

$$R_0 = 0,999935$$
$$Q_{ZB1} = 4,9998 \cdot 10^{-5} \tag{22}$$
$$Q_{ZB2} = 1 \cdot 10^{-5}$$
$$Q_B = 5 \cdot 10^{-6}$$

The presented process of surveillance systems analysis allows us to determine the level of reliability of the proposed integrated system. This is possible by using the different three systems which can work both together and separately and that can ensure an adequate level of reliability indicators. So far, Authors presented some material about cooperation of systems and possibility to replace each of them. This time, the article focuses on integration of system into one with appropriate reliability indicator.

4 Summary

Analyzing the functionality of the integrated airport surface surveillance system in which there are three subsystems of surveillance turned out to be very important in case of safety in air traffic. Authors illustrated the relationships in its structure in terms of reliability. This allowed to determine the relations defining probability of system staying in the states of full operational capability, partial operational capability, no operational capability. This is illustrated graphically and analytical having each subsystem failure. In its analysis, authors omitted functionality of the repair process related to subsystems. This process authors plan to consider in future studies. Presented article turned out to be basis for discussion on sense of implementation the integrated system of surface surveillance at the airport where such a surveillance is carried out only visually and there is a bigger risk of causing a dangerous situation.

References

1. Doc 9870 AN/463 Manual on the Prevention of Runway Incursions, International Civil Aviation Administration, Montreal (2007)
2. Lewitowicz, J.: Podstawy eksploatacji statków powietrznych. Tom 3, Wydawnictwo Instytutu Technicznego Wojsk Lotniczych, Warszawa (2006)
3. Malarski, M.: Inżynieria ruchu lotniczego, Oficyna Wydawnicza Politechniki Warszawskiej, Warszawa (2006)
4. Kierzkowski, A., Kisiel, T.: An impact of the operators and passengers behavior on the airport's security screening reliability. In: Source of the Document Safety and Reliability: Methodology and Applications - Proceedings of the European Safety and Reliability Conference, ESREL 2014, pp. 2345–2354 (2015)

5. Kierzkowski, A., Kowalski, M., Magott, J., Nowakowski, T.: Maintenance process optimization for low-cost airlines. In: 11th International Probabilistic Safety Assessment and Management Conference and the Annual European Safety and Reliability Conference 2012, PSAM11 ESREL 2012, vol. 8, pp. 6645–6653 (2012)

6. Laskowski, D., Łubkowski, P., Kwaśniewski, M.: Identyfikacja stanu zdatności usług sieci bezprzewodowych/Identification of suitability services for wireless networks. Przegląd Elektrotechniczny 89(9), 128–132 (2013)

7. Siergiejczyk, M., Krzykowska, K., Rosiński, A.: Reliability assessment of cooperation and replacement of surveillance systems in air traffic. In: Zamojski, W., Mazurkiewicz, J., Sugier, J., Walkowiak, T., Kacprzyk, J. (eds.) Proceedings of the Ninth International Conference on DepCoS-RELCOMEX. AISC, vol. 286, pp. 403–411. Springer, Heidelberg (2014)

8. Siergiejczyk, M., Krzykowska, K., Rosiński, A.: Parameters analysis of satellite support system in air navigation. In: Selvaraj, H., Zydek, D., Chmaj, G. (eds.) Proceedings of the Twenty-Third International Conference on Systems Engineering. AISC, vol. 330, pp. 673–678. Springer, Heidelberg (2015)

9. Siergiejczyk, M., Rosiński, A., Krzykowska, K.: Reliability assessment of supporting satellite system EGNOS. In: Zamojski, W., Mazurkiewicz, J., Sugier, J., Walkowiak, T., Kacprzyk, J. (eds.) New Results in Dependability & Comput. Syst. AISC, vol. 224, pp. 353–363. Springer, Heidelberg (2013)

10. Sumiła, M.: Evaluation of the drivers' distraction caused by dashboard MMI interface. In: Mikulski, J. (ed.) TST 2014. CCIS, vol. 471, pp. 396–403. Springer, Heidelberg (2014)

11. Rosiński, A.: Reliability-exploitation analysis of power supply intransport telematics system. In: Nowakowski, T., Młyńczak, M., Jodejko - Pietruczuk, A., Werbińska – Wojciechowska, S. (eds.) Safety and Reliability: Methodology and Applications - Proceedings of the European Safety and Reliability Conference, ESREL 2014, pp. 343–347. CRCPress/Balkema (2015)

A Group Decision Support Technique for Critical IT Infrastructures

Inna Skarga-Bandurova, Maxim Nesterov, and Yan Kovalenko

Technological Institute of East Ukrainian National University, 59-a Radyansky Avenue, Severodonetsk, Luhansk Region, 93010, Ukraine
skarga_bandurova@ukr.net, nesxam@gmail.com, czech16@email.ua

Abstract. The paper presents a formal semantics of decision-making based on Dempster-Shafer belief structures. We introduced a method of decision support taking into account the subjective expert information formalized in the form of family of estimations based on the combination of hypotheses and ordered weighted average operators. The task is formulated in terms of the belief structures and allows evaluating the minimum and maximum objectives through different types of aggregation operators. In the context of critical infrastructure management, our research shows that the matrix of possible solutions can be represented as a payoff matrix including performance indicators or in the form of a risk matrix corresponding losses on the specific combinations of decisions. To ensure variation in the goals we use different types of ordering alternatives depending on the type of the specific problem. Finally, an illustrative example was given to selecting strategies and prioritizing decisions to mitigate targeted cyber intrusions fit to effective IT-security risk management in different critical application.

Keywords: group decision-making, belief structure, ordered weighted averaging operator, critical infrastructure.

1   Introduction

With a growing of computer society, the problem of cyber security has emerged increasingly. Computer systems and networks tailored to automatic control, monitoring and analyzing industrial and business processes became an exceptional component exercises a significant influence on safety and security of companies, governments and organizations.

To solve the cyber security problems, much of the effort has been focused on the development of better hardware and software solutions with little thought to the human factors of cyber security [1]. However, the humans play a major role in cyber security as they interact with computer system and, for example, in case of industrial application (on nuclear power plants, chemical plants, transportation systems, etc.) they control the inputs and outputs of complex machines. At that rate, allocating human effort to activity is critical since inappropriate allocation can result in hit attacks when they will be unchallenged or human time being wasted. Time pressure, the presence of ambiguous

© Springer International Publishing Switzerland 2015   445
W. Zamojski et al. (eds.), *Theory and Engineering of Complex Systems and Dependability,*
Advances in Intelligent Systems and Computing 365, DOI: 10.1007/978-3-319-19216-1_42

information and the high objectives involved can aggravate the judgments associated with the allocation process. Sometimes a situation gets more complicated when human needs analyze information from different sources under information redundancy (or lack of information) and under confronting with the contradictory facts. To reduce a risk of decision-making and ensure the reliability and accuracy of the decisions the group decision making techniques can be suitable.

Group decision-making is a situation where two or more decision makers are involved in the decision of a joint problem whereas each of them has their own understanding of the problem and the decision consequences (competing hypotheses). Formally, competing hypotheses or conflict set is considered as a set of objects concerning which there is no consensus among at least two experts. Conceptual model M of a typical situation assessment problem in the presence of competing hypotheses

$$M = <A, S, P, D>$$

where A is a set of possible conclusions about the situation (alternatives), a generalization of logic experts; S is a set of baseline data on the situation which is measured in quantitative and qualitative scales; P are the analytical dependences, which provide formation of conclusions $a \in A$ according to the data S; and D are the techniques that allow to select the most important information from S.

2   Statement of the Problem of Decision-Making under Competition

Let A be a set of alternatives $\{A1; A2; ...; Aq\}$ whose values describe variants of the decision; S be a set of object states $\{S1; S2; ...; Sn\}$, characterizing the possible scenarios; values c_{11}; c_{12}; c_{1n}; c_{21}; c_{22}; c_{2n}; c_{n1}; c_{n2}; ...; c_{ln} – are the specific level of effectiveness of the solution corresponding to a specific alternative in a certain situation. Knowledge of the safety conditions fixed in terms of belief structure m. B_1, ..., B_r are the focal elements of m and $m(B_k)$ are the associated weights.

The task involves finding the best alternative that delivers the payoff to the decision makers.

Moreover, to solve the problem, consider the following conditions:

— the presence of subjective quality expert information, characterized by a set of competing hypotheses and requiring aggregation;
— form of the matrix of solutions may vary depending on the selected performance indicators;
— the method should provide support for decision-making, in order to lookup minimal losses as well as for the problem of finding maximum efficiency.

The next sections present the theoretical provisions based on the extended Dempster-Shafer belief structure and the method for automated decision support based on evidence-based reasoning applicable for critical IT infrastructure. We have implemented our method on top of decision-support software tool, so it can be easily adopted to the different IT security risk management tasks. Dempster-Shafer theory has unique advantages in handling uncertainty in critical IT-infrastructures analysis, namely, a means to explicitly account for unknown possible causes of observational data and the ability to deal with the lack of prior probabilities for all events and the ability to combine beliefs from multiple sources [2], [3], [4].

3   The Problem of Decision-Making Using Dempster-Shafer Belief Structures

The Dempster-Shafer belief structure is defined in the space X consisting of a set of n nonzero subsets B_j, $j=1,...,n$, called the focal elements and basic belief assignment m called the mass function or the probability of mass [4] which is denoted as m. It is a mapping function defined as $m: 2^X \rightarrow [0,1]$, satisfying

$$\sum_{j=1}^n m(B_j) = 1, \ \forall \ B_j \subseteq X,$$

$$m(A) = 0, \forall \ A \neq B_j.$$

Model of the belief structure [5] is a distributed evaluation with the levels of believes to represent an effectiveness of alternative for the selected criteria.

Suppose that the criterion is evaluated by a full range of possible situations with n estimated classes, $H = \{H_1; H_2; ..., H_j, ...,H_n\}$, where H_j is the j-th evaluation class.

Without loss of generality, we may assume that H_n is preferred H_{n+1}. This assessment criterion can be represented by the following distribution

$$S(c) = \{F(H_j, m(B_j))\}, \ j = 1, ..., n, \tag{1}$$

where $m(B_j) \geq 0$, $\sum_{j=1}^N m(B_j) \leq 1$.

The function (1) denotes that the criterion is assessed for the class H_n with the level of confidence $m(B_j)$.

Estimation S(s) is complete if $\sum_{j=1}^N m(B_j) = 1$ and incomplete if $\sum_{j=1}^N m(B_j) < 1$. A special case is $\sum_{j=1}^N m(B_j) = 0$ which means a complete disregard for the criterion.

There are two measures associated with the belief structures – plausibility (Pls) and belief (Bel) or similarity [6].

Pls is defined as the measure $Pls: 2^X \rightarrow [0,1]$, such that

$$Pls(A) = \sum_{A \cap B_j \neq \emptyset} m(B_j).$$

Similarly, the confidence measure is defined as $Bel: 2^X \rightarrow [0,1]$, such that

$$Bel(A) = \sum_{B_j \subseteq A} m(B_j).$$

Bel represents precise support, while *Pls* is a possible support. Through these measures is possible to submit confidence interval A as [*Bel*(A), *Pls*(A)]. This interval is considered respectively as the lower and upper levels of trust.

Schafer model defines distinguishing frame, Θ, as the space of all possible solutions.

Dempster rule allows for each set of initial subsets (focal elements) on the entire set of input data to generate the resulting subsets and calculate their confidence level (combined measure of confidence (probability mass)). Dempster's rule for combining hypotheses X and Y is performed by orthogonal summing corresponding confidence measures m_1 and m_2

$$m_{12}(A) = \frac{\sum_{X \cap Y=A} m_1(X) m_2(Y)}{1-k_{12}} \tag{2}$$

where

$$k_{12} = \sum_{X \cap Y=\emptyset} m_1(X) m_2(Y). \tag{3}$$

The main problem with this approach in the design of automated decision support systems is the presence of a normalizing factor $(1 - k_{12})$ which completely ignores the conflict. Practically, when k_{12} equal 1, the combination rule of evidence (2) is not determined mathematically.

To solve this problem, a number of models combining different hypotheses were developed, among them models of D. Dubois et al. [7], E. Lefevre et al. [8], C. Murphy [9], P. Smets [10], R. R. Yager et al. [11].

In this work we use calculation rule [12] by selecting $\omega_m(\Theta) = 1$ and $\omega_m(A \neq \Theta) = 0$: $m(\emptyset) = 0$

$$m(A) = \sum_{X \cap Y=A} m_1(X) m_2(Y), \tag{4}$$

$$m(\Theta) = m_1(\Theta)m_2(\Theta) + \sum_{X \cap Y=\emptyset} m_1(X) m_2(Y) = \omega(\Theta) + \omega(\emptyset),$$

if $= \Theta$, where $\forall A \in 2^\Theta, A \neq \emptyset$.

In critical applications (for distributed team decision-making, or under interdisciplinary incomprehension, for example) the individual solutions can be compared to formal aggregation procedures to select a general consensus. The final solution must be obtained from the synthesis of performance degrees of criteria. To this end, the aggregation of information is fundamental.

One of the most common methods of aggregation is a method using the ordered weighted averaging (OWA) operator, introduced by Ronald R. Yager in [13]. Since its description, the given operator has been used in a wide range of applications [14], [15], [16], [17], [18]. It provides a parameterized family of operators, including arithmetic mean, geometric mean (the ordered weighted geometric (OWG) operator); harmonic mean (the ordered weighted harmonic (OWH) operator); a set of nonadditive integrals (Sugeno integral, Choquet integral); weighted minimum; weighted maximum, as well as enhanced operators of ordered weighted average [19],

[20]. Here, we consider two orders of the OWA operator – ascending and descending, as well as some of the main results of their use in the decision-making model.

Assume that X is a set of information sources, $f(x_i)$ is a value supplied $x_i(c)$, σ and s are the permutations such that $a_{\sigma(i)} \geq a_{\sigma(i+1)}$, $a_{s(i)} \leq a_{s(i+1)}$.

Then, according to [13], the OWA operator in ascending order is calculated by (5), descending one by (6).

$$OWA_\sigma = \sum_{i=1}^{n} w_i a_{\sigma(i)} \tag{5}$$

$$OWA_s = \sum_{i=1}^{n} w_i a_{s(i)} \tag{6}$$

where w is a weight vector such that: $w_i \in [0,1]$, $\sum_{i=1}^{n} w_i = 1$.

If the characteristics of the individual values represent the relative values of the dynamics, for example, describe the average growth rate, it is advisable to use the geometric mean. In this case, the operators of the weighted geometric mean [21] in ascending and descending order are calculated by (7) and (8), respectively.

$$OWG_\sigma = \prod_{i=1}^{n} a_{\sigma(i)}^{w_i} \tag{7}$$

$$OWG_s = \prod_{i=1}^{n} a_{s(i)}^{w_i} \tag{8}$$

4   The Procedure of Group Decision Making

To get the best alternative in the group decision-making, the following steps are involved:

Step 1. Formation of a decision matrix
Depending on the type of the problem, the matrix of possible solutions can be represented as a payoff matrix including performance indicators, or in the form of a risk matrix consists of financial loss indexes. It corresponds to certain combinations of alternatives to decision-making and possible scenarios

	S_1	S_2	$\cdots$	S_n
A1	c_{11}	c_{12}	$\cdots$	c_{1n}
A2	c_{21}	c_{22}	$\cdots$	c_{2n}
$\vdots$	$\vdots$	$\vdots$	$\vdots$	$\vdots$
Al	c_{l1}	c_{l2}	$\cdots$	c_{ln}

Step 2. Definition of a set focal elements $B \subseteq \Theta$ and the appointment of the main mass of probability to subsets

$$B^1 = \left(B_1^1, B_2^1, \dots, B_i^1, \dots, B_q^1\right),$$

$$B^2 = \left(B_1^2, B_2^2, \dots, B_j^2, \dots, B_\tau^2\right).$$

Step 3. Calculation of belief function for the combined sets using (4)

$$m(B_k) = \sum_{B^1 \cap B^2 = B} m_1(B_i^1) m_2(B_j^2).$$

Step 4. Determination of the weight coefficients collection used in the aggregation functions for the individual sets of focal elements: $w = (w_1, w_2, ..., w_n)$ such that $w_j \in [0,1]$; $\sum_{j=1}^{n} w_j = 1$.

To calculate weighting values w_j ($I=1$, ..., n) we use formula [22]. Each weight can be obtained by

$$w_j = Q\left(\frac{j}{n}\right) - Q\left(\frac{j-1}{n}\right) \tag{9}$$

Where Q is a function of fuzzy linguistic quantifiers, proposed by Zadeh [23] and defined as

$$Q(r) = \begin{cases} 0, \ if \ r < \alpha, \\ \frac{r-\alpha}{\beta-\alpha}, \ if \ \alpha \leq r \leq \beta \\ 1, \ if \ r > \beta. \end{cases} \tag{10}$$

1) $Q(0) = 0, \ Q(1) - 1$;

2) $r < t \Rightarrow Q(r) \leq Q(t)$;

3) $\sum_{j=1}^{n} w_j = \sum_{j=1}^{n} \left(Q\left(\frac{j}{n}\right) - Q\left(\frac{j-1}{n}\right) \right) = Q(1) - Q(0) = 1.$

Quantifier Q in (10) is defined as a linear membership function for all $\alpha, \beta, r \in [0,1]$.

The values α, β are determined depending on the linguistic meaning of the quantifier.

Step 5. Calculating a set N_{ik}, which is formed when the i-th alternative has selected and k-th focal element, $\forall i, k: N_{ik} = \{c_{ij} | s_j \in B_k\}$.

Step 6. Ordering N_{ik} sets for each of the criteria

$$OWA_\sigma, OWG_\sigma: s_1 > s_2 >, ..., > s_j >, ..., > s_{n-1} > s_n,$$

$$OWA_s, OWG_s: s_1 < s_2 <, ..., < s_j <, ..., < s_{n-1} < s_n,$$

$$\forall s_j \in N_{ik}, \quad j = 1, ..., n.$$

Step 7. Calculation of aggregated values M_{ik}

$$M_{ik} = \sum_{j=1}^{n} w_j \cdot s_j.$$

Step 8. Calculation of the expected value of the overall index for each alternative

$$C_i = \sum_{k=1}^{r} M_{ik} \cdot m(B_k).$$

Step 9. Ordering and selection of an alternative in accordance with the objectives and the current rules.

5   Numerical Example

To illustrate the method, let's suppose there is a problem selecting strategies to mitigate targeted cyber intrusions. This problem can be solved by combining subjective threat judgment information received from the decision makers based on their professional experience.

Planning team has to identify the best mitigation actions that can be readily implemented but because of funding, technical support, and other causes may not be immediately available for every action. Therefore, it is necessary to prioritize the most suitable mitigation actions to implement in the target system.

1. Assume that the decision problem has four mitigation strategies (alternatives A1, A2, A3, A4). To each strategy, we attribute generalized metrics, which allow assessing the mitigation actions in four process areas: s_1 - vulnerability management, s_2 - patch management, s_3 - configuration management, and s_4 - incident management as a base for analysis.

	s_1	s_2	s_3	s_4
A1	10	40	20	30
A2	15	20	25	30
A3	40	30	10	20
A4	40	50	10	30

2. Assume further that there are two groups of experts, each of that defined its own judgment concerning the best mitigation actions and used the model of the belief structure for each alternative on each criterion as follows.

Group 1: ({s1, s2, s4}, 0,8; {s2, s3, s4}, 0,1; {s2, s4}, 0,1).
Group 2: ({s1, s2, s4}, 0,5; {s2, s3, s4}, 0,4; {s2, s4}, 0,1).

Then the set of focal elements to merging sets can be represented as follows:

	$\{s_1,s_2,s_4\}$ 0,8	$\{s_2,s_3,s_4\}$ 0,1	$\{s_2,s_4\}$ 0,1
$\{s_1,s_2,s_4\}$ 0,5	$\{s_1,s_2,s_4\}$ 0,4	$\{s_2,s_4\}$ 0,05	$\{s_2,s_4\}$ 0,05
$\{s_2,s_3,s_4\}$ 0,4	$\{s_2,s_4\}$ 0,32	$\{s_2,s_3,s_4\}$ 0,04	$\{s_2,s_4\}$ 0,04
$\{s_2,s_4\}$ 0,1	$\{s_2,s_4\}$ 0,08	$\{s_2,s_4\}$ 0,01	$\{s_2,s_4\}$ 0,01

3. Calculation of the belief function carried out by (4):

B1	$\{s_1,s_2,s_4\}$	0,4
B2	$\{s_2,s_3,s_4\}$	0,04
B3	$\{s_2,s_4\}$	0,56

4. Determination of weight coefficients w, which are used for aggregation functions for the individual sets of focal elements. Let $w_1 = (0,4; 0,6)$, $w_2 = (0,3; 0,4; 0,4)$.
5. Defining sets N_{ik}.
6. Ordering sets $N_{ik}: a_{\sigma(i)} \geq a_{\sigma(i+1)}$, and $a_{s(i)} \leq a_{s(i+1)}$
7. Calculation of aggregated values M_{ik} performed by (11), the results are presented in Table 1.

Table 1. The results of the calculation of aggregate values

Operator	Aggregate value											
	M_{11}	M_{12}	M_{13}	M_{21}	M_{22}	M_{23}	M_{31}	M_{32}	M_{33}	M_{41}	M_{42}	M_{43}
OWA$_s$	31	34	36	24,5	28	26	34	23	26	45	35	42
OWA$_\sigma$	28	32	34	23	27	24	32	21	24	43	31	38
OWG$_s$	24,2	29,8	35,6	21,6	25,1	25,5	29,8	19,1	25,5	40,1	26,5	40,7
OWG$_\sigma$	21,1	27,8	33,6	20,1	24,1	23,5	27,8	17,1	23,5	38,1	22,5	36,8

8. Calculation of the overall index is performed by (12). The results are summarized in Table 2.

Table 2. The results of the calculation of the generalized index

	OWA$_s$	OWA$_\sigma$	OWG$_s$	OWG$_\sigma$
A1	33,92	31,52	30,81	28,37
A2	25,48	23,72	23,92	22,16
A3	29,08	27,08	26,96	24,96
A4	42,92	39,72	39,89	37,87

9. The choice of an alternative is performed in accordance with the preference rule presented in Table 3.

Table 3. The results ordering alternatives and prioritizing the most suitable mitigation actions in the target system

Operators	The order of preference alternatives
OWA$_s$, OWG$_s$	$A2 > A3 > A1 > A4$
OWA$_\sigma$, OWG$_\sigma$	$A4 > A1 > A3 > A2$

If the main goal of intrusion mitigation programs formulated for improving system security or increasing confidence of security we will use descending order of operators, otherwise for example for risk reduction, ordered weighted operators in ascending order will applicable. For OWA$_\sigma$, OWG$_\sigma$ operators, as the best solution chosen A4 because it gives the highest expected value. For OWA$_s$ and OWG$_s$ operators selected variant A2, since in these cases it is believed that the best result is the lowest.

6 Conclusions

We propose a group decision support technique permits the use of subjective expert information formalized in the form of family of estimations and based on the combination of hypotheses and ordered weighted average operators. The task is formulated in terms of the belief structures and allows evaluate the minimum and maximum objectives. As an example, the problem of prioritization for cyber intrusion mitigation programs is considered. Another interesting issue to consider in the context of critical IT-infrastructures is the group decision support in following areas:

- Prediction the situation change trends, when mitigation actions have not undertaken.
- Prediction safety/security trends in case the decisions were not taken.
- Selection factors that have maximum impact on the critical IT-infrastructure attributes (safety, security, reliability, etc.).
- Evaluation of the impact of individual measures or groups of measures on the critical IT-infrastructure attributes.
- Factoring intrusion-sensitive activity.
- Failure effects evaluation and consequence analysis for individual mitigation programs measures or groups of measures, etc.

The proposed method is effective for decision support under competing hypotheses and helps in enhancing cyber security team performance. The formal basis for automated reasoning based on the theory of Dempster-Schafer has been successfully extended and applied by authors to a number of problems including multisensory data fusion and analysis of process data. Where those decisions relate to the involvement of human analyst resources to activities, this technique essentially improves the efficiency of group decision-making in critical environments.

References

1. Finomore, V., Sitz, A., Blair, E., Rahill, K., Champion, M., Funke, G., Mancuso, V., Knott, B.: Effects of Cyber Disruption in a Distributed Team Decision Making Task. In: Proceedings of the Human Factors and Ergonomics Society Annual Meeting, vol. 58, pp. 415–418 (2014)
2. Zhanga, Y., Huanga, S., Guob, S., Zhu, J.: Multi-sensor Data Fusion for Cyber Security Situation Awareness. Procedia Environmental Sciences 10, 1029–1034 (2011)
3. Zomlot, L., Sundaramurthy, S.C., Luo, K., Ou, X., Rajagopalan, S.R.: Prioritizing Intrusion Analysis Using Dempster-Shafer Theory. In: AISec 2011 Proceedings of the 4th ACM Workshop on Security and Artificial Intelligence, pp. 59–70 (2011)
4. Hall, D.L., McMullen, S.A.H.: Mathematical Techniques in Multisensor Data Fusion, 2nd edn. Artech House, Inc. (2004)

5. Yang, J.B., Singh, M.G.: An evidential reasoning approach for multiple attribute decision making with uncertainty. IEEE Transactions on Systems, Man, and Cybernetics 24(1), 1–18 (1994)
6. Shafer, G.: A mathematical theory of evidence. Princeton University Press, Princeton (1976)
7. Dubois, D., Prade, H.: Representation and combination uncertainty with belief functions and possibility measures. Computation Intelligence 4, 244–264 (1988)
8. Lefevre, E., Colot, O., Vannoorenberghe, P.: Belief functions combination and conflict management. Information Fusion 3(2), 149–162 (2002)
9. Murphy, C.: Combining belief functions when evidence conflicts. Decision Support Systems 29(1), 1–9 (2000)
10. Smets, P.: The combination of evidence in the transferable belief model. Pattern Analysis and Machine Intelligence 12, 447–458 (1990)
11. Yager, R.R., Kacprzyk, J., Fedrizzi, M. (eds.): Advances in the Dempster-Shafer theory of evidence. John Wiley & Sons, Inc., NY (1994)
12. Yager, R.R.: On the Dempster-Shafer framework and new combination rules. Information Science 41(2), 93–137 (1987)
13. Yager, R.R.: On ordered weighted averaging aggregation operators in multi-criteria decision making. IEEE Trans. Systems, Man and Cybernetics 18, 183–190 (1988)
14. Zhou, S.-M., Chiclana, F., John, R.I., Garibaldi, J.M.: α-Level Aggregation: A Practical Approach to Type-1 OWA Operation for Aggregating Uncertain Information with Applications to Breast Cancer Treatments. IEEE Transactions on Knowledge and Data Engineering, 1–14 (2011)
15. Merigó, J.M., Gil-Lafuente, A.M.: Decision making with the OWA operator in sport management. In: XIV Congreso Español Sobre Tecnologías y Lógica Fuzzy, ESTYLF 2008, pp. 1–7 (2008)
16. Merigó, J.M.: Probabilistic decision making with the OWA operator and its application in investment management. In: European Society for Fuzzy Logic and Technology, EUSFLAT, pp. 1364–1369 (2009)
17. Deng, X., Li, Y., Deng, Y.: A group decision making method based on Dempster-Shafer theory of evidence and IOWA operator. Journal of Computaional Information Systems 8(9), 3929–3936 (2012)
18. Miller, S., Appleby, S.: Evolving OWA operators for cyber security decision making problems. In: IEEE Symposium on Computational Intelligence in Cyber Security (CICS), pp. 15–22 (2013)
19. Merigó, J.M., Gil-Lafuente, A.M.: A method for decision making with the OWA operator, http://www.doiserbia.nb.rs/img/doi/1820-0214/2012/1820-02141100044M.pdf (access date: January 2015)
20. Torra, V., Narukawa, Y.: Modeling decisions: information fusion and aggregation operators. Springer (2007)
21. Yager, R.R., Xu, Z.S.: The continuous ordered weighted geometric operator and its application to decision making. Fuzzy Sets and Systems 157, 1393–1402 (2006)
22. Yager, R.R.: Quantifier guided aggregation using OWA operators. Int. J. of Intelligent Systems 11, 49–73 (1996)
23. Zadeh, L.A.: A computational approach to fuzzy quantifiers in natural languages. Computer and Mathematics with Applications 9, 149–184 (1983)

Planning of Electric Power Distribution Networks with Reliability Criteria

Ioannis I. Stamoulis, Agapios N. Platis, and Vassilis P. Koutras

Department of Financial and Management Engineering, University of the Aegean,
41 Koudouriotou Street, Chios, GR-82100, Greece
john.stamoulis@gmail.com,
platis@aegean.gr,
v.koutras@fme.aegean.gr

Abstract. With the development of the Distributed Generation (DG), the role of power distribution networks becomes more and more important, and their configuration tends to change from radial to meshed. In this paper, algorithms for planning meshed networks are proposed. The problem of network planning has been transformed to an operational research problem. The algorithms are automated to reduce the planning time. Additional checks have been added to reduce the total length of the final network. A reliability assessment has been proposed to locate the weak spots of a network and to compare different solutions. Finally an application of the proposed algorithms is demonstrated.

Keywords: Network reliability, Network planning, Electric power distribution systems.

1 Introduction

Even today, power distribution networks are used as passive terminations of transmission networks, simply supplying customers efficiently and reliably [1]. This use tends to change with the development of Distributed Generation (DG): Small and dispatched generation units, usually connected to the distribution network [2].

The main three factors leading to the development of the DG are the protection of the environment, the saturation of the majority of the transmission networks and the liberation of the electric power market [3]. In [2], an analysis of the DG is presented, and in [4], the superiority of the meshed arrangements in the presence of DG is underlined.

So, the distribution networks tend to change, by adding connections (expansion), from radial, where each load point can be supplied by only one supply route, to meshed, which allows alternative routes in case of emergency [5].

A distribution network is planned maximizing the ability of the network to meet the forecasted loads, for a given level of service, while at the same time minimizing its cost [12]. In [6], an algorithm is proposed for assessing reliability indexes of general distribution systems. In [1], mathematic expressions for the calculation of the costs applying on distribution networks are presented.

455

W. Zamojski et al. (eds.), *Theory and Engineering of Complex Systems and Dependability,*
Advances in Intelligent Systems and Computing 365, DOI: 10.1007/978-3-319-19216-1_43

Network planning is a complex problem, with many variables and constraints. Assumptions have to be made carefully, so the problems solution comes in a reasonable time and as close to the optimum as possible. There are plenty of algorithms proposed for network planning; most of them can be applied for planning radial networks, such as [7] – [11], and only a few for planning meshed networks, as [1], [5], [12] and [13].

The algorithms presented in this paper are based on the meshed network building algorithms MNB-1 and MNB-1V [13]. First they were programmed and tested on Matlab software. Then, these automated algorithms have been improved in order to reduce the total length of the final network.

2   Problem Statement

The goal is to connect all the given load points in a meshed network, while at the same time minimizing the total cost. So, algorithms for meshed network planning have to be developed.

These algorithms have to be programmed and automated, in order to eliminate the drawbacks originating from the involvement of the planner (human factor). Then, they have to be applied for network expansion. Also, a reliability assessment of the network has to be made, to assure the level of service and to compare different solutions.

In this paper, the positions and the load of all the load points are considered known. After being placed in a Cartesian system x0y, the distances between the load points are calculated [13]. The connection cost between two load points i and j equals to the Cartesian distance $l_{i,j}$ between them [13], in order to simplify the problem. In other cases, $l_{i,j}$ could be considered as the real or natural distance between i and j.

If we set $k_{i,j}$ as the average connection cost per length unit, the connection cost can be calculated as $k_{i,j} \cdot l_{i,j}$ [13]. That way, the problem of minimizing the total cost can be transformed to a problem of minimizing the total length and vice-versa [13].

In order to plan the network as an operational research problem, a non-oriented graph is considered, depicting the load points as nodes and the connections as edges. Also, we consider as degree of a node, the number of edges incident in this node [13].

3   Proposed Algorithms

3.1   The Algorithms MNB-1 and MNB-1V

The Meshed Network Building (MNB) algorithms are presented in [13]. In the beginning, the network consists of all the feasible edges. Then, the algorithm repeatedly eliminates edges from the longest to the shortest, that connects nodes with a degree of three (3) or higher. But this constraint is not efficient for securing the inexistence of radial arrangements.

An example is shown in Fig. 1. According to the constraint, one of the two dotted lines could be eliminated, but in that case there would be a Loop Radial Supplied (LRS), on the left, which should not be present in meshed networks. So, before eliminating an edge, the formation of a LRS has to be checked.

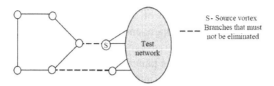

Fig. 1. Loop Radial Supplied 13

According to [13], the steps of the MNB-1 are:

— Step 1: In a Cartesian system x0y the placements of power sources and loads are presented. All the placements will define the vortices set V.
— Step 2. The distances between each two placements are calculated;
— Step 3: The edges with length smaller than the threshold are retained. The edges will define the set LR;
— Step 4: The condition grad(vi) ≥ 2, ∀ v$_i$∈V is evaluated. If the condition is not fulfilled, the algorithm stops;
— Step 5: Two sets, LD and LE, are defined. The set LD contains the available edges. The set LR contains the eliminated edges. Initially, LD = LR and LE = ∅};
— Step 6: As long as LD ≠ {∅}, are executed the operations:
— Step 6.1: The edge with maximum length is determined Lmax∈LD;
— Step 6.2: The incidence of Lmax at vortices with degree at least 3 is verified. If not, Lmax is declared unavailable, is excluded from LD and the algorithm returns to Step 6.1;
— Step 6.3: The construction, after eliminating Lmax, of a loop radial supplied (LRS) is verified. If not, the algorithm goes to Step 6.4. If yes, Lmax is declared unavailable, is excluded from LD and the algorithm returns to Step 6.1;
— Step 6.4: The edge Lmax, which is included in LE and excluded from LD, is eliminated. The degree of the vortex where Lmax is incident is decreased by 1;
— Step 7: The set LB of the edges that forms the meshed network is determined, LB = LR − LE.

In step 6.1, between two or more edges with equal length, which to choose first is important, since may result to different solutions. In this paper, the edge that incidents in the node with the smaller index number is chosen first.

In order to increase the reliability of the network, in the MNB-1V algorithm an extra constraint is added: Until the formation of a meshed network is completed, the elimination of an edge incident in a source load is forbidden. At a next step, if a local loop includes a source, edges incident in nodes with a degree of three (3) or higher can be eliminated [13].

3.2   Automation of the Algorithms MNB-1 and MNB-1V

In order to reduce the side effects of the human factor in running these algorithms, the automation of them is proposed. The first step is to automate the step 6.3, the check for LRS, which can be transformed into the check for radial connection of the nodes i and j, that Lmax is incident. For each of the two nodes, the algorithm searches two independent routes to a source. Independent routes don't share edges, but can share nodes.

For the search of the routes, in this paper, Dijkstra's algorithm (problem 2) [15] is used, after extracting the Lmax. After finding the first route, the used edges are temporarily extracted, and then the search is repeated to find the second route. The same process is followed for both nodes. If these four (4) routes are found, there is no radial connection, so the Lmax can be eliminated.

In rare cases, for the tested node, there is only one route found, even though it is not radial connected. This is because when the first route was extracted, all the other routes were cut off. To overcome this problem, another property of the meshed networks is used: If a node stops being supplied when only one edge fails, this node is radial connected.

When two independent routes from the tested node to a source are not found, an extra check (Fig. 2) is applied. After extracting Lmax, every edge is temporarily and successively extracted one by one. If the tested node is supplied in every loop, it is meshed connected, so the Lmax can be eliminated.

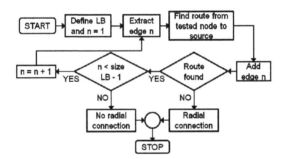

Fig. 2. Check for radial connection of a node

3.3   Improvements of the Algorithms MNB-1 and MNB-1V

Meeting the Demand of All Load Points
In some cases, the proposed solution does not have the form of a unique network. The separation of the network does not violate any of the constraints of the algorithm MNB-1(V), but in a sub-network the power demand may exceed the supply. To avoid this, before the separation, a capacity check in every sub-network is recommended.

In order to notice a separation, the algorithm keeps track of the connections among the sources, by finding routes between them. If the sources separate into two groups when an edge is eliminated, the network is separated too, and every sub-network corresponds to one group of sources and vice-versa. Then, the algorithm sums up the capacity of the sources in the same group and compares them to the load of the nodes that are connected to this group.

Consequently, after the verification of the inexistence of radial connection, in case of separation, follows the capacity check of the sub-networks. If there is no lack of power in any of them, the edge Lmax can be eliminated.

Reduction of the Total Length
Eliminating edges from the longest to the shortest, in the algorithm MNB-1(V), there is the possibility to eliminate an edge (Lmax) resulting in maintaining two others with greater total length. An example is illustrated in Fig. 3.

When the Lmax (3-5) is eliminated the edges 3-9 and 5-9 have to remain to avoid radial connections. But, if the Lmax is not eliminated, then the edges 3-9 and 5-9 can be eliminated. In this case the total length of the remaining edges is smaller, since ($l_{3,5}$ < $l_{3,9}$ + $l_{5,9}$).

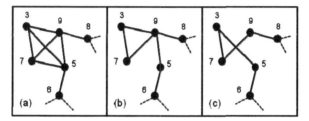

Fig. 3. Special case of (no) elimination of an edge

To avoid this unpleasant situation, the case of what would have happened if the edge Lmax was not eliminated should be examined. If the total length of the final network is shorter in the second case, the Lmax should not be eliminated.

In order to apply this check, for every edge elimination during the first run of the algorithm, the algorithm should run again without this elimination. Also, in the MNB-1V algorithms, the extra constraint of the connections to the sources has to be transformed. The new form of the constraint is that in the final network the degree of each source must equal to three (3) or higher.

In this step, the order of the checks is important. It is impossible to know which order will give the best result, and it is inefficient to run the check in every possible order. So, there are two main options:

- If the check runs from the longest edge to the shortest, as the main algorithm, it could result in keeping long edges instead of short ones. (Reduction a)
- If the check runs from the shortest edge to the longest, the decision about an edge may turn wrong after the decision about a longer edge, since the check runs in a reverse order to the main algorithm. (Reduction b)

3.4   Using the Algorithms for Expansion Planning

All the algorithms above can be used for the planning of a network that does not currently exist. In order these algorithms to be applicable for the expansion of an existing network, the length (or the cost) of every existing edge is considered 0.

4   Reliability Assessment

First, the failure probabilities are considered: For an edge it is considered proportionate to its length and for a node (source or not) is considered 0. Also, according to the properties of the meshed networks, there is the possibility of alternative routes. So the power supply to a node is interrupted when at least two (2) edges of the same loop fail. Thus, all possible pairs of edges are found, and their length is summed up.

For each node, the algorithm checks the elimination of which pairs of edges causes interruption of service for this node (by searching for a route from the node to a source), and their length is summed up. Then, the length of the pairs that cause interruption is divided by the length of all possible pairs. Finally, the result is multiplied with 1000, to calculate the probability per thousand of an interruption of service for the node, when exactly two edges fail (Fig. 4).

In the reliability assessment it could be helpful to locate the weak spots of the network. So, an estimation of the importance of each edge is recommended. Thus, the algorithm counts how many times each edge appeared at the interruptions of the process described above. Then, it multiplies the times of appearance with the length of the edge (Fig. 4). In addition, it could be multiplied with the energy not supplied or the importance of the node not supplied (e.g. public services versus residences).

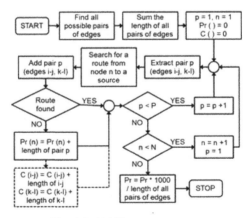

Fig. 4. Reliability assessment

5   Application - Example

All the proposed algorithms were used for planning a network of 72 nodes, with 3 sources and 69 load points [12], same as [13]. Initially, all edges longer than 4km (threshold) are eliminated (Fig. 5). Fig. 6-11 illustrate the final meshed networks after applying each of the proposed algorithms.

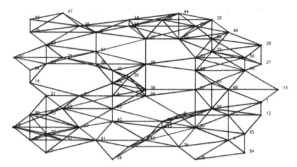

Fig. 5. Initial network after threshold [13]

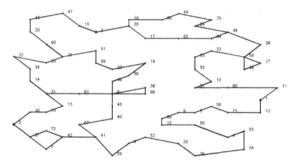

Fig. 6. Automated MNB-1 (78 edges, 152.70km)

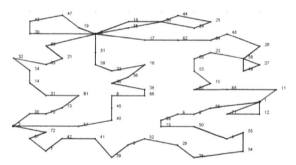

Fig. 7. Automated MNB-1V (78 edges, 164.75km)

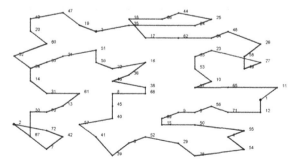

Fig. 8. MNB-1 length reduction a (73 edges, 150.27km)

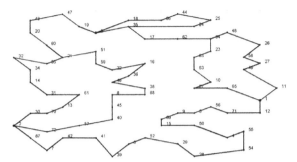

Fig. 9. MNB-1V length reduction a (75 edges, 152.90km)

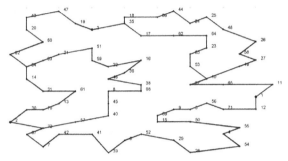

Fig. 10. MNB-1 length reduction b (74 edges, 145.80km)

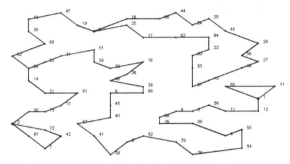

Fig. 11. MNB-1V length reduction b (74 edges, 149.89km)

Table 1. Solutions given by the proposed algorithms

Algorithm	Total length (km)	Number of edges	Average running time (sec)
MNB-1	157.90	79	Unknown
MNB-1V	167.70	80	Unknown
MNB-1 automated	152.70	78	3
MNB-1V automated	164.75	78	3
MNB-1 reduction a	150.27	73	2200
MNB-1V reduction a	152.90	75	2200
MNB-1 reduction b	145.80	74	2200
MNN-1V reduction b	149.89	74	2200

For all the networks above, the probability of interruption of every node has been calculated, as described in chapter 4. The results are illustrated in the chart (Fig. 12), which shows the number of nodes with a probability of interruption that exceeds each limit.

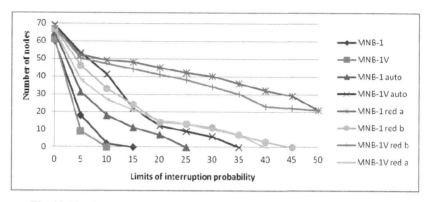

Fig. 12. Number of nodes with probability of interruption exceeding each limit

6   Conclusions

In this paper, algorithms for meshed network planning are proposed. The planning is described as an operational research problem; a non-oriented graph is considered, as well as the load points as nodes and the connections as edges.

The algorithms MNB-1 and MNB-1V from [13] are presented, and the automation of them is proposed, in order to speed up the planning process. Then, two checks are added; a check for the meeting of the demand of all load points, and a check to reduce the total length of the final network. After that, the use of the algorithms for expansion planning is described.

The network planning not only reduces the total length (cost), but it ensures a high level of service (less energy not supplied - ENS). Thus, a reliability assessment is proposed, for locating the week spots of a network and for comparing two networks.

The example illustrates that a network can be planned in a few seconds, by having the algorithm programmed and automated. In addition, it shows that the added checks indeed reduce the total length of the final network. It also shows that reducing the number of the edges weakens the network, increasing the probability of interruptions.

References

1. Celli, G., Ghiani, E., Loddo, M., Pilo, F.: An Heuristic Technique for the Optimal Planning of Meshed MV Distribution Network. In: IEEE Russia Power Tech, PowerTech, St.Petersburg, June 27 (2005)
2. Dondi, P., Bayoumi, D., Headrli, C., Julian, D., Suter, M.: Network integration of distributed power generation. Journal of Power Sources 106(1-2), 1–9 (2002)

3. Hadjsaid, N., Alvarez-Hérault, M.-C., Caire, R., Raison, B., Descloux, J.: Novel architectures and operation modes of Distribution Network to increase DG integration. In: PES General Meeting, PES 2010 (2010)
4. Celli, G., Pilo, F., Pisano, G., Allegranza, V., Cicoria, R., Iaria, A.: Meshed vs. Radial MV Distribution Network in Presence of Large Amount of DG. In: IEEE PES Power Systems Conference and Exposition 2, pp. 709–714 (2004)
5. Božić, Z., Hobson, E.: Urban underground network expansion planning. IEE Proc. -Gener. Transm. Distnb. 144(2), 118–124 (1997)
6. Wang, Z., Shokooh, F., Qiu, J.: An Efficient Algorithm for Assessing Reliability Indexes of General Distribution Systems. IEEE Transactions on Power Systems 17(3), 608–614 (2002)
7. Aoki, K., Nara, K., Satoh, T., Kitagawa, M., Yamanaka, K.: New Approximate Optimization Method for Distribution System Planning. IEEE Transactions on Power Systems 5(1), 126–132 (1990)
8. Nara, K., Satoh, T., Kuwabara, H., Aoki, K., Kitagawa, M., Ishihara, T.: Distribution Systems Expansion Planning by Multi-stage Branch Exchange. IEEE Transactions on Power Systems 7(1), 208–214 (1992)
9. Nara, K., Kuwabara, H., Kitagawa, M., Ohtaka, K.: Algorithm for Expansion Planning in Distribution Systems Taking Faults into Consideration. IEEE Transactions on Power Systems 9(1), 324–330 (1994)
10. Miranda, V., Ranito, J.V., Proença, L.M.: Genetic Algorithms in Multistage Distribution Network Planning. IEEE Transactions on Power Systems 9(4), 1927–1933 (1994)
11. Jian, L., Wenyu, Y., Jianming, Y., Meng, S., Haipeng, D.: An Improved Minimum-cost Spanning Tree for Optimal Planning of Distribution Networks. In: Proceedings of the 5th World Congress on Intelligent Control and Automation, Hangzhou, P.R. China, June 15-19, pp. 5150–5154 (2004)
12. Glamocanin, V., Filipovic, V.: Open loop distribution system design. IEEE Transactions on Power Delivery 8(4), 1900–1906 (1993)
13. Dumbrava, V., Ulmeanu, P., Duquenne, P., Lazanoiu, C., Scutariu, M.: Expansion Planning of Distribution Networks by Heuristic Algorithms. In: 2010 45th International IEEE Proceedings of the Universities Power Engineering Conference (UPEC), pp. 1–6 (2010)
14. Dialynas, E.: Reliability Analysis of Technological Systems, in Greek, Simeon, Athens (1998)
15. Dijkstra, E.W.: A Note on Two Problems in Connexion with Graphs. Numerische Mathematik 1(1), 269–271 (1959)

Selected Aspects of Modeling the Movement of Aircraft in the Vicinity of the Airport with Regard to Emergency Situations

Paulina Stańczyk and Anna Stelmach

Warsaw University of Technology, Faculty of Transport
paulina.stanczyk@gmail.com,
ast@wt.pw.edu.pl

Abstract. Due to the steadily increasing air traffic, the problem of selection of the appropriate landing of the aircraft in the absence of the possibility of a touchdown at the destination airport is becoming increasingly important. The article presents the results of the analysis of elements affecting safety and traffic flow during the operation of an aircraft emergency landing at the airport. The article presents the essential elements of airport infrastructure affecting the safe touchdown the aircraft, as well as presents the basic elements that affect the occurrence of an emergency landing.

Keywords: airport area, emergency landing, aircraft crash.

1   Introduction

In the era of the dynamic development of the sciences, uninterrupted security is a priority of each of the transport sector. Security depends on many factors that determine the quality, the number of operations performed and the performance of the air transport system whereas the means of transport which is the aircraft. Factors should be understood by any action case, condition or situation whose existence or non-existence increases the probability of unsuccessful termination of the flight.

Technical and organizational complexity of the air transport system, a multitude of aviation personnel, aircraft operation in all weather conditions are the source of a variety of factors affecting flight safety. Detailed calculation taking into account all the factors of complexity air transport system is practically impossible [7].

This article is an introduction to research aimed at developing a concept model for simulation studies, designating a place of touchdown of the aircraft crash-landed at the airport using the developed mathematical model. It sets out the essential elements that aim to provide the required level of safety during flight. The first element affecting the quality of the flight is to determine the airspace of the flight.

2   Characteristics of Airspace, Traffic, The Airport and Flight

In the early days of aviation, aircraft movement took place at the designated routes in the air freely. With increasing interest in the air transport, congestion has increased

© Springer International Publishing Switzerland 2015
W. Zamojski et al. (eds.), *Theory and Engineering of Complex Systems and Dependability,*
Advances in Intelligent Systems and Computing 365, DOI: 10.1007/978-3-319-19216-1_44

significantly in the sky. In order to ensure adequate separation of aircraft, established relevant organizations to ensure flight safety. These organizations have developed adequate rules and standards by which they have been defined:

- Restricted airspaces,
- Danger areas and endangered,
- Control and advisory services,
- Check of the areas, areas of aerodromes and airports, and
- Routes and airways.

In Poland distinguished airspace controlled and uncontrolled.

2.1   Controlled Airspace

Controlled airspace is separated in the area of flight information region, space of defined dimensions in which air traffic control service is provided in relation to controlled flights. The Polish controlled airspace includes:

- Air routes (fixed and contingent) -AWY,
- Flights level,
- Control zones of public and military airports available for air transport - CTR,
- Control areas of airport - TMA.

Figure 1 presents the division of airspace [10].

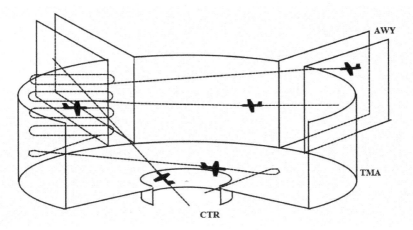

Fig. 1. Division of airspace

2.2   Uncontrolled Airspace

This is the space over the area of land, inland and the territorial sea of the Republic of Poland, entailing the controlled airspace. It is divided into:

- Space of coordinated flights. It is part of the space, not entailing the free flight of space and other specified space, within which the flight of civil aircraft, prior to their commencement, subject to coordination,
- Space of free flights. It is part of the airspace reaching an altitude of 400 m above the ground or water, not covering space of coordinated flights and other specified space in which civil aircraft may operate VFR flights (Visual Flight Rules), with speeds up to 300 km / h and are not subject to coordination prior to the beginning [12].

Another element which is necessary for the safe and efficient flight is to determine the conditions of the air.

2.3   Air Traffic

The specificity of air traffic, space requirements and the condition of the aircraft maintain a certain speed and flight level causes the airways require special design. Air traffic as a movement of all aircraft operating or moving on the maneuvering area is divided into:

- Air traffic controlled as a movement of aircraft covered by the action of air traffic services, held in controlled airspace and airports on the maneuvering area controlled,
- Traffic monitored as the movement of aircraft covered by the action of the air traffic surveillance services taking place in the area of supervised airports and airports on the maneuvering area supervised, as well as in any other area or part of another airport, temporarily assigned to the movement.

Each aircraft flight begins and ends at the specified (in different ways) airport. Type airports also affects the flight, so it becomes essential to also specify the conditions for its implementation.

2.4   Airport

The airport is on the ground or surface water (or any buildings, installations and equipment) intended in whole or in part for the arrival, departure and surface movement of aircraft [4]. Distinguished:

- Controlled airport where the air traffic control service is provided in relation to airport traffic,
- Airports supervised on which air traffic surveillance service is provided in relation to airport traffic,
- Spare airport, airports which aircraft can fly, if it is not possible or not appropriate to flight to the destination airport or landing on it, stands out:
  - After takeoff airport - at which an aircraft can land, if it is necessary shortly after take-off, and it is not possible to use the start airport,
  - On route airport - airport where the aircraft on the route under abnormal or dangerous conditions could land. and
  - Destination airport - airport at which an aircraft can fly, if the landing at the airport of intended landing becomes impossible or impracticable.

2.5   Emergency Landing

Emergency landing is the unscheduled landing, made for technical, meteorological or other reasons.

When choosing an airport as a supplementary special role of the airport operating minima. The carrier is required to determine its use according to the principles recognized by the civil aviation authority, operating minima for each airports used by the aircraft, and to approve the method of determining such minima. For their selection take into account:

- Flight crew,
- Dimensions and characteristics of the runways (as required for different categories of airports)
- Type, maneuverability and performance of the aircraft,
- Technical characteristics and quality of visual installed at the airport and other operational support to the approximation, approach and landing,
- Aircraft equipment for navigation and control of the track during approach to landing and the missed approach procedure for landing
- Barriers to the take-off area, the approach to landing and go-around,
- Barriers in the area of initial climb,
- The means for measuring and methods of administration meteorological conditions.

Along with the occurrence of an emergency situation may lead to a situation in which reality becomes the impossibility of continuing the flight. This situation entails the need to ensure maximum safety and minimum risk to the crew and passengers of the aircraft and, where possible, to maintain the integral whole of its hull. There are many emergency situations, ranging from slight limitations to the performance of the aircraft (often catastrophic) aircraft structural damage, or damage to the powertrain. It is also possible when the speed at which deteriorates performance of the aircraft will determine the kind of action that will be taken. Nevertheless, any discussion on taking appropriate steps should be done early enough to be able to land the plane at full (or with small performance constraints) pilots control [13].

During a dangerous emergency, the most part of the aircraft is safe drive, or structural destruction of the aircraft. This situation forces the immediate landing of the need to ensure the correct orientation of the aircraft in space. If there is no possibility of landing on a dedicated field or airstrip located in the coverage area of the aircraft, it must land "in the adventitious".

In the history of the emergency landing occurred not only on land but also on the river (eg. US Airways Flight 1549 January 15, 2009, due to the precipitation of Canadian geese flock during the climb, the emergency landing on the river Hudson)

3   Reasons for Aircraft Accident near Airport

Aviation accidents are the result of many causes. Typically, there are a result of a number of interconnected with each other reasons. Each of these reasons, considered separately may seem trivial, but in combination with the other, generates a sequence of events seemingly insignificant, which inevitably lead to an accident [2].

Aviation safety affects the interplay of the three main elements of the aviation system (Fig. 2):

- Man (human factor),
- Aircraft (technical factor),
- The environment (impact of natural and artificial environment).

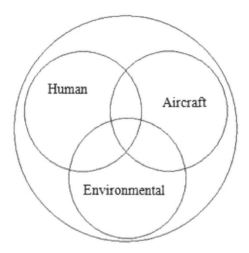

Fig. 2. Model of systematical - causal aviation safety

In most cases, the risks associated with the exercise of air, can be attributed to human activities despite the interaction of the other two categories. An example would be performing some type of tasks that can trigger increased tension in the remote control or enhanced attention, contributing to him committing errors. You may also find yourself in a situation which was not adequately prepared and trained.

This leads to premature wear of components of the aircraft, which in turn increases the load on the pilot and the likelihood of him committing an error.

Factors failure aviation always present numerous problems, particularly when their position is located at the junction of the above-mentioned elements. In view of the fact that man has a stake in each of these three elements, it is suspected that most of the causes of air accidents are always different human errors.

For ease of classification of the causes of air accidents, the International Air Transport Association IATA has established the following division, comprising five main categories of air accident:

- Category I (HUM) - with the participation of the human factor. Covers only accidents, which is the main reason for doing the flight crew error,
- Category II (TEC) - Technical. Include fault plane - malfunctioning installation and systems, errors in the operating instructions, manufacturer errors, the errors in maintenance, and when troubleshooting,

- Category III (ENV) - external conditions. Include the whole of the conditions within which work crew, such as:
  - Meteorological conditions,
  - Air traffic control,
  - Airport facilities,
  - Collisions with birds,
  - Communication: crew-cabin crew, crew-ATC controller, crew - ground personnel,
- Category IV (ORG) - organizational. Include accidents that cause was:
  - Wrong selection or training of the crew,
  - Errors in the management and administration of aviation organization,
  - Inadequate control,
  - Poor flow of information, or improper purposes,
- Category V (O) - Other. Additional group of accidents are, those to whom could not establish sufficient facts [1].

Whatever the reason causing the accident, as well as the type of aircraft, the primary source of information are data from the records of on-board flight recorders. On the basis of these data, obtained from airlines flight operations can be modeled using a method of computer identification or modeling using artificial neural network. Examples presented in [7 - 8] and [10 - 11]. An important fact of the presented method is the need to verify the accuracy of the mapping and analysis of the actual flight independently for each phase of flight

4   Development of Model Concept for Simulation Research

As shown in the previous chapter, air transport and flight safety should be considered multifaceted. First of all, it must be ensured that the high degree of flight service was performed by the relevant services, flight took place in an appropriate airspace, in other words to all the technical elements were provided.

Another very important aspect, in accordance with chapter 3, which has an impact on the fluency and safety of operation is the selection of appropriate staff. Any person participating in the transport process should be thoroughly selected, on trained and qualified.

Another factor that significantly affects the course of the flight is the environment. It is the only element in the midst of the others, as described above, for which we have no control. In the event of an emergency landing can not clearly determine which is the most important aspect. Both the technical condition, as well as training and weather conditions play an important role in the decision-making process, undertaken mainly by aircraft pilots. All these elements have an impact on the trajectory of the flight, so while carrying out research on the simulation model, they are taken into account.

The research aims to develop a mathematical model using artificial neural networks, serving determining the trajectory of an emergency flight, taking into account all the necessary elements[5].

One of the essential elements of the studies using the model and simulation techniques will designate the kinematic equations of motion. These equations will be described the flight path, which characterizes the behavior of the aircraft in the airspace, depending on the space (described in Section 2) as well as the type of geographical coordinates destination.

Will be prepared algorithm of computer program to determine the coordinates of the flight path using the models developed by computer identification and an artificial neural network shown in Table 1.

Table 1. The algorithm for determining the coordinates of flight path of aircraft

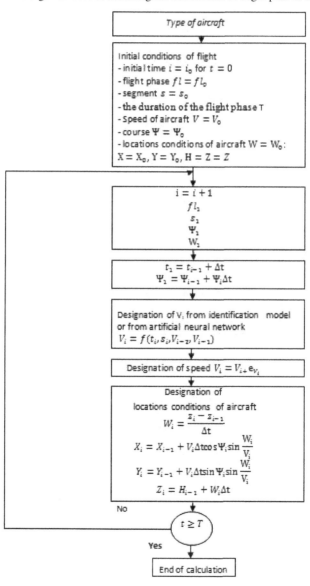

The decisive element with an accuracy of mapping the actual flight by the model is the selection of a suitable form of the mathematical model. Anticipating the results of modeling, assumed the form of the model as follows (Fig. 3):

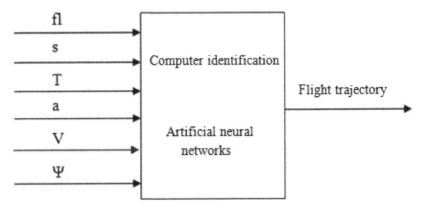

Fig. 3. Simulated model of the trajectory

In the first place will be referred the type of aircraft. Then will be designated the initial conditions of flight, ie .:

- fl - flight phase,
- s - flight segment,
- T - the duration of the flight phase,
- a - the type of accident,
- V - airspeed ($V = V_0$), where V_0 is the velocity at time i = 0,
- Ψ - course
- X, Y, Z – location coordinates of the aircraft ($X = X_0$, $Y = Y_0$, $H_i = Z_i = Z_0$).

Then they set the parameters controlling the aircraft at the time of i + 1, ie. the phase of flight (fl_i) flight segment (s), rate of turn ($Ψ_i$) and vertical rate of climb (W_i). The next step will be the appointment of the algorithm of duration of the flight in the time of i, and depending on the exchange rate $\psi_i = \psi_{i-1} + \dot{\psi}_i \, \Delta t$ as well as the altitude $Z_i = H_{i-1} + W_i \Delta t$.

Then the speed will be determined in the i-th moment of the mathematical model developed by computer identification [11], or optionally with an artificial neural network [12] - $V_i = f(t_i, s_i, a_i, V_{i-2}, V_{i-1})$. The position coordinates of the aircraft will be determined from the relation $X_i = X_{i-1} + V_i \, \Delta t \cos \psi_i \sin \vartheta_i$,

$$Y_i = Y_{i-1} + V_i \, \Delta t \sin \psi_i \sin \vartheta_i \quad Z_i = H_i \text{ wherein } \vartheta_i = \arcsin \frac{W_i}{V_i}. \text{ In the next step}$$

will be determined velocity, including interferences $V_{ZAK_i} = V_i + e_{V_i}$. Wherein e_{V_i} will be calculated from the interference pattern [12].

Using the described algorithm of calculations will be developed computer program TL, which will be used for the calculation and graphical representation of the trajectory. Fig. 4 shows an example of the trajectories of the two flights for the landing phase.

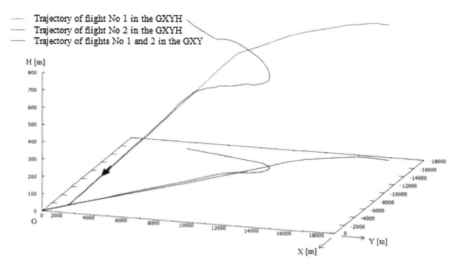

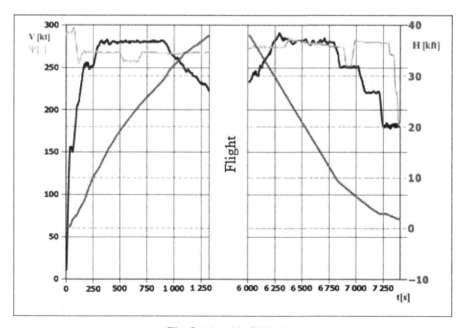

Fig. 4. Flight trajectory in the landing phase

On the other hand, Fig. 5 shows an example of the time course of the aircraft's flight from the take-off and landing phase ending. This figure shows the variation in time of speed, course and height.

Fig. 5. Aircraft's flight time

5   Conclusion

Problem of aircraft emergency landing appeared with the beginning of aviation. That is why it is the subject of many research centers.

The article indicates the basic elements of air traffic affecting flight safety, the main causes of air accidents, as well as presents the basic elements that affect the flight path of the aircraft.

The aim of the paper was to develop an initial concept phase simulation research model aircraft emergency landing at the airport, an indication of the selected flight parameters have a direct impact on the traffic flow, which in turn made it possible to create an algorithm for determining the coordinates of the flight path of the aircraft.

Tests were carried out using artificial neural networks, which are becoming increasingly popular in modern science. They are used not only in technology, but also in economics, medicine, and other fields of science. The studies serve to create a model that can be implemented to flight simulators. These in turn are used to train candidates for the pilots.

With the presented model can be modeled multiple flights, with varying technical conditions of the aircraft, weather conditions and evaluate the decisions taken by the user. This solution will eliminate, already in the process of training pilots, decisions having tragic consequences.

References

[1] Accident Prevention Manual Doc 9422-AN/923 ICAO

[2] JAA Airline Transport Licence Theoretical Knowledge Manual, Oxford Aviation Services Limited (2001)

[3] Chuang, C., Su, S., Hsiao, C.: The Annealing Robust Backpropagation (ARBP) Learning Algorithm. IEEE Transactions on Neural Networks 11 (September 2000)

[4] Cieślak, G.: Infrastruktura krytyczna lotnisk – kuracja ze skutkami ubocznymi: Terroryzm, nr 1 (2010)

[5] Information on safety of flights and parachute jumps in civil aviation of Poland in 2009, the Civil Aviation Authority (April 2010)

[6] Kierzkowski, A., Kisiel, T.: An impact of the operators and passengers behavior on the airport's security screening reliability. In: Source of the Document Safety and Reliability: Methodology and Applications - Proceedings of the European Safety and Reliability Conference, ESREL 2014, pp. 2345–2354 (2015)

[7] Siergiejczyk, M., Rosiński, A., Krzykowska, K.: Reliability assessment of supporting satellite system EGNOS. In: Zamojski, W., Mazurkiewicz, J., Sugier, J., Walkowiak, T., Kacprzyk, J. (eds.) New Results in Dependability & Comput. Syst. AISC, vol. 224, pp. 353–363. Springer, Heidelberg (2013)

[8] Stańczyk, P., Stelmach, A.: Modelowanie ruchu samolotu podczas operacji startu i lądowania z wykorzystaniem sztucznych sieci neuronowych. In: Skorupski, J. (red.) Współczesne problemy inżynierii ruchu lotniczego - modele i metody, pp. 48–57. Oficyna Wydawnicza Politechniki Warszawskiej, Warszawa (2014)

[9] Stańczyk, P., Stelmach, A.: Artificial Neural Networks Applied to the Modeling of Aircraft Landing Phase. In: 10th European Conference of Young Research and Scientists - Proceedings, Zilina, pp. 169–173 (2013) ISBN:978-80-554-0690-9

[10] Stańczyk, P., Stelmach, A.: Analiza wpływu elementów infrastruktury krytycznej na gotowość operacyjną portu lotniczego. Prace Naukowe Politechniki Warszawskiej - Transport, zeszyt nr 92, Warszawa, str. 211–219 (2013)

[11] Stelmach, A.: Modeling of the selected aircraft flight phase using data from Flight Data Recorder. Archives of Transport XXIII(4), 541–555 (2011)

[12] Stelmach, A.: Identyfikacja modeli matematycznych faz lotu samolotu. Oficyna Wydawnicza Politechniki Warszawskiej, Warszawa (2014)

[13] Annex 2 ICAO

[14] Annex 14 ICAO

Identifying and Simulation of Status of an ICT System Using Rough Sets

Marek Stawowy and Zbigniew Kasprzyk

Faculty of Transport, Warsaw University of Technology, Koszykowa 75,
00-662 Warsaw, Poland
{mst,zka}@wt.pw.edu.pl

Abstract. This paper presents a method to identify the operational state of the data communications system (ICT). The article presents the support of the management and operation of process simulation. The presented method based on rough set theory. Presented calculation of coefficient supporting decision about operational state of system ICT. At the end of the article presents the results of a simulation program written by the author. Computer simulations have shown the scope of the analysis method and scope of coefficient supporting decision.

Keywords: operation, identifying status, rough sets, information and communications, simulation.

1 Introduction

In the process of operation of ICT systems (eg: presented in [2]), often there is doubt whether operated system needs repair. Particularly in ICT systems transmitting live images [9]. It is caused by an increasingly effective methods of error correction [5].

This paper presents a method to identify the operating state of the communication system that can assist the management or operation of the simulation process. When the information of system state is incomplete. Especially when the amount of transmitted data and amount of errors are unknown.

The article uses a method which is based on rough set theory [3,4].

Increasingly, decision support method uses rough set theory to detect the state of the system, such as in this publication [1,7]. This publication uses the uncertainty modelling using rough sets. This article is a continuation of the study of 2014 years on methods of decision support using rough set theory [8].

2 Object and Model Description

In many cases, the analytical description of the operating states of the objects or systems, and operational model is very complicated or even impossible. Often this happens with the states of ICT systems when operating in the process can not identify whether the system is capable of unfitness [5]. Especially in situations where the

© Springer International Publishing Switzerland 2015   477
W. Zamojski et al. (eds.), *Theory and Engineering of Complex Systems and Dependability,*
Advances in Intelligent Systems and Computing 365, DOI: 10.1007/978-3-319-19216-1_45

unknown is the effectiveness of corrective methods, which may depend on the effectiveness of the transmitted data. Through the use of rough set such a description can be made available. The method described here allows to support the decision of the state of the object in terms of operational and can be helpful for object management or the management of computer simulation. When the information of system state is incomplete.

Fig. 1 shows a simplified operational state model. Nodes Z1 and Z2 represent states of: operation and repair. Because the case is considered real, not marked on the intensity of the transitions. However transition between state of operation and state of repair was considered given deliberations relate to provided herein [5,10] considerations.

Fig. 1. State diagram of repairable model. Own development on the basis of [7,8].

You can create decision table of decision making for simple, two-state operational model where it is doubtful only transition in an airworthy condition to a state of repair.

Table 1. Decision table for transition into state of repair (failure). Own development.

Datagram	Correct	Retransmission	Correction	Failure
1	YES	NO	NO	NO
2	NO	YES	NO	NO
3	NO	YES	YES	NO
4	NO	NO	YES	NO
5	NO	YES	NO	YES
6	NO	YES	YES	YES
7	NO	NO	YES	YES
8	NO	NO	NO	YES

From Table 1 it can be concluded that:

• Datagram 1 (dg1) indicates a fully operational ICT system.
• Datagram 8 (dg1) indicates a non-operational ICT system.

The rest of the datagram indicates a contradiction (inconsistency of information).

This is a preliminary decision on the basis of the elimination of the same processes which give different results. As a result of deductive reasoning. Can also be seen that the efficiency is influenced by any factor that action is not known. For a full and proper deductions should know the impact of this unknown factor to the analyzed ICT system. However, with the use of inductive inference can be presented to this unnecessary.

3   Basic Definitions

Using the method of rough sets [3,4][1], you can define the lower approximations (1) and upper approximations (2) for a set of datagrams.

$$B_*(X) = \{x \in U : B(x) \subseteq X\}$$

(1)

$$B^*(X) = \{x \in U : B(x) \cap X \neq 0\}$$

(2)

Where:
   U – universe (non-empty set of finite objects, set of datagrams from analysed example),
   X – set, non-empty subset of the universe,
   x – object of the set X,
   B(x) – abstract class containing object x from full relation (B-elementary set),
   $B^*(X)$ – upper approximation of set X,
   $B_*(X)$ – lower approximation of set X,

The following formula describes the difference between upper and lower approximation (3).

$$BN_B(X) = B^*(X) - B_*(X)$$

(3)

$BN_B(X) = \emptyset$ only when upper and lower approximations are equal. Then the set is exact set. In another case, as in the here considered, set is the approximate and exact B-rough set.

   Quantitative measurement of approximation was determined using formula (4).

$$\alpha_B(X) = \frac{|B_*(X)|}{|B^*(X)|}$$

(4)

Where:
   $\alpha_B(X)$ - accuracy of approximation,
   $|B_*(X)|$ - number of lower approximation elements,
   $|B^*(X)|$ - number of upper approximation elements.

When $BN_B(X) \neq \emptyset$ the approximation accuracy rate will be set to 1. As mentioned above, then we are dealing with a set of exact. This factor, in the present case, enable decision support system regarding the operational status of ICT system.

---

[1]   Given multiplicity of rough set definitions, the author was drawing on publications by prof. Zdzislaw Pawlak [3,4].

4   Calculations and Results

Using the definitions of the previous section and description section 2 can be derived inference two lines that lead to calculating accurate approximations for fully operational and non-operational ICT system. It can be assumed that the sum of these factors need not be equal to 1.

We assume that the universe U is the set of datagrams from 1 to 8. Subsets of the universe Xs and Xn are, respectively, a subset of fully operational and non-operational ICT system. Abstraction class B (x) is described in dependency decision table shown in Table 1. As the table describes the universe will be applicable to both fully operational and non-operational ICT system, which are described Xs and Xn subsets of the same universe U.

For the lack of failures to the rough sets can be defined:

- Lower approximation of the state of no failures is a set of datagrams, consisting of datagram 1 i.e. $B*(Xs)=\{$ dg1$\}$.
- Upper approximation of the state of no failures is a set of datagrams, consisting of datagram 1, 2, 3 and 4 i.e. $B^*(Xs)=\{$dg1,dg2,dg3,dg4$\}$.

For instance failure rough sets can be defined:

- Lower approximation of failure is a set of datagrams, consisting of datagram 8 i.e. $B_*(Xn)=\{$ dg8$\}$.
- Upper approximation of the state of no failures is a set of datagrams, consisting of datagram 5, 6, 7 and 8 i.e. $B^*(Xn)=\{$dg5,dg6,dg7,dg8$\}$.

On this basis, using the formula (4) can be calculated accurate approximation coefficients for each of the cases: $\alpha_B(Xs)=0,25$ i $\alpha_B(Xs)=0,25$. This value can serve as an indicator of the accuracy of decision.

To complete the picture, you can influence the decision-making tables of changing it according to the present state of datagrams. Only the most interesting cases were considered.

Table 2. Decision table of transition to state of repair for some datagrams from table 1. Own development.

Datagram	Correct	Retransmission	Correction	Failure
1	YES	NO	NO	NO
2	NO	YES	NO	NO
3	NO	YES	YES	NO
4	NO	NO	YES	NO
5	NO	YES	NO	YES
6	NO	YES	YES	YES
7	NO	NO	YES	YES

For a class abstraction described in Table 2 when removed dg8 provider of system failure, these results were obtained:

Fully operational system:
B∗(Xs)={dg1}; B*(Xs)={dg1,dg2,dg3,dg4};

$$\alpha_B(Xs) = \frac{1}{4} = 0,25$$

(5)

Non-operational system:
B∗(Xn)={ }; B*(Xn)={dg5,dg6,dg7};

$$\alpha_B(Xn) = \frac{0}{3} = 0$$

(6)

For a class abstraction described in Table 3 when removed dg7 provider of system failure, these results were obtained:

Fully operational system:
B∗(Xs)={ dg1,dg2}; B*(Xs)={dg1,dg2,dg3,dg4};

$$\alpha_B(Xs) = \frac{2}{4} = 0,5$$

(7)

Non-operational system:
B∗(Xn)={dg8 }; B*(Xn)={dg5,dg6,dg8};

$$\alpha_B(Xn) = \frac{1}{3} = 0,33(3)$$

(8)

Table 3. Decision table of transition to state of repair for some datagrams from table 1. Own development.

Datagram	Correct	Retransmission	Correction	Failure
1	YES	NO	NO	NO
2	NO	YES	NO	NO
3	NO	YES	YES	NO
4	NO	NO	YES	NO
5	NO	YES	NO	YES
6	NO	YES	YES	YES
8	NO	NO	NO	YES

For all the possibilities can be calculated in the decision accuracy of the indicators [7]:

$$A = \alpha_B(Xs)/\alpha_B(Xn)$$

(9)

5   Simulation

The simulation was performed by one of the authors of all 256 possibilities. The simulation program uses a mask bit datagrams. For example, the table 2 contains mask 127 (01111111) and Table 3 contains 191 (10111111). There are 256 possibilities but Table 4 shows only the results of simulations that are possible for the calculation (without division by zero).

Grouping results as in Table 5, the possibility of carrying the graph shown in Figure 2. This graph shows how to select the right choice.

Table 4. Result of simulation. Own development.

Mask of datagrams	$B^*(Xs)$	$B_*(Xs)$	$B^*(Xn)$	$B_*(Xn)$	$\alpha_B(Xs)$	$\alpha_B(Xn)$	A
00010001	1	1	1	1	1,00	1,00	1,00
00010100	1	1	1	1	1,00	1,00	1,00
00110010	0	1	1	2	0,00	0,50	0,00
00110011	1	2	1	2	0,50	0,50	1,00
00111011	2	3	1	2	0,67	0,50	1,33
01110010	0	1	2	3	0,00	0,67	0,00
01110011	1	2	2	3	0,50	0,67	0,75
01110110	0	2	1	3	0,00	0,33	0,00
01110111	1	3	1	3	0,33	0,33	1,00
10011111	3	4	1	2	0,75	0,50	1,50
10111011	2	3	2	3	0,67	0,67	1,00
10111111	2	4	1	3	0,50	0,33	1,50
11110010	0	1	3	4	0,00	0,75	0,00
11110011	1	2	3	4	0,50	0,75	0,67
11110111	1	3	2	4	0,33	0,50	0,67
11111110	0	3	1	4	0,00	0,25	0,00
11111111	1	4	1	4	0,25	0,25	1,00

Table 5. Result of simulation after grouped. Own development.

$A = \alpha_B(Xs)/\alpha_B(Xn)$	0,25	0,33	0,50	0,67	0,75	1,00
0,25	1,00					
0,33		1,00	0,67			
0,50		1,50	1,00	0,75	0,67	
0,67			1,33	1,00		
0,75			1,50			
1,00						1,00

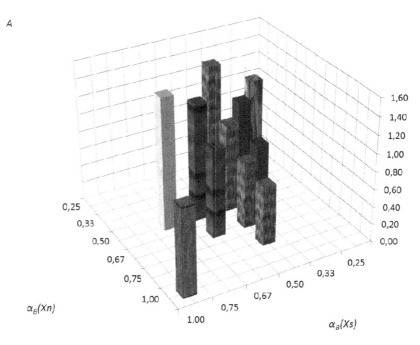

Fig. 2. Factors of decision correctness made on base of result of simulation in table 5. Own development.

Higher value of the coefficient *A* allows you to make better decisions: the system is fully operational. In the case of the lower value of the coefficient *A* better decision is: a system in need of repair and will not operate efficiently. It can be expected that the value of the coefficient *A* will be very high for low values α_B *(Xn)* and lower for low values α_B *(Xs)*.

6   Summary

The results indicate the possibility of the use of coefficient *A* in the decision correctness of management and ICT systems management simulation. This is possible even when the system information is incomplete. Even when the number of transmitted bits or amount of errors are unknown. It was possible after applying the uncertainty modelling using the theory of rough sets.

Presented calculation of coefficient *A* in the decision correctness has supported only when cannot be easily estimated eg: when information about system are not full.

In the course of the continuation of this studies the authors intend to expand the scope of the variables affecting decision-making.

References

1. Galar, D., Gustafson, A., Tormos, B., Berges, L.: Maintenance Decision Making Based on Different Types of Data Fusion. Eksploatacja i Niezawodnosc – Maintenance and Reliability 14(2), 135–144 (2012)
2. Kasprzyk, Z., Rychlicki, M.: Analysis of Physical Layer Model of WLAN 802.11g Data Transmission Protocol in Wireless Networks Used by Telematic Systems. In: Zamojski, W., Mazurkiewicz, J., Sugier, J., Walkowiak, T., Kacprzyk, J. (eds.) Proceedings of the Ninth International Conference on DepCoS-RELCOMEX. AISC, vol. 286, pp. 265–274. Springer, Heidelberg (2014)
3. Pawlak, Z.: Rough Sets. Research Report PAS 431, Institute of Computer Science, Polish Academy of Sciences (1981)
4. Pawlak, Z.: Zbiory przybliżone nowa matematyczna metoda analizy danych. Miesięcznik Politechniki Warszawskie (May 2005)
5. Rosinski, A., Dąbrowski, T.: Modelling Reliability of Uninterruptible Power Supply Units. Eksploatacja i Niezawodność – Maintenance and Reliability 15(4), 409–413 (2013)
6. Siergiejczyk, M., Paś, J., Rosiński, A.: Application of Closed Circuit Television for Highway Telematics. In: Mikulski, J. (ed.) TST 2012. CCIS, vol. 329, pp. 159–165. Springer, Heidelberg (2012)
7. Stawowy, M.: Comparison of Uncertainty Models of Impact of Teleinformation Devices Reliability on Information Quality. In: Nowakowski, T., Młyńczak, M., Jodejko-Pietruczuk, A., Werbińska–Wojciechowska, S. (eds.) Proceedings of the European Safety and Reliability Conference, ESREL 2014, pp. 2329–2333. CRC Press/Balkema (2015)
8. Stawowy, M.: Identifying Status of an ICT System Using Rough Sets. Archives of Transport System Telematics 7(1), 50–53 (2014)
9. Stawowy, M.: Zastosowanie analizy obrazu do rozwiązywania zagadnień transportowych, nr 862, Raport Prace IPI PAN (1998)
10. Ważyńska-Fiok, K.: Podstawy teorii eksploatacji i niezawodności systemów transportowych. Wydawnictwo Politechniki Warszawskie (1993)

Popular FPGA Device Families in Implementation of Cryptographic Algorithms

Jarosław Sugier

Wrocław University of Technology, Department of Computer Engineering
Janiszewskiego St. 11/17, 50-372 Wrocław, Poland
jaroslaw.sugier@pwr.edu.pl

Abstract. This work evaluates implementation efficiency of different crypto-graphic algorithms in selected hardware organizations and in different FPGA devices. The tests included AES symmetric cipher and two more contemporary hash algorithms: Salsa20 and Keccak-f[400] permutation function. Each algo-rithm was realized in hardware in five organizations: the basic iterative one, two with the loop unrolled and two with the loop unrolled and pipelined, then automatically implemented in two popular-grade FPGA devices from Xilinx: Spartan-3 and Spartan-6. Results of 30 test cases allowed for evaluation of par-ticular strengths and weaknesses of the ciphers, the organizations and the FPGA architectures. In particular, the evaluation took into account implementation ef-ficiency offered by the two device families, scalability of the ciphers with the loop unrolling factor and specific routing problems which came out in some configurations.

Keywords: AES, Salsa20, Keccak, loop unrolling, pipelining, Spartan FPGA.

1 Introduction

Cryptographic algorithms are ubiquitous in contemporary information systems in applications like data protection, authentication methods, digital fingerprinting, etc. If high data throughput is required their implementation in hardware, often in configu-rable devices, is of fundamental importance.

Numerous positions in the literature propose particularly efficient hardware im-plementations of the ciphers ([3-6], [8-10], [13]) but the aim of this work is different: by providing consistent test methodology we want to compare implementation effi-ciency of selected three cryptographic algorithms (AES, Salsa20 and Keccak-f[400]) in five hardware organizations (iterative, with loop unrolled and pipelined) and in two different FPGA devices (Spartan-3 and Spartan-6 from Xilinx) – in a total of 30 de-signs. The text is presented in just two essential parts: the second chapter introduces the selected five hardware organizations viable for round-based ciphers and discusses their particular realizations for the three algorithms, and the third chapter includes the results of implementation and presents their evaluation with regard to efficiency of-fered by the two platforms, scalability of the ciphers with the loop unrolling factor and specific routing problems of some configurations.

© Springer International Publishing Switzerland 2015   485
W. Zamojski et al. (eds.), *Theory and Engineering of Complex Systems and Dependability,*
Advances in Intelligent Systems and Computing 365, DOI: 10.1007/978-3-319-19216-1_46

2   Test Suite: Different Organizations of the Ciphers

2.1   Implementing a Round-Based Cipher in Hardware

If processing of an algorithm is expressed in a series of rounds executed repeatedly n_r times over the same bits of cipher state, the simplest approach is to implement it in a plain iterative organization where there is one cipher round instantiated in hardware and the state data is propagated through it in a loop of n_r iterations. This concept is illustrated in Fig. 1a. With each round completed in one clock cycle the total latency of computations is n_r clock ticks.

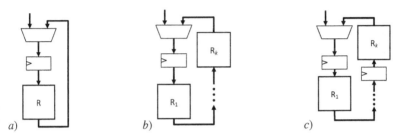

Fig. 1. This paper considers three concepts of hardware organization for a round-based cipher: a) basic iterative (x1);   b) with loop unrolled (xk);   c) loop unrolled and pipelined (PPLk)

Taking such an architecture as a starting point, the two opposing techniques can be used to create various derivate organizations with different area vs. speed trade-offs: loop unrolling or round folding ([3]). Round folding aims at minimizing size of the hardware at the cost of reduced performance and will not be included in this analysis. In loop unrolling (Fig. 1b) more than one round is instantiated in hardware so the number of loop iterations is reduced and thus the speed of data processing can be increased. With k instances of rounds the state is propagated through all of them in each clock cycle so the latency is reduced to n_r / k clock ticks.

Pipelining is a generic technique where long combinational propagation paths in the hardware are split into shorter segments by dividing it with registers: the clock cycle can be shorter (higher frequency of operation) and in every cycle each segment processes its own chunk of data so with multiple data being processed simultaneously overall throughput increases remarkably. In a case of a cipher, in its unrolled organization the long combinational cascade of rounds can be split by introducing registers at the output or input of each round; with k round instances this gives k pipeline stages and k blocks of data processed autonomously in parallel. Latency of computation of the entire cipher for each data block returns to n_r.

Pipelining is possible only if the cipher scheme allows to encode subsequent blocks of data in the input stream independently. It is worth furthermore noting that this technique usually does increase area of ASIC implementations but in FPGA arrays, where each logic cell is equipped with a flip-flop register regardless of the fact whether it is used or not, the increase in design size is often negligible and it is often complemented by substantial improvement in routing thanks to separation of long combinational propagation paths into shorter segments.

2.2   The Test Suite

In this analysis each of the three tested ciphers was implemented in its basic iterative organization (denoted later as "x1"), in two organizations with loop unrolled with factor k = 2 and 5 ("x2" and "x5") and in their pipelined variants ("PPL2" and "PPL5"). Because the considered ciphers have either 10 or 20 rounds this is the complete common set of possible unrolling factors. With every of the 3 ciphers being implemented in these 5 organizations there is a total of 15 designs which were prepared.

In further text we will use the x1 organization as a reference in comparison of the remaining ones because it takes the least amount of hardware resources and instantiates just one copy of cipher round. Having the x1 parameters, size and speed of the unrolled and pipelined cases can be estimated as follows. The size (e.g. number of logic cells used in the FPGA array) should increase approximately in proportion to the number of rounds implemented in hardware thus:

$$Size_{xk} \approx Size_{x1} \cdot k \tag{1}$$
$$Size_{PPLk} \approx Size_{PPL1} \cdot k \tag{2}$$

As it was noticed above, additional registers which are present in the pipelined organizations usually do not introduce any extra burden in the FPGA arrays and therefore the above estimations are identical for both xk and PPLk cases. What is not considered in these simple equations is the input multiplexer visible in the Fig. 1a which is counted in $Size_{x1}$ but is not replicated k times - therefore actual size parameters for the unrolled or pipelined cases can be somewhat smaller.

Maximum frequency of operation – or the minimum clock period – depends on the other hand on the number of rounds the state must go through in one clock cycle:

$$Tclk_{xk} \approx Tclk_{x1} \cdot k \tag{3}$$
$$Tclk_{PPLk} \approx Tclk_{x1} \tag{4}$$

Again, $Tclk_{x1}$ includes propagation delay of the input multiplexer which is not duplicated in xk/PPLk organizations, hence the clock periods may also be little overestimated.

2.3   Particularities of Cipher Organizations

AES is the cipher with 10 regular rounds preceded by a simple initial one ([7]). Each round processes the 128b data and needs a key which is computed from the user-supplied external key by so called key expansion routine running in parallel to the encryption path with another 128b of data. Uniformity of iteration is questioned by the two factors: a) the initial round is significantly different form the following ones; b) the last round is slightly modified with one elementary transformation omitted. Due to the first aspect, all the AES architectures need to have an extra simple hardware implemented before the regular loop with just 128b xor logic but it makes estimation of eq. (1) even more unnecessary restrictive. Moreover, its execution needs a separate clock cycle (which is needed anyway for preparation of the key for the first round) so the total computation latency is 11 clock cycles for x1 and PPLk architectures and 6 or 3 (1 + 10/2 or 1 + 10/5) cycles for x2 and x5 cases. The second aspect – different processing in the last round – required special multiplexers for bypassing column mixing inside the round hardware when it is executing the last iteration, which again weakened the estimations (1) ÷ (4).

Salsa20 comprises 20 rounds executed over 512b data ([1]). The rounds can be of two kinds: column round and row round and these are executed interchangeably with the only difference between them in applying different permutations to the input data. There is no separate key expansion routine because the user key is processed as a part of the 512b input. The iteration scheme is overall much more uniform but the actual fragment of the cipher being iterated is a double round (executed 10 times) rather than a single round. In this situation, implementation of a strict iterative scheme "20 repetitions of a single round" would lead to a 512b wide multiplexer which would switch between column and row round inputs, impairing both size and speed of the hardware. In [10] we have shown that a better alternative is to consider a double round as an elementary unit of the iteration and such an organization – "10 repetitions of a double round" – was adopted in this work to be the basic "x1" architecture with $n_r = 10$. Therefore, the "x2" organization computes the result in 5, while "x5" – in 2 clock cycles. This is on par with latencies of the AES variants but (nearly) doubles the sizes.

Processing of Keccak-f[400] permutation consists in application 20 identical rounds to the data 400b wide ([2]) with the only irregularity in using 20 different 16b constants as an auxiliary round parameter. These constants could be computed on-the-fly by LFSR registers independently for each round instance but it was simpler to tabularize them in distributed ROM modules which, being relatively small, do not add noticeably to the total size but (compared to the LFSR operation) conveniently simplify timing of data distribution. This solution was optimal in both xk and PPLk architectures. It should be noted that Keccak has the highest number of rounds in our comparison; with $n_r = 10$ AES and Salsa20 require half of them.

3   Evaluation of Results

Each of the 15 designs was implemented and tested on two hardware platforms: Spartan-3 and Spartan-6 from Xilinx – the inventor of the FPGA devices and still their leading manufacturer. Every architecture was described in the VHDL language at register transfer level (RTL), using consistent coding style for modelling the plain standard specification. No instances of architecture specific elements were inserted in order to keep the code as portable as possible. Then, the code was automatically synthesized and implemented by Xilinx ISE software ver. 14.7 with XST synthesis tool, and targeted for two devices – XC3S2000-5 (Spartan-3, [11]) and XC6SLX150-3 (Spartan-6, [12]), both in FGG676 package. Having each design implemented twice in different chips gave the final total of 30 implementations under the tests. Exactly the same code was used for both platforms.

Table 1. Parameters of the implementations

		Spartan-3					Spartan-6			
	min T_{clk} [ns]	Longest path - routing [%]	Levels of logic	Avg. non-clk fanout	Size (LUTs)	min T_{clk} [ns]	Longest path - routing [%]	Levels of logic	Avg. non-clk fanout	Size (LUTs)
AES										
x1	13.1	73.1	6	3.99	8 755	6.26	76.7	3	5.22	1 400
x2	20.9	69.5	14	4.44	10 757	9.74	78.2	6	6.72	2 349
x5	43.8	72.1	29	5.32	15 734	23.9	81.4	14	8.45	4 956
PPL2	13.5	73.4	7	4.39	11 610	5.70	74.5	4	6.20	2 289
PPL5	13.6	73.1	7	4.95	20 591	5.58	75.0	3	8.36	4 873
Salsa20										
x1	51.9	47.3	102	2.37	3 535	22.7	62.0	50	2.86	3 367
x2	99.8	48.5	199	2.47	5 575	54.8	74.4	75	2.80	5 391
x5	271	55.9	425	2.47	11 708	138	75.6	188	2.78	11 528
PPL2	49.7	45.4	105	2.28	5 575	28.7	72.5	45	2.58	5 555
PPL5	56.7	52.4	97	2.28	11 726	27.6	68.2	57	2.60	12 140
Keccak										
x1	8.92	64.3	4	3.45	1 777	4.89	71.1	3	4.18	1 339
x2	15.9	74.0	6	3.47	2 792	10.7	84.4	4	4.67	2 166
x5	37.9	69.6	21	3.45	5 908	39.6	89.8	15	5.16	3 601
PPL2	8.67	69.3	4	2.98	2 984	8.42	85.7	3	3.44	2 107
PPL5	8.78	61.2	5	3.18	6 140	7.03	86.1	2	4.48	3 428

Parameters of the implementations are given in Table 1. Minimum clock period (T_{clk}) was estimated by the implementation tools in static timing analysis of the final, completely routed design. Two additional parameters describe the longest (lengthiest) path of combinational propagation (this path determined T_{clk}): percentage of the delay attributed to routing resources (the rest is generated by logic elements) and the number of logic levels. Also the next parameter – average fan-out of non-clock nets – is related to the routing structure . Finally, size of the complete design is given as the number of occupied Look Up Tables (LUT) which are the elementary components in every logic cell generating combinational functions in Xilinx arrays. Number of registers is not included in this analysis because it depends predominantly on internal functionality of the algorithm and not on efficiency of the implementation. Moreover, as it was noted at the end of chapter 2.1, in FPGA arrays there is abundance of registers and, in case of ciphers (which are first of all logic-intensive), their utilization is not as critical as consumption of LUT generators.

3.1   Efficiency of Implementation Offered by the Two FPGA Platforms

Fig. 2 presents ratios of the basic speed (T_{clk}) and size (LUT) metrics between the two platforms: parameter of Spartan-3 implementation was divided by the value for Spartan-6 and the quotient is displayed in percent.

Using the device with more advanced architecture and manufactured in faster technology, Spartan-6 implementations are unquestionably expected to use significantly less LUT elements and offer shorter clock periods, thus both the quotients should reach high values. This indeed is the case in most of the cases, although with some notable exceptions.

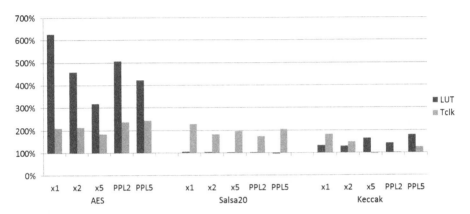

Fig. 2. Ratios of Spartan-3 : Spartan-6 parameters for speed (T_{clk}) and size (LUT) metrics

Looking at the size comparison (the left-hand bars) we can see that AES is the only cipher that benefits remarkably from moving to the newer Spartan-6 platform: the size is reduced from 6.3 to 3.2 times. In Keccak the reductions are still noticeable although only by factors 1.3 ÷ 1.8. In Salsa20, on the other hand, number of LUT elements remains virtually unchanged with PP5 case being the only one when this number actually increases – and this despite the fact that 6-input LUT generators in Spartan-6 are *much* more powerful than their 4-input counterparts in Spartan-3. This indicates that this potential of the new platform remains useless in implementation of atomic operations defined for this particular cipher.

Speed comparison (the right-hand bars) indicate severe problems that plague Keccak implemented in Spartan-6. While both AES and Salsa20 organizations reduce their clock periods by approx. 2.4 ÷ 1.7 on the new platform, Keccak demonstrate significant problems with scaling when its size (the number of unrolled rounds) increases. For the x1 and x2 designs the T_{clk} reduction is approx. by 1.8 and 1.5, but in x5 the calculated ratio is only 96%, i.e. the clock period in Spartan-3 is actually shorter that in Spartan-6. It is a surprising and unusual situation that this design is slower in the newer FPGA device than in its predecessor.

3.2   Scalability of the Ciphers

Straight comparisons of parameters taken from Tab. 1 would not lead to legitimate conclusions: being dependent on intrinsic specifics, absolute values are not comparable between different ciphers, organizations and hardware platforms. Therefore this study will be limited to evaluation of different organizational cases in relation to their basic iterative architecture for specific cipher / FPGA combination. Fig. 3 presents these relations: speed and size parameters of each derivative organization (x2, x5, PPL2 and PPL5) were divided by their estimations calculated form respective x1 parameters by applying the equations (1) ÷ (4). The lower the displayed bar, the faster (shorted T_{clk}) or the smaller (number of LUT) was the actual design in comparison to what could be expected from the x1 case implemented in the same hardware. The value of 100% is the threshold separating "worse than" from "better than expected".

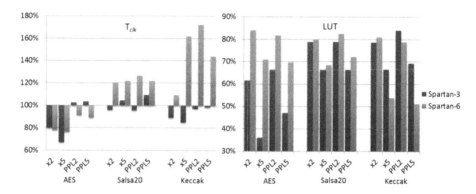

Fig. 3. Actual speed (T_{clk}, left) and size (LUT, right) of the scaled designs versus predictions based on respective x1 architectures

Analyzing the graphs once again we should distinguish the AES algorithm as the one which behaves in the most predictable way and achieves results which are in most of the cases better than the expectations (which harmonizes with comments from chapter 2.2 about the estimations (1) ÷ (4) being too pessimistic). It is the AES which offers the greatest reductions in T_{clk} (increases in speed): the x5 organization in Spartan-3 reaches 67% of the expected T_{clk} and only the pipelined organizations on this platform achieve slightly worse results than the estimations. Reductions of T_{clk} on the Spartan-6 platform, on the other hand, are not as spectacular but are more consistent because they include also PPLk cases.

As for the size metric, we can see that long combinational paths which are present in x2 and x5 AES organizations were particularly suitable for efficient optimizations in partitioning of the logic into LUT generators in Spartan-3 arrays. Such an optimization significantly reduced their use: the record is 36% actually used in the x5 case while in Spartan-6 optimizations are not as remarkable: at most down to 70%.

For Salsa20 and Keccak the reductions in T_{clk} are not so undisputable. While in Spartan-3 Salsa20 designs actually do not offer any noticeable improvement over the

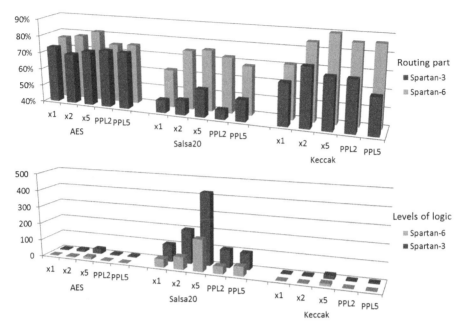

Fig. 4. Parameters of the longest propagation path: the part generated by the routing resources (above) and levels of logic (below)

estimations (ratios $0.96 \div 1.09$), for Keccak at least the x2 and x5 designs can reduce clock period to $85 \div 89\%$.

Probably the most striking observation from Fig. 3 is that, in contrast to AES, in the newer (and potentially much faster) Spartan-6 family reductions in T_{clk} are negative for both Salsa20 and Keccak. For Salsa20 the actual clock periods are $21 \div 27\%$ longer than expected even though at the same time the optimization in LUT usage remains quite good (down to $68 \div 82\%$ vs. estimations). This negative result becomes dramatic in Keccak: clock periods are by 62% longer than expected in the largest x5 organization and, notably, pipelining added in the PPL5 case was only a partial solution (this case scored still an increase by 44%, not seen in any implementation of the two other ciphers).

3.3   Routing Problems

The problems of Keccak in Spartan-6 should be attributed to routing congestion which is confirmed by looking at reported logic vs. routing ratio of the longest path. These data from Table 1 are presented graphically in the upper part of Fig. 4. With the values of 71 $\div$ 90%, Keccak designs in Spartan-6 have by far the highest routing components amongst all the tested cases. Such high values – and, consequently, small values for logic components – indicate that schemas of configurable connections implemented in the new Spartan-6 family do not fit well particular requirements of propagation rules in case of Keccak individual bits. Neither AES nor Salsa20 presented such problems although indeed this is a general rule that in the much bigger array of the Spartan-6 device the routing components are noticeably higher (this difference is most notable in all Salsa20 implementations).

Routing problems can usually be alleviated by pipelining which splits long propagation paths into shorter, better routable segments. In terms of combinational length PPL2 and PPL5 designs have the propagation paths as long as the x1 design (i.e. they span one round) so the T_{clk} and also routing part should be roughly the same. It should be noted that this effect is completely absent in routing parts of Keccak designs implemented in Spartan-6.

The most significant observation form the lower part of Fig. 4, on the other hand, is that elementary operations of Salsa20 are worst suited for aggregation in LUT elements: processing of one double round needs 102 (Spartan-3) and 50 (Spartan-6) levels of logic versus $3 \div 6$ levels in AES or Keccak. In the unrolled organizations these numbers increase to $199 - 425$ (Spartan-3) and to $75 - 188$ (Spartan-6) while the two other ciphers reach at most 29 and 15 so the difference remains fundamental. This also affects performance (by far the lowest operating frequency) and explains why the LUT usage in Salsa is much higher than in Keccak on both platforms.

4   Conclusions

This work discussed implementations of 3 ciphers in five organizations, each implemented in two different FPGA devices. The uniform approach applied to all the 30 test cases provided a comprehensive testbed for evaluating particular strengths and weaknesses of the ciphers, the organizations and the FPGA architectures.

The results showed that the oldest AES cipher is best handled by both FPGA devices in spite of its complex set of elementary transformations and irregular structure. The newer hash algorithms, although were especially designed with hardware realizations in mind and do not include, for example, wide substitution functions, are based on large and irregular combinational networks which are a challenge for FPGA implementation particularly at the routing level. In this context the older architecture of Spartan-3 in some cases may turn out to be the better option than its more advanced successor.

References

1. Bernstein, D.J.: The Salsa20 family of stream ciphers. In: Robshaw, M., Billet, O. (eds.) New Stream Cipher Designs. LNCS, vol. 4986, pp. 84–97. Springer, Heidelberg (2008)
2. Bertoni, G., Daemen, J., Peeters, M., Van Assche, G.: The Keccak sponge function family, http://keccak.noekeon.org (access date: March 2015)
3. Gaj, K., Homsirikamol, E., Rogawski, M., Shahid, R., Sharif, M.U.: Comprehensive evaluation of high-speed and medium-speed implementations of five SHA-3 finalists using Xilinx and Altera FPGAs. In: The Third SHA-3 Candidate Conference. Available: IACR Cryptology ePrint Archive, 2012, 368 (2012)

4. Gaj, K., Kaps, J.P., Amirineni, V., Rogawski, M., Homsirikamol, E., Brewster, B.Y.: ATHENa – Automated Tool for Hardware EvaluatioN: Toward Fair and Comprehensive Benchmarking of Cryptographic Hardware Using FPGAs. In: 20th International Conference on Field Programmable Logic and Applications, Milano, Italy (2010)
5. Junkg, B., Apfelbeck, J.: Area-efficient FPGA implementations of the SHA-3 finalists. In: 2011 International Conference on Reconfigurable Computing and FPGAs (ReConFig), pp. 235–241. IEEE (2011)
6. Liberatori, M., Otero, F., Bonadero, J.C., Castineira, J.: AES-128 Cipher. High Speed, Low Cost FPGA Implementation. In: Proc. Third Southern Conference on Programmable Logic, Mar del Plata, Argentina. IEEE Comp. Soc. Press (2007)
7. National Institute of Standards and Technology: Specification for the ADVANCED ENCRYPTION STANDARD (AES). Federal Information Processing Standards Publication 197, http://csrc.nist.gov/publications/PubsFIPS.html (2001; access date: March 2015)
8. Sugier, J.: Implementation of symmetric block ciphers in popular-grade FPGA devices. J. Polish Safety and Reliability Association 3(2), 179–187 (2012)
9. Sugier, J.: Implementing Salsa20 vs. AES and Serpent Ciphers in Popular-Grade FPGA Devices. In: Zamojski, W., Mazurkiewicz, J., Sugier, J., Walkowiak, T., Kacprzyk, J. (eds.) New Results in Dependability & Comput. Syst. AISC, vol. 224, pp. 431–438. Springer, Heidelberg (2013)
10. Sugier, J.: Low-cost hardware implementations of Salsa20 stream cipher in programmable devices. J. Polish Safety and Reliability Association 4(1), 121–128 (2013)
11. Xilinx, Inc.: Spartan-3 Family Data Sheet. DS099.PDF, http://www.xilinx.com (access date: March 2015)
12. Xilinx, Inc.: Spartan-6 Family Overview. DS160.PDF, http://www.xilinx.com (access date: March 2015)
13. Yan, J., Heys, H.M.: Hardware implementation of the Salsa20 and Phelix stream ciphers. In: Proc. Canadian Conf. Electrical and Computer Engineering, CCECE 2007, pp. 1125–1128. IEEE (2007)

Aspect-Oriented Test Case Generation
from Matlab/Simulink Models

Manel Tekaya[1], Mohamed Taha Bennani[2], Mohamed Abidi Alagui[3],
and Samir Ben Ahmed[4]

[1] University of Carthage, TELNET Innovation Labs, Tunisia
manel.tekaya@telnet-consuling.com
[2] University of Tunis El Manar, Tunisia
taha.Bennani@enit.rnu.tn
[3] University Paris Dauphine, France
mohamedlabidi.allagui.14@campus.dauphine.fr
[4] University of Tunis El Manar, Tunisia
samir.benahmed@fst.rnu.tn

Abstract. Matlab/Simulink is a widely used modeling notation for control systems design in automotive industries. Safety standards, such as ISO 26262, are emphasizing model-based testing, in which, test cases derived from the design model are used to show model-code conformance. In this paper, we propose a new aspect-oriented test case generation approach called "MB-ATG" from Simulink models. This approach exploits model checking technique capability to generate counterexamples that constitute test cases. We experiment a real automotive Simulink model with MB-ATG prototype to show its performance. Experimental results show that MB-ATG approach is compliant with standard structural coverage criteria and does not provide redundant test cases.

Keywords: Model-based testing, Embedded Systems, ISO 26262-6, Aspect-oriented models.

1   Introduction

Testing transportation systems is tedious and may leads to untrustworthy solutions. Safety standards have defined risk reduction levels to quantify the customer's trust. For instance, (IEC-61508[1]) and railway (CENELEC 50126/128/129 [2]) conventions have proposed five levels scale safety integrity. However, aviation (DO-178/254 [3]) standard has identified five design assurance levels. For Automotive applications, International Organization for Standardization "ISO" has introduced a functional safety standard for road vehicles (ISO 26262). ISO specifies a vocabulary, provides an overall organizational safety management of the development life cycle and defines an abstract classification of inherent safety risk in an automotive system called ASIL (Automotive Safety Integrity Level). ASIL argues the safety of the automotive system and proposes four testing levels: ASIL-A/B/C/D [4].

The automotive standard tightens test inputs among the system implementation by using classical structural coverage criteria. Despite the relevance of the test data set,

© Springer International Publishing Switzerland 2015   495
W. Zamojski et al. (eds.), *Theory and Engineering of Complex Systems and Dependability*,
Advances in Intelligent Systems and Computing 365, DOI: 10.1007/978-3-319-19216-1_47

which could cover 100% of source code, it is insufficient to detect both omission and non-conformance faults that happens willfully or accidentally. Omission faults occur when the application developer implements partially the system model and non-conformance ones occurs when the developer misunderstands the application model. In the model-based testing (MBT), models are the cornerstone of test cases generation, which can be used to show model-code conformance.

Automotive controllers are usually designed using the Simulink/Stateflow [22] (SL/SF) language. Simulink design verifier (SLDV) tool [5] integrates a test case generation functionality that is suitable for discrete Matlab/Simulink models. The generator suffers from random strategy drawbacks. Therefore, this functionality does not provide an efficient set of test inputs that comply with standard requirements. Hamon et al. [6] have been established that model checking could be used to generate test cases according to structural coverage criteria. SLDV integrates, also, model checking feature. We focus in this paper how could we use the integrated model checker to generate test cases that comply with standard requirements.

This paper introduces an aspect-oriented approach, which feeds the Prover plug-in model checker [7] with a processed model to generate counterexamples. These counterexamples are considered as test inputs. The paper is structured as follows: Section 2 analyses the related works. Proposed oriented-aspect approach is described in section 3. Section 4 and 5 present the protocols of our approach. Section 6 shows the experimental results.

2   Related Works

Researchers have proposed various approaches of generating test sequences from embedded systems models using model checking according to test coverage criteria [9, 17 and 18]. In [17], authors present an approach to generating test case from SCADE models. They use sal-atg tool [22], and the SCADE model is transformed into SAL Specification language. MC/DC coverage criterion was adopted for generating the test goals to be covered with generated test cases. Appropriate trap variables are defined for the translated model in SAL language based on MC/DC criterion. Sal-atg is then used to derive the test cases for the defined boolean trap variables. Ambar et al. [9] present a flow for automatic test-case generation (ATG) from SL/SF model based on the model checker sal-atg. SL/SF model is transformed into SAL specification interpretable by sal-atg model checker. Structural coverage criteria such as block, condition and decision and MC/DC lookup-table coverage, states and transitions coverage are defined by test specification that includes different coverage goals. In addition, "SmartTestGen" tool proposed in [18], integrates several test case generation using sal-atg tool, Random testing, local constraint solving and guided Heuristics based guided coverage.

These methods have some limitations [19] such as difficulties in handling floating point data types in SAL and translating SL/SF to SAL specification. All the above tools generate test cases only for their relative input model and do not accept any other modeling language as input. However, [20] shows that such transformation process entails the increase of the project cost owing to the need of developing a tool that transform each modeling language into a specification language.

In [23], authors detail tools framework. It allows test case generation from a behavioral model of the system in the fully formal and executable specification language (i. e. RSML-e). After the specification is validated, the analyst can translate the specification to the PVS or NuSMV input languages for verification. Test sequences are generated according to MC/DC coverage criterion expressed in LTL format.

The common denominator of the approaches introduced above is the model transformation into a model checker internal representation. Thus, test coverage criteria are expressed as LTL format or trap properties. Unlike, these approaches that transform the application design to comply with the model checker language; in this paper we propose a homogenous approach that retains the original application model. It creates a set of properties in the same design language. These latter are weaved automatically into the original model through an aspect-oriented protocol. Weaved properties are non-functional, and they have no influence on the original behavior of the system. Checking the transformed model generates a set of test cases. Properties generator implements the ISO 26262-6 coverage criteria that release test case in accordance with the standard requirements.

3   Aspect Oriented Approach

Figure 1 details oriented-aspect proposed approach overview. This approach exploits SLDV formal verification by verifying properties constrained by assumptions [10]. It performs exhaustive formal analysis to confirm Matlab/Simulink models correctness with respect to given properties and assumptions. This approach is composed of three steps: (1) Properties and assumptions generator (2) Properties and assumptions weaving protocol (3) Model verification.

The first component requires transformed Matlab/Simulink model parsing according to ASIL coverage criteria. Section 4 details this generation. Component (2) consists of two steps: (a) model transformation into a tree structure and (b) New nodes creation which tie in properties and assumptions position weaving. These properties and assumptions must be expressed using the same language used in system model design. The property ψ, respectively hypothesis H, of a model is called "property observer", respectively "Assumption". These observers, as the model M, are implemented with Simulink operators. These latter are called "Proof objective" and "Assumption"; they are accessible through the SLDV library [5]. "Proof Objective" operator (respectively "Assumption") is identified by the letter P; respectively A. Section 5 details the weaving process.

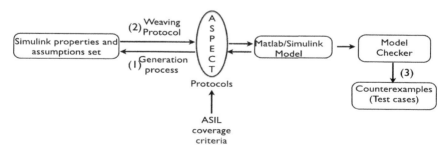

Fig. 1. MB-ATG Oriented-Aspect approach overview

Component (3) consists in the verification of properties constrained by assumptions. The SLDV model checker performs this verification. As described in the user's guide [5], SLDV uses the model checker Prover Plug-In [7] which is developed and maintained by Prover Technology. In order to prove properties, this model checker searches for all possible values of a Simulink functions to prove these properties. It aims to find a simulation that satisfies the objective. Prover Plug-in key idea is to model properties as propositional logic formulas. In addition, it applies proof methods to decide whether the formulas are valid or not. Prover Plug-In uses a "reductio ad absurdum" argument to solve this problem [11]. This method uses a combination of different analysis to check whether the formula negation is not satisfied, and the assumptions are valid (i. e. true) throughout property analysis. In this case, a counter-example is generated, and it is equivalent to a test case. We proved the equivalence between test cases and counterexamples in [12]. Our proposal is fully automated by the implementation of two main components: properties and assumptions generator and properties and assumptions weaver.

4   Properties and Assumptions Generator

In this section, we introduce our properties and assumptions generation approach according to ASIL coverage criteria. First, we expose ASIL coverage criteria correspondence with structural model coverage. After that, we present our properties and assumptions generation algorithm. Properties and assumptions generator takes as inputs Matlab/Simulink model and a structural coverage criterion. It generates a set of Matlab/Simulink properties and assumptions in order to be weaved.

4.1   Structural Model Coverage Correspondence

In [13], authors present four classes of structural model coverage criteria that are: Control-flow-oriented coverage criteria, Data-flow-oriented coverage criteria, Transition-based coverage criteria and UML-based coverage criteria. We focus on the first criterion that is based on the statements, decision (or branch), loops, and paths in source code. In Matlab/Simulink logical operator notations, there are the same concepts of statements and decisions. However, there are no loops. To apply code-based coverage criteria to Matlab/Simulink logical models: Firstly, for the statement coverage, we consider model blocks as statements. Each block (statement) needs to be tested only once. Secondly, branch coverage is equivalent to decision coverage; we consider model global decision and sub-decisions as if-then-else expression. Thirdly, for MC/DC coverage criterion for models has the same definition as the code.

Chilenski investigated three different notions of MC/DC [14]: Unique-Cause MC/DC, Masking MC/DC and the combination of the former two notions. In masking MC/DC, a basic condition is masked if varying its value cannot affect the outcome of a decision. This is due to the structure of the decision and the value of other conditions. To satisfy masking MC/DC for a basic condition, we must have test states in which the condition is not masked and takes on both "true" and "false" values. Masking MC/DC is the easiest of the three forms of MC/DC to satisfy since it allows for more independence pairs per condition and more coverage test sets per expression than the other forms. Chilenski's analysis showed that Masking MC/DC could allow fewer tests than Unique-Cause MC/DC. However, its performance in terms of probability of error detection was nearly identical to the other forms of MC/DC. This led Chilenski to conclude in [14] that Masking MC/DC should be the preferred form of MC/DC. In this paper, we will use masking MC/DC [15].

4.2   Properties and Assumptions Generation Approach

For properties and assumptions generation according to masking MC/DC criterion, we have proposed a new approach inspired by the hypothesis presented in [15]. In the original proposition, authors suggest an MC/DC approach that aims to determine if existing requirement based test case provide the level of rigor required to achieve masking MC/DC of the source code. The cornerstone of this approach is to create a schematic representation of the source code. Basic building blocks are fundamental to the MC/DC approach such as AND, OR, XOR and NOT. They represent the base of this approach. Compliance verification is based on the determination of masking MC/DC for each basic block. Thus, a basic block represents a decision.

According to this hypothesis, we proposed to verify masking MC/DC for each basic block. To satisfy this verification, we parse Boolean expression, which is equivalent to Simulink model and extract each decision, depicted by a basic block. For this reason, we have to generate two counterexamples for each decision (Basic block) that must cover it once to "true" and once to "false". According to the definition of masking MC/DC for basic blocks detailed in [15], we have to assign an assumption for each block to ensure generation of the required test cases. This approach is applied to both decisions with common logical operators and decisions with mixed logical operators.

4.3   Properties and Assumptions Generation Algorithm

Algorithm 1 details properties and assumptions generation process.

Algorithm 1 Properties and assumptions generation algorithm

1. **Input**: XMLFILE: File which corresponds to the model Matlab/Simulink
   ASILCOVERAGE: String
2. **Output**: PROPERTIESFILE: Text file which contains properties and assumptions list
3. **Begin**
4. Path: XML file path
5. Tr: Binary Tree
6. RootTree: Block
7. VBlocks, LogicalExpression, Operators, subExpressions, ListProperties: Vector
8. VBlocks ← ListBlocks(Path)
9. RootTree ← getoutport_name(Path)

10. Tr ← TreeConstruction(RootTree,VBlocks)
11. LogicalExpression ← ExpressionConstruction(Tr)
12. **switch** (ASILCOVERAGE)
13. **case** "Blocks": ListPropertiesAsp ← GeneratePropertiesAspB(LogicalExpression)
    break
14. **case** "Decisions": ListPropertiesAsp ← GeneratePropertiesAspDec(LogicalExpression)
    break
15. **case** "MC/DC": Operators ← Operators(LogicalExpression)
16. SubExpressions ← Decomposition(LogicalExpression)
17. ListPropertiesAsp ← GeneratePropertiesAsp(LogicalExpression,Operators)
    break
18. **default:**
19. write("Please choose a coverage criterion")
20. **Endswitch**
21. **End**

Lines 1 and 2 present respectively algorithm's input and output. The inputs (Line 1) correspond to the Matlab/Simulink model transformation into XML format using SimEx tool [16] and coverage criterion chosen to generate test cases. The output (Line 2) consists of a text file containing properties and assumptions list. Lines 4, 5, 6 and 7 are variables declaration. VBlocks vector (Line 8) and RootTree (Line 9) contains all blocks names and the model output respectively. They are obtained by parsing XML file. Line 10 corresponds to tree construction. This construction uses RootTree and VBlocks variables. Expression construction needs tree model as input (Line 11) and generate LogicalExpression vector as output that contains all the expression elements with parenthesis.

We assign a specific treatment for each structural coverage criterion using a switch loop (Line 12). For blocks coverage criterion (Line 13) we generate only one property that correspond to the LogicalExpression or its negation. Line 14 presents decisions coverage criterion case. To ensure this criterion, we generate two properties. The first one corresponds to the LogicalExpression and the second to its negation. For each of these coverage criteria, properties are saved into ListPropertiesAsp vector (Line 13, Line 14). Line 15 presents properties and assumptions generation according to MC/DC coverage criterion.

Operators vector contains all operators within the LogicalExpression. Line 16 corresponds to the logical expression decomposition into sub-expressions that present decisions. After the decomposition, properties and assumptions list is generated using sub-expression and operators lists (Line 17). This generation consists of two steps. The first one consists in generating properties for the model expression and each sub-expression by writing in ListPropertiesAsp (Line 17) the expression and sub-expressions and their negations. The second one consists of parsing operators list and generates an assumption for each operator.

If the user does not choose a structural coverage criterion, an error message will be shown. For each structural coverage criterion, ListPropertiesAsp will be saved into a text file and will present properties and assumptions weaver input.

After analyzing the time complexity of all instructions, we conclude that the algo-rithm 1 complexity is O(n log n); where n is the number of nodes in the binary tree.

5   Properties and Assumptions Weaver

In this section, we present weaving algorithm. Our properties and assumptions weaver takes as inputs: (1) Matlab/Simulink model and (2) properties and assumptions file depicted in Table 1. As output, it generates Matlab/Simulink model in which Sim-ulink properties and assumptions are weaved.

Algorithm 2 shows weaving process algorithm. The steps: Model output extraction into `RootTree` variable and transformation of XML file into a tree structure are common with the previous algorithm 2.

Algorithm 2 Weaving process algorithm
1. **Input**: PathFile: path of PROPERTIESFILE, Path: XML file path, Tr: Binary tree which corresponds to the XML file
2. **Output**: XMLFILEOutput: Transformed XML file after properties and assumptions weaving
3. **Begin**
4. Verify: boolean
5. ListPropAs ← parse(PathFile)
6. **For** j ← 0 **to** size(ListPropAs) **do**
7. Verify ← VerifyPlacement(ListPropAs[j],RootTree)
8. **Endfor**
9. **IF** Verify == true
10. ListPropAsOuptput ← LocatePlacement(ListPropAs,RootTree)
11. XMLFILEOutput ← weave(Path, ListPropAsOuptput)
12. **ELSE**
13. write("Properties and assumptions list in invalid")
14. **End**

Properties and assumptions weaving algorithm takes as inputs (Line 1): properties and assumptions file path, model structure tree `Tr`, `RootTree` and XML file path. The three former inputs are generated from properties and assumption generator. As an output, it generates an XML file modified with properties and assumptions weaved (Line 2). This file will be transformed into .mdl file using SimEx tool. `ListPropAs` vector (Line 5) contains properties and assumptions list to be verified.

Weaving algorithm consists of three main steps. (1) Verifying if the list of properties and assumptions is valid (Line 7). If it is the case, "Verify" variable takes "true" value (Line 9). If the properties and assumptions list is not valid, an error message is sent to the user (Line 13). (2) Placement and weaving position identification for each property and assumption to be weaved (Line 10). (3) Weaving properties and assumptions on the initial XML file and generates a modified one (Line 11).

`LocatePlacement` function ensures second step (Line 10). It takes as inputs initial properties and assumptions file and the modified binary. It generates a new properties and assumptions file saved in `ListPropAsOuptput` vector. For each property and assumption, we identify each sub-expression and locate the operator. Then we replace it with the annotation inserted in the binary tree. This annotation represents position insertion in the model. The time complexity of the algorithm 2 is $O((\log n)^n)$; where n is the binary tree nodes number.

After properties and assumptions weaving according to a structural coverage criterion, we apply SLDV property proving option on the transformed model. This verification allows counterexamples generation. These counterexamples are considered as test case suite.

6   Experimental Results

Our oriented-aspect approach has been implemented in J2EE platform and used Matlab scripting language m-script. We will apply our tool to a real automotive system that adjusts the critical area of automobile engine knocking. We have sliced, manually, the primary model into nine sub-models according to the approach proposed by [8]. The unit sub-models sizes vary from 2 blocks up to 20 SL/SF blocks and three input variables. These models contain logical blocks, multi-dimensional inputs, non-linear blocks like multiplication and division, dynamic lookup tables and hierarchical triggering of blocks.

Figure2, below, contains three charts (a), (b) and (c) which show, respectively, Blocks coverage, Decisions, and Masking MC/DC test case generation results. Each chart includes three curves: sal-atg results (i.e. diamond line), SLDV results (i.e. square line) and MB-ATG results (i.e. triangle line). The abscissa identifies the nine sub-models, and the ordinate shows the test input generated.

Sal-atg tool produces a higher number of test cases for all the models. Thus, it shows redundant test cases for SM1, SM4, SM5, SM6 and SM8. Redundant elements generation is due to the setting to "true" trap variable when a corresponding SAL module for a block is activated. For Decision coverage criterion (chart b), sal-atg generates three or four test cases with redundancy for models SM1, SM2 and SM8. Also, we observed that SLDV always generate three test cases without redundancy. The higher test case number for SLDV and sal-atg is due to their scope language to express properties and assumptions. Redundancy and higher test cases number generate a heavyweight test driver and causes a time overhead of test phase. Chart c presents Masking MC/DC criterion coverage. For sal-atg tool, the corresponding SAL modules are more complex because they have to take into account all the combination of changes in the conditions and decisions to capture the semantics of this criterion. Thus, SAL scope language does not ensure properties and assumptions expression. SLDV tool generates three test cases and four for SM7 model. These test cases do not present a redundancy. All this contributes to a lower coverage criterion. For all these criteria, MB-ATG achieves 100% coverage for the nine sub-models. Using MB-ATG tool helps to harvest a minimal and efficient test case suite according to structural coverage criteria. In another way, it contributes to reduce time and cost of unit testing.

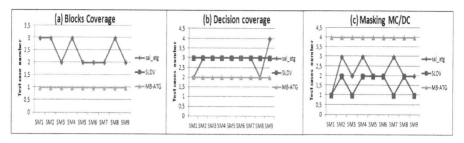

Fig. 2. Comparison of MB-ATG results with those SLDV and sal-atg for (a) Blocks coverage and (b) Decision coverage (c) Masking MC/DC coverage

7   Conclusion and Outlook

This paper has presented an aspect-oriented approach that generates test cases for Simulink models according to blocks, decision coverage and Masking MC/DC structural coverage. This approach depends on two main protocols: Properties and assumptions generator and properties and assumptions weaver. The former analyses the original model of the automotive system and generates a set of properties and assumptions that are compliant with a structural coverage criteria. The weaving protocol introduces the generated elements into the original model. The model checker processes the modified model and produces an efficient and minimal test suite. We applied our prototype to a real automotive model, and test case generation is done in a nested way that limits the size of the systems under test. The first outcome has given promising results in the automatic test case generation. Our future work aims to automate the whole process by adapting the proposition of Reicherdt, R. et al. [8] which introduced an approach to slicing Simulink models.

Acknowledgment. This research and innovation work is carried within a MOBIDOC thesis funded by the European Union under the PASRI project. The authors would like to thank Mr. Anis YOUSSEF and the automobile department team in TELNET.

References

1. Functional safety of electrical/electronic/programmable electronic safety-related systems, http://www.iec.ch/functionalsafety/
2. CENELEC: European Committee for Electro-technical Standardization, http://www.iec.ch/functionalsafety/
3. Hilderman, V., Baghi, T.: Avionics certification: a complete guide to DO-178 (software), DO-254 (hardware). Avionics Communications (2007)
4. ISO - International Organization for Standardization, ISO 26262 Road vehicles Functional safety Part 10: Guideline on ISO 26262 (2012)
5. Mathworks, Inc. Simullink Design Verifier 1: User's Guide (2012)

6. Hamon, G., De Moura, L., Rushby, J.: Generating Efficient Test Sets with a Model Checker. In: Software Engineering and Formal Methods, SEFM, pp. 261–270. IEEE (2004)
7. Sheeran, M.: Prover plug-in documentation (2000)
8. Reicherdt, R., Glesner, S.: Slicing MATLAB simulink models. In: 34th International Conference on Software Engineering (ICSE), pp. 551–561. IEEE (2012)
9. Gadkari, A., Mohalik, S., Shashidhar, K.C., Yeolekar, A., Suresh, J., Ramesh, S.: Automatic generation of test cases using model checking for SL/SF models. In: 4th Model-Driven Engineering, Verification and Validation Workshop, pp. 33–46 (2007)
10. Bochot, T., Virelizer, P., Waeselynck, H.: Paths to property violation: a structural approach for analyzing counter-examples. In: 12th International Symposium on High-Assurance Systems Engineering (HASE), pp. 74–83. IEEE (2010)
11. Sauders, P.T.: Reduction. Reductio ad absurdum. Formal Aspects of Cognitive Processes 22(3), 118–119 (2007)
12. Tekaya, M., Bennani, M.T., Ben Ahmed, S., Youssef, A.: Equivalence entre Propriétés Simulink et Critères de Couverture. In: Conférence francophone sur l'Architecture Logicielle (CAL). ACM (2014)
13. Utting, M., Legeard, B.: Practical model-based testing: a tools approach (2007)
14. Chilenski, J.: An investigation of three forms of the modified condition decision coverage (MCDC) criterion. DTIC Document (2010)
15. Hayhurst, K., Veehusen, D.S., Chilenski, J., Rierson, L.: A practical tutorial on modified condition/decision coverage. National Aeronautics and Space Administration, Langley Research Center (2001)
16. ITPower SimEx, http://www.itpower.de/102-1-SimEx-Bi-directional-conversion-of-MATLABSimulink-models-to-XML.html
17. Wakankar, A., Bhattacharjee, A.K., Dhodapkar, S.D., Pandya, P.K., Arya, K.: Automatic test case generation in model based software design to achieve higher reliability. In: 2nd International Conference on Reliability, Safety and Hazard (ICRESH), pp. 493–499. IEEE (2010)
18. Peranandam, P., Raviram, S., Satpathy, M., Yeolekar, A., Gadkari, A., Ramesh, S.: An integrated test generation tool for enhanced coverage of Simulink/Stateflow models. In: Design, Automation & Test in Europe (DATE), pp. 308–311. IEEE (2012)
19. Venkatesh, R., Shrotri, U., Darke, P., Bokil, P.: Test generation for large automotive models. In: IEEE International Conference on Industrial Technology (ICIT), pp. 662–667. IEEE (2012)
20. Schlich, B., Kowalewski, S.: Model checking C source code for embedded systems. International Journal on Software Tools for Technology Transfer 11(3), 187–202 (2009)
21. Mathworks, Inc. Getting started guide: R2014b
22. Hamon, G., De Moura, L., Rushby, J.: Automated test generation with SAL. CSL Technical Note (2005)
23. Rayadurgam, S., Heimdahl, M.: Coverage based test-case generation using model checkers. In: Eighth Annual IEEE International Conference and Workshop on the Engineering of Computer Based Systems, ECBS, pp. 83–91. IEEE (2001)

Ranking and Cyclic Job Scheduling
in QoS-Guaranteed Grids

Victor Toporkov[1], Anna Toporkova[2], Alexey Tselishchev[3],
Dmitry Yemelyanov[1], and Petr Potekhin[1]

[1] National Research University "MPEI",
ul. Krasnokazarmennaya, 14, Moscow, 111250, Russia
{ToporkovVV,YemelyanovDM,PotekhinPA}@mpei.ru
[2] National Research University Higher School of Economics,
ul. Myasnitskaya, 20, Moscow, 101000, Russia
atoporkova@hse.ru
[3] European Organization for Nuclear Research (CERN), Geneva, 23, 1211, Switzerland
Alexey.Tselishchev@cern.ch

Abstract. In this work, we describe approaches to creation of a ranked jobs framework within the model of cycle scheduling in Grid virtual organizations with such quality of service (QoS) indicators as an average job execution time and a number of required scheduling cycles. Two methods for job selection and scheduling are proposed and compared: the first one is based on the knapsack problem solution, while the second one introduces a heuristic parameter of job and computational resources "compatibility". Along with these methods we present experimental results demonstrating the efficiency of proposed approaches and compare them with random job selection.

Keywords: Grid, virtual organization, scheduling, resource management, job, flow, batch, knapsack problem, quality of service.

1 Introduction

The complexity of resource management and scheduling in distributed computing environment like Grid is determined by geographical distribution, resource dynamism and inhomogeneity of jobs and execution requirements defined by users of virtual organizations (VO) [1, 2]. A matter of the utmost importance for the VO is to efficiently manage available computational resources with such QoS indicators as an average job execution time and a number of required scheduling cycles while fulfilling requirements of all stakeholders: users, resource owners and VO administrators. The fact that resources are non-dedicated makes the efficient scheduling problem even more complex. In distributed computing with a lot of different participants and contradicting requirements the most efficient approaches are based on economic principles [3-6]. Different approaches to job scheduling can be classified based on job-dispatching methods. When job-dispatching process is decentralized, schedulers usually reside and work on the client side and fulfill end-user requirements (AppLes [7], PAUA [8]). Centralized

© Springer International Publishing Switzerland 2015
W. Zamojski et al. (eds.), *Theory and Engineering of Complex Systems and Dependability,*
Advances in Intelligent Systems and Computing 365, DOI: 10.1007/978-3-319-19216-1_48

job-dispatching implies that a meta-scheduler ensures the efficient usage of all the resources. While managing the scheduling process the meta-scheduler works with meta-jobs that are accompanied by a resource request, that contains resource characteristics required for the job execution. Such a hierarchical model is used in X-Com [9], GrADS [10] and other systems. It is also possible to evaluate job resource requirements by other means: statistically or by using expert systems [11]. Generally, the job-flow scheduling problem is solved using standard methods or algorithms [12-14], which include First-Come-First-Served, backfilling, user ranking mechanisms and resource separation. Within these approaches it is important to maintain the queue order and user priorities when executing these jobs. Even more "honest" queue forming is based on economic principles [6], which takes into account single job features and their impact on the queue.

Cycle job-flow scheduling [15] allows fulfilling VO requirements to a greater extent. Such scheduling is based on the set of dynamically updated information about the load of available resources. In that way three problems are being solved within each scheduling cycle: 1) job selection from a global job-flow; 2) forming jobs framework; 3) jobs framework scheduling and allocation based on the selected VO policy. During the job batch execution the VO policy, as a rule, has higher priority than single batch jobs preferences. This allows optimizing overall job batch execution parameters. For example, in a similar solution [16], it is described how a problem of minimizing the total energy consumption is solved during the job batch execution. However, at the same time queue order can be affected. There are two main steps in the cycle scheduling scheme (CSS) [15] for a single job batch: firstly, several execution options (alternatives) are found for each job for a given scheduling interval and, secondly, the set of alternatives (one alternative for each job) is chosen following the VO policy [17]. The fact that several execution alternatives are taken into account allows optimizing the schedule for a batch of independent jobs. In order to fulfill VO user requirements the job batch is populated with the jobs with the highest priority (e.g. those in the beginning of a standard queue). Execution alternatives allocation is also performed sequentially for each job, which, in its turn, guarantees, that the priorities are followed. When additionally, a user optimization criteria are used, one can guarantee a "fair" scheduling of the whole job batch [17, 18]. However, it is worth noting, that job selection using simple user priorities can negatively impact the scheduling efficiency of the whole job batch. In other words, in order to increase the whole job batch scheduling efficiency according to the VO requirements and QoS indicators one should evaluate different methods of job framework ranking.

In this paper, we review common problems of job batch forming for the cycle scheduling process. A heuristic job and resource domain "compatibility" parameter is proposed for the job-flow distribution and job batch selection. Two job batch forming approaches are proposed: the first one is based on the knapsack problem, and the second one is using the heuristic "compatibility" parameter.

The rest of this paper is organized as follows: section 2 contains proposed approaches to form a job framework, section 3 describes the experimental results, final results and next steps are defined in the summary section.

2   Job Framework Forming

For the hierarchical structure of job flow distribution key problems include configuring and composing of computational environment and also deliberate user jobs allocation between available resource domains. The distribution process may be based on meta-information from a user resource request, which represents user QoS expectations. High level of heterogeneity and diversity of distributed and parallel computing systems results in existence of a variety of resource request representation forms.

The CSS model [15] has the following basic resource request requirements to computational nodes: minimal performance p, required for job execution, maximum total job execution cost (budget) S, number n of computing nodes needed for the job and resource reservation time t (estimated for a resource with performance p). Thus, the model is based on a user estimate of the job execution time. If the user estimate is inaccurate, either the job can be interrupted after resource allocation time is over, or the resources can be released prematurely. The framework of independent jobs at each scheduling cycle is represented as a job batch in a certain manner formed from the job flow. Such selection makes it possible to increase overall scheduling efficiency in the VO compared to scheduling each job individually due to optimization of the general criterion formalizing the VO policy and fair resource sharing based on preferences of key stakeholders [2-6, 11, 15].

2.1   Job Batch Size Restrictions

An important step bearing at the first glance no relation to job flow cyclic scheduling efficiency is determining the job batch size during each scheduling cycle. By varying the job batch size limit (which can be expressed, for example, in number of the jobs in the batch or their cumulative execution budget) scheduling efficiency can be increased according to one or several different criteria. There are following scheduling efficiency criteria considered in our model.

- Overall resource domain computing nodes utilization level.
- Optimization criterion formalizing the VO policy. For example, job flow execution time minimization, with restriction on the total execution cost.
- Number of scheduling cycles required to complete job flow processing. Minimizing this factor provides a higher throughput of the distributed computing environment.

Specifying the batch size directly, for instance, by VO administrators, is not reasonable. Under conditions when local schedules of computing nodes change dynamically and parameters of incoming jobs differ, it's impossible to specify a limit that would contribute to the scheduling efficiency according to the VO policies in advance. A more flexible batch size limiting mechanism is based on relation between job requirements and computing environment parameters. In the context of economic principles, it is logical to choose resource utilization cost and resource reservation time as base characteristics for such a relation. When using time limit, total time of slot occupation is evaluated for each job. To execute a job a set of suitable slots has to be allocated. Each of the slots is characterized by start time, length and utilization cost [17]. A slot set forms a "window", for which total time and cost of slot utilization can be calculated. Note that for the purpose of normalization these values need to be calculated to a resource of base performance. Thus total time of slot utilization by single job can be evaluated as ptn. For resource domain we evaluate cumulative slot length. The job batch should be composed so that a total time of the slot utilization by the batch jobs is not greater than the cumulative slot length of the resource domain. Additionally a limit coefficient $limitCoefficient \in (0;1]$ is introduced to control job flow execution process in computing environment. Cost limit can be calculated similar to time limit. As opposed to a batch with a fixed number of jobs, time and cost limits introduction allows adjusting batch size under conditions of dynamically changing nodes utilization and heterogeneity of the job flow. Experimental study of this approach is conducted and presented in section 3.

2.2   Job and Computing Environment Compatibility Indicator

Job batch grouping schemes proposed in this paper form batch based on job and computing environment characteristics compatibility. Thus the batch is composed of jobs which resource requests are most fitted for executing in current scheduling interval. As a compatibility measure of an individual job and a resource domain an empirical coefficient D_Q (Distribution Quality) is proposed. D_Q coefficient describes chances for a job to be scheduled and executed successfully during the present resource domain scheduling interval utilization level. D_Q can have positive (high chance to be executed) or negative values (low chance to be executed). To figure out the structure of D_Q coefficient experimental studies were conducted. As a result, the following environment characteristics and resource request parameters that most influence the probability of a successful scheduling outcome were discovered.

1. A "price/quality" ratio of the domain computational nodes Q_0 and user jobs Q. For an individual computing node Q_0 is calculated as the ratio of the specified utilization cost (per time unit) to its performance factor c / p. For an individual job the factor is evaluated in a similar manner: $Q = S / ntp$.

2. The number of the available resources n_0 in the resource domain and the number of the computing nodes n required for job execution respectively.

3. An average slot length l_s in the environment and resource reservation time tp. The characteristics are calculated with regard to base resource performance.

4. Total domain available processor time V_s (cumulative length of available slots) and processor time required for the job execution tpn.

D_Q coefficient consists of four summands, corresponding to the mentioned characteristics. For each of the summands adjusting parameters are introduced: K_q, K_n, K_l, and K_v – weight coefficients of the summands; C_q, C_n, C_l, and C_v – threshold values, approximately determining the value at which at least one alternative for the job is likely to be found. The values of the adjusting parameters can be formed based on statistics of the previous scheduling cycles or expert estimates. Thus, D_Q coefficient is defined as the sum of the following terms.

Each of the summands can be presented in the following form:

$$D_{Q_r} = K_r\left(C_r - r/r_0\right)/C_r. \tag{1}$$

In (1), r is a job characteristic and r_0 is the corresponding resource domain parameter.

For example, the following term characterizes the ratio of slot utilization time required to execute the job and total processor time available during the considered scheduling cycle:

$$D_{Q_v} = K_v\left(C_v - tpn/V_s\right)/C_v. \tag{2}$$

Using D_Q coefficient it is possible to form the job batch in different ways. One possible approach consists in selecting jobs with the maximum value of D_Q at each scheduling cycle. However, in this case, after successful scheduling of the most "valuable" jobs at the first cycles, scheduling efficiency may reduce abruptly for jobs left in the queue at subsequent scheduling cycles.

2.3   Job Batch Grouping Methods

Two fundamentally different batch generation methods are proposed in the paper. In the first method, job batch grouping process reduces to solving the well known dynamic programming problem of the optimal knapsack filling. This approach seems to be most natural as it allows formalizing the job selection procedure under characteristics of job and resource domains known in advance. The second method is based on the flat D_Q coefficient and allows flexibly adjusting scheduling process to a dynamically changing structure of the resources and jobs of the flow.

The idea of the knapsack problem application when organizing scheduling is not new, however in the known approaches [16-18] it is usually used for optimal jobs allocation to non-dedicated resources. On the other hand, we propose using it to fill the job batch, as a preparatory step before scheduling. In our task, a knapsack is represented by a job batch, weight limit of the batch and weight of an individual job can be either time or cost depending on the chosen limit type. Weight limit is chosen based on summary resource characteristics with limitCoefficient as described in section 2.1. The value of a job is calculated as $1/D_Q$: that is, selecting the most "problem" jobs out of those that can be executed successfully at the current scheduling interval. This assumption is aimed for balancing the job flow execution during many cycles. Note that jobs with value less than or equal to zero will never be put into batch, since they make no positive contribution to total batch value but occupy some "useful weight".

Another approach uses the flat D_Q indicator to select jobs into the batch. However, the indicator is modified to take into account the jobs already put into the batch. For instance, when time limit is used, total processor time required by the jobs already put into the batch is taken into account and (2) is replaced in the following way:

$$D_{Q_v} = K_v \left(C_v - \left(tpn + \sum_{i=1}^{N} t_i p_i n_i \right) / V_s' \right) / C_v \ . \tag{3}$$

In (3), D_{Q_v} includes parameters of all N jobs, already put into the batch, and V_s' is total processor time for all the jobs.

As was mentioned earlier, the jobs with the minimal a positive value of D_Q coefficient have the highest priority during the selection process. As the number of jobs in the batch increases, the value of D_{Q_v} reduces and takes negative values. Batch generation process continues until there are any jobs with positive value of D_Q coefficient left in the job flow. In this batch grouping method the limiting coefficient VO administrators operate is the threshold parameter C_v .

Note, that when solving the knapsack problem batch size limit is strict and cannot be exceeded. However, when using D_Q method exceeding the limit will result in D_{Q_v} taking a negative value while entire D_Q coefficient can still be positive, and then the job will be put into the batch.

3   Simulation Studies

The efficiency of the considered job batch grouping techniques was studied using a Grid simulator [19]. The following job batch grouping methods were studied:

- **Random** – each time the batch is filled with a constant number of jobs randomly selected from the job flow;
- **KnapsackT** – the knapsack problem with a restriction on total reservation time is solved to fill the job batch;
- **KnapsackC** – the knapsack problem with a restriction on total reservation cost is solved to fill the job batch;

- **D_QT** – the job batch is filled according to jobs D_Q indicator with a restriction on jobs reservation time;
- **D_QC** – the job batch is filled according to jobs D_Q indicator with a restriction on jobs reservation cost.

Job flow scheduling in a single simulation experiment is performed cyclically: before the beginning of each scheduling cycle the job batch is filled with jobs from the job flow, and then scheduling and execution simulation are performed. Jobs failed to run during the scheduling cycle are returned to the job flow (job execution decline) and can be selected for the following scheduling cycles. The cyclic scheduling continues until all the jobs of the flow are completed. The computing environment consists of 24 nodes with performance distributed uniformly between 2 and 15 of relative performance units. Job length and the number of processors required by a job were also distributed uniformly at segments [50, 150] and [2, 5] respectively. The job flow contains 150 user jobs, scheduling was performed at a time interval of 600 units of time.

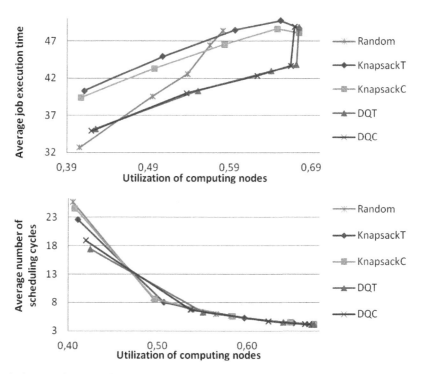

Fig. 1. Average job execution time and number of required scheduling cycles depending on the nodes utilization level

Figure 1 shows a 15% advantage of **$D_QT(C)$** approach over **KnapsackT(C)** by job execution time (main VO scheduling criterion). However, an average number of scheduling cycles required to complete the job flow execution was approximately the same for each grouping method with the same observed resources utilization level. Note, that the **Random** approach provided a relatively lower resource utilization level for the same job flows: its graphs in Figure 1 terminate earlier.

Figure 2 contains graphical data on average execution alternatives number and job batch size depending on the scheduling cycle number. As was mentioned earlier, the $D_Q T(C)$ algorithm aims to select jobs with the minimal positive D_Q compatibility indicator value. On the other hand, the **KnapsackT(C)** approach maximizes the sum of the batch jobs D_Q values with a strict restriction on total weight. As it turns out, this policy tends to select more relatively small jobs since they make a smaller contribution to the total batch weight. Thus the job flow scheduling is uneven: relatively small jobs are selected for the first scheduling cycles while jobs with higher resource demands remain till the last cycles.

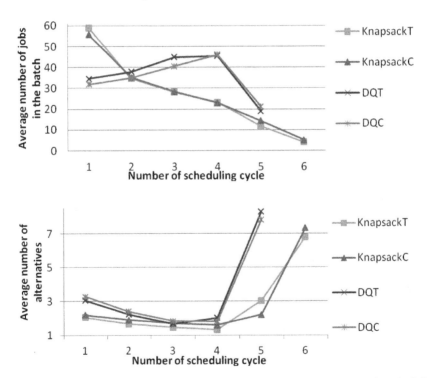

Fig. 2. Average execution alternatives number and job batch size depending on the scheduling cycle number

As a result, at the first cycles **KnapsackT(C)** forms batches of a large size, while at later cycles with the same weight limit the batch size decreases by several times. Similarly, at the first scheduling cycles a large number of relatively small jobs are competing for a limited resources and each reserved alternative further "granulates" available processor slots. During the following cycles it becomes difficult to allocate even several execution alternatives for a batch of much more resource demanding jobs. At the same time, the scheduling with $D_Q T(C)$ is uniform from cycle to cycle: there is now skew in the batches toward more or less resource demanding jobs.

The **Random** approach provided a lower average resource utilization level, a greater number of job execution declines and required a larger number of scheduling cycles to execute the job flow compared to other approaches. This means an inefficient resources usage by this regular approach.

4   Summary

In this paper, the problem of job selection for resource scheduling in virtual organization of QoS-guaranteed Grids is considered. In order to increase the cycle scheduling efficiency, we propose the general compatibility parameter D_Q for a job and a chosen resource domain. Two methods based on different job selection approaches for job batch forming using D_Q are proposed and analyzed.

The first **KnapsackT(C)** method forms job batch based on the knapsack problem solution with a strict limit for a total execution time or the cost of the job batch execution. The second $D_Q T(C)$ method uses D_Q for job selection, which in its turn dynamically changes based on jobs already selected and their characteristics.

Experiment results show significant advantage of $D_Q T(C)$ over **KnapsackT(C)**. This is due to the fact that when local resources and their schedules are changing dynamically and jobs in a flow are inhomogeneous, it is very important to take into account as many dynamic environment parameters as possible.

Thus, softer constraints for a job set size and dynamically changing D_Q allows $D_Q T(C)$ to ensure the best efficiency criteria value. One should mention that both methods secured better scheduling results and higher efficiency of resource distribution compared to the traditional random job distribution method.

Further research is aimed at developing methods for job allocation between several resource domains and forming a job framework while fulfilling requirements of all VO participants.

Acknowledgements. This work was partially supported by the Council on Grants of the President of the Russian Federation for State Support of Young Scientists and Leading Scientific Schools (grants SS-362.2014.9 and YPhD-4148.2015.9), RFBR (grants 15-07-02259 and 15-07-03401), and by the Ministry on Education and Science of the Russian Federation, task no. 2014/123 (project no. 2268).

References

1. Garg, S.K., Konugurthi, P., Buyya, R.: A Linear Programming-driven Genetic Algorithm for Metascheduling on Utility Grids. Par., Emergent and Distr. Systems 26, 493–517 (2011)

2. Cafaro, M., Mirto, M., Aloisio, G.: Preference-Based Matchmaking of Grid Resources with CP-Nets. Grid Computing 11(2), 211–237 (2013)
3. Buyya, R., Abramson, D., Giddy, J.: Economic Models for Resource Management and Scheduling in Grid Computing. Concurrency and Computation 14(5), 1507–1542 (2002)
4. Toporkov, V.V., Yemelyanov, D.M.: Economic Model of Scheduling and Fair Resource Sharing in Distributed Computations. Programming and Computer Software 40(1), 35–42 (2014)
5. Ernemann, C., Hamscher, V., Yahyapour, R.: Economic Scheduling in Grid Computing. In: Feitelson, D.G., Rudolph, L., Schwiegelshohn, U. (eds.) JSSPP 2002. LNCS, vol. 2537, pp. 128–152. Springer, Heidelberg (2002)
6. Mutz, A., Wolski, R., Brevik, J.: Eliciting Honest Value Information in a Batch-queue Environment. In: 2007 8th IEEE/ACM International Conference on Grid Computing, pp. 291–297. IEEE Computer Society (2007)
7. Berman, F., Wolski, R., Casanova, H., et al.: Adaptive Computing on the Grid Using AppLeS. IEEE Trans. on Parallel and Distributed Systems 14(4), 369–382 (2003)
8. Cirne, W., Brasileiro, F., Costa, L., et al.: Scheduling in Bag-of-task Grids: The PAUÁ Case. In: 16th Symposium on Computer Architecture and High Performance Computing, pp. 124–131. IEEE (2004)
9. Voevodin, V.: The Solution of Large Problems in Distributed Computational Media. Automation and Remote Control 68(5), 773–786 (2007)
10. Dail, H., Sievert, O., Berman, F., et al.: Scheduling in the Grid Application Development Software Project. In: Nabrzyski, J., Schopf, J.M., Weglarz, J. (eds.) Grid Resource Management. State of the Art and Future Trends, pp. 73–98. Kluwer Acad. Publ. (2003)
11. Kurowski, K., Oleksiak, A., Nabrzyski, J., et al.: Multi-criteria Grid Resource Management Using Performance Prediction Techniques. In: Gorlatch, S., Danelutto, M. (eds.) Integrated Research in GRID Computing, pp. 215–225. Springer (2007)
12. Moab Adaptive Computing Suite, http://www.adaptivecomputing.com/products/moab-adaptive-computing-suite.php (access date: November 2014)
13. Kannan, S., Roberts, M., Mayes, P., et al.: Workload Management with LoadLeveler. IBM (2001)
14. Tsafrir, D., Etsion, Y., Feitelson, D.: Backfilling Using System-generated Predictions Rather than User Runtime Estimates. IEEE Trans. on Parallel and Distributed Systems 18(6), 789–803 (2007)
15. Toporkov, V., Toporkova, A., Tselishchev, A., Yemelyanov, D., Potekhin, P.: Preference-Based Fair Resource Sharing and Scheduling Optimization in Grid VOs. Procedia Computer Science 29, 831–843 (2014)
16. Zhou, Z., Lan, Z., Tang, W., Desai, N.: Reducing Energy Costs for IBM Blue Gene/P via Power-Aware Job Scheduling. In: 17th Workshop on Job Scheduling Strategies for Parallel Processing, Boston, pp. 96–115 (2013)
17. Toporkov, V., Toporkova, A., Tselishchev, A., Yemelyanov, D.: Slot Selection Algorithms in Distributed Computing. J. of Supercomputing 69(1), 53–60 (2014)
18. Soner, S., Özturan, C.: Integer Programming Based Heterogeneous CPU-GPU Cluster Scheduler for SLURM Resource Manager. In: 14th IEEE International Conference on High Performance Computing and Communication & 9th IEEE International Conference on Embedded Software and Systems, pp. 418–424. IEEE, Liverpool (2012)
19. Toporkov, V., Tselishchev, A., Yemelyanov, D., Bobchenkov, A.: Composite Scheduling Strategies in Distributed Computing with Non-dedicated Resources. Procedia Computer Science 9, 176–185 (2012)

Web Based Engine for Processing and Clustering of Polish Texts

Tomasz Walkowiak

Wroclaw University of Technology, Wybrzeze Wyspianskiego 27, 50-320 Wroclaw
tomasz.walkowiak@pwr.edu.pl

Abstract. The paper presents a service oriented, online engine for processing and clustering texts in the Polish language. The engine, designed according to Web-Oriented Architecture paradigm, allows to run a large number of different language tools (like tagger, named entity recognizer, feature extractor) and clustering tools (like CLUTO or R) from almost any type of applications including HTML/JavaScript's ones. It allows constructing of a complex workflow, not only a simple chain of tools. To meet high availability requirements, the engine is deployed in a private cloud.

Keywords: natural language processing, clustering Polish texts, web application.

1   Introduction

Language technology is under development for many years. However, it faces several barriers blocking its widespread [10], especially for languages other than English. It is also a case of technologies for Polish. First of all, large number of language tools and resources are not accessible in the network, secondly descriptions are not available or are very limited. Moreover, tools are very hard to be installed and integrated since they are developed in different technologies (C++, Python, and Java). In addition, processing of large texts requires huge computational power.

These problems are overcoming by making linguistic tools available online. This idea is a background of CLARIN[1] project and related initiatives such as multilingual WebLicht [6] and Multiservice [3] for Polish. However, these tools phases some limitations, like allowing only serial chain of tools and lacking of data mining and machine learning tools. Therefore, authors aimed to develop an open access, scalable and highly available engine with the various types of interfaces that allows to integrate and run different language and machine learning tools and construct a complex work-flow.

To allow usability of designed solution the Web-Oriented Architecture (WOA) [9] paradigm was used which aims at building systems consisting of modular, distributable sharable and loosely coupled components. WOA follows the SOA paradigm and regards entire world as resources accessed via the representational state transfer (REST).

---

[1] http://clarin.eu/

© Springer International Publishing Switzerland 2015   515
W. Zamojski et al. (eds.), *Theory and Engineering of Complex Systems and Dependability,*
Advances in Intelligent Systems and Computing 365, DOI: 10.1007/978-3-319-19216-1_49

The paper is organized as follows. Firstly, system assumptions are analyzed and the performance analysis of preliminary solution is given. Next, the engine interface functionality is defined. It is followed by an architecture overview. Next, techniques of developing web applications with a use of the described engine is given with examples of results of real texts clustering. Finally, the short summary is presented.

2 Architecture Assumptions

The design of any software architecture requires a number of project decisions to be taken. To make them justified we have performed a set of experiments which aim was to construct a chain of language processing tools available by web services. The criterion for selection the proper solution was processing speed and usability.

2.1 Chain of Polish Language Tool

The chain consisted of three basic language tools: morphological analyzer and converter (MACA [7]), tagger (WCRFT [8]) and named entity recognizer (Liner2 [5]). The chain input consist of texts in Polish, the results are annotated texts in CCL[2] format.

Two different architectures were analyzed. The first one, based on existing SOAP web services available for MACA, WCRFT and Liner2 and the second on (named NLPServer), based on WOA paradigm, horizontal scaling and a use of tokens for input/output file identifications.

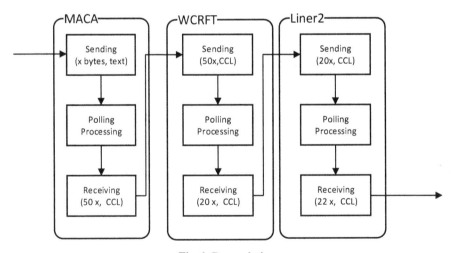

Fig. 1. Base solution

---

[2] http://nlp.pwr.wroc.pl/redmine/projects/corpus2/wiki/
CCL_format

The base SOAP web services (see Fig. 1) work in a polling style and require to download results from one web service as an output to the next one. It results in a long processing time. From maximum 14 sentences per second for client localized in the same network down to 2 sentences per second for clients with slow internet connection. The main reason of slow performance is resending files from and to web services. As it is marked on Fig. 1, files after MACA tool are 50 times larger than the original input size. The experiments showed that an overhead for data transfer is from 15% to 500%. To solve this problem we propose to split the processing and file transfer into separate functionality of a web service and to store tool results on a server side. It allows running a tool chain without a need of uploading results from following tools.

Next reason of slow performance are sizes of models used by tagger and named entity recognizer. Therefore, smaller models with a small decrease in recognition rate were developed. Finally, to allow vertical and horizontal scaling we propose to use a central dispatcher. Language and data mining tools are collecting tasks from the dispatcher defining its functionality by a name for each type of tool. It allows running several instances of tools on the same host (if it is technically possible) or other hosts.

Developing a web application for base web service requires server side scripts due to a need to store intermediate data, moreover a use of SOAP web services in JavaScript is not straight-forward.

Nowadays the user interface is mostly based on web browsers. WOA paradigm allows developing cross platform applications that runs on almost any device (personal computers, tablets, and smartphones) connected to internet. Therefore, the possibility of accessing language tools from JavaScript is a very important aspect. It can be achieved by a usage of REST protocol and JSON for transmission of data.

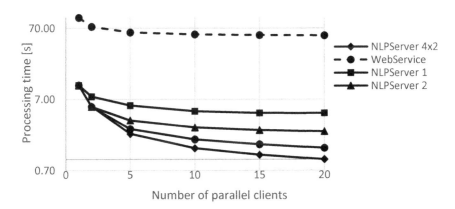

Fig. 2. Speed up of a chain of language tools for different solutions. Base – the base solution, NLPserver *NxM* – the preliminary solution that applies proposed assumptions for different types of scaling, *N* – number of hosts on which tool were deployed, *M* – number of tool instances run on one host.

2.2   Performance Tests

We have performed load tests to compare performance of the base solution with the proposed one (named NLPServer) and it's scaling capabilities. The processing time of the test case texts in a function of parallel clients for the base solution and different configuration of NLPServer: without scaling (NLPServer 1), with vertical scaling (NLPServer 2, 4) and vertical and horizontal scaling (NLPServer 4x2) is presented in Fig 2. The average performance was from 132 sentences/s for a single client, up to 1689 sentences/s for 100 clients for four servers each running four instances of language tools on four core processors. Load tests showed that proposed solution is much faster and easier to scale then base one.

2.3   Clustering of Texts

The clustering of texts in an important technique of natural language processing [2,4]. Like all machine learning methods it requires a pre-processing stage that generates a feature vector representing each of texts. In case of Polish language and language tool mentioned in section 2.1, it could be done by Fextor tool [5].

Moreover, it requires a bit different way of processing then chain described in section 2.2. An input to Fextor tool consists of a configuration file and a list of annotated files in CCL format. It requires an additional functionality of the engine: aggregation of results of language tools into one file (directory) and for a purpose of parallel processing a method for synchronization, that starts next tool when all input files are ready. There exist a large number of open-source tool that allows data clustering like CLUTO or R. But they could be run is a similar way as language tools.

3   Engine Functionality

The results of analysis given in the previous section was a base for the design and deployment of an effective engine for processing and clustering of Polish texts.

3.1   File Transfer

As mentioned is section 2.1 there is a need to split tool processing from file transfer to allow a construction of processing chains without a need of intermediate file transfer to and from a client. Therefore, the file transfer web service has to have at least two functionalities:

- file download - with file content sent as input, it returns a file token;
- file upload - with file token as an input and returns the file content.

The separation of loading/downloading from the task execution allows calling a sequence of tools without a need of loading/downloading large amount of date from and to a client.

3.2   Language and Clustering Tools

Language tools (for example tagger or name entity recognizer) or clustering tools (like CLUTO) could be treated as activities that has one input file and configuration parameters (for example defining the used model) and produces one file on output. To allow asynchronous processing (in a polling style) following functionalities are required:

- task start -   with file token and tool name as an input and returns the ID of a task;
- check task - with the task   ID as an input;   returns the status of processing; when status is equal to "DONE" it returns also the ID of output file;

The asynchronous mechanism prevents the clients from crashing due to time-outs caused by long time processing (for example due to a system overload).

3.3   Aggregation of Results

As mentioned is section 2.3 there is a need of special web service to deal with tools that requires a set of input files. Therefore a simple web service was designed that as input has a list file tokens and places files in one directory or makes a zip archive of them, the token of resulting file or directory is retuned.

4   Engine Architecture

4.1   Architecture Overview

The described functionalities and the scaling requirement result in the engine architecture presented in Fig 3. The core of the system consists of the NLPTasker. It consists of the simple asynchronous REST service, task queues, data storage and a set of workers. The workers executes language and clustering tools (LCT).

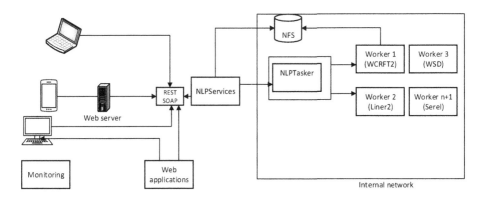

Fig. 3. The engine architecture

The data to be processed are stored on NFS. The REST service is called that creates the task to be processed (described by the LCT name and input file token) and stores it in the queue. A client of the REST service can check if the task was finished and get the token of the result file. The data processing is done by LCTs working as workers. Each worker collects a task from the queue, loads data from the NFS, processes them and returns the result to NFS. The workers and the queue system allows an effective scaling.

Additional NLPServices server grants the access from the Internet. It works as a proxy for the core system delivering a large set of different APIs. Different techniques for accessing the NLPServices including synchronous as well as asynchronous services, SOAP and REST, as well as XML and JSON are available. Such approach allows an easy integration with almost any kind of application (starting with mobile ones, through desktops, servers and JavaScript ones) built nearly in any programming language.

4.2   High Availability

To achieve high availability requirements the engine was deployed on a scalable hardware and software architecture that can be easily optimized to deliver high performance. The hardware consists of eight Cisco UCS B-Series Blade Servers based on Intel® Xeon® processors E5 product families. Servers are connected by fast fiber channel connection with highly scalable midrange virtual storage designed to consolidate workloads into a single system for simplicity of management (the IBM Storwize V7000). XEN Citrix creating a private cloud controls each server. It make the virtual infrastructure management more convenient and efficient since operating systems are independent from the hardware. Each language tool is deployed on separate virtual machine. Therefore, it is easy to scale up the system just by duplicating virtual machines as a reaction to a high number of requests for a given type of LCT. In addition, virtual machines can be easily moved to another server as a reaction to any failure or resource shortage. Moreover, virtualization provides a disaster recovery mechanism ensuring that when virtualized system crash, it will be restored as quickly as possible.

```
var show=new NLPSHOW(); var lock=new NLPLOCK();
var cluto=new NLP('cluto',show, cluto_params);
var fextor=new NLP('fextor',cluto,fextor_params);
var dir=new NLPAGREGATE('dirFiles',fextor);
for (i=0;i<files.length;i++)
{ var liner=new NLP('liner2',dir);
  var wcrft=new NLP('wcrft2',liner);
  var any2txt=new NLP('any2txt',wcrft);
  any2txt.inits(files[i].serverid,files[i].name,lock);
}
lock.start();
```

Fig. 4. Listing of exemplar NLP choreography in JavaScript

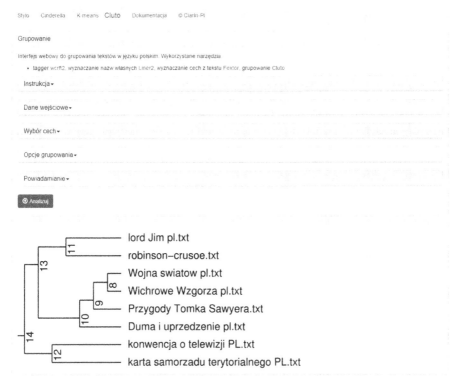

Fig. 5. GUI of web based application for Polish text clustering

5   Web Applications

To allow an easy development of web applications a JavaScript library for accessing the engine was developed. It uses   REST and AJAX to perform asynchronous communication. It allows an easy construction of flexible workflow of LCTs as presented in a code example shown in Fig. 4. The NLPSHOW class is used for showing the processing results. The NLPAGREGATE class allows the aggregation of multiple files into one and provides synchronization that starts further processing only when previous tools finished its processing. The NLPLOCK starts the processing for a set of tools. Due to a usage of AJAX and events, the library allows to run LRTs in parallel on a server side.

To show a functionality of the engine a set of web applications for defined workflows were developed. They are available online at http://ws.clarin-pl.eu. These applications were developed in a pure HTML5 and JavaScript technology using REST web service to run and control LCTs on a server side. The exemplar graphical interface with a result of clustering of a set of texts in Polish is presented in Fig. 5.

6   Summary

The paper presented the online and scalable engine for processing and clustering texts in the Polish language. The engine allows to construct and run a flexible workflow (not only a chain) of language (like tagger, named entity recognizer, feature extractor) and clustering tools (like CLUTO or R). The workflow is controlled and monitored by a set of web services that allows a fast development of HTML and JavaScript applications. To meet high availability requirements, the engine is deployed in a private cloud.

Work financed as part of the investment in the CLARIN-PL research infrastructure funded by the Polish Ministry of Science and Higher Education

References

1. Broda, B., Kędzia, P., Marcińczuk, M., Radziszewski, A., Ramocki, R., Wardyński, A.: Fextor: A feature extraction framework for natural language processing: A case study in word sense disambiguation, relation recognition and anaphora resolution. In: Przepiórkowski, A., Piasecki, M., Jassem, K., Fuglewicz, P. (eds.) Computational Linguistics. SCI, vol. 458, pp. 41–62. Springer, Heidelberg (2013)
2. Eder, M.: Rolling stylometry. DSH: Digital Scholarship in the Humanities, vol. 30 (in press, 2015)
3. Hinrichs, M., Zastrow, T., Hinrichs, E.: WebLicht: Web-based LRT Services in a Distributed eScience Infrastructure. In: Proceedings of the International Conference on Language Resources and Evaluation, pp. 489–493. European Language Resources Association (2010)
4. Kuta, M., Kitowski, J.: Clustering Polish Texts with Latent Semantic Analysis. In: Rutkowski, L., Scherer, R., Tadeusiewicz, R., Zadeh, L.A., Zurada, J.M. (eds.) ICAISC 2010, Part II. LNCS, vol. 6114, pp. 532–539. Springer, Heidelberg (2010)
5. Marcińczuk, M., Kocoń, J., Janicki, M.: Liner2 — A Customizable Framework for Proper Names Recognition for Polish. In: Bembenik, R., Skonieczny, Ł., Rybiński, H., Kryszkiewicz, M., Niezgódka, M. (eds.) Intell. Tools for Building a Scientific Information. SCI, vol. 467, pp. 231–254. Springer, Heidelberg (2013)
6. Ogrodniczuk, M., Lenart, M.: A multi-purpose online toolset for NLP applications. In: Métais, E., Meziane, F., Saraee, M., Sugumaran, V., Vadera, S. (eds.) NLDB 2013. LNCS, vol. 7934, pp. 392–395. Springer, Heidelberg (2013)
7. Radziszewski, A., Śniatowski, T.: Maca: a configurable tool to integrate Polish morphological data. In: International Workshop on Free/Open-Source Rule-Based Machine Translation, pp. 29–36 (2011)
8. Radziszewski, A.: A tiered CRF tagger for polish. In: Bembenik, R., Skonieczny, Ł., Rybiński, H., Kryszkiewicz, M., Niezgódka, M. (eds.) Intell. Tools for Building a Scientific Information. SCI, vol. 467, pp. 215–230. Springer, Heidelberg (2013)
9. Thies, G., Gottfried, V.: Web-oriented architectures: On the impact of web 2.0 on service-oriented architectures. In: Asia-Pacific Services Computing Conference, pp.1075–1082 (2008)
10. Wittenburg, P., et al.: Resource and Service Centres as the Backbone for a Sustainable Service Infrastructure. In: Proceedings of the International Conference on Language Resources and Evaluation, pp. 60–63. European Language Resources Association (2010)

Effectiveness of Providing Data Confidentiality in Backbone Networks Based on Scalable and Dynamic Environment Technologies

Radosław Wielemborek, Dariusz Laskowski, and Piotr Łubkowski

Institute of Telecommunications, Faculty of Electronics,
Military University of Technology Gen. S. Kaliskiego 2 Street, 00-908 Warsaw
radoslaw.wielemborek@student.wat.edu.pl,
dlaskowski@wat.edu.pl

Abstract. Along with the dynamic evolution of wide area networks and network technology development, data security is becoming increasingly important. This issue is particularly important in organizations or companies with branch offices in many distant places all over the world where communication is required in real-time often and data leakage may even cause the collapse of the company. In order to guarantee the security of transmitted data between remote sites such as the central division office of the branches and mobile workers is proposed to apply technologies that use secure, encrypted tunnels such as virtual private network. The latest solution is a dynamic multipoint virtual private network technique that eliminates defects of previous versions. This paper verifies the efficiency of data privacy and protection afforded by IT.

Keywords: AISC, DMVPN, efficiency, VPN, data security.

1 Introduction

This paper contains the results of research aimed at examining the effectiveness of mechanisms ensuring the security offered by the dynamic multipoint virtual private network (DMVPN) in Hub & Spoke configuration. [1] Data transmitted between remote nodes and the central nodes are vulnerable to unauthorized access and "leakage" of private data (information). [2] Furthermore using IT devices in LAN nodes poses a threat of electromagnetic leakage of private data. In order to protect the private data the IT devices have to be protected using electromagnetic shielding [3]. It is important to ensure the highest data security during data transport. In order to ensure the confidentiality of transmitted information, created data tunnels are encrypted and keys used with algorithms should be as "strongest" as possible. In addition to the complexity of the data encryption algorithm it is also important that the method of encryption shouldn't consume too much power and don't overload the processor. Overloading can cause errors in encryption or other processes performed by this device. In consideration of multiple determinants in the process of compilation

W. Zamojski et al. (eds.), *Theory and Engineering of Complex Systems and Dependability,*
Advances in Intelligent Systems and Computing 365, DOI: 10.1007/978-3-319-19216-1_50

of confidential channels within the DMVPN it was found that it is advisable to test this innovative and forward-looking technology also following the use of various encryption algorithms (DES, 3DES, AES256).[4] The reference will be simple local area network. Some issues can be affected by vibration occurring in close environment of the network. It can be new area of the further research [5-6].

2   DMVPN Essence

Network topologies used in most companies are based on combination of remote branch offices (mobile workers) with central division (data server or specific services). Often it is necessary to have a connection between remote offices to share information. [7-9]. Such connections are compiled mostly through the public Internet. In order to increase security, the most commonly used protocol- IPSec allows you to create secure, encrypted connections between end-points. However, it has certain limitations. Created tunnels can only be point-to-point. The solution to this problem is to combine IPSec tunnels topologies in full / partial mesh or hub & spoke. Full meshed VPN topology allows direct compilation of VPN tunnels between all VPN devices. However, this technique works only in small networks where the number of VPN tunnels is small. Configuring all network devices in larger networks would be impractical. The Hub & Spoke topology, where the central node mediates in the compilation of tunnels to other VPN devices, configuring and compiling direct connections with all devices with a large number of VPN tunnels VPN is also not very effective. In contrast, the basic problem in using IPSec VPN topologies is the inability to use dynamic routing protocols due to the fact that they use multicast or broadcast communication and IPSec does not allow transfer such packets. These problems are solved by GRE tunnels in conjunction with IPsec. This protocol requires static addressing ends of the tunnel and the existence of routing between endpoints. In large networks, configuration of the central node would be very long and complicated, therefore full mesh topologies for direct connections between Spoke branches are used. This technique provides a greater scalability and flexibility for VPN IPSec. The size of routers concentrating traffic from multiple locations configuration is minimized. Encryption between routers that initially do not know about each other is automatic. DMVPN enables the connection pooling with routers, which IP address is assigned dynamically by DHCP, which is undoubtedly a simplification. The DMVPN technique is an extension of a VPN. It offers several modes of operation. The main of them is the basic configuration Hub & Spoke (star topology) based on the transport of data between remote branch offices through central point- Hub. This is illustrated on the following figure. Each remote unit is connected to a central node by static IPSec tunnel. Each client is registering as a client to a NHRP server located in Hub which assures finding best path from source to destination. This is followed by the data transmission between remote branch offices through a central hub node using IPSec tunnels connecting the remote branches to the central division.

For the above described configuration can be added Spoke-to-Spoke functionality, that allows for direct communication between remote offices. In this case, the basic mode configuration described above, at a time when communication occurs from one branch to another client initiated call will query the NHRP server located in the central node for actual IP address of the branch destination. Then an IPSec tunnel will be created directly between the branches. Spoke -type devices will receive information about the LAN networks in all branches of the panel through a dynamic routing protocol updates (OSPF, BGP, EIGRP, or RIP), acting in GRE tunnels between headquarters and remote nodes, sent by the central node Hub. GRE protocol makes it possible to transmit multicast traffic over the network and the operation of routing protocols between nodes VPN.

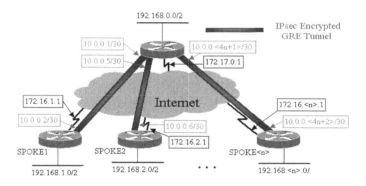

Fig. 1. Hub & Spoke Topology [14]

In order to improve network reliability DMVPN technology allows you to extend the basic topology Hub & Spoke of additional node or hub. It can be connected to another DMVPN network and it could even cooperate with another ISP provider. NHRP protocol provides to Spoke devices the ability to dynamically obtain information about the actual addresses of the interfaces from other branches. This means that each router can dynamically obtain information that is sufficient to direct statement of an IPSec tunnel with another branch of Spoke. This is possible through the cooperation of dynamic routing protocols, mapping and solving interface addresses to real addresses tunnels physical interfaces (NHRP). DMVPN technology allows for connections IPSec nodes with IP addresses obtain dynamically. DMVPN also supports split tunneling for Spoke devices, which means simultaneously directing unencrypted packets to the Internet and encrypted by IPSec tunnel. An important advantage of this technique is a stencil Spoke configuration nodes (change occurs only in IP addresses), and the hub node configuration unchanged even when adding new devices Spoke. This allows for quick and easy configuration of the nodes using the same pattern and adding new devices to an existing topology DMVPN. For a new type of Spoke branch, should be perform basic configuration as for each node, and add the information necessary for the authorization of a new branch in the ISAKMP configuration of a central device. Dynamic routing protocol will distribute information about the new destination network to all endpoints. DMVPN technology greatly reduces the size of the configuration file on each node VPN. On the basis of information contained above has been prepared test environment concept art DMVPN. Selected mode used to test was the Hub & Spoke topology.

3   The Concept of the Research Environment

DMVPN testing environment scope covers a wide variety of topics ranging from network devices via the mechanisms of transmission media network configurations such as routing protocols and VPN tunnels (Fig. 2.). Devices, used for the implementation of the research were software and equipment from renowned manufacturers i.e. Drytek Access Point, Cisco routers and firewalls, LANForge generates stateful network traffic and monitors packets for throughput and correctness, Hewlett Packard servers, 3Com switches, Wireshark and JPerf software. [10,11]

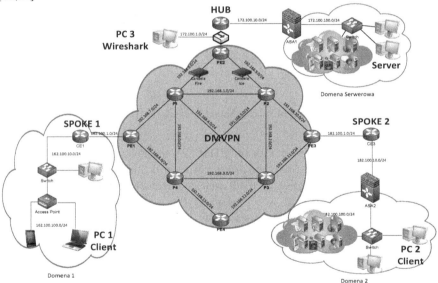

Fig. 2. Testbed

LANForge platform was used as traffic emulator allowing the most accurate representation of the actual conditions occurring in telecommunication networks by regulation of network parameters. The efficiency and quality of DMVPN service was examined basis on determining the impact of changes in latency and packet loss on the network throughput as compared with the ordinary telecommunications network LAN without security mechanisms. LANForge platform has been incorporated in the backbone. Desired type of network traffic was generated by the program JPerf installed on the network end devices (PC1 and Server). In contrast, traffic was collected by Wireshark protocol analyzer installed on computer PC3, connected to the network through a hub. Topology of the test network is presented in the figure below.

After the hardware configuration, and setting the LANForge platform and program JPerf on termination equipment it was possible to test ordinary LAN to obtain a reference point for subsequent performance testing techniques. It is worth mentioning that any of the measurements was not free from the influence of distorting results. Even if the DMVPN technique was not implemented, the cable installation affected the quality of the network, the errors resulting from long-term operation of devices, IP traffic load on the network and interference from all other working devices. However, the impact of the above factors on the test results was negligible and works on all measurements equally.

4   Web Performance Test

Network Testing was done by sending 1000 packets about the size of 256 kB over the network. These tests were performed for the local area network and DMVPN technique using encryption algorithms DES, 3DES and AES256. This study was designed to test the effectiveness of the network for the transmission of its many small packets and determine the total loss rate. The test results are shown in the following table (Table 1) and chart (Fig. 3.).

Table 1. Stream data rate units for 1000 packets 256 kB

Interval [s]	C [Mb/s]			
	LAN	DMVPN AES256	DMVPN 3DES	DMVPN DES
1	88,1	86,0	88,1	92,7
2	90,2	88,1	88,1	87,9
3	90,2	90,2	90,2	90,7
4	91,8	91,8	90,2	90,0
5	90,7	88,6	88,1	87,9
6	90,2	89,6	88,1	88,7
7	90,2	88,6	90,2	89,5
8	90,2	90,1	90,2	88,2
9	90,2	90,2	88,0	90,9
10	88,1	88,1	88,1	87,6
11	90,2	90,2	90,1	88,8
12	90,2	90,2	90,2	88,9
13	90,2	90,2	88,1	88,8
14	91,8	90,2	90,2	87,9
15	90,7	90,2	88,1	89,8
16	90,2	88,1	90,2	88,4
17	90,2	90,2	88,1	89,0
18	90,2	90,2	88,1	89,2
19	90,2	88,0	90,2	89,2
20	90,2	88,1	90,2	88,4
21	90,2	90,1	88,1	88,3

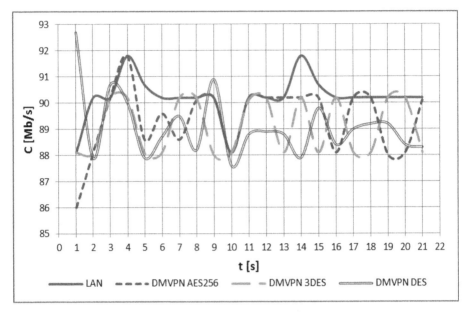

Fig. 3. Graph stream data rate units' 1000 packet size 256 kB

LAN after running DMVPN technique, using any encryption algorithm is characterized by a lower rate. The reason for this is taking place in the processes of securing data stream IPSec. The data are sent from the transmitting station to the receiving station. On the way the packets first encounter router Spoke 1. The data packets are encapsulated using the protocol mGRE to enable IPSec security through encryption of transmitted packets, because the protocol does not support multicast traffic. For the purpose of this Paper used protocols are DES, 3DES and AES with 256 bit key. Then, as the encrypted packets enter the router Hub, which performs the same steps in reverse order and then decoded packets arrive to the destination. Processes associated with compiling and preparing DMVPN tunnel packets to pass through the network (the above-mentioned encryption and encapsulation) are performed in routers enabled technology DMVPN (Hub and Spokes). This makes the routers must devote part of their computational power on these processes. It may also cause declines in rate and increase the packet loss associated with the performance of these routers also other processes.

From the above statements can be seen that the network with DMVPN technique implemented with AES 256 encryption algorithm is characterized by the highest average rate. The disadvantage is a large variation ranging from 86 Mb/s to 91.8 Mb/s for 1000 packets of size 256 KB. DES and 3DES algorithms are characterized by a lower average, but their constancy of rate is higher (it amounts to 2.2 to 3.3 for 3DES and DES at 5.8 for AES256). In comparison with usual LAN better is DMVPN technique using AES encryption with 256 bit key length. It is worse than the ordinary LAN, only about 0.8% for the transmission 1000 packet size 256 kB. However, these differences compared to the usual LAN are so small to feel the inconvenience of working with a tunnel or without him. These values according to the theory can be up to 10%, so the results with interest are in the theoretical data. Implementation DMVPN technique slightly prolongs the process of obtaining the desired data in the terminal station, but does not cause a problem, since these are too small to be appreciable.

In order to investigate network performance with the DMVPN technique carried out a second test of which consists on testing the network throughput between the client and the server PC1 located in the domain of the server in terms of increasing the delay generated by the device LANForge. The study consisted of 250 MB of data transferred over the network and pulling the average of all samples collected rate every second, the network of tunnels and different without him. The measurement results are presented below. The study consisted of transferred 250 MB of data over the network and averaged all collected samples rate taken every second, the network with different tunnels and without him. The measurement results are presented below.

Table 2. Throughput and delay as a function of variation

Latency [ms]	C [Mb/s]			
	LAN	DMVPN AES256	DMVPN 3DES	DMVPN DES
10	43,7	42,0	42,2	42,2
20	43,0	41,8	42,1	42,1
30	43,5	40,9	41,9	41,9
40	43,3	41,1	41,8	41,8
50	41,8	41,4	40,4	40,4
80	38,9	40,8	37,1	37,4
90	37,3	29,1	36,1	36,6
100	39,9	20,2	38,2	38,5
200	35,9	17,6	38,7	28,9

From Figure 3 it can be observed that implements a LAN with DMVPN technique is more susceptible to the increasing delay value. For a delay of about 80 ms practical differences are imperceptible (loss up to 2 Mb/s). Above this value, there was a huge drop in the value rate for the DMVPN tunnel with AES256 encryption algorithm. Outperformed the other algorithms as the decrease rate of the entire range of measurements (delay value selected from 10 to 200 ms) is not greater than 2 Mb/s.

It is worth noting that the existing public network delays do not reach values above 80 ms. Average their size are below 30 ms. This means that the work of the tunnel and use any encryption algorithm does not cause any difficulty, as occurs 2% difference in the rate for all tested encryption algorithms for delays up to 80 ms compared to the usual LAN is insignificant. In real network conditions - when the aim is to provide the best conditions for network usage, the effect is quite noticeable DMVPN, and it provides security and scalability in any network conditions are higher than for ordinary LAN.

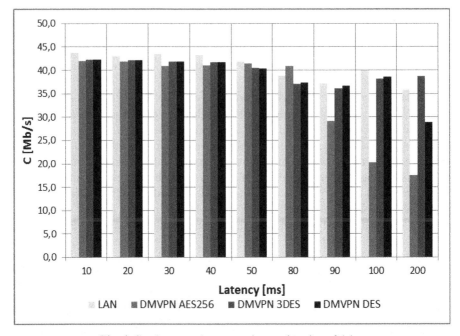

Fig. 4. Graph stream data rate units as a function of delay

5   Conclusion

Described DMVPN technique is the perfect solution for users who appreciate data security provided by IPSec, together with the provision of access to corporate resources from anywhere in the world via a secure encrypted tunnels. This provides a quick and easy reconfiguration of the networks administrator in case of, for example, the expansion of new branches or new users. It is a very convenient tool to give each user access rights needed for the position occupied by the work that each employee has access to only the necessary documents to him. In this way DMVPN provides great flexibility and confidentiality associated with the use of IPSec tunnels made from a variety of encryption algorithms and hash functions. That will greatly influence the ability to create dispersed organizational structures, remote workstations and will facilitate the operation of the intercontinental companies. [12]

In this paper was described DMVPN network solution to check Hub and Spoke type, where the tunnels are compiled on the road Spoke-Hub. This mode causes the Hub - the central node is fully functional supervisor and mediator in each data transmission. In order to assess the impact of dynamic, scalable virtual private networks on the quality of the connection is attempted data transmission via an IPSec tunnel using the emulator platform LANForge actual network conditions. [13] The data presented above demonstrate a minimum, practically imperceptible impact of this technology on the time of receipt of the data. DMVPN technique in any noticeable way hinders the work of a normal user, it is the imperceptible. In conclusion DMVPN technique is very good and convenient solution for large companies, where the mobility of workers and security play a major role. This technique allows you to secure data transmission and it is not forcing you to constantly maintain active VPN connections. Additionally CCP software makes anyone able to quickly and effectively establish DMVPN connection even in home conditions.

References

1. Jankuniene, R., Jankunaite, I.: Route Creation Influence on DMVPN QoS. In: Proceedings of the ITI 2009 31ST International Conference on Information Technology Interfaces, Dubrovnik, pp. 609–614 (2009)
2. Roebuck, K.: Internet Privacy of Data and Information. Emereo Publishing, Vendors (2012)
3. Nowosielski, L., Łopatka, J.: Measurement of Shielding Effectiveness with the Method Using High Power Electromagnetic Pulse Generator. In: Progress In Electromagnetics Research Symposium, PIERS 2014 Conference Proccedings, China, pp. 2687–2691 (2014)
4. Chen, H.: Design and Implementation of Secure Enterprise Network Based on DMVPN, Business Management and Electronic Information (BMEI), pp. 506–511 (2011)
5. Burdzik, R., Konieczny, Ł., Figlus, T.: Concept of On-Board Comfort Vibration Monitoring System for Vehicles. In: Mikulski, J. (ed.) TST 2013. CCIS, vol. 395, pp. 418–425. Springer, Heidelberg (2013)
6. Burdzik, R.: Research on the influence of engine rotational speed to the vibration penetration into the driver via feet - multidimensional analysis. Journal of Vibroengineering 15(4), 2114–2123 (2014)
7. Siergiejczyk, M., Paś, J., Rosiński, A.: Application of closed circuit television for highway telematics. In: Mikulski, J. (ed.) TST 2012. CCIS, vol. 329, pp. 159–165. Springer, Heidelberg (2012)
8. Siergiejczyk, M., Paś, J., Rosiński, A.: Evaluation of safety of highway CCTV system's maintenance process. In: Mikulski, J. (ed.) TST 2014. CCIS, vol. 471, pp. 69–79. Springer, Heidelberg (2014)
9. Siergiejczyk, M., Krzykowska, K., Rosiński, A.: Reliability assessment of cooperation and replacement of surveillance systems in air traffic. In: Zamojski, W., Mazurkiewicz, J., Sugier, J., Walkowiak, T., Kacprzyk, J. (eds.) Proceedings of the Ninth International Conference on DepCoS-RELCOMEX. AISC, vol. 286, pp. 403–412. Springer, Heidelberg (2014)
10. Lubkowski, P., Laskowski, D., Pawlak, E.: Provision of the reliable video surveillance services in heterogeneous networks. In: Safety and Reliability: Methodology and Applications - Proceedings of the European Safety and Reliability Conference, ESREL 2014, pp. 883–888 (2015)
11. Laskowski, D., Łubkowski, P., Pawlak, E., Stańczyk, P.: Anthropo-technical systems reliability. In: Safety and Reliability: Methodology and Applications - Proceedings of the European Safety and Reliability Conference, ESREL 2014, pp. 399–407 (2015)
12. Sławińska, M., Butlewski, M.: Efficient Control Tool of Work System Resources in the Macro-Ergonomic Context. In: Applied Human Factors and Ergonomics, AHFE Conference, pp. 3780–3788 (2014)
13. Laskowski, D., Łubkowski, P.: Confidential transportation of data on the technical state of facilities. In: Zamojski, W., Mazurkiewicz, J., Sugier, J., Walkowiak, T., Kacprzyk, J. (eds.) Proceedings of the Ninth International Conference on DepCoS-RELCOMEX. AISC, vol. 286, pp. 313–324. Springer, Heidelberg (2014)
14. http://www.cisco.com/c/en/us/support/docs/security-vpn/ipsec-negotiation-ike-protocols/41940-dmvpn.html

Models for Estimating the Execution Time of Software Loops in Parallel and Distributed Systems

Magdalena Wróbel

Maritime University of Szczecin, ul. Wały Chrobrego 1-2, 70-500 Szczecin, Poland
m.wrobel@am.szczecin.pl

Abstract. Presented are new methods of loop execution time estimation for parallelized and distributed systems. The proposed solutions take account of data transfer time, data locality and synchronization of threads. In this way the methods have been adjusted to modern parallel and distributed systems, which permits to estimate execution times of loops compatible with FAN, PAR and PIPE transformations. The model-based estimates have been compared to real measurements of program loops parallelized in the OpenMP standard and adapted to distributed systems satisfying the MPI standard. The presented approach can be used for optimized allocation of tasks in multithreaded processors and in distributed systems.

Keywords: time estimation, loops program, fan, par, pipe, the time estimation with limited bandwidth.

1   Introduction

The development of computer multithreading architecture has allowed to accelerate computations by parallelizing them. However, even multithreaded processors are not sufficient to perform complex computational tasks. In such cases distributed computing systems come to assistance. In these systems, tasks are distributed to be executed by individual nodes, and collected at the end. Even a number of powerful computers may execute certain tasks more slowly than a single computer running in the parallelized mode. This requires task distribution, data transmission, synchronization of calculations and data locality. Data transmission is a particularly important issue, crucial in systems with a limited link, such as those occurring in maritime transport and aviation. In spite of existing efficient broadband connections, e.g. satellite systems, the cost of using them for transmission of large portions of data becomes unprofitable. To date, the only form of data transfer are slow short-range systems, mainly based on radio transmission, such as VHF [1]. This essentially restricts possible implementation of the distributed system. This remark refers mainly to systems requiring high computing power, often exceeding the capabilities of user's systems, such as those carried by ships. Examples of such systems are decision support systems, which today can calculate optimal courses of the ship, or warn the navigator against a risk of collision, taking into account the current navigational situation and relevant Collision Regulations.

This article presents methods of estimating execution time of program loops compatible with FAN, PAR and PIPE transformations. Development of these methods

© Springer International Publishing Switzerland 2015

W. Zamojski et al. (eds.), *Theory and Engineering of Complex Systems and Dependability,*
Advances in Intelligent Systems and Computing 365, DOI: 10.1007/978-3-319-19216-1_51

applicable to parallel and distributed systems is a necessary step towards further research   aimed at improving methods of optimal task assignment to individual computers engaged in computations and the determination whether given type of computations should be made parallel or distributed. The presented solutions take into consideration data transfer time and data locality, essential in systems with limited bandwidth.

2   Program Loops Compatible with FAN, PAR and PIPE Transformations

In high-level programming languages, such as C and C++, there may exist different types of dependencies that make parallelization impossible. The character of these dependencies defines the type of loop and the parallelization method. In the book [2] the author proposes to divide loops depending on their transformation and presents three types: those consistent with the FAN transformation (with dependencies in the loop body or lack of them), PAR (dependencies between iterations) and PIPE (mixed dependencies).

FAN loops may have dependencies inside the body but cannot have data dependencies between iterations. Each thread inside a parallel section performs the same operations, but with different data, e.g. with different indexes of the array [2, 3]. The threads perform calculations returning the results to the main thread that assigns another task to each thread, or a thread is waiting for the next task. FAN transformation is a classic model of a fork-join. The main thread divides the task into smaller parts, executed at the same time. Dependencies inside the loop are retained because operations inside the body are executed by only one thread. The fork-join model does not guarantee the execution of *j-th* iteration before *j+1* iteration, so dependencies between iterations cannot exist inside the FAN transformation.

$$T_k = (r_0 + r_1 + w_0 + w_1) + r_1 F(S/k) + (k - 1)(w_0 + w_1(S/k)) \qquad (1)$$

The estimation model (1) for the FAN transformation presented in [2] calculates the time of program execution on k processors from environment data and a function defining computational complexity $F(S/k)$ for a specified size of data S divided by the number of processors k. Environment data include operation preparation time (r_0), operation execution time (r_1), time of data preparation for transmission (w_0) and time of data transfer to threads (w_1). The time of data transfer from the main thread to the other threads $(k-1)(w_0+w_1(S/k))$ is added to the computing time.

Loops, compatible with PAR transformation can have data dependencies between loop iterations, but cannot have dependencies inside the loop body. For a loop to be performed in parallel, it must have at least two statements inside the loop body. The created threads execute independent instructions for the whole loop. PAR and FAN transformations alike allocate tasks to individual threads, but the difference between them is that threads in the FAN transformation perform identical instructions on different data, while threads in the PAR execute different instructions. Additionally, in the PAR transformation each iteration must be followed by a barrier for data synchronization. This prevents   downloading outdated data to the threads. The lack of dependencies between instructions inside the loop body allows to distribute instructions to each thread. The more instructions the loop has, the more threads this loop can be divided into [2].

$$T_n = M_{i=2}^n \{[r_0 + r_1 F(S_1)]; [(i-1)(w_0 + w_1 S_i) + r_0 + r_1 F(S_i) + w_0 + w_1]\} \quad (2)$$

The model (2) estimating the time of executing a PAR loop [2] is a maximum function determining a critical path. The result of the maximum function may be a path $[r_0 + r_1 F(S_1)]$, being a sequential execution or one of the other n paths $(i-1)(w_0 + w_1 S_i) + r_0 + r_1 F(S_i) + w_0 + w_1$. According to the PAR transformation, the loop is divided into independent clauses, the model tests each i-th clause and chooses the longest time. Data S in the model (2) are divided into threads of the size S_i. The division depends on data dependencies inside the loop.

The PIPE transformation of a loop is used when there are dependencies between instructions inside the body and dependencies between iterations. Such dependencies make it impossible to use another transformation, FAN or PAR. Unlike FAN or PAR, the PIPE transformation does not synchronize threads after each iteration, but asynchronously executes the instructions inside the loop body. Each iteration of the loop is performed in parallel, but these iterations are shifted in time to eliminate the dependency. Independent instructions in different iterations can be performed at the same time [2]. For implementation of the PIPE transformation the following conditions have to be met:

— the statements inside the loop body are numbered and executed by the order;
— iterations do not have to be synchronized, but iteration j where $j<k$ is always executed before k-iteration;
— the same instructions from different iterations can be performed at the same time by different threads;
— dependencies called backward loop-carried $[i-1]$ are safe, while forward loop-carried $[i+1]$ dependencies must be changed to remove such dependency;
— the number of iterations must be greater than the number of threads.

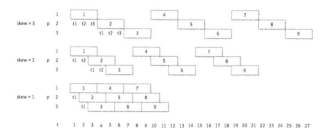

Fig. 1. Parallel execution for skew = 1,2,3 [2]

The exclusion of reverse dependencies is done by adding the delay skew, which delays the work of a thread by the skew value. The delay eliminates the problem of data synchronization by delaying the execution of a dependent instruction. The skew parameter is calculated from an analysis of data dependencies in the program loop.

The larger the skew, the lower acceleration can be obtained. Figure 1 schematically shows the execution of each iteration in parallel by three threads. The longer delay, the longer execution time of the loop was. In the worst case, the execution time of a parallel loop can be compared to sequential execution.

$$T_k = r_0 k + r_1 \{(N-1) \sum_{i=1}^{skew} t_i + \sum_{i=1}^{k} t_i\} + k(N-1)(w_0 + w_1 S) \qquad (3)$$

In the model (3) [2] estimating the execution time of a PIPE loop for each thread, we introduce the skew removing dependencies. The sum of all delays $\sum_{i=1}^{skew} t_i$ is added to the sum of instruction execution time $\sum_{i=1}^{k} t_i$ and is multiplied by the number of iterations $(N-1)$ and the operation execution time (r_1). Thus calculated time is added to the preparation time of operations on k processors $(r_0 k)$ and the time of data transfer from the main thread to the remaining threads for $N-1$ iterations.

3   Parallelization Methods of Program Loops Compatible with the FAN, PAR and PIPE Transformations

Dependencies that may occur in the loops have impact on the methods used for parallelization [4]. In the simplest case – the lack of data dependencies – the relevant additional clauses should be added to the programming clause *#pragma omp parallel for*. This is done to reach the best acceleration and to get correct values of variables after leaving the parallel region [5]. Data dependencies have to be analyzed in program loops with flow-dependence, anti-dependence or output-dependence inside the loop body (FAN transformation) to determine the data sharing attribute clauses. Appropriate use of the clauses in the standard OpenMP provides a correct result even though dependencies occur between instructions in the loop body [6,7].

Data dependencies between iterations (PAR transformation) can be divided into two types: dependencies that refer to subsequent indexes of variables $(i+1)$ and those referring to previous indexes $(i-1)$. The former dependencies can be removed for readout by making a copy of this variable, in which dependencies exist between loop iterations. The copy of the variable will guarantee that each thread reads unchanged by other threads values of the original variable. Dependencies of the $i-1$ type can be removed by adding the skew parameter to the index array, which removes the reference to the previous iteration. Additionally, loop skewing requires an analysis of the order of executed instructions, aimed at establishing a new order of their execution, complying with the instruction execution sequentially. Besides, before and after the program loop the instructions omitted by using the skew parameter should be added. The dependencies of the type $i-1$ in multidimensional arrays can be parallelized by the selected independent dimension of the array [6].

Mixed dependencies, that is those existing inside the loop body and between iterations (PIPE transformation) can be converted into smaller loops, where only dependencies inside the body or between iterations occur. Further removal of the dependencies is done similarly to the methods for FAN or PAR transformations.

4   Models for Estimating the Execution Time of FAN, PAR and PIPE Loops

This author proposes new models for estimating the execution time of loops compatible with FAN, PAR and PIPE transformation. The estimation for a FAN loop can be a basis for all types of loops. The reason for this is that at present the loops PAR and PIPE are mainly transformed fragmentarily to a loop compatible with FAN transformation. For the PAR loop, instructions from the loop are divided separately into smaller loops executed in accordance with FAN transformation. In case of the PIPE loop, FAN or PAR loops are obtained by adding a loop index skew.

An additional element that has been added to the models for estimating the execution time of the loops is a data locality parameter. Data locality is important in the calculation in multidimensional arrays, where the loop retrieves and/or writes data scattered in the memory. For example, a loop that calculates by iterating subsequent lines in a two-dimensional array is performed longer than a loop iterating column by column in the same array.

This author proposes a new form of the model for estimating the execution time of loops compatible with the FAN transformation. The model is expanded in comparison to the ones developed in author's earlier works [3]. The new model (4) takes into consideration data localities and features more detailed information about how to take measurements. New models for other types of loops are also herein presented.

The following model can be used for estimating the execution time of loops compatible with FAN transformation running on multithreaded computers:

$$T(n) = \sum_{k=1}^{K} \frac{(r_k \cdot li \cdot z_k)}{lp \cdot n} + (w \cdot md) + cw + ti, \tag{4}$$

where:

r – execution time of single operation
w – communication time
li – number of iterations (in nested loops adds up all iterations)
md – number of data needed for the calculations by one thread

z – the number of operations inside the loop body
lp– pipeline stages in processor
cw – synchronization time of threads
ti – initialization time of measurement
k – the type of data locality
n – number of threads

The model (4) is divided into three parts. In the first part, dependent on parameter r, the execution time of instructions inside the loop body is estimated. The sign of the sum relates to a locality where parameter r_k should be set for all types of data k (locality) taken for calculations. Their total number is K. The execution time of a single operation (r) is multiplied by the number of operations inside the loop body (z) and by the number of iterations (li) of the loop (in nested loops all iterations are added up). The whole formula is divided by the product of the number of threads and the number of pipeline stages. That number determines the number of operations to be executed by a single pipeline stream inside the processor. Given the construction and speed of modern processors, it is assumed that instructions inside a pipeline are executed approximately in the time unit.

The second part of the model refers to communication time, that is time required for data delivery to the threads. Each thread works with a different portion of data. All data needed for calculations are expressed by the parameter md. The number of data was multiplied by the transmission time (w) calculated for a single data item sent to a thread.

Loop execution in parallel requires the creation and synchronization of $lw\text{-}1$ number of threads. Based on the delivered data, the thread performs calculations and the results are transferred to the main thread. Time cw is the sum of the time of the initiation of the region and the time of final synchronization of calculations.

Finally, the author has introduced parameter $ti,$ which represents a measurement error resulting from the method inaccuracy in measuring the real time of loop execution. Methods for determining the parameters r, w, md, cw and ti are presented in [3].

For distributed systems, this model obtains the following form presented model (5).

$$T(n) = \sum_{i=1}^{m} \sum_{k=1}^{K} \frac{r_{ki} \cdot li_i \cdot z_k}{n_i lp_i} + \sum_{i=1}^{m} w_i \cdot md_i + \sum_{i=1}^{m} cw_i + ti \qquad (5)$$

The notations of parameters remain the same as for the model (4). They have been supplemented with the index i, which indicates a number of computers in a distributed system for m computers. In this case, m is a number of threads available in the distributed system. The parameter denoted as cw indicates the synchronization time for all items of information.

The author proposes for loops compatible with PAR transformation the following new model for estimating the execution time of these loops on multithreaded systems presented model (6)

$$T(n) = M_{l=1}^{L} (\sum_{k=1}^{n} \frac{(r_{kl} \cdot li \cdot z_{kl})}{lp \cdot n} + (w_l \cdot md_l) + cw_l) + ti \qquad (6)$$

Where l indicates the critical path determined by the maximum function that is here necessary. For example, if a loop compatible with the PAR transformation has two instructions inside the loop body, they will be executed by separate threads. Then the execution time is equal to the longest executed instruction.

For distributed systems, this model (7) takes the following form:

$$T(n) = M_{l=1}^{L} (\sum_{i=1}^{m} \sum_{k=1}^{K} \frac{r_{kil} \cdot li_i \cdot z_{kl}}{n_i lp_i} + \sum_{i=1}^{m} w_{il} \cdot md_{il} + \sum_{i=1}^{m} cw_{il}) + ti \qquad (7)$$

For the loops compatible with the PIPE transformation it is proposed to divide the loop into the FAN and PAR parts, for which the models for execution time estimation are shown above.

The models proposed in [2] for FAN and PAR transformations include a function defining computational complexity. As the use of computational complexity is too general, which leads to inaccuracies, the author proposes new models in which general computational complexity has been replaced by a product of iteration number and the number of operations inside the loop body. Additionally introduced parameters permit to more accurately estimate loop execution time. These include: lp identifying the pipeline stage in a processor, cw – time of synchronization between threads during the operation of loops in parallel, and ti – calculating real measurement error. Extra parameters allow to adjust the models to the current architecture of multistage pipeline processors.

Apart from models estimating time in the parallel environment, this author has presented models for a distributed environment and for each transformation. A separate method of calculating environment parameters is presented for these models.

5   Tests of Presenting Models

The tests of loops from NAS-Parallel-Benchmark [8] made use of loops with all types of dependencies. All the loops were parallelized to OpenMP [5] and MPI [9] standards. The testing environment had the following specification: Intel Core i7 Q760 processor (8 threads), 8GB RAM, operating system Windows 7 64 bit. The tests were performed for different numbers of threads: 2, 4 and 8. For a distributed system an additional computer was used, with Intel Core i5-2450M processor (4 threads), 4GB RAM connected with the base computer by a 10/100 Mbps switch. Selected loops are shown table 1.

Table 1. Programming loops used to tests

MG_mg_f2p_1	for(ax = 1; ax <= 3; ax++){
	    for(k = N1; k <= N2; k++){
	        ng[ax][k] = ng[ax][k+1]/2;
	    }}
		CG_cg_f2p_6 with 2D array	for (i=2; i< N2;i++){
			for (j=2; j< N2; j++){
			A[i][j]= A[i][j] + A[i][j-1];
			}}

MG_mg_f2p_4 with 2D array	for (i=2; i< N2; i++){
	for (j=2; j< N2;j++){
	    A[i][j]= A[i][j+1] + m1[i][j+1] * m2[i][j+1] *
	m3[i][j+1];
	}}
		CG_cg_f2p_6	for(j = 2; j <= N1; j++){
			    rowstr[j] = rowstr[j] + rowstr[j-1];
			}

Seidel	for (i=0; i<MATRIX_SIZE-1; i++){
	for (j=0; j<MATRIX_SIZE-1; j++){
	    a[i][j] = (a[i-1][j-1] + a[i-1][j] +
	a[i-1][j+1]+ a[i][j-1] + a[i][j] + a[i][j+1]+
	a[i+1][j-1] + a[i+1][j] + a[i+1][j+1])*0,11;
	}}
		MG_mg_f2p_4	for(j = N1; j <= N2; j++){
			    ir[j]=ir[j+1]+m1[j+1]*m2[j+1]*m3[j+1];
			}

The loops in Table 1 have data dependencies and dependencies between loop iterations. All of them are loops compatible with the PIPE transformation. Each loop was analyzed and transformed to FAN or PAR loop using OpenMP and MPI standards.

The method of measuring the parameters is shown in Figure 2.

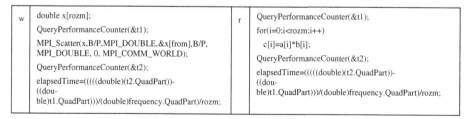

Fig. 2. Method of estimating the values of parameters r and w for the parallelized environment
Source: author's study

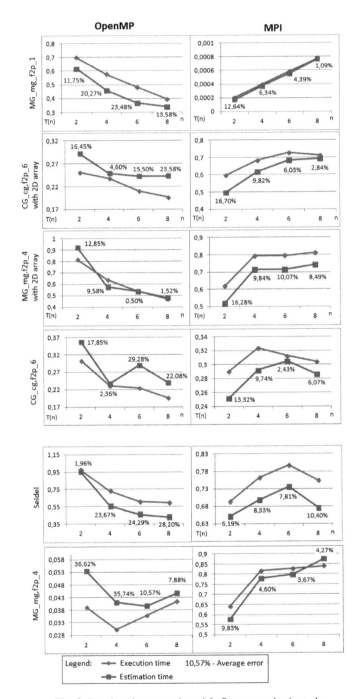

Fig. 3. Results of presented models Source: author's study

The measurement results and times estimated using the models are given in Figure 3. In table 2 compared approximation error (the same program loops) models (1),(2),(3) presented in [2] with proposed models (4),(6).

The models used for the estimation of execution times of parallelized programming loops in OpenMP standard show a characteristic similar to real measurements. The maximum average error of these models was 22%, whereas the minimum error reached 6%. The maximum average error of Lewis models (1),(2),(3) was 780,23%, whereas the minimum error reached 22,79%. All approximation errors shown in Table 2.

Table 2. Approximation error [%] of Lewis models (1),(2),(3) and proposed models (4),(6) on OpenMP standard

Threads num. / Loop name	Models (1), (2), (3)				Models (4), (6)			
	2	4	6	8	2	4	6	8
MG_mg_f2p_1	13,82	22,06	52,70	91,53	11,75	20,28	23,48	13,58
CG_cg,f2p_6 2D	69,34	83,02	109,07	124,21	16,46	4,60	15,50	23,58
MG_mg_f2p_4 2D	1,55	57,37	100,34	134,87	12,85	9,59	0,50	1,53
CG_cg,f2p_6	56,30	120,64	132,07	168,01	17,85	2,36	29,28	22,08
Seidel	49,61	26,35	9,05	6,13	1,96	23,67	24,29	28,20
MG_mg.f2p_4	454,67	862,20	711,25	1092,80	36,63	35,74	10,57	7,88

The models estimating the time show a strong positive linear Pearson correlation (79 - 99%), which indicates good conformity of the models to real measurements.

Models for estimating the execution time of software loops in the distributed environment also depict very good characteristics. The maximum average error was 11%, while the minimum average error was 3.5%. This proves high conformity of the presented models to real measurements, additionally confirmed by strong and positive linear correlation (82 - 99%).

6   Summary

The presented new methods of estimating loop execution times are applicable to all types of loops in various programming languages. The described models allow to estimate loop execution time on multithread processors and in distributed systems. The introduced models have been adjusted to current architecture of processors and distributed systems. This could be achieved by improving measurement methods of model components and by introducing new parameters, including a parameter accounting for data localities, synchronization of threads and time of data transfer to the threads. For verification, the models have been tested using loops from NAS-Parallel-Benchmark. The tests have revealed the correctness of the models of loop execution time estimation, compared to real time measurements. In the future, the models may be used for building a system of efficient task allocation to maximize hardware capabilities in parallel and distributed systems.

References

1. Pietrzykowski, Z., Nozdrzykowski, Ł.: Technologie informacyjno-komunikacyjne w transporcie morskim – od udostępniania informacji do udostępniania usług In: TTS Technika Transportu Szynowego (October 2013)
2. Lewis, T.: Foundations of Parallel Programming: A Machine-Independent Approach. IEEE Computer Society Press (1992)
3. Wróbel, M., Nozdrzykowski, Ł.: Model for estimating the execution time of the loop program with limited connection. Logistyka 4, 3425–3435 (2014)
4. Siedlecki, K.: Algorytmy wyszukiwania drobno- i gruboziarnistej równoległości w pętlach programowych z zależnościami afinicznymi, PhD thesis, Szczecin (2008)
5. OpenMP, http://www.openmp.org (access date: February 2015)
6. Chandra, R., Dagum, L., KohrD., M.D., McDonald, J., Menon, R.: Parallel Programming in OpenMP. Morgan Kaufmann (2001)
7. Czech, Z.: Wprowadzenie do obliczeń równoległych, Wydawnictwo Naukowe PWN (2010)
8. NAS Parallel Benchmarks, http://www.nas.nasa.gov/publications/npb.html (access date: February 2015)
9. Message Passing Interface (MPI), http://www.mpich.org/ (access date: February 2015)

Analysis of Different Aspects
of Infomobility for Public Transport in Latvia

Irina Yatskiv (Jackiva)[1], Vaira Gromule[2], and Irina Pticina[1]

[1] Transport and Telecommunication Institute, Riga,
Latvia, Lomonosova 1, LV 1019
{Jackiva.I,Pticina.I}@tsi.lv
[2] JSC "Riga International Coach Terminal", Riga, Latvia, Pragas 1, LV 1050
vaira@autoosta.lv

Abstract. The concept of information mobility and its application spheres have been overviewed in the article. The main task in the way of infomobility implementation is to create and manage the information systems that facilitate the use of the transport system by its potential users. The information systems for public transport in Latvia have been analysed on the information services offered. Based on the received results of the development, the recommendations of the infomobility service level improvement in Latvia have been formulated.

Keywords: mobility, public transport, information, services, systems.

1 Introduction

Public transport plays an important role in social and economic life of the community. There are 150 annual public transport (PT) journeys per urban inhabitant in the EU. In other words, the 'regular' urban denizen in the EU undertakes an average of some three journeys every week using PT [1]. The provision of PT services is a complex process and in order to attract travellers to use PT, there is a need to know and satisfy their requirements. On the other side, PT is an area in which the implementation of information technologies will contribute to radical changes over the 21 century. The dissemination of information technologies in that area is characterized, first of all, by technological factors: widespread deployment of the new generation smart cards technologies and the development of spatial information systems. The integration information systems (IS) and integrated ticketing, introduction of other Information and Communication Technology (ICT) and ITS services for the customer are essential for enhancing the use of public transport. According to UITP [1], ICT is crucial for the study, design, development, implementation, support or management of computer-based IS, particularly software applications and computer hardware. One of the key user needs is "Door to door information", which require the information to be sufficiently detailed, accurate and covering all aspects of the journey, in forms that users understand (e.g. language for tourists, maps for sensory impaired travellers) etc.

© Springer International Publishing Switzerland 2015   543
W. Zamojski et al. (eds.), *Theory and Engineering of Complex Systems and Dependability,*
Advances in Intelligent Systems and Computing 365, DOI: 10.1007/978-3-319-19216-1_52

The implementation of customized functions that are now characteristic of IS shall become feasible. User wants to buy a ticket for "the journey" and not for "the means of transport". He wants to order a journey: from the place of current residence, with the best possible choice of means of transport and routes, and "invisible to the passenger" problems and their solution if something happens during the trip. In the case of information services in transport, the changes will refer to:

— travel payments system: the transport system will be multimodal, with more flexible and individualized offer, taking into account the specific needs of different groups of users, in particular with disabilities;
— tickets ordering: the development of spatial IS, which will entail the development of monitoring: global, regional or local. Access to the data of such systems will become widespread. In combination with the development of mobile and satellite technologies, it will create a new quality in planning, organizing, and the whole course of the journey [2].

The state of information service for PT in Latvia have been analysed in the article and it is organized as follows. In section two we will make a literature review on the concept of "infomobility". Section 3 overviews the main stakeholders in the infomobility market in Latvia and information services offered to passengers. Section 4 presents the results of the analysis and develops recommendations. Finally, research paths for further developments in the field of infomobility are addressed.

2   Literature Review on "Infomobility" Concept

"Infomobility" is a term increasingly discussed and analysed in academic literature and practical areas. The term "infomobility" has different meanings relating to the specific area of analysis: technical or social. In the linguistic sense it is "information-for-mobility". Mobility is a core element in any urban and regional development and can be defined as the ability to reach the destination at the time and cost that are satisfactory. Information is the data that has been processed in the way to be meaningful to the person who receives it. So, we can define infomobility similarly to all information services for users of transport system.

From the technological aspects of view, it was Kotsakis [3] who first determined the "infomobility as a term used to describe the set of technologies and the applications that allow the "user on the move" to access positioning and geographic information anytime, anywhere". Paganelli and Giuli denoted more service aspects and describe the term as "Information delivery for user mobility support" [4]. In [5] it was described "with the term "Infomobility" we refer to the broad range of information services which provide users with information and transactions required for supporting mobility of persons and goods – both in private and public transport – in Intelligent Transport Systems (ITS)". One of the definitions is: "Infomobility systems provide access to information and services for the support of user mobility. Bidirectional communication between the client devices and the system that can travel by several different transportation means, ranging from cars to trains and foot" [6]. In [7] it is defined as "the set of information management systems that aim at improving the mobility systems, providing the required information to managers and users". M. Arena et.al. determine main application fields: traffic management, fleet management, mobility payments, safety and security and information service. These fields contain overlapping information areas, but the areas that have the greatest impact on others, are the spheres of information services since information services permeate all other areas and are an integral part of all others. Sterle C. considering this term in the frame of city logistics defines "info mobility" as "a theme increasingly debated due to its potentiality of making the mobility system more efficient and effective in meeting users' needs" and considers the following application fields [8]:

- planning the vehicles routing and scheduling,
- vehicles localization,
- travel guide,
- exchange of the information among the different urban logistic actors,
- management of the logistic flows and tracking of the freights,
- traffic light regulation,
- payment services (road and park pricing),
- optimization of parking places.

The wide range of various information services successfully functioning in the systems of PT in the countries of the European Union gives topicality to the problem of transferring of the best practices on infomobility services within the EU. Many EU projects have investigated this concept in the last decade: Quantis, Connect, eMOTION, POLITE [9 – 12], etc. In TTRANS report [13] it has been mentioned: "Reliable, personalized, and "anytime-anywhere" based real-time travel and traffic information (RTTI) is a key element of intelligent mobility services envisioned for the future. The activities in the Infomobility sector mainly focus on: traffic and traveller information; geo-localization; freight and logistics, access and demand management".

So, the term infomobility basically refers to the adoption of information technologies to support the mobility and movement of people and goods, through continuous interactive and intelligent access to multimedia information supporting the needs that may occur in the field of transportation, during working activities and in a person's spare time. Infomobility, on the one hand, helps normal citizens who move through traffic (using a car, motorcycle, or even by bicycle or on foot), and on the other hand it helps those who use means of public transportation (with real-time information regarding bus and train times, or the location of stops), as well as logistic operators and those who transport goods.

Infomobility concept was described in [14] as a set of the following conditions:

- seamless real-time travel and traffic information including multi-modal journey planning and IS;
- freight IS combining operators' freight-flow and public authorities traffic flow requirements contributing to the optimum use of road capacity and the reduction of negative impact on the environment;
- eCall leading to a reduction in facilities;

- electronic Toll Collection as a key instrument for internalization of external costs;
- traffic demand management leading to cleaner road transport and less congestions;
- integration of several core applications on an open in-vehicle Telematics platform.

But, first of all, the quality of an infomobility system refers to its capability to provide managers and customers with the "right" information.

3   Main Stakeholders and Services in the Latvian Infomobility Market

As a part of EU Latvia shall follow and harmonize its national policy with the EU directives. In 2010 EU published the directive 2010/40/EU. This directive refers to the framework for the deployment of Intelligent Transport Systems in the field of road transport and for the interfaces with other modes of transport [15]. In the Latvian transportation development guidelines (2014 – 2020), there are noted the following objectives [16]:

- available and affordable PT which ensures access throughout the whole country by providing a convenient and unified PT system which can ensure good connections between bus and rail services;
- implementation of unified PT planning in intercity and regional routes;
- unified policy on tariffs;
- unified tickets.

The main parties which are involved in this process from state institutions:

- Ministry of Transport, which supports ITS by providing planning, management and monitoring of the EU funds and activities;
- Latvian State Roads, managed by 188 data collection points all over Latvia and available on-line and the historic data about traffic intensity on public webpage (http://www.lvceli.lv/traffic);
- Ltd "Road Transport Administration" (RTA);
- City Councils (two cities are active in the field of ITS: Riga and Jelgava);
- Planning Regions and local authorities.

Buses and train are the most popular PT service in the Latvian regional and interregional markets. It operates properly in conventional infrastructures, and offers different opportunities in terms of size/passenger capacity, speeds, combustion technology, etc. The State "Road Transport Administration" Ltd (RTA) is a unified state policy implementer in the field of licensing international transport and commercial road transport businesses in Latvia. The main function of the Road Transport Administration in PT development is to maintain and develop an effective, safe, competitive, environment friendly and flexible PT system. As we can see in Fig.1, there is the inter-city bus coach network tightly covering the territory of Latvia. The problems faced by RTA in the last decade: insufficient state subsidies; service level variations from authority to authority; lack of information for decision making: only few surveys, data not reliable;

no network planning at the strategic level, etc. The main goal of the RTA in the Latvian infomobility market, through the introduction of a single ticket in Latvia, is to solve the problems with the control of subsidies, to simplify the payment system for passengers and to improve the efficiency of passenger operators.

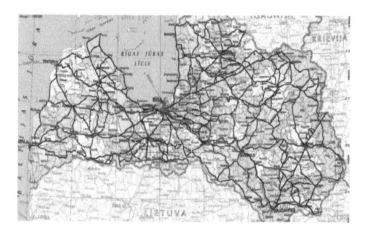

Fig. 1. Inter-city bus coach network in Latvia

There are four main information systems for the PT in Latvia (see Fig.2): VIPUS (JSC Passenger Train); Baltic Lines (33 Latvian bus and Coach Terminals), including www.bezrindas.lv, www.buseurope.eu; VBTS (State Ltd "Road Transport Administration"); ATLAS (Rīgas karte). Let us describe the services offered by systems.

The State Joint-Stock Company Latvijas Dzelzceļš (Latvian Railway) manages expansive and varied infrastructure including rail tracks, engineering structures, rail traffic management systems, rail telecommunications network, radio communication, power supply and contact lines. The Joint-Stock company "Passenger Train" was founded in 2001 on the basis of two former branches of the State Corporation Latvijas dzelzceļš; now it is the only local carrier which provides public rail transport services. On 10 March 2005, JSC "Passenger Train" and BTI Ltd signed a contract for the purchase of hardware and automatic ticket sales, joint revenue and passenger tracking system VIPUS installation and implementation of all Latvian railway stations.

Interactive complex solutions services for railway passengers are intended for the multi-purpose audience and consider the basic requests of passengers in Latvia. These services allow passengers to manage and plan the trip, the needed resources, and to determine the level of comfort in their travels. Information from 150 cash registers in a single system in the online mode has been collected and there has been offered the employed interactive schedule temporal model. Using interactive IS, passengers can buy tickets via Internet and mobile phone (SMS); reserve seats in the wagons, use wireless Internet, carry bulky luggage; use of special equipment for passengers with special needs.

There is possibility to plan the journey in the portal of the national Railway Roads - www.ldz.lv and www.pv.lv. There, travellers will find an interactive map that shows all most important roads in Latvia, and it is possible to plan the journey with the help of this map that shows the modes of transportation in the biggest cities of Latvia.

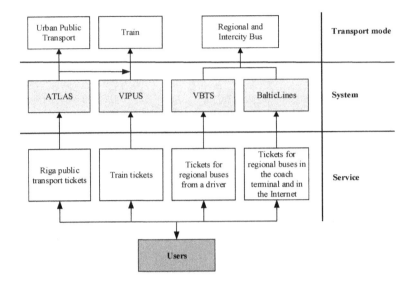

Fig. 2. The main information systems for the PT ticketing in Latvia

Riga International Coach Terminal and other bus and coach terminals in Latvia use the integrated system for tickets purchase and trip management on the bases of the IS "Baltic Lines". The structure of the IS is formed by 10 modules with the continuous inter-exchange of information flows. Organization of the outward and inward information flows of the IS provides necessary connection between the users of the system and other ISs, such as the bookkeeping accounting in the IS Microsoft Dynamics, selling tickets at www.bezrindas.lv, etc. The information services include:

- coach route schedule and operational information on changes;
- coach traffic information (arrival/departure time, location on platforms, delays);
- system of ticket reservation and sale, including route planning, using services of many carriers and means of transports (multimodal and intermodal principles of transportations) and different ways of payments and communications (cashless settlements with credit cards, payments through Internet, using mobile telephone);
- connection with other services (urban transport, luggage transportation, hotel services, etc.);
- development of the control system and the coach station services process;
- communication among dispatcher service, ticket sale and information service;
- 24/7/365 service.

Both systems (for train and bus passengers) are in the process of constant development and upgrading, have the plans widespread to the whole Latvian market and, also, have the similar disadvantages:

- not possible to pay on board with the contactless payment cards (bank cards without cash);
- difficult to use interface for the elderly people;
- no additional functionalities.

In December 2008, the RTA created the VBTS database, which was based on the bus ticket sales IS "Baltic Lines". By 1 September 2009, the RTA had concluded contracts with all regional intercity bus route carriers on the VBTS use and on 18 September 2009 launched the VBTS database separation from IS "Baltic Lines" and on 18 January 2010 VBTS a new database software version was put into operation. VBTS ensures the following data:

- routes in all routes networks (length, maps, etc.);
- public transport services tariffs;
- calendar (possibility to make corrections in tariffs);
- travel times, stops, roads and streets;
- list of administrative territories, carriers, list of public transport services contracts;
- vehicles involved in carriage of passengers, GPS/GPRS devices in vehicles;
- carriers' performance information on voyages, tickets, drivers, etc.;
- electronic map for monitoring of vehicles movement;
- performance reports (sold tickets, cash registers, number of passengers in stops, etc.).

The problems with the VBTS results concern three lacks of: complete information from system users (carriers and terminals); necessary equipment in the vehicles (not all carriers ensure all data for the system); and the most important – motivation to use VBTS (control instrument for contracting authorities).

Data about passengers throughput only cover for buses – 40 % for intercity busses and 5% for local busses, due to infrared connection failure and for trains – 0 %, because VBTS does not support railway.

In June 4 2014, at a practical seminar the conception of PT integrated database STIFS, its development goals and working model were presented. It was announced as the integrated database and modern tool for traffic planning, agreements control, and financial losses calculation with the following realization stages:

1. STIFSS database with the information about buses and trains transportation, passengers' throughput (tickets information offline mode); reports based on BI Tool.
2. GPS/GPRS systems development and combining with STIFSS and controlling on its basis distances, fuel, speed, trips time, etc.
3. Developing of unified tickets reservation portal.

But information about the start of this project realization is absent.

Another big stakeholder in this market – SIA "Rigas Karte", which on Feb.2013 became an Electronic Money Institution. In Riga public transport, e-tickets were fully introduced on May 1, 2009, and they are valid in all PT vehicles of "Rīgas satiksme". E-ticket uses Atlas Public Transport Ticketing System. Atlas systems and services are used by over 1,000 municipal, regional and national operators to run more than 150,000 pieces of equipment, including: automated ticket vending machines, validators, booking office machines, portable inspector terminals, access gates. Atlas equipment enables 50 million passengers per day to use buses, trolleybuses, trams and trains. And the Atlas tailor-made solutions allow central management of equipment across different modes of transportation. E-ticket has solved the problem of the avail-ability of paper tickets in retail trade, and requires no cash. A ticket loaded in the e-form is valid for 12 months from the moment of its purchase, except when the tariff of the ticket type changes or the ticket type is cancelled and Riga municipal company "Rīgas satiksme" sets a transition period for the validity of the ticket type. Electronic validators are located in PT vehicles – buses, trolleybuses and trams and register pas-sengers paying for the trip.

The vision of the conception of Riga PT development is promotion of the accessi-bility to PT on the basis of connection of regional routes with Riga city network; application of unified tariffs in the regional transport system carried out under the principles of state procurement and implementation of unified ticketing in all modes of PT (city transport, rail, bus coaches).

Another city (Jelgava) develops their projects in ITS usually in the frame of the EU funds or by participating in different programmes like INTERREG. For example, the Jelgava City council, in the frame of the Latvia – Lithuania cross border cooperation program, has established a traffic control centre.

Two widely used travel ISs to manage PT journeys (web pages) are available for public use:

- www.1188.lv – information service (trip duration, price, distance, number of stops and route itself),
- www.bezrindas.lv – ticket selling (possibility to buy tickets on-line).

Company Janis Solution & Solutions offers android free software "Trains and bus in Latvia" with schedules in the 3 languages (the same information as on www.1188.lv).

4   Analysis and Next Steps

At present, the transport authorities in Latvia focus their attention on ticketing and real-time information. Real-time status information on PT (e.g. bus and rail) already exists. However, the existing services do not offer travellers real-time information across all the stages of a multimodal trip. Travellers increasingly expect real-time vehicle location, predictions (arrival times) and notifications of travel disruptions, particularly while the journey is taking place, and on mobile devices.

There are also limitations of ticketing. For many destinations, it is not possible to book an integrated ticket that includes the long-distance part and the first/the last part of the journey. It is usually difficult to compare different transport operator offers for the same itinerary.

Information services in Latvia remain very fragmented in what they offer, both in the geographical scope and the coverage of the modes of transport; they rarely pro-vide cross-border travel information, let alone EU-wide or door-to-door coverage. Such services as car-sharing, car-pooling and demand-responsive transport, which provide more environmentally-friendly modes of transport, have not yet been inte-grated into travel planning at a practical level.

The next steps in the Latvian infomobility market development mean to be more integrated with the help of the same technological platforms: ticket validation systems, based on rechargeable and contactless e-cards, and real-time IS for all modes of transport, either in panels at stops or stations, or via smart phone applications. In our opinion, the directions of the Latvian infomobility market development are the following:

1. To consider the actual information quality attributes that can make an infomobility effective.
2. To develop multimodal journey planner. Existing services do not offer travellers real-time information across all stages of a multimodal trip. Information provided to travellers is incomplete.
3. The same applies to ticketing: while there are many examples of electronic and smart ticketing, it is still not possible to buy a single ticket for a multimodal journey in Latvia and across national boundaries in EU. Based on the good results of the e-ticketing implementation in Riga, there appears a need to develop smart ticketing solutions for more cities and for intercity transportation in Latvia. Integrated ticketing is a key part of an attractive, user-friendly multimodal transport system and a prerequisite for a seamless journey. The ability to travel by multiple modes of transport, while only needing to purchase one ticket for the whole journey, is a valuable incentive to encourage travellers to combine several modes of transport.
4. The cities and the government of Latvia are eager to better control their subsidies to public transports companies, and they need automatic passenger counter systems widely accepted across the country. The information can be used also by transit agencies and operators mainly to derive ridership rates and to deal with the revenue management. Passenger counts can be applied in the assignment model calibration and validation (for planning).
5. Automated fare collection (AFC) systems, besides keeping track of passenger ticket payments, may store valuable information concerning the used stops and lines, time of the trip, personal information. This data is valuable to observe the travellers' behaviour, to estimate Origin-Destination matrices, etc. However, usually passengers are required to validate their electronic tickets/cards only at the beginning of their journey. This makes the use of AFC datasets more complicated.
6. Implementation of the concept "Open Data". Data collected through the AFC systems can apply at the strategic level – to network planning; at the tactical level – to derive information for the schedule adjustment; network monitoring and the detection of irregularities in the smart card system itself at the operational level.

5   Conclusions

Use of ICT in PT improve efficiency, safety, quality and reliability of the service; and in whole the attractiveness of PT; reduce maintenance costs and offer possibilities of increasing revenues. However, ICT have been and remain the tools which will never replace clear policy and efficient management. Taking into account the technical and technological evolution of the new ICT tools in the field of ITS, it is possible to provide real-time and accurate information to travellers and to better monitor and manage the PT system and to reach a unique challenging goal – to create the integrated Latvian PT system. The integration of PT can be considered at four levels: integrated information -> integrating ticketing and fares -> coordinated transport services -> coordination with other modes of transport (integration) and other policies (land planning, environmental and social policies). The transfer from policy of coordination to integration is another step of the development of the Latvian PT system on the way to sustainable cities and regions.

References

1. UITP, Information Technology and Innovation Commission (accessed 2014), http://www.uitp.org/Public-Transport/technology
2. Communication from the European Communities Commission. Freight transport logistics action plan. COM, 607, Brussels (2007)
3. Kotsakis, E., Caignault, A., Woehler, W., Ketselidis, M.: Integrating Differential GPS data into an Embedded GIS and its Application to Infomobility and Navigation. In: 7th EC-GI & GIS Workshop, Managing the Mosaic, Potsdam, Germany (2001)
4. Paganelli, F., Giuli, D.: An Evaluation of Context Aware Infomobility Systems. In: Context-Aware Mobile and Ubiquitous Computing for Enhanced Usability, Cap, vol. 15 (2009)
5. Giuli, D., Paganelli, F., Cuomo, S., Cianchi, P.: A systemic and cooperative approach towards an integrated infomobility system at regional scale. In: 11th International Conference on ITS Telecommunications, pp. 547–553. IEEE (2011)
6. Canali, C., Lancellotti, R.: A distributed architecture to support infomobility services. ACM Digital library (2006)
7. Arena, M., Azzone, G., Franchi, F., Malpezzi, S.: An Integrated Framework for Infomobility. In: 3rd International Engineering Systems Symposium, CESUN 2012, DelftUT (2012)
8. Sterle, C.: Location-Routing models and methods for Freight Distribution and Infomobility in City Logistics, PhD Thesis, Università degli Studi di Napoli 'Federico II' (2009)
9. Quantis project, http://www.quantis-project.eu
10. Connect project, http://www.connect-project.org/index.php?id=9
11. eMOTION project, http://www.emotion-project.eu
12. POLITE Project, http://www.polite-project.eu/polite-project
13. T-TRANS Project. Report Deliverable 3.1 – ITS state of the art assessment, http://www.ttransnetwork.eu/ttrans/wp-content/uploads/2013 (access date: February 2015)
14. Ambrosino, G., Boero, M., Nelson, J.D., Romanazzo, M.: Infomobility Systems and Sustainable Transport services. Italian National Agency for New Technologies, Energy and Sustainable Economic Development: ENEA (2010)
15. Directive 2007/2/EC of the European Parliament and the Council establishing an Infrastructure for Spatial Information in the European Union, INSPIRE (2007)
16. LR Ministru kabinets, Transporta attīstības pamatnostādnes 2014. – 2020. gadam. Riga, http://www.mk.gov.lv (access date: December 2013)

Data Actualization Using Regression Models in Decision Support System for Urban Transport Planning

Irina Yatskiv (Jackiva) and Elena Yurshevich

Transport and Telecommunication Institute Riga, Latvia, Lomonosova 1, LV 1019
{Jackiva.I,Jurshevicha.J}@tsi.lv

Abstract. A systematic approach to urban transport system planning and managing means the inclusion of a systematic monitoring system to collect the necessary data and periodically updating the DSS databases, as well as updating of models in their repositories. This should be supported by introduction of new data and information without changing and deleting the old. The authors proposed the application of regression analysis for data actualisation and new obtaining, and considered several task settings for realization of such approach. The proposed methodology focuses on the issues of data updating and preparation for modelling, consideration of model preparation and simulation scenarios including the analysis of the influence of the new solution implementation on the neighbouring fragments of the network. The approach has been approved using the simulation model for a fragment of Riga City. The offered procedures can be used in the frame of model-driven DSS and give the possibility to fulfill the process of the model actualisation faster and less expensive without loss of accuracy.

Keywords: transportation system, decision-making, modelling, credibility, regression, procedure.

1   Introduction

In the field of urban transport system (UTS) planning and management over the last 10 years, the decision support system (DSS) based on different concepts approach to modelling has been actively implemented. There are DSS designed for planning and management at different levels of decision-making. At the strategic level, decision-making is most often used in DSS through the macroscopic meta-model, as well as expert evaluation. In this case, DSS is used to furnish politicians with UTS-related solutions at a high abstraction level, without any particularization at a local level. At the tactical and operational level, decision-making is the most reasonable DSS based on the microscopic and mesoscopic modelling. Such kind of DSS is intended for planning at the level of individual fragments of UTS, traffic control, traffic light control, high-speed mode, etc. In addition, there are DSS designed for control and prevention of accidents and incidents on the roads and traffic control in specific situations.

© Springer International Publishing Switzerland 2015

W. Zamojski et al. (eds.), *Theory and Engineering of Complex Systems and Dependability,*
Advances in Intelligent Systems and Computing 365, DOI: 10.1007/978-3-319-19216-1_53

Especially, using of microscopic models (MM) repositories in model-driven DSS allows one to get a comprehensive idea of the upcoming changes on the global and local level of UTS and assess the impact on their subsequent implementation, and as a result to improve decision-making by the responsible person [1-3]. However, despite the obvious advantages, the using of MM as a part of DSS is studied poorly, and they are rarely used due to the complexity of the model structure and its application, as well as high requirements to the data [4].

The authors' research [5, 6, 7] is devoted to several aspects of MM application as a part of DSS. There are provided several procedures that can be used for simplifying the MM application and that support the semi-automated processes of data preparation, actualisation and transition from databases to MM and back to DSS repositories. It will make the MM based DSS more attractive for decision makers at tactical and operational level of decision making regarding UTS planning, development and operation.

The main attention in this article is paid to the problem of the data actualisation. There is demonstrated the idea of that problem solving by means of the models of different kind of applications and presented the results of this idea approbation with the example of the regression model (RM) application.

2   Literature Review of UTS Model-Driven DSS and the Experience of MM Application

The idea of model-driven DSS application for UTS planning, management and control was considered by different scientists (see the review in [8]). Juan de Dios Ortuzar in his monograph [2] describes the role of DSS in transportation system management and planning; he paid special attention to modeling application as a part of decision-making. Barcelo in [1] considered the DSS for the Madrid UTS management by using the models implemented in AIMSUN and GERTRAM software. Ulied and Esquius in [9] considered the framework of DSS meant for the European transport system management and control. Soo et.al in [10] presented the DSS framework providing a holistic framework to perform analytical assessments of integrated emergency vehicle pre-emption and transit priority systems.

The leaders in DSS implementation are the U.S., Spain, Germany, and the UK. For instance, a new adaptive control DSS was described in the paper published in the magazine "Traffic Engineering & Control" [11]. This system is used in New York City and combines adaptive control of traffic lights in real-time mode and some operators who are involved in the process. Another example is the activities of ICM (Integrated Corridor Management) in San Diego (California, USA). Within the framework of the ICM implementation, a DSS is created which integrates both the inclusive communication system and the formulation of inter-jurisdictional agreements. At the heart of the DSS is simulation-based prediction system AIMSUN Online [12].

DSS used in today's practice of decision-making in transport planning are mainly based on macroscopic models – the use of microscopic models in this context is given little attention. There are only several examples of DSS that incorporate MM: CAPITALS/MADRID [1], SCOOT [13], SCATS [3], ICM AMS [12], TRIM [14], ROMANSE [15]. The reason of these is the so called "no popularity" - there are still a lot of unsolved and non-completely investigated problems, such as:

- combined use of different types of models and the MM integration as a part of DSS;
- inevitable obsolescence of the data and the models stored in repositories and DSS databases;
- organization of the data and model actualisation in case of fragmentary measurements of the data in a real system (usually, there is considered the system with systematic data collection in a real system) or in case of their absence (for instance, consideration of a new solution);
- etc.

In authors' opinion, these unresolved issues are the main reasons of why MM is rarely used in DSS. This paper is devoted to the last issue of those listed above.

3   Conception of MM Traffic Volume Actualisation on the Base of Fragmentary Data Measuring

The main tasks of DSS are the following:

- provision of storing and access to the data required for UTS planning and control tasks;
- knowledge storing in knowledge bases and providing access to it;
- models storing in the model repositories and organizing access to them;
- ensuring the user interface to provide for data input by users;
- ensuring manipulation with the models and data;
- presentation of the output results in graphic and textual forms or in the form of generalized reports.

MM used as a part of DSS for UTS planning provides for the following benefits and capabilities:

- an opportunity of a more careful investigation of the causes of traffic congestion occurrence, and the possibility of locating UTS bottlenecks more precisely;
- a possibility of conducting a more detailed analysis of the consequences of the proposed strategic decisions at a local level and refining them more in detail as follows: selecting interchanges of engineering solutions, optimizing the operating plan of traffic light controllers and organizing the traffic flow, regulating operation of public transport etc.

Careful reproduction of UTS properties by MM involves with itself using of different kinds of data: about UTS network, traffic properties and its organization, public transport scheduling and routing, pedestrian flows organization etc. DSS should provide the communication between databases, data warehouses, data marts on the one hand and MM on the other hand. A possible scheme of its organization is presented in Fig.1.

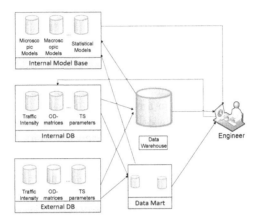

Fig. 1. Organization of access to the data and models

Unlike the macroscopic model, the MM application means the usage of wide range of different data. Taking into account that at the micro-level the traffic properties and UTS network configuration are changing faster than at macro-level, both the data in databases and the models in repositories are needed for a frequent actualisation. The traffic volumes especially require more frequent updating. Let us consider two common situations:

1. It is necessary to estimate the influence of new solutions on one UTS fragment on the traffic volume of neighboring. The researcher can have the microscopic models of considered UTS fragments, but he has not the possibility to connect them together to transmit the new volumes of traffic.
2. It is necessary to repeat the investigation of UTS fragment based on a previously created simulation model and with preliminary knowledge about the changed situation.

In the first case the traditional solution is either expensive or does not exist at all. In the second case, it is possible either to conduct the repeated UTS survey implementation, or to apply the UTS macroscopic model as a source of data. The disadvantages of these approaches are the following: the first approach is very expensive and there is no possibility to implement it in UTS on constant basis, but the second one is possible only in case of such model existing. The paper presents the alternative conception, which is based on three main stages:

1. The conduct the fragmentary traffic flow volume measuring on some fragments of UTS.

2.   To predict the new traffic flow volume on others UTS fragments using the fragmentary measurements of traffic and special approximation model.
3.   To transfer the updating data regarding traffic volume to the input of MM and apply the simulation.

The key question of that approach: to find the approximation model that can describe the dependences between UTS fragments traffic flow volumes without detailed reproduction of UTS properties (Fig.2). The second requirement is very important because it is supposed that that kind of model will not be needed in actualisation as often as MM.

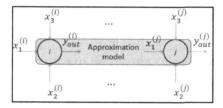

Fig. 2. Illustration of the idea of approximation model application for the dependence between traffic volumes approximation

The authors proposed the idea to apply for that either regression models or artificial neural networks and there is expounded the idea of RM application in the offered paper. This approach has been developed and approved on simulation model that constructed in the frame of the project "Pedestrian and Transport Flows Analysis for Pedestrian Street Creation in Riga City" in 2011[5]. The holistic set of procedures for the use of RM for updating the existing traffic data and obtaining the new one were developed in [6].

4   Data Actualisation for MM on the Basis of Regression Models

Let us have at the time moment t_C in the DSS models repositories M the subset of models $\{M_l\}$, $l = 1..8$ presented on Fig.3 and the data $D(t_p)$ for these models collected at the fixed time moments $t_p, t_p < t_C$. At the time moment t_C the data $D(t_p)$ may be needed in actualisation. Let us have a model $M_i(t_p)$ for some fragment i of UTS created/updated at time t_p. It is necessary to analyse the influence of throughput capacity of the fragment of UTS i on the other fragments lying at the same distance from the considered one. Let us denote the other fragment as j, $j = i \pm k$, where k – is a number of crossroads located between   crossroad i and j, $k=2..n$ (Fig.3). It is necessary to simulate the new output traffic volumes from UTS fragment i and to estimate the influence of this traffic flow volume on the level of congestion of the other UTS fragment j using the corresponding simulation model.

Several alternatives of possible situations have been formulated for planning the experiments and there was determined the set of factors for regression modelling. There were considered different types of streets, the distance between crossroads, the number of RM that can be used for that problem, etc. In general the plan of experiments was included 16 different set of factors (experiment scenario). The implementation of each experiment for regression modelling includes the following steps:

- to form a sample of dependent and independent variables based on the average values of traffic flow volumes and the existing simulation model;
- to check samples for homogeneity;
- to separate the generated samples into two subsamples: for RM creation and for checking their approximation quality;
- to estimate the parameters of RM and to analyse the model quality;
- to make the decision about the possibility of its application for the data actualisation based on the quality analysis results.

The following statistics were used for the RM quality analysis: $\overline{R^2}$ – adjusted coefficient of multiple determination, F-test – Fisher test, SEE – standard error of estimation; for the approximation quality testing: the root mean square error (RMSE), the root mean square normalized error (RMSNE), the 95% confidence interval (CI) and the 95% tolerance interval (TI). Let us consider two typical examples.

The *first example* concerns the crossroads disposition which presented in Fig.3. The dependent variable is $y^{(3)}_{out\ to\ Chaka}$ – the volume of output traffic flow from intersection N3. The independent variables are $x_1^{(1)}$ and $x_3^{(1)}$, $x_1^{(2)}$ and $x_4^{(2)}$, $x_1^{(3)}$ and $x_4^{(3)}$ – the volume of input traffic flows of intersections N1, N2 and N3, accordingly. The analysis demonstrated that the number of independent variables can be decreased and the obtained results are presented in Table 1.

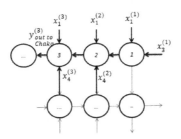

Fig. 3. Scheme of the crossroads location

Table 1. Characteristics of the quality of obtained RM for the case on Fig.3

$$y^{(3)*}_{out\ to\ Chaka} = 78.70 + 0.268 \cdot x_3^{(1)} + 0.402 \cdot x_4^{(3)}$$

$\overline{R^2}$	F	p-level	SEE	RMSE	RMSNE	Obtained/ Predicted	95% CI	95% TI
0.73	106.55	<0.05	13.02	12.94	0.033	386/400	(396;404)	(374;427)

Thus it can be concluded by the values of $\overline{R^2}$, F-test and p-level that the RM generally is qualitative. The values of RMSE (12.94), and RMSNE (0.033) are satisfactory. The RM was tested for data prediction. In all cases the tolerance interval covers the obtained values of output traffic volume $y^{(3)}_{out\ to\ Chaka}$. The obtained results are satisfactory and on the basis of estimated RM and fragmentary data measurements new values of the volume of input traffic flow arriving at neighboring intersection can be obtained.

The *second example* illustrates RM application for approximation of the dependence of the input traffic flow on the other traffic flows located at some distance from each other at the different levels (roads) according to the main street of UTS (case 2, Fig.4).

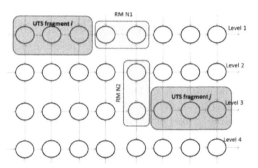

Fig. 4. Two-stage application of RM for UTS fragment j input traffic flow volume prediction

There are two fragments of UTS: the *i*-th fragment consists of intersections N3 and N4, the *j*-th one includes intersection N2; fragments were located at a distance from each other and at different levels. It is assumed that the simulations models of these UTS fragments exist. It is necessary to test the hypothesis for RM approximating the influence of the output traffic of the *i*-th UTS fragment to the value of the input traffic flow of *j*-th UTS fragment of the transport network. It was suggested that this problem can be solved by using two RM (Fig.4):

• the RM N1 – the dependence of the output traffic of intersection N6 on the values of the output traffic of UTS fragment i and other input flows;

• the RM N2 – the dependence of the input traffic flow of UTS fragment j on the value of the output traffic intersection N6 and other input flows.

The results of created models are presented in Table 2. The experiment demonstrated the positive results of two RM applications to approximate the traffic flows of two fragments of UTS, located at a distance from each other and at different levels.

Table 2. Characteristics of obtained RM quality

RM	$\overline{R^2}$	F	p-level	SEE	RMSE	RMSNE	Obtained/ Predicted	95% CL	95% TL
N1	0.88	161.19	<0.05	13.244	12.02	0.041	336/331	(326;335)	(304;358)
N2	0.907	202.55	<0.05	6.067	5.88	0.022	315/312	(309;316)	(300;325)

5   Results Discussion

The following findings of the implemented experiments were obtained:

1. The RM can be used to approximate the dependence of output traffic flow of one of the intersections on the value of traffic flow volumes of other intersections. It can be used for two, sometimes three intersections (no more). The intersections or group of intersections can be located at some distance and either at the same or different level on the streets covered by UTS.

2. A preliminary analysis of the structure of traffic flows affecting the situation on the given intersection is necessary for building RM, and it's desirable to include into the model the variables corresponding to all input and output flows of the intersection, with the stepwise elimination of insignificant factors from the model.

3. The results of the output flows prediction can be used as input data of another RM for evaluation of the output of another traffic intersection.RM being easy-to use and not time-consuming for updating data can be mentioned as the advantages of RM application.

4. Some limitations with respect to operation can be mentioned, too – in particular: the proposed procedures work good if the approximation is performed with regard to dependence of traffic flows located not far from each other. The greater the distance, the worse the result. It is reasonable to apply data updating by using RM if intersections are located at one level, i.e., at a distance not exceeding 2-3 crossroads; if intersections are located at different levels – at a distance not exceeding 6 intersections.

6   Conclusions

The research was dedicated to the problem of MM application as a part of model-driven DSS aimed at decision-making support according to UTS planning. The rare use of microscopic models in the planning stage is explained by the complexity of the technology simulation, the additional data requirements, as well as additional requirements to the user of these models. However, it can be assumed that the use of microscopic simulation in the planning stage can enrich decision-makers with information enabling one consider a plan of solving the existing problems and a more detailed organization of UTS descended from abstract-making at the strategic level to a more detailed and specific design study at the tactical level.

On substantiation of one of the major problems of MM credibility – i.e., the data quality on the basis of the theoretical and practical analysis, – the possibility of using a DSS based on data updating procedures relying on regression models have been proposed and experimentally investigated.

The authors proposed the RM application for the data actualisation and new obtaining, and considered several task settings for realization of such approach:

- data actualisation for MM of UTS fragments that are on one street and on different ones;
- data obtaining for analysis of new solutions' influence on the neighbouring fragments of UTS (both for UTS fragments located at the same level (road) and the different one).

References

1. Barcelo, J.: Advanced Traffic Control Strategies in Madrid. ITS, Berlin (1997)
2. Ortuzar, J.D., Willumsen, L.G.: Modelling Transport. John Wiley & Sons, England (2006)
3. Kergaye, C., Stevanovic, A., Martin, P.T.: Comparative Evaluation of Adaptive Traffic Control System Assessments through Field and Microsimulation. Intelligent Transportation Systems 14(2), 109–124 (2010)
4. Wood, S.: Traffic microsimulation – dispelling the myths. Tecmagazine, 339–344 (2012)
5. Yurshevich, E., Yatskiv, I.: Consideration of the Aspects of the Transportation Systems Microscopic Model Application as Part of a Decision Support System. Transport and Telecommunication 13(3), 209–218 (2012)
6. Yurshevich, E.: Methodology of Decision Making Support Based on Urban Transportation System Microscopic Models Repositories. PhD Thesis, Riga, TSI (2013)
7. Yurshevich, E., Yatskiv, I.: Decision Support System for Transport System Management And Microscopic Traffic Modelling. In: 7th International Scientific Conference TRANSBALTICA, Vilnius, Lithuania, pp. 74–81 (2011)
8. Brand, C., Moon, D.: Integrated Modelling and Decision Support for Environmental Transport Policy and Technology Choices Across Europe. In: European Transport Conference, Cambridge, pp. 189–202 (2000)
9. Ulied, A., Esquius, A.: Developing Advanced Decision-Support Systems (DSS). An Open and Networked Transport DSS for Europe, Agora Jules Dupuit – Publication AJD-52 (2000), http://www.e-ajd.net/source-pdf/AJD-52-Ulied-Esquius-Pap_DSS.pdf
10. Soo, H.Y., Teodorovic, D., Collura, J.: A DSS Framework for Advanced Traffic Signal Control System Investment Planning. Journal of Public Transportation 9(4), 87–106 (2006)
11. Development of ACDSS and its implementation to New York City arterials. Traffic Engineering & Control, pp. 367–371 (September 2009)
12. Alexiadis, V., Cronin, B., Mortensen, S., Thompson, D.: Integrated Approach: Analysis, Modelling, and Simulation Results for the ICM Test Corridor. Traffic Technology International Annual, 46–50 (2009)
13. Thomas, G., Baffour, K., Brown, T.: Simulation and Implementing a SCOOT UTC Strategy for a Planned Event. Traffic Engineering & Control, 118–122 (March 2010)
14. Mitrovich, S., Valenti, G., Mancini, A.L.: Decision Support System for Traffic Incident Management in Roadway Tunnel Infrastructure. Thesis from European Transport Conference (September 18-20, 2006)
15. Wylie, M.: Modelling for the future. Traffic Engineering & Control, 173–175 (April 2009)

New Parallel Algorithm
for the Calculation of Importance Measures

Elena Zaitseva, Vitaly Levashenko, Miroslav Kvassay, and Jozef Kostolny

Faculty of Management Science and Informatics,
University of Zilina, Univerzitna 8215/1, 01026 Zilina, Slovakia
{Elena.Zaitseva,Vitaly.Leavshenko,Miroslav.Kvassay,
Jozef.Kostolny}@fri.uniza.sk

Abstract. A lot of mathematical approaches are used in importance analysis, which permits to investigate influence of system component state changes on the system reliability or availability. One of these approaches is Logical Differential Calculus, in particular Direct Partial Boolean Derivatives. A new algorithm for the calculation of Importance Measures with application of Direct Partial Boolean Derivatives is proposed in this paper. This algorithm is developed based on parallel procedures.

Keywords: Importance measures, Direct Partial Boolean Derivatives, matrix procedures, parallel algorithm.

1 Introduction

Consider a system of n components. From reliability point of view, the system and all its components can be in one of two possible states: functional (presented as 1) and failed (presented as 0). The mathematical dependency between the system state and states of its components can be defined by the structure function [1]:

$$\phi(x_1, x_2,\ldots, x_n) = \phi(\boldsymbol{x}): \{0, 1\}^n \to \{0, 1\}, \tag{1}$$

where x_i is the state of component i, for $i = 1,2,\ldots,n$, and $\boldsymbol{x} = (x_1, x_2,\ldots, x_n)$ is a vector of components states (state vector).

Every system component is characterized by probability p_i (represents the availability of component i) and probability q_i (defines its unavailability):

$$p_i = \Pr\{x_i = 1\}, \quad q_i = \Pr\{x_i = 0\}, \quad p_i + q_i = 1. \tag{2}$$

When the system structure function and availabilities of all system components are known, then system availability/unavailability can be computed as follows [2, 3]:

$$A = \Pr\{\phi(\boldsymbol{x}) = 1\}, \quad U = \Pr\{\phi(\boldsymbol{x}) = 0\}, \quad A + U = 1. \tag{3}$$

The availability is one of the most important characteristics of any system. It can also be used to compute other reliability characteristics, e.g. mean time to failure,

© Springer International Publishing Switzerland 2015
W. Zamojski et al. (eds.), *Theory and Engineering of Complex Systems and Dependability,*
Advances in Intelligent Systems and Computing 365, DOI: 10.1007/978-3-319-19216-1_54

mean time to repair, etc. [2, 3]. But they do not permit to identify the influence of individual system components on the proper work of the system. For this purpose, there exist other measures that are known as *Importance Measures* (IM). The IMs are used in part of reliability analysis that is known as importance analysis. The comprehensive study of these measures has been performed in work [4]. IMs have been widely used for identifying system weaknesses and supporting system improvement activities from design perspective. With the known values of IMs of all components, proper actions can be taken on the weakest component to improve system availability at minimal costs or effort.

There exist a lot of IMs, but the most often used are the *Structural Importance* (SI), Birnbaum's Importance (BI), *Criticality Importance* (CI) and *Fussell-Vesely Importance* (FVI) (Table 1).

Table 1. Basic Importance Measures

Importance Measure	Meaning
SI	The SI concentrates only on the topological structure of the system. It is defined as the relative number of situations in which a given component is critical for the system activity
BI	The BI of a given component is defined as the probability that the component is critical for the system work.
CI	The CI of a given component is calculated as the probability that the system failure has been caused by the component failure, given that the system is failed.
FVI	The FVI of a given component is defined as the probability that the component contributes to the system failure probability.

Different mathematical methods and algorithms can be used to calculate these indices. Ones of them are *Direct Partial Boolean Derivatives* (DPBDs) that have been introduced for importance analysis in paper [5]. In paper [1], the mathematical background of DPBDs application has been considered. But efficient algorithm for computation of DPBDs has not been proposed. In this paper, a new parallel algorithm for the calculation of a DPBD is developed.

This paper has the next structure organization. In section 2, the general definition of a DPBD is provided. The definition and calculation aspects of IMs (Table 1) based on DPBDs are considered in this section too. In section 3, the development of the new parallel algorithm for DPBD calculation is considered. This algorithm is developed by the transformation of initial definition of a DPBD into matrix form. Based on the matrix definition of the DPBD, the new parallel algorithm is proposed.

As alternative result for the new algorithm, algorithms in [6] can be considered. The authors of the paper [6] proposed algorithms for calculation of a DPBD based on the structure function representation by a Binary Decision Diagram (BDD) that includes parallel procedure too. But the algorithms in [6] need a special transformation of initial representation of the structure function into a BDD, and this increases the computation complexity.

2   Importance Analysis

2.1   Direct Partial Boolean Derivatives

A DPBD is a part of Logical Differential Calculus [1, 7, 8]. In analysis of Boolean functions, a DPBD allows identifying situations in which the change of a Boolean variable results the change of the value of Boolean function. In case of reliability analysis, the system is defined by the structure function (1) that is a Boolean function. Therefore, a DPBD can be used for the structure function analysis too. In terms of reliability analysis, a DPBD allows investigation the influence of a structure function variable (=component state) change on a function value change (=system state). Therefore, a DPBD of the structure function permits indicating components states (state vectors) for which the change of one component state causes a change of the system state (availability). These vectors agree with the system boundary states [1, 5].

DPBD $\partial \phi(j \to \bar{j})/\partial x_i(a \to \bar{a})$ of the structure function $\phi(x)$ with respect to variable x_i is defined as follows [8]:

$$\frac{\partial \phi(j \to \bar{j})}{\partial x_i(a \to \bar{a})} = \left\{ \phi(a_i, x) \leftrightarrow j \right\} \wedge \left\{ \phi(\bar{a}_i, x) \leftrightarrow \bar{j} \right\}, \tag{4}$$

where $\phi(a_i, x) = \phi(x_1, x_2, \ldots, x_{i-1}, a, x_{i+1}, \ldots, x_n)$, $a, j \in \{0, 1\}$ and $\leftrightarrow$ is the symbol of equivalence operator (logical bi-conditional).

Clearly, there exist four DPBDs for every variable x_i [1, 7, 8]:

$$\frac{\partial \phi(1 \to 0)}{\partial x_i(1 \to 0)}, \quad \frac{\partial \phi(0 \to 1)}{\partial x_i(0 \to 1)}, \quad \frac{\partial \phi(1 \to 0)}{\partial x_i(0 \to 1)}, \quad \text{and} \quad \frac{\partial \phi(0 \to 1)}{\partial x_i(1 \to 0)}.$$

In reliability analysis, the first two DPBDs can be used to identify situations in which a failure (repair) of component i results system failure (repair). Similarly, the second two DPBDs identify situations when the system failure (repair) is caused by the i-th component repair (failure). The second two derivatives exist (are not equal to zero) for a noncoherent systems [1]. In this paper, coherent systems are taken into account only. These systems meet the next assumptions [4]: (i) the structure function is monotone, and (ii) all components are independent and relevant to the system.

For example, consider a system of three components ($n = 3$) in Fig. 1 with structure function:

$$\phi(x) = \text{AND}(x_1, \text{OR}(x_2, x_3)). \tag{5}$$

The influence of the first component failure on the system can be analyzed by DPBD $\partial \phi(1 \to 0)/\partial x_1(1 \to 0)$. This derivative has three nonzero values for state vectors $x = (x_1, x_2, x_3)$: ($\underline{1 \to 0}$, 1, 1), ($\underline{1 \to 0}$, 0, 1) and ($\underline{1 \to 0}$, 1, 0). Therefore, the failure of the first component causes a system breakdown for working state of the second and the third component or working state of one of them. The system is not functioning if the second and the third components are failed and, therefore, a failure of the first component does not influence system availability.

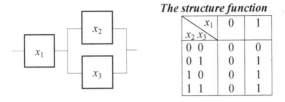

Fig. 1. The system example with its structure function

2.2   Importance Measures and Direct Partial Boolean Derivatives

In reliability analysis, the structure function and the system components are used instead of the Boolean function and the Boolean variables, respectively. Using this coincidence, the authors of the papers [1] have developed techniques for analysis of influence of individual system components on system failure/functioning using DPBDs. Let us summarize the definitions of IMs (Table 1) for the system failure based on DPBDs.

The SI of component i is defined as the relative number of situations, in which the component is critical for system failure. Therefore, the SI of component i can be defined by DPBD $\partial \phi(1{\rightarrow}0)/\partial x_i(1{\rightarrow}0)$ as the relative number of state vectors for which the considered DPBD has nonzero values [1, 5]:

$$SI_i = \frac{\rho_i^{(1\rightarrow0)}}{2^{n-1}}, \tag{6}$$

where $\rho_i^{(1\rightarrow0)}$ is a number of nonzero values of DPBD $\partial \phi(1{\rightarrow}0)/\partial x_i(1{\rightarrow}0)$ and 2^{n-1} is a size of the DPBD.

Similarly, the modified SI, which takes into account the necessary condition for component being critical, can be defined as follows [1, 5]:

$$MSI_i = \frac{\rho_i^{(1\rightarrow0)}}{\rho_i}, \tag{7}$$

where ρ_i is a number of state vectors for which $\phi(1_i, x) = 1$.

The BI of component i defines the probability that the i-th system component is critical for system failure. Using DPBDs, this IM can be defined as the probability that the DPBD is nonzero [1]:

$$BI_i = \Pr\{\partial\phi(1 \rightarrow 0)/x_i(1 \rightarrow 0) \leftrightarrow 1\}. \tag{8}$$

A lot of IMs are based on the BI, e.g. the CI, Barlow-Proschan, Bayesian, redundancy, etc. For example, the CI is calculated as follows [4]:

$$CI_i = BI_i \cdot \frac{q_i}{U}, \tag{9}$$

where q_i is component state probability (1) and U is the system unavailability.

The FVI takes into account the contribution of a component failure to system failure [4]. This contribution is computed based on *Minimal Cut Sets* (MCSs), which correspond to minimal sets of components whose simultaneous failure result system failure. Using MCSs, the FVI of the i-th component is defined as the probability that at least one MCS containing component i is failed, given that the system is failed [4]:

$$FVI_i = \Pr\{MCS(x_i) \mid \phi(\mathbf{x}) = 0\} = \frac{\Pr\{MCS(x_i) \cap \phi(\mathbf{x}) = 0\}}{\Pr\{\phi(\mathbf{x}) = 0\}} = \frac{\Pr\{MCS(x_i)\}}{U}, \tag{10}$$

where $MCS(x_i)$ represents the event when at least one MCS that contains component i is failed.

MCSs can also be expressed in the form of state vectors. These vectors are known as *Minimal Cut Vectors* (MCVs). A state vector $\mathbf{x}$ is a MCV if $\phi(\mathbf{x}) = 0$ and $\phi(\mathbf{x}') = 1$ for any $\mathbf{x}' > \mathbf{x}$ [2, 9]. Based on the investigation in paper [9], the FVI can be defined using MCVs as follows:

$$FVI_i = \frac{\Pr\{\mathbf{x} \le MCV(x_i)\}}{U}, \tag{11}$$

where $MCV(x_i)$ is a set of all MCVs for which $x_i = 0$.

According to [9], a state vector $\mathbf{x}$ is a MCV, if every variable meets one of the following two conditions: (*a*) DPBD $\partial\phi(0 \to 1)/\partial x_i(0 \to 1)$ has a nonzero value for state vector $\mathbf{x}$, or (*b*) DPBD $\partial\phi(0 \to 1)/\partial x_i(0 \to 1)$ does not exist for this state vector.

The condition (*a*) agrees with the definition of a MCV that supposes that $\phi(\mathbf{x}) = 0$ and $\phi(\mathbf{x}') = 1$ for any $\mathbf{x}' < \mathbf{x}$. The condition (*b*) assumes that $x_i = 1$ and, therefore, DPBD $\partial\phi(0 \to 1)/\partial x_i(0 \to 1)$ for the state vector $\mathbf{x}$ does not exist.

Therefore, DPBDs $\partial\phi(0 \to 1)/\partial x_i(0 \to 1)$ with respect to every variable of the structure function are calculated for the indication of MCVs. The intersection of these derivatives identifies the state vectors that are MCVs.

Consider some computational aspects of MCVs calculation based on DPBDs. It is known that the DPBD with respect to variable x_i does not depend on this variable [1, 8]. The derivative $\partial\phi(0 \to 1)/\partial x_i(0 \to 1)$ is defined only for state vectors $(x_1, x_2, \dots, x_{i-1}, 0, x_{i+1}, \dots, x_n)$ and cannot be computed for state vectors that have the form of $(x_1, x_2, \dots, x_{i-1}, 1, x_{i+1}, \dots, x_n)$. According to conditions (*a*) and (*b*), analysis of non-existing values of the DPBD is supposed in the calculation of MCVs. Definition of DPBD (4) is transformed and non-existing values of DPBD will be marked using a special symbol "*":

$$\partial\phi(0 \rightarrow 1)/\partial x_i (0 \rightarrow 1) = \begin{cases} 1 & \text{if } x_i = 0 \text{ and } \phi(1_i, \boldsymbol{x}) \neq \phi(0_i, \boldsymbol{x}) \\ 0 & \text{if } x_i = 0 \text{ and } \phi(1_i, \boldsymbol{x}) = \phi(0_i, \boldsymbol{x}) \\ * & \text{if } x_i \neq 0 \end{cases} \tag{12}$$

For example, consider the system of three components in Fig. 1. All derivatives $\partial\phi(0 \rightarrow 1)/\partial x_i (0 \rightarrow 1)$ for this structure function are calculated according to (12) in Table 2. There are two MCVs for this system: (0, 1, 1) and (1, 0, 0). These MCVs are identified based on the intersection of all DPBDs. The rule for the intersection of two DPBDs (12) with respect to two different variables x_i and x_j is defined in Table 3 [9].

Table 2. Calculation of MCVs using DPBDs (12)

$x_1\ x_2\ x_3$	$\partial\phi(0{\rightarrow}1)/\partial x_1(0{\rightarrow}1)$	$\partial\phi(0{\rightarrow}1)/\partial x_2(0{\rightarrow}1)$	$\partial\phi(0{\rightarrow}1)/\partial x_3(0{\rightarrow}1)$	**The intersection**
0 0 0	0	0	0	**0**
0 0 1	1	0	*	**0**
0 1 0	1	*	0	**0**
0 1 1	1	*	*	**1**
1 0 0	*	1	1	**1**
1 0 1	*	0	*	**0**
1 1 0	*	*	0	**0**
1 1 1	*	*	*	*****

Table 3. Defining the intersection of two DPBDs (12)

Value of $\partial\phi(0{\rightarrow}1)/\partial x_i(0{\rightarrow}1)$	Value of $\partial\phi(0{\rightarrow}1)/\partial x_j(0{\rightarrow}1)$		
	*	0	1
*	*	0	1
0	0	0	0
1	1	0	1

To illustrate the calculation of all IMs using DPBDs consider the system in Fig. 1. Values of IMs for this system are computed in Table 4. According to these IMs, the first component has the most influence on the system failure from point of view of the system structure, because the values of the SI, MSI and BI are greatest for this component. The CI is maximal for the second and third components and, therefore, it indicates the first component as non-important taking into account the probability of failure of this component (it is minimal for this component, i.e. $q_1 = 0.10$). The FVIs implies that the second and third components contribute to system failure with the most probability.

So, DPBDs are one of possible mathematical approaches that can be used in importance analysis, and they allow us to calculate all often used IMs (Table 1). Mathematical background of its application for the definition of IM has been considered in papers [1, 5]. In this paper new algorithm for the calculation of DPBD based on a parallel procedure is developed.

Table 4. IMs for the system in Fig.1

Component	Probability of component state, p_i	SI_i	MSI_i	BI_i	CI_i	FVI_i
x_1	0.90	0.75	1.00	0.90	0.46	0.52
x_2	0.70	0.25	0.50	0.32	0.49	0.54
x_3	0.65	0.25	0.50	0.27	0.49	0.54

3   Parallel Algorithm for the Calculation of Direct Partial Boolean Derivatives

One of possible way for the formal development of parallel algorithms is transform mathematical background into matrix algebra. Therefore, consider DPBD (4) in matrix interpretation. As the first step in such transformation, the initial data (structure function) has to be presented as a vector or matrix.

The structure function is defined as a *truth vector* (Fig. 2) in matrix algorithm for calculation of DPBD. It is column of a truth table of function $\mathbf{X} = [x^{(0)}\ x^{(1)}\ \ldots\ x^{(2^n-1)}]^\mathrm{T}$, where $x^{(i)}$ is value of a function $\phi(x)$ for state vector $x = (x_1, x_2, \ldots, x_n) = (i_1, i_2, \ldots, i_n)$ $((i_1 i_2 \ldots i_n)$ is binary representation of the parameter i, $1 \le i \le 2^n-1$).

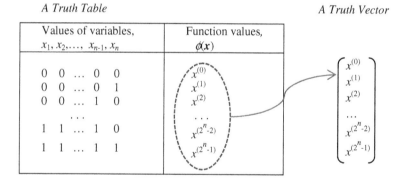

Fig. 2. Truth vector of the structure function

For example, the truth vector of the structure function for the system in Fig. 1 is:

$$\mathbf{X} = [x^{(0)}\ x^{(1)}\ x^{(2)}\ x^{(3)}\ x^{(4)}\ x^{(5)}\ x^{(6)}\ x^{(7)}]^\mathrm{T} = [0\ 0\ 0\ 0\ 0\ 1\ 1\ 1]^\mathrm{T}.$$

The value of the function can be defined by the truth vector unambiguously. Consider the truth vector element $x^{(5)} = 1$. The state vector for this function value is defined by the transformation of the parameter $i = 5$ into binary representation: $i = 5$ $\Rightarrow (i_1, i_2, i_3) = (1, 0, 1)$. Therefore, the truth vector element $x^{(5)} = 1$ agrees with the function value $\phi(1, 0, 1) = 1$.

The truth vector of DPBD (derivative vector) is calculated based on the truth vector of the structure function as:

$$\partial \mathbf{X}(j \to \bar{j})/\partial x_i(a \to \bar{a}) = \left(\mathbf{P}^{(i,a)} \cdot (j \leftrightarrow \mathbf{X})\right) \wedge \left(\mathbf{P}^{(i,\bar{a})} \cdot (\bar{j} \leftrightarrow \mathbf{X})\right), \tag{13}$$

where $\mathbf{P}^{(i,l)}$ is the differentiation matrix with size $2^{n-1} \times 2^n$ that is defined as:

$$\mathbf{P}^{(i,l)} = \mathbf{M}^{(i-1)} \otimes \left[l\bar{l}\right] \otimes \mathbf{M}^{(n-i)}, \tag{14}$$

and $\mathbf{M}^{(w)}$ is diagonal matrix with size $2^w \times 2^w$, $\left[l\bar{l}\right]$ is the vector for which $l = s$ for the matrix $\mathbf{P}^{(i,a)}$ and $l = \bar{a}$ for matrix $\mathbf{P}^{(i,\bar{a})}$, and $\otimes$ is the Kronecker product [10].

Note that the calculation $(j \leftrightarrow \mathbf{X})$ and $(\bar{j} \leftrightarrow \mathbf{X})$ in (13) agrees with the definition of state vectors for which the function value is j and $\bar{j}$, respectively. The matrices $\mathbf{P}^{(i,a)}$ and $\mathbf{P}^{(i,\bar{a})}$ allows indicating variables with values a and $\bar{a}$, respectively. The operation AND ($\wedge$) integrates these conditions.

DPBD $\partial \phi(j \to \bar{j})/\partial x_i(a \to \bar{a})$ ($\partial \mathbf{X}(j \to \bar{j})/\partial x_i(a \to \bar{a})$) does not depend on the i-th variable [8]. Therefore, the derivative vector (13) has size of 2^{n-1}.

Consider an example for calculation of derivative vector $\partial \mathbf{X}(1 \to 0)/\partial x_1(1 \to 0)$ for the structure function with the truth vector $\mathbf{X} = [0\ 0\ 0\ 0\ 0\ 1\ 1\ 1]^T$ (it is the truth vector of the structure function of the system depicted in Fig. 1). According to (14), the rule for the calculation of this derivative is:

$$\partial \mathbf{X}(1 \to 0)/\partial x_1(1 \to 0) = \left(\mathbf{P}^{(1,1)} \cdot (1 \leftrightarrow \mathbf{X})\right) \wedge \left(\mathbf{P}^{(1,0)} \cdot (0 \leftrightarrow \mathbf{X})\right) = [0\ 1\ 1\ 1]^T, \tag{15}$$

where matrices $\mathbf{P}^{(1,1)}$ and $\mathbf{P}^{(1,0)}$ are defined based on the rule (14) as:

$$\mathbf{P}^{(1,1)} = \mathbf{M}^{(0)} \otimes \begin{bmatrix} 1 & 0 \end{bmatrix} \otimes \mathbf{M}^{(2)} \text{ and } \mathbf{P}^{(1,10)} = \mathbf{M}^{(0)} \otimes \begin{bmatrix} 0 & 1 \end{bmatrix} \otimes \mathbf{M}^{(2)}.$$

The derivative vector $\partial \mathbf{X}(1 \to 0)/\partial x_2(1 \to 0)$ has three nonzero values that imply that the change of the structure function value from 1 to 0 is caused by change of the first variable value from 1 to 0 if the values of the second and third variable are 1, or one of these variables has 0-value and other has value 1. In term of components states and the system availability, the DPBD indicate three state vectors $x = (x_1, x_2, x_3)$: $(1 \to 0, 1, 1)$, $(1 \to 0, 0, 1)$ and $(1 \to 0, 1, 0)$. Therefore, the failure of the first component causes a system breakdown for working state of the second and the third components or working state of one of them. This result is equal to result that has been calculated by definition (4) for DPBD $\partial \phi(1 \to 0)/\partial x_1(1 \to 0)$.

A matrix procedure can be transform in parallel procedure according to [10]. Therefore the equation (13) can be interpreted by parallel procedure. For example, the flow diagrams for the calculation of the derivative vectors $\partial \mathbf{X}(1 \to 0)/\partial x_1(1 \to 0)$, $\partial \mathbf{X}(1 \to 0)/\partial x_2(1 \to 0)$ and $\partial \mathbf{X}(1 \to 0)/\partial x_3(1 \to 0)$ for the structure function (5) according (13) are presented in Fig. 3. These diagrams illustrate the possibility to use parallel procedures for the calculation of DPBD.

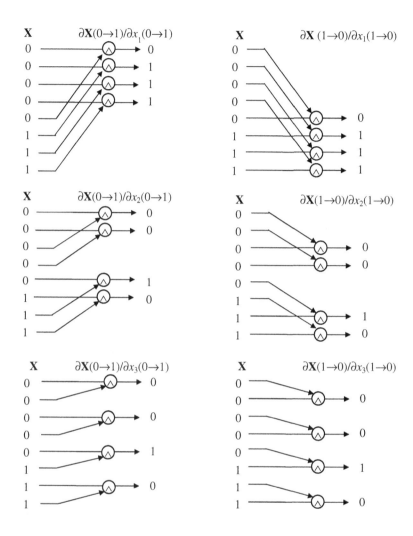

Fig. 3. Calculation of DPBDs based on parallel procedures

4   Conclusion

In this paper the new algorithm based on the parallel procedures is proposed. All most often-used IMs (Table 1) can be calculated based on this algorithm according (6) – (11). The computational complexity of the proposed algorithm is less in comparison with algorithm based on the typical analytical calculation (Fig. 4).

The proposed algorithm for the calculation of IMs based on the parallel procedures can be used in many practical applications. The principal step in these applications is representation of the investigated object by the structure function. As a rule the structure function is defined based on analysis of the structure of investigated object.

For example, this algorithm can be used for calculation of importance of service points in service system in [9].

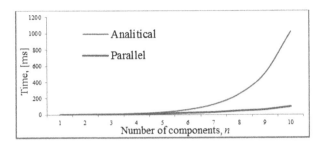

Fig. 4. Computation time for calculation of DPBDs based on analytical and parallel procedures

Acknowledgement. This work was partially supported by the grant VEGA 1/0498/14 and the grant of 7th RTD Framework Program No 610425 (RASimAs).

References

1. Zaitseva, E., Levashenko, V.: Importance Analysis by Logical Differential Calculus. Automation and Remote Control 74(2), 171–182 (2013)
2. Barlow, R.E., Proschan, F.: Importance of system components and fault tree events. Stochastic Processes and their Applications 3(2), 153–173 (1975)
3. Schneeweiss, W.G.: A short Boolean derivation of mean failure frequency for any (also non-coherent) system. Reliability and Engineering System Safety 94(8), 1363–1367 (2009)
4. Kuo W., Zhu X.: Importance Measures in Reliability, Risk and Optimization. John Wiley & Sons, Ltd. (2012)
5. Zaitseva, E.: Importance Analysis of a Multi-State System Based on Multiple-Valued Logic Methods. In: Lisnianski, A., Frenkel, I. (eds.) Recent Adv. Syst. Reliab. Signatures, Multi-State Syst Stat Inference, pp. 113–134. Springer (2012)
6. Zaitseva, E., Kostolny, J., Kvassay, M., Levashenko, V.: A Multi-Valued Decision Diagram for Estimation of Multi-State System. In: Proc. of the IEEE International Conference on Computer as a tool (EUROCON 2013), Zagreb, Croatia, pp. 645–650 (2013)
7. Moret, B.M.E., Thomason, M.G.: Boolean Difference Techniques for Time-Sequence and Common-Cause Analysis of Fault-Trees. IEEE Trans. Reliability R-33, 399–405 (1984)
8. Bochmann, D., Posthoff, C.: Binary Dynamic Systems. Academic Verlag, Berlin (1981)
9. Kvassay, M., Zaitseva, E., Levashenko, V.: Minimal cut sets and direct partial logic derivatives in reliability analysis. In: Safety and Reliability: Methodology and Applications - Proceedings of the European Safety and Reliability Conference, pp. 241–248. CRC Press (2014)
10. Kukharev, G., Shmerko, V., Zaitseva, E.: Multiple-Valued Data Processing Algorithms and Systolic Processors. Minsk, Nauka and Technica (1990)

An Unloading Work Model at an Intermodal Terminal

Mateusz Zajac and Justyna Swieboda

Wroclaw University of Technology,
Wyb. Wyspianskiego 27, 50-370 Wroclaw, Poland
{mateusz.zajac,justyna.swieboda}@pwr.edu.pl

Abstract. The article contains a method for distributing containers on the yard while unloading a container train. The method assumes a definition of the sequence of operations and their adjustment to the time of the container stay on the yard. As compared to the traditional approach to the issue, the proposed method allows for saving the load service time and further savings connected with the machine service cost.

Keywords: service process, container transport, container terminal.

1   Introduction

Intermodal transport can be defined as the successive use of various modes of transportation (road, rail, air and water) without any handling of the goods themselves during transfers between modes. It is a form of freight transport, which is becoming more and more popular. Such a method of transport in intercontinental relations is a very complex process, which involves over ten means of transport, several cases of overloading and a total storage time of over ten days at container ports and land terminals. Container terminals are an important element of this process.

Marine terminals should function in a way that allows for reducing the time of loading units at the terminal to the necessary minimum. This is caused by the necessity to obtain a high throughput as a result of infrastructural conditions and the assumed container turnover. The use of refined overloading technologies, such as full automation of the process, makes it possible to significantly reduce the load service time, eliminate errors and increase the safety level of the process. There are very expensive technologies which are not broadly used. At smaller ports, a high throughput is obtained by rationalization of actions. Growing rates for storage of loads at the port are a factor, which forces load recipients to look for savings.

Land terminals, on the other hand, combine transport and storage functions. In this case, the rates for storing empty or loading containers are degressive. Both types of containers are stored in one storage space. The problem, which occurs in intermodal overloading nodes, is the adoption of an appropriate storage method for intermodal units, i.e. the implementation of the container storage process so that it is not necessary to translocate them to another storage place. Such a situation occurs very often at intermodal modes if containers are stacked, there is a high turnover volume and decisions are made intuitively. This is the cause of the generation of additional costs and sometimes it makes it necessary to move a container even several times.

© Springer International Publishing Switzerland 2015   573
W. Zamojski et al. (eds.), *Theory and Engineering of Complex Systems and Dependability,*
Advances in Intelligent Systems and Computing 365, DOI: 10.1007/978-3-319-19216-1_55

Both technologies and terminal design principles do not provide any clues as to the method of management of the flow of load units. However, the operation of land container terminals is much different from typical container ports, about which a lot of information can be found in the literature, e.g. [3].

In recent years, research has been conducted on the functioning of intermodal terminals. Attention is drawn to the effectiveness of the terminal functioning in study [12]. It was emphasized that the effectiveness of the entire service is the most important in the competition with road transport. Simulation tests of the terminal operation were conducted, including the selection of the appropriate type and quantity of equipment to serve a specific number of containers. In some studies, attention is paid to quite new technologies. The study [16] presents the functioning of an automated container terminal, which is designed to reduce the number of operations at the terminal. This should result in the reduction of costs and in increased reliability and reduced service times. Detailed issues on this area are contained in study [5].

The problem of planning the unloading of a container train was discussed in the article [3]. The authors considered the problem of planning the operation of cranes for various tasks (storage order, wagon-semi-trailer overloading). The issue of unloading from parallel tracks on the rail terminal was also analyzed in [1]. The schedule of the crane operation at an intermodal overloading node was considered in [2] and [14]. The train loading planning using the "robust approach" was presented in [4]. The slot optimization in trains was also dealt with in [11]. In this study, it was assumed that it was optimization connected with the organization and costs. It was emphasized that the use of optimization methods in slot allocation is the basis for effective use of space in wagons.

Literature pertaining to the organization of intermodal terminal operation also deals with the problem of scheduling road transport in intermodal transport [12].

A simulation analysis was conducted in [10], which was aimed at finding balance between the required service level, costs and train (service) delays. Interesting simulation algorithms are presented in [6] and [8].

The literature also includes studies on the optimization of the load service process in container-type installations. This optimization pertains to, amongst other things, to the energy intensity of the logistic process [15]. An important topic undertaken in scientific studies includes the operation, reliability and safety of the load service process. As far as safety level estimation methods are concerned, [7] and [9] are an interesting proposal, where an analysis of the Petri net lies in the transport process. An interesting approach was also proposed in [14] and [15], where fuzzy logic was used to estimate the system susceptibility to damage.

Despite of various analyses of the work modelling issue at an intermodal overloading node, there are no studies, which would deal with the aspect of the sequence of unloading operations and define the sequence of operations and select the container storage place. The author performed his own analysis of the sensitivity of several most important arguments (according to container terminal operators), which influence the selection of the loading unit storage place. However, one should take into consideration the fact that a few dozen such arguments were formulated during previous research conducted by the author. Simulation tests confirmed that the storage process conducted according to specific rules may result in a much lower energy demand while servicing containers. Depending on the size of the overloading node, the degree of its use and the rate of the intermodal unit flow, these savings may reach up to 50%.

The aim of the article is to present a method for performing work connected with the wagon-yard overloading of containers, which would allow for reducing the train service time. This method takes into account the time of container stay at the terminal and, by assigning service priority to them, allows for work rationalization and reduction of the number of operations connected with the movement of loads in the afore-mentioned direction.

Next, the article presents the principles of operations while unloading containers at the terminal and information to be provided when loads are placed on the yard. A method rationalizing overloading work in an intermodal overloading node in the wag-on-yard direction; parameters of the operation of overloading machines and calculations comparing load service times in reality and those obtained using the method.

2   The Procedure Diagram for Container Unloading

The condition of containers is checked after the train enters the terminal. Next, after the list of containers received at the terminal is entered into the system, the load units are physically moved. This is done traditionally using reachstackers or container cranes. Containers after being taken off wagons are transported and placed on the container yard. Decisions on the selection of the storage place for containers are made by machine operators. Based on their own experience, operators make decisions on where a given container should be placed and in what the sequence of the performance of the next transport order should be.

The proposed approach to the load unit storage problem is significantly different. The procedure diagram is presented in Fig. 1.

N - set of containers on wagons,
N_w – set of free spaces in the storage yard,
P_{ki} – priority value of the ith container to be unloaded,
P_{ki} – priority value of the ith container,
$T_{ti\text{-}>j}$ - time of placing the ith container at the jth place
$T_{ti\text{-}>j}$ - set of times of placing the ith container at the jth place

After wagons are rolled on the terminal, information is entered into the system about the sequence of container placement on the wagons. On the basis of prior notification or the known round trip, i.e. information about the date of releasing the container for road transport, the priority of further overloading operations is defined. Containers which will be kept at the terminal for the longest time are served first. The containers which will stay at the terminal for the shortest time are the last to be overloaded. In this way, containers which were unloaded later will be situated in higher layers and will be available for loading onto a semi-trailer immediately.

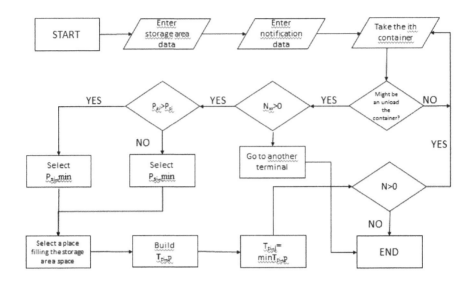

Fig. 1. Diagram of the method of the container train unloading process

K-levels of container service priority are established. After specifying a list of containers with first-order priority, it is defined where these containers should be placed by checking the forecast storage time of a load unit at available locations that are several rows away from their place in the wagon in each direction. For each container, a place is selected, for which the transport together with putting it away will take the shortest time. It does not need to be a place, which is the closest to the place on the wagon because the time of lifting may take more time than taking the container several dozen metres away and placing it on a lower layer.

Thus, input information for the method includes:

- the list of overloading tasks,
- the list of free spaces on the yard,
- the list of occupied spaces on the yard,
- the expected time of the stay of containers in the highest layers.

On the basis of combined information about occupied spaces and the time of container stay in the highest layers, the storage areas are assigned priority $P\{1,2,3,4\}$ - the lower the value is, the sooner the unit will leave the terminal. Containers to be unloaded from a train are marked in a similar manner. Containers with a lower value may not be stored on ones with lower priority value.

After establishing the priorities, i.e. available spaces, the load placement time analysis is performed. In this analysis, total operation times are compared, which include (from the time the crane reaches the wagon). These include:

t_{1i} - setting the crane in the right position for lifting the container,

t_{2i} - lifting the container,

t_{3i} - setting the crane in the right position for driving,

t_{4i} - driving to the storage location,
t_{5i} - setting the crane in the right position for placing the container,
t_{6i} - placing the load on a container stack,
t_{7i} - setting the crane in the right position for driving.

The total time of the ith container placement is:

$$T_{t_i} = \sum t_{1i} + t_{2i} + \cdots + t_{ni}$$

(1)

The total time of unloading N-containers from wagons is:

$$T_{i=1}^{N} = \sum T_{t_i}$$

(2)

This procedure does not include the driving time to the wagon to take another container. The problems of the performance of this issue are a separate issue, which is outlined below. After designating places for all containers with a given priority, the sequence of operations is determined. Each transport order can be regarded as a cycle of activities from setting the crane in the right position for lifting the container until the grapple is lowered after the load unit was placed in the storage area. Hence, the establishment of the sequence of operations involves the verification of subsequent passages from the container placement area to another container on the wagon. After establishing the unloading sequence for containers with the same priority, the sequence for units with the next priority level is defined. After all relations have been established, load units are overloaded. The next part of the study presents calculations focusing on the container transport process from the wagon to the storage area on the yard.

3   Tests and Observations

During research conducted on one of container terminals in Poland in 2014, several dozen processes of train unloading were observed. The aim of the research included:

- identification of components of this type of unloading process,
- assigning time values to individual components of the process,
- becoming familiar with arguments taken into consideration while establishing the sequence of container operation during the wagon-yard unloading work,
- becoming familiar with arguments taken into consideration while making decisions about the container storage place.
- proposing a knowledge-based method for storing load units.

It was noticed during the research that there is a fundamental difference between load unit service ties, depending on the floor on which they are stored.

During more than 800 observations, the distance of transportation was measured and time recorded. Table 1 shows summary information for the group of observations.

Table 1. The overall results of the observation

Operation	Mean value	Std deviation	Variance
Setting up the drive [s]	17.98	8.48	71.94
Drive [s]	28.45	20.49	419.87
Distance (without.) [m]	54.04	38.37	1472.23
Setting to download and download [s]	15.57	6.00	36.02

As can be seen from the data presented in the table 1, most of the time during the execution of the discharge cycle takes a ride with the cargo, immediately after driving without a load. The total cycle takes an average 133s reloading, while driving is more than 62s. It can be seen here that one of the fundamental option of shortening the cycle is a shortened drive to and with the container. Reloading cycle-times differ depending on the relationship of handling. Table 2 and Table 3 present results for handling relationships and track - first floor, and track - the fourth floor.

Table 2. Results of the unloading in relation I track – I floor

Operation	Mean value	Std deviation	Variance
Setting up the drive [s]	16.76	8.86	78.44
Drive [s]	25.29	20.18	407.22
Distance (without load) [m]	45.47	31.99	1023.14
Setting to download and download [s]	14.71	5.16	26.60
Setting up the drive [s]	20.88	17.20	295.74
Drive [s]	30.12	18.98	360.11
Distance (with load) [m]	62.00	42.39	1796.50
The setting for unloading and unloading [s]	15.29	7.74	59.97

In relationships I track - I floor, the average operating time was 123s. Like the collective results, driving time was also the largest (total 55 s). Due to the experience of the operators, it turned out to be a very short time - away containers. Due to the visibility of the container, and no need for very precise withdrawal of the unit load on the already standing - another - this time was the shortest of all possible relationships. Unloading on the fourth floor proved to be far more time-consuming. The average duration of the operation was over 150s; in this case, for almost one third the time, it took defer the load on a high level. The driving time with load was similar to those obtained with the unloading of the first floor: a little over 55 s.

Table 3. Results of the unloading in relation I track – IV floor

Operation	Mean value
Setting up the drive [s]	22.00
Drive [s]	31.50
Distance (without load) [m]	50.00
Setting to download and download [s]	11.00
Setting up the drive [s]	14.00
Drive [s]	26.50
Distance (with load) [m]	20.00
The setting for unloading and unloading [s]	45.50

The studies lead to the following general conclusions:

- if taking account the time of operations, it does not matter whether the container is deposited on level 1 or level 2.
- postponing the level 3 lasts longer than half the level of 1-2. Putting the container on level 4 in comparison to the level 1 is three times longer. the difference between putting on levels 1 - 2 and level 3 allows for passing around 5-10m.
- the difference between putting on levels 1 - 2 and level 4 allows the passing of about 50-60m.

Observations of duration times of the elements of the process allowed for determining values needed to calculate the operation time according to the method.

4   Results of Observations and Comparison

One of container unloading example is presented in a further part of the study. The first column contains the number of the next container to be serviced. The second column contains the actual (measured) service time while the next column contains the calculated service value. The results achieved by method are presented in chapter 2 and compared with observed data.

As it can be read from the table, time saving for single movements was as high as over 50%. The last column in this table presents the accumulated time saved owing to the proposed method.

The proposed method, however, does not always ensure a reduction in the container unloading time. In the worst case for profitability, it allowed for saving only less than 2 minutes. The comparative results for actual measurements and the method are presented in table 5.

In this case, nearly 42% of loads allowed for reducing the container service time. However, in nearly 30% of the cases, worse service times were obtained when the procedure was followed, the remaining part includes times, which are identical for the theory and practice. It should be noticed that, despite obtaining similar container unloading times, the places indicated by the method and actually selected by the operators were different. The operators did not take into consideration priorities assigned to the containers. As a result, in several cases, it was necessary to relocate containers so it was necessary to use additional technical and human resources and additional time. It was not necessary if the method was used.

Table 4. Sample results for train service

Container number	Container service time, actual, [s]	Container service time according to the method [s]	Time-saving for one movement	Service time saving, accumulated value [min]
1	94.4	90.54	4.09%	
2	114.81	89.00	22.48%	0.49
3	89.81	89.00	0.90%	0.51
...	...	...	...	...
25	150.42	90.54	39.64%	11.71
26	165.11	89.00	46.06%	12.98
27	180.16	89.00	50.56%	14.49

Table 5. Results for the minimum saving-time solution

Container number	Container service time, actual, [s]	Container service time according to the method [s]	Time-saving for one movement	Service time saving, accumulated value [min]
1	114.81	90.53	21.14%	
2	89.81	92.07	-2.52%	0.37
3	89.4	90.53	-1.27%	0.35
4	114.4	92.07	19.51%	0.72
5	94.1	93.61	0.41%	0.73
...	...	...	...	...
21	89.44	90.54	-1.27%	1.47
22	94.12	90.54	3.68%	1.53
23	90.21	92.07	-2.07%	1.50
24	89.4	90.54	-1.27%	1.48

Figure 2 presents a case, in which the highest and the lowest savings were obtained while servicing loads for the sake of comparison. As can be seen in this figure, the service did not provide such promising results for the greatest savings, it was even lower than the lowest saving course.

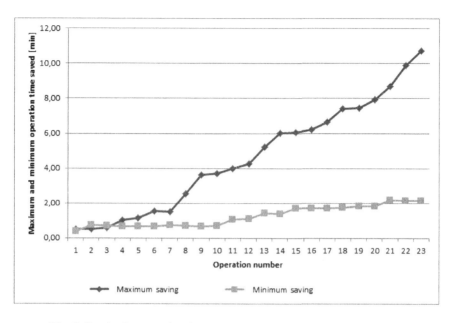

Fig. 2. Graph of progressing time saving together with the operation number

5   Conclusions

The article presents an unloading work method at an intermodal overloading node The proposed method is based on a comparison of the expected storage time for containers on the yard and intended for unloading. Next, the method provides for the verification of the time needed for placing the container in individual storage areas and makes it possible to choose the shortest service time. The research conducted allowed for specifying the parameters of overloading equipment operation, which were then used to determine theoretical load service times. The comparison of the service time for a single container usually makes it possible to save time. For a train with 50 containers, up to 30 minutes can be saved.

The next stage of research involves the development of the method to include an element indicating the sequence of loading operations. The assignment of the sequence of unloading operations will make it possible to gain further benefits, such as reduced train service time and reduced fuel consumption.

Acknowledgements. The results presented in this paper have been obtained within the project "The model of operations in intermodal terminal" (contract no. POIG.01.03.01-02-068/12 with the Polish Ministry of Science and Higher Education) in the framework of the Innovative Economy Operational Programme 2007-2013.

References

1. Ballis, A., Golias, J.: Comparative evaluation of existing and innovative rail – road freight transport terminals. Transportation Research Part A 36, 593–611 (2002)
2. Bostel, N., Dejax, P.: Models and algorithms for container allocation problems on trains in a rapid transshipment shunting yard. Transportation Science 32(4), 370–379 (1998)
3. Boysen, N., Fliedner, M.: Determining crane areas in intermodal transshipment yards: The yard partition problem. European Journal of Operational Research 204(2), 336–342 (2010); Smith, T.F., Waterman, M.S.: Identification of Common Molecular Subsequences. J. Mol. Biol. 147, 195–197 (1981)
4. Bruns, F., et al.: Robust load planning of trains in intermodal transportation. OR Spectrum, 1–38 (2013)
5. Bujak, A., Zając, P.: Can the increasing of energy consumption of information interchange be a factor that reduces the total energy consumption of a logistic warehouse system? In: Mikulski, J. (ed.) TST 2012. CCIS, vol. 329, pp. 199–210. Springer, Heidelberg (2012)
6. Kierzkowski, A.: Reliability models of transportation system of low-cost airlines. Reliability, Risk and Safety: Back to the Future, pp. 1325–1329 (2010)
7. Kierzkowski, A., Kisiel, T.: An impact of the operators and passengers behavior on the airport's security screening reliability (2015). Source of the Document Safety and Reliability: Methodology and Applications - Proceedings of the European Safety and Reliability Conference, ESREL, pp. 2345–2354 (2014)
8. Kierzkowski, A., Kowalski, M., Magott, J., Nowakowski, T.: Maintenance process optimization for low-cost airlines. In: 11th International Probabilistic Safety Assessment and Management Conference and the Annual European Safety and Reliability Conference 2012, PSAM11 ESREL 2012, vol. 8, pp. 6645–6653 (2012)
9. Kowalski, M., Magott, J., Nowakowski, T.: Werbińska-Wojciechowska Sylwia: Analysis of transportation system with the use of Petri nets. Eksploatacja i Niezawodność - Maintenance and Reliability 1, 48–62 (2011)
10. Kozan, E.: Optimum capacity for intermodal container terminals. Transportation Planning and Technology 29(6), 471–482 (2006)
11. Liu, D., Yang, H.-L., Zhang, Y.: Slot Allocation Optimization Model for Multi-node Container Sea-Rail Intermodal Transport. Journal of Transportation Systems Engineering and Information Technology 2, 26 (2013)
12. Nossack, J., Pesch, E.: A truck scheduling problem arising in intermodal container transportation. European Journal of Operational Research 230(3), 666–680 (2013)
13. Rotter, H.: New operating concepts for intermodal transport: The mega hub in Hanover/Lehrte in Germany. Transportation Planning and Technology 27(5), 347–365 (2004)
14. Vališ, D., Pietrucha-Urbanik, K.: Utilization of diffusion processes and fuzzy logic for vulnerability assessment. Maintenance and Reliability 16(1), 48–55 (2014)
15. Vališ, D., Fuchs, P., Saska, T., Soušek, R.: How to Calculate the Accident Probability of Dangerous Substance Transport. The Archives of Transport 24(3), 273–284 (2012) ISSN 0866-9546
16. Zając, P.: Transport-storage system optimization in terms of exergy. In: Bruzzone, A. (ed.) The 13th International Conference on Harbor Maritime Multimodal Logistics Modeling & Simulation, HMS 2010, Fes, Morocco, October 13-15, pp. 143–148. Laboratoire des Sciences de l'Information et des Systèmes, Marseille (2010)

Cost-Aware Request Batching
for Byzantine Fault-Tolerant Replication

Maciej Zbierski

Institute of Computer Science
Warsaw University of Technology, Poland
m.zbierski@ii.pw.edu.pl

Abstract. One of the most commonly applied optimizations to Byzantine fault-tolerant replication is batching. Such approach involves packing multiple client requests into a single instance, thus reducing the per-request overhead and providing a potential increase in throughput. Existing solutions use either constant-sized batches or determine their sizes based on the performance of the underlying replication protocol. In this article we propose a different approach and introduce a method for selecting batch sizes that minimize the cost of cryptographic operations performed by replication protocols. The results of performed experiments show that our method can obtain up to 50% increase in throughput when compared to existing batching schemes. The proposed approach can be applied not only in modern BFT replication protocols, but also in solutions using other fault models, as long as they use some form of message authentication.

Keywords: Byzantine fault tolerance, batching, cryptography, state machine replication, distributed systems, dependability.

1   Introduction

State Machine Replication [15] is a technique widely used for improving the availability of distributed systems in the presence of faults. In such approach the actual service is replicated among a set of servers and enhanced with a protocol for coordination and consistency management. While originally proposed to tolerate benign faults, such as server crashes, recent studies have stressed the importance of extending the model to arbitrary faults. Such systems, often referred to as Byzantine Fault Tolerant (BFT), provide a correct service even despite the arbitrary behavior of a fraction of nodes, due to for instance malicious attacks [4], transient hardware errors [14, 19] or bugs in code [16]. As a response to an increased demand to tolerate such unexpected behavior, BFT protocols have recently been gaining on popularity, being deployed not only in safety critical environments, but also in other solutions, such as replicated databases, distributed services or in the cloud (see for instance [10, 7–9, 1, 23]).

One of the most commonly applied optimizations to fault-tolerant replication is request batching [2, 13]. The goal of such approach is to group multiple client requests

© Springer International Publishing Switzerland 2015
W. Zamojski et al. (eds.), *Theory and Engineering of Complex Systems and Dependability,*
Advances in Intelligent Systems and Computing 365, DOI: 10.1007/978-3-319-19216-1_56

into larger entities. This allows to reduce the number of messages exchanged in the system, leading to smaller per-request overhead and potentially higher throughput. Batching becomes particularly important when considering replicated systems tolerating Byzantine faults. This is because the underlying replication protocols perform additional cryptographic operations required to authenticate messages exchanged between nodes in order to guarantee their integrity [2]. While these operations are required to prevent faulty replicas from disrupting the service, they might take even several times longer than the transfer of the corresponding message between nodes. Consequently, optimizing the cost of cryptographic operations could reduce the amount of time required for a message exchange between nodes, leading to an increase in throughput of fault-tolerant replication protocols.

While batching mechanisms are currently used by the majority of Byzantine fault-tolerant replicated state machine protocols [10, 9], the existing approaches do not focus directly on optimizing the cost of cryptographic operations. Instead, the batch sizes are usually selected based on the short-term performance analysis of the underlying replication protocols, using for instance a sliding window mechanism [2]. Since this approach requires fine-tuning the batching policy, some implementations simplify it by using fixed batch sizes [10]. We on the other hand propose to select batch sizes that would minimize the time required to authenticate the batched requests.

In this article we contribute with the first batching method targeted for minimizing the cost of cryptographic operations performed by BFT replication protocols. Additionally, we propose an algorithm for adaptive batch size estimation, combining our method with the historical data on incoming requests. We verify our approach by implementing it in two different Byzantine replication protocols, PBFT [2] and MinBFT [20]. The performed tests have demonstrated that our approach can increase the throughput of the underlying replication protocol up to 50% over other batching strategies used in existing BFT protocols. Finally, although in this article we focus mostly on protocols tolerating Byzantine faults, our approach can also be applied in solutions using different fault models, as long as they use some form of message authentication.

The rest of the paper is constructed as follows. Section 2 provides the background on Byzantine fault-tolerant replication protocols and their batching strategies. Our model of cost-aware batching is introduced in section 3. Section 4 proposes adaptive batch size estimation combined with the introduced batching model. The description of performed experiments and the evaluation of results is presented in section 5. Finally, section 6 presents the related work and section 7 concludes the paper.

2   Byzantine Fault-Tolerant Replication Protocols

This section presents two approaches to Byzantine fault-tolerant replication and describes the way they take advantage of request batching to improve their overall throughput. PBFT [2] is considered to be the first Byzantine replication protocol whose safety does not depend on synchrony assumptions and is usually treated as a baseline solution in that field. The protocol requires 3f + 1 machines in order to tolerate up to f faulty replicas.

The normal operation mode of PBFT consists of three all-to-all communication rounds, required to reach an agreement, plus an additional round to relay the result to the client. To guarantee communication integrity, the exchanged messages are enhanced with a vector of message authentication codes (MAC), containing one element for every recipient of the message. Consequently, the cost of authentication increases linearly with the number of replicas in the system, requiring an overall of $8f + 3$ message authentication codes for processing a single message. This number can be reduced b times through the use of batching, where b is the batch size, i.e. the number of requests grouped together. The batching is performed before the first communication round and requests grouped in one batch are treated as a single message whenever they are exchanged between replicas.

The batch size in PBFT is dynamically selected using a sliding-window mechanism. The size of the window represents the number of previous request batches that have been assigned with an identifier, but have not yet been processed by the system. If the current window size is greater than the maximum accepted threshold, incoming requests are enqueued until one of the previous batches has been processed. When this is satisfied, the awaiting messages are grouped in a single batch and relayed to other replicas. Additionally, the maximum number of messages in a batch is limited by a certain threshold. It has been later suggested that the efficiency of batching in PBFT can be further improved by using batches of a constant size for all incoming requests [10].

MinBFT [20] is a recent approach to designing a Byzantine fault-tolerant replication protocol using lower number of replicas and fewer communication rounds. Authors use dedicated trusted service (USIG) to assign unique identifiers to outgoing messages. USIG is also used for authenticating the communication process using a HMAC algorithm. The application of a trusted component allows to reduce the overall number of replicas to $2f + 1$ and remove one communication round, as compared to PBFT. This reduces the per-request overhead produced by cryptographic operations, although the cost of authentication still scales linearly with the number of replicas. Finally, MinBFT reuses the batching policy of PBFT described above. MinBFT is often used as a reference protocol whenever new replication system with a trusted service is constructed [9, 5].

3   Cost-Aware Batching Model

Unlike the methods described in the previous section, we propose to optimize request batching with regard to the cost of performed cryptographic operations, i.e. authentication of the exchanged messages. In this section we present a method for creating a batching model, which can be later used to determine how a sequence of requests should be divided into batches to minimize the cost of performed cryptographic operations.

Let M be a sequence of n client requests with overall payload size s. Assume that the sequence is divided into a number of batches of constant size x, which in turn are sequentially authenticated using the corresponding algorithm (e.g. HMAC). We denote the largest time required to authenticate a message from M using batches of size x as $T_{max}(n, s, x)$. Additionally, $T_{avg}(n, s, x)$ represents the average amount of time spent on authenticating a message from M. The batch size optimal with regard to the cost of cryptographic operations can be selected as the one minimizing the time required to provide authentication for all messages in M. Consequently, the optimal batch size for a sequence of messages can be selected using the function presented as equation 1, with W_m and W_a being the

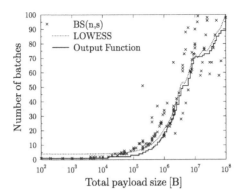

Fig. 1. The cost-aware batching model for HMAC function

weights of $T_{max}(n, s, x)$ and $T_{avg}(n, s, x)$ respectively. Although these weights can be selected arbitrarily, we have observed the best results for $W_m = W_a = 0.5$ and consequently use them in the remainder of this article.

$$BS(n,s) = \arg\min_x (W_m \times T_{max}(n,s,x) + W_a \times T_{avg}(n,s,x)) \tag{1}$$

The process of creating a batching model for an authentication algorithm involves iteratively selecting input parameters n and s, and calculating the respective value of $BS(n,s)$. This is performed experimentally by authenticating messages from a sequence generated according to analyzed parameters, and selecting the batch size minimizing the equation 1 based on values T_{max} and T_{avg} obtained for different batch sizes x. The result model is created based on the values of $BS(n,s)$ obtained for different combinations of parameters n and s. Finally, the output function is obtained through local regression and sampling. A model for HMAC algorithm, computed using the approach presented above is shown in figure 1. A similar model needs to be computed once for every message authentication algorithm used by the protocol prior to its deployment.

4   Adaptive Batch Size Estimation

In this section we show how the previously created cost-aware model can be used to optimize batching in BFT replication protocols. We do so by proposing a simple adaptive algorithm for selecting the batch size based on both the historical data about incoming requests and the batching cost model. Please note that other algorithms, such as for instance tuned sliding-window [13] or optimized adaptive request batching [4], can also be combined with our solution to possibly obtain even further increase in throughput.

The proposed adaptive algorithm operates in two separate modes to provide a distinction between high and low traffic. For a large inflow of requests, their number to arrive in a selected time interval and their overall size are approximated based on historical data. Based on that estimation, the cost model is used to select the size of the next batch. This estimate can be updated if the number of incoming requests turns out to differ significantly from the prognosis.

Algorithm 1. Adaptive batch size estimation.

Initialization:
1: *count_hist* := allocate_memory(HISTORY_LENGTH)
2: *size_hist* := allocate_memory(HISTORY_LENGTH)
3: *next_batch_size* := FALLBACK_BATCH_SIZE

Procedure update_estimates(V_{count}, V_{size})
4:    *count_hist* := *count_hist*.shift() ∪ {V_{count}}
5:    *size_hist* := *size_hist*.shift() ∪ {V_{size}}
6:    **if** average(*count_hist*) < T_{count} **or** average(*size_hist*) > T_{size} **then**
7:       *next_batch_size* := FALLBACK_BATCH_SIZE
8:    **else**
9:       *next_batch_size* := batch_model(average(*count_hist*), average(*size_hist*))
10:   **end if**

Procedure build_batch()
11:   *batch* := ∅
12:   **while** size(*batch*) < *next_batch_size* **and** age(*batch*) < Δ_B **do**
13:      **if** *req_queue* not empty **then**
14:         *batch* := *batch* ∪ {*req_queue*.first()}
15:         *req_queue*.shift()
16:      **end if**
17:   **end while**

Procedure main()
18:   **while** *true* **do**
19:      *batch* := build_batch()
20:      process(*batch*)
21:      update_estimates(count(*batch*) / age(*batch*), size(*batch*) / age(*batch*))
22:   **end while**

If the incoming traffic is considered low, or the received messages are very large, the algorithm provides a fallback to constant sized batching. This approach reduces the latency for low number of small-sized requests, since the cost model would suggest batching many more messages and the algorithm would constantly wait for the timeout before reaching the estimated batch size. Additionally, we have observed that cost-aware batching does not provide a significant performance increase for very large messages. The algorithm can switch between both modes of operation if the number of incoming requests or the size of their payload exceeds the assumed thresholds.

The pseudocode for request processing with the adaptive batch size estimation is presented as algorithm 1. The batching process continuously collects the incoming requests to create a new batch (line 19), signs it and transfers to other replicas (line 20), and uses its properties to update the estimates (line 21). The process of creating a batch consists of collecting incoming client requests from the queue (lines 13-16) until either the number of messages exceeds the estimated batch size or the creation time becomes greater than some assumed timeout Δ_B (line 12). The function estimating the next batch size stores the parameters of the previous batch (lines 4 and 5). If the number of incoming requests is too small or their payload becomes too large, as determined by the average of the parameters of previous batches (line 6), the constant size batching is used (line 7). Otherwise, the next batch size is estimated using the cost-aware model with the average parameters of previous batches (line 9).

5   Experiments and Evaluation

In order to verify our approach we have created an implementation of both PBFT and MinBFT according to their original descriptions. The protocols were additionally modified to take advantage of different batching schemes. Although PBFT originally uses MD5 algorithm to create message authentication codes, in our implementation we have replaced it with SHA-512, since MD5 is generally discouraged due to known safety issues [17,18]. SHA-based HMACs are already used in newer Byzantine replication protocols [20]. Finally, the variant of MinBFT protocol used in our tests bases on NS-USIG service, although in general the obtained results also apply to other types described in the original article.

The performed tests compare the cost-aware batching method proposed in this article with two other approaches present in the literature. The first one is the sliding-window mechanism of PBFT, implemented according to the description in [2]. Similarly to the original solution, we have set the window size to one and assume that the next batch is created immediately after the previous one has been processed. The other tested approach is the constant size batching, with the batch size set to 10 messages, as suggested by Kotla et. al. [10]. Finally, wherever applicable, the results without batching are supplied as a reference.

The goal of the first experiment was to compare the efficiency of the cost-aware batching model with other approaches for fixed sizes of request payloads. The clients involved in the experiment periodically issued requests of a previously assumed constant size. The results for PBFT with different batching methods are presented in figure 2, while figure 3 shows the results for MinBFT.

The cost-aware batching has demonstrated the greatest increase in throughput in both protocols for all tested message sizes. The highest difference between the cost-aware approach and other methods can be observed for smaller messages, with a maximum achieved throughput higher by around 50% than for constant size batching. This is because the cost-aware method batches up to several orders of magnitude more small-sized requests than both sliding window and constant size batching. The results for request with 4kB payload are less spectacular, although the cost-aware method was still faster by around 5%. It is worth noting however that modern replication protocols usually transmit only hashes of the payload, and therefore rarely require to transfer such large messages during the agreement. As a result, when using SHA-512 hashing function, the typical size of a single message would rarely exceed 64 bytes.

While the experiments for constant request sizes allow to analyze how batching mechanisms generally behave, in real systems payload sizes usually vary from request to request. The second experiment analyzed the efficiency of the proposed batching model for requests with varying payload sizes. Similarly as before, the clients periodically issued requests, however the sizes of their payloads were selected using random number generators with different probability distributions. In the first case, the payload sizes in bytes were generated using a uniform distribution $U(0,256)$, while in the other they were selected as absolute values of Cauchy distribution with parameters $C(0,128)$. The corresponding results are presented in figures 4a and 4b respectively.

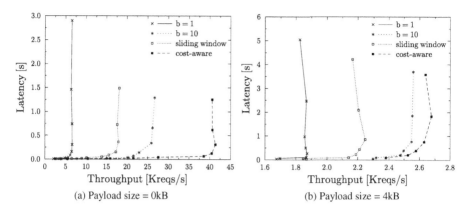

Fig. 2. Latency vs. throughput for PBFT with different batching methods

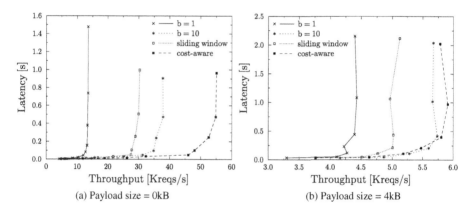

Fig. 3. Latency vs. throughput for MinBFT with different batching methods

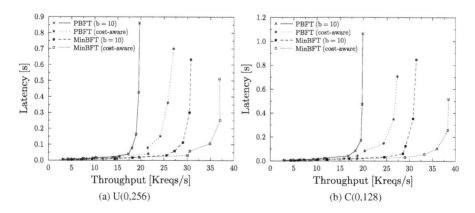

Fig. 4. Latency vs. throughput for varying request payload sizes

Despite the inability to exactly determine the optimal batch size, the performed approximation method allowed for an increase in throughput over constant batching of around 40% for PBFT and 20% for MinBFT. What is interesting is that, unlike for cost-aware approach, the latencies of constant size batching for payloads generated using Cauchy distribution exceeded those uniformly distributed by around 25%. This shows that while the proportions between the throughputs obtained by both methods for tested distributions remain the same, the latency of cost-aware batching reacts better to occasional requests of very large size.

6   Related Work

Practical Byzantine Fault Tolerance, or PBFT [2], while not being the first protocol tolerating Byzantine faults, is nevertheless considered by many to be the first practical enough to be implemented in real distributed systems. The two main reasons for this are the safety guarantees in asynchronous environments and a number of applied optimizations such as batching or the usage of MACs for message authentication. PBFT is still considered a baseline in the field of Byzantine fault-tolerant replication.

The authors of Zyzzyva [10] replaced the sliding window mechanism with a constant size batching. The optimal batch size has been obtained by a trial-and-error to maximize the throughput of the protocol. Authors of protocols deriving from PBFT or Zyzzyva (such as [20-22,11,7]) also very often apply constant size batching, usually using the same batch size as in [10].

Apart from the constant size batching techniques, various adaptive methods have been proposed. Friedman and Hadad [6] suggested a protocol-agnostic adaptive batching algorithm for replicated services. In their proposition multiple messages are batched together whenever a node wishes to broadcast them among other replicas at roughly the same time.

Santos and Shiper [13] focused on crash-fault tolerant services and provided an analysis of the sliding-window mechanism present in Paxos [12]. The authors optimized the parameters of the Paxos batching mechanism to fully utilize the network bandwidth. As the sliding-window approach in Paxos is very similar to the one in PBFT, their findings can be extended to Byzantine fault tolerant systems.

Sá et. al. [4] designed an adaptive request batching algorithm targeted specifically for Byzantine replication protocols. Their solution, designed as an extension to PBFT, takes into account the dynamic parameters of the computational environment, such as mean time between arrivals, to calculate the optimal batch size. Please note that our approach is non-exclusive with these adaptive batching methods and can be applied as an additional decision factor to potentially further increase the efficiency of those solutions.

A different approach to request batching has been proposed in Zzyzx [8]. Authors implement a locking mechanism which uses an underlying replication protocol to extract a selected portion of the state and assign it to the request originator for exclusive access. Zzyzx batches requests from a single client by allowing multiple operations on a set of objects after a single setup phase.

7   Conclusion

In this article we have presented a different approach to request batching in Byzantine fault tolerant replication protocols. Instead of determining batch sizes based on the speed of processing or using batches of fixed size, we have proposed to focus on the cost of cryptographic operations. When compared to constant size batching, an approach widely used in the literature, our method achieved over 50% increase in the throughput while processing requests without additional payload and around 30% increase for requests with randomly generated payload smaller than 256 bytes. Furthermore, the proposed approach has proven to provide better results even for larger messages.

In the future we plan to combine our approach with other adaptive batching methods present in the literature and use factors other than total message size while determining the batch size. Additionally, we intend to analyze the behavior of our method in wide area networks, characterized by high transmission latency and varying bandwidth.

References

1. Behl, J., Distler, T., Kapitza, R.: Scalable BFT for multi-cores: Actor-based decomposition and consensus-oriented parallelization. In: 10th Workshop on Hot Topics in System Dependability (HotDep 2014), Broomfield, CO. USENIX Association (2014)
2. Castro, M., Liskov, B.: Practical Byzantine fault tolerance. In: Proceedings of the Third Symposium on Operating Systems Design and Implementation, OSDI 1999, pp. 173–186. USENIX Association, Berkeley (1999)
3. Crypto++ library 5.6.2., http://www.cryptopp.com
4. Santos de Sá, A., Freitas, A.E.S., de Araújo Macêdo, R.J.: Adaptive request batching for Byzantine replication. SIGOPS Oper. Syst. Rev. 47(1), 35–42 (2013)
5. Dettoni, F., Lung, L.C., Correia, M., Luiz, A.F.: Byzantine fault-tolerant state machine replication with twin virtual machines. In: 2013 IEEE Symposium on Computers and Communications (ISCC), pp. 398–403 (July 2013)
6. Friedman, R., Hadad, E.: Adaptive batching for replicated servers. In: 25th IEEE Symposium on Reliable Distributed Systems, SRDS 2006, pp. 311–320 (2006)
7. Guerraoui, R., Knežević, N., Quéma, V., Vukolić, M.: The next 700 BFT protocols. In: Proceedings of the 5th European Conference on Computer Systems, EuroSys 2010, pp. 363–376 (2010)
8. Hendricks, J., Sinnamohideen, S., Ganger, G.R., Reiter, M.K.: Zzyzx: Scalable fault tolerance through byzantine locking. In: 2010 IEEE/IFIP International Conference on Dependable Systems and Networks (DSN), pp. 363–372 (2010)
9. Kapitza, R., Behl, J., Cachin, C., Distler, T., Kuhnle, S., Mohammadi, S.V., Schröder-Preikschat, W., Stengel, K.: CheapBFT: Resource-efficient Byzantine fault tolerance. In: Proceedings of the EuroSys 2012 Conference (EuroSys 2012), pp. 295–308 (2012)
10. Kotla, R., Clement, A., Wong, E., Alvisi, L., Dahlin, M.: Zyzzyva: Speculative Byzantine fault tolerance. In: Symposium on Operating Systems Principles, SOSP (2007)
11. Kotla, R., Dahlin, M.: High throughput Byzantine fault tolerance. In: Proceedings of the 2004 Conference on Dependable Systems and Networks, pp. 575–584 (2004)

12. Lamport, L.: The part-time parliament. ACM Trans. Comput. Syst. 16(2), 133–169 (1998)
13. Santos, N., Schiper, A.: Tuning Paxos for high-throughput with batching and pipelining. In: Bononi, L., Datta, A.K., Devismes, S., Misra, A. (eds.) ICDCN 2012. LNCS, vol. 7129, pp. 153–167. Springer, Heidelberg (2012)
14. Schiffel, U., Schmitt, A., Süßkraut, M., Fetzer, C.: ANB- and ANBDmem-Encoding: Detecting hardware errors in software. In: Schoitsch, E. (ed.) SAFECOMP 2010. LNCS, vol. 6351, pp. 169–182. Springer, Heidelberg (2010)
15. Schneider, F.B.: Implementing fault-tolerant services using the state machine approach: a tutorial. ACM Computing Surveys 22(4), 299–319 (1990)
16. Singh, A., Fonseca, P., Kuznetsov, P., Rodrigues, R., Maniatis, P.: Zeno: Eventually consistent Byzantine-fault tolerance. In: Proceedings of the 6th USENIX Symposium on Networked Systems Design and Implementation, NSDI 2009, pp. 169–184. USENIX Association (2009)
17. Sotirov, A., Stevens, M., Appelbaum, J., Lenstra, A., Molnar, D., Osvik, D.A., de Weger, B.: MD5 considered harmful today – creating a rogue CA certificate (December 2008)
18. Stevens, M., Lenstra, A.K., De Weger, B.: Chosen-prefix collisions for MD5 and applications. Int. J. Appl. Cryptol. 2(4), 322–359 (2012)
19. van Renesse, R., Ho, C., Schiper, N.: Byzantine chain replication. In: Baldoni, R., Flocchini, P., Binoy, R. (eds.) OPODIS 2012. LNCS, vol. 7702, pp. 345–359. Springer, Heidelberg (2012)
20. Veronese, G.S., Correia, M., Bessani, A.N., Lung, L.C., Verissimo, P.: Effcient Byzantine fault tolerance. IEEE Transactions on Computers 62(1), 16–30 (2013)
21. Wood, T., Singh, R., Venkataramani, A., Shenoy, P., Cecchet, E.: Zz and the art of practical BFT execution. In: Proceedings of the Sixth Conference on Computer Systems, EuroSys 2011, pp. 123–138 (2011)
22. Yin, J., Martin, J.-P., Venkataramani, A., Alvisi, L., Dahlin, M.: Separating agreement from execution for Byzantine fault tolerant services. In: Proceedings of the Nineteenth ACM Symposium on Operating Systems Principles, pp. 253–267. ACM Press (2003)
23. Zbierski, M.: Iwazaru: The Byzantine sequencer. In: Kubátová, H., Hochberger, C., Daněk, M., Sick, B. (eds.) ARCS 2013. LNCS, vol. 7767, pp. 38–49. Springer, Heidelberg (2013)

Symbolic Analysis of Timed Petri Nets

Wlodek M. Zuberek

Department of Computer Science, Memorial University,
St. John's, NL, Canada A1B 3X5
wlodek@mun.ca

Abstract. In timed Petri nets temporal properties are associated with transitions as transition firing times (or occurrence times). Specific properties of timed nets, such as boundedness or absence of deadlocks, can depend upon temporal properties and sometimes even a small change of these properties has a significant effect on the net's behavior (e.g., a bounded net becomes unbounded or vice versa). The objective of symbolic analysis of timed nets is to provide information about the net's behavior which is independent of specific temporal properties, i.e., which describes preperties of the whole class of timed nets with the same structure.

Keywords: timed Petri nets, symbolic analysis, boundedness, absence of deadlocks, producer–consumer model.

1 Introduction

Petri nets are formal models of systems which exhibit concurrent activities [10], [7], [4]. Communication networks, multiprocessor systems, manufacturing systems and distributed databases are simple examples of such systems. As formal models, Petri nets are bipartite directed graphs, in which the two types of vertices represent, in a very general sense, conditions and events. An event can occur only when all conditions associated with it (represented by arcs directed to the event) are satisfied. An occurrence of an event usually satisfies some other conditions, indicated by arcs directed from the event. So, an occurrence of one event causes some other event to occur, and so on.

In order to study performance aspects of systems modeled by Petri nets, the durations of modeled activities must also be taken into account. This can be done in different ways, resulting in different types of temporal nets [2], [3], [14], [8]. In timed Petri nets [17], firing times or occurrence times are associated with events, and the events occur in real–time (as opposed to instantaneous occurrences in other models [1]). For timed nets, the state graphs of nets are Markov chains (or embedded Markov chains), so the stationary probabilities of states can be determined by standard methods [13], [5]. Stationary probabilities are used for the derivation of many performance characteristics of the model [12].

In timed nets, all firings of enabled transitions are initiated in the same instants of time in which the transitions become enabled. If, during the firing

© Springer International Publishing Switzerland 2015
W. Zamojski et al. (eds.), *Theory and Engineering of Complex Systems and Dependability*,
Advances in Intelligent Systems and Computing 365, DOI: 10.1007/978-3-319-19216-1_57

period of a transition, the transition becomes enabled again, a new, independent firing can be initiated, which will overlap with the other firing(s). There is no limit on the number of simultaneous firings of the same transition (sometimes this is called "infinite firing semantics").

The firing times of transitions can be either deterministic or stochastic (i.e., described by a probability distribution function); in the first case, the corresponding timed nets are referred to as D–nets [15], in the second, for the (negative) exponential distribution of firing times, the nets are referred to as M–nets (Markovian nets) [16]. In both cases, the concepts of states and state transitions have been formally defined and used in the derivation of different performance characteristics of the models [15], [16], [17].

In Petri nets with deteministic firing times, the propoerties such as boundedness or absence of deadlocks can depend upon specific values of firing times and sometimes even a small changes of firing times can have a significant effect on the behavior of a net (e.g., a bounded net becomes unbounded or a deadlock is created).

This paper proposes symbolic analysis of timed Petri nets which analyzes the behavior for the whole spectrum of temporal properties, so the results do not depend upon specific temporal properties.

Section 2 recalls a few basic concepts of Petri nets and timed Petri nets. Section 3 introduces symbolic analysis while an illustrative example is presented in Section 4. Several concluding remarks are in Section 5.

2   Petri Nets and Timed Petri Nets

Place/transition Petri nets are bipartite directed graphs in which the two types of vertices are called places and transitions. Place/transition nets are also known as condition/event systems.

A Petri net (sometimes also called net structure) $\mathcal{N}$ is a triple $\mathcal{N} = (P, T, A)$ where:

- P is a finite set of places (which represent conditions);
- T is a finite set of transitions (which represent events), $P \cap T = \emptyset$;
- A is a set of directed arcs which connect places with transitions and transitions with places, $A \subseteq P \times T \cup T \times P$, also called the flow relation or causality relation (and sometimes represented in two parts, a subset of $P \times T$ and a subset of $T \times P$).

For each transition $t \in T$, and each place $p \in P$, the input and output sets are defined as follows:

$$
\begin{aligned}
Inp(t) &= \{p \in P \mid (p, t) \in A\}, \\
Inp(p) &= \{t \in T \mid (t, p) \in A\}, \\
Out(t) &= \{p \in P \mid (t, p) \in A\}, \\
Out(p) &= \{t \in T \mid (p, t) \in A\}.
\end{aligned}
$$

The dynamic behavior of nets is represented by markings, which assign non-negative numbers of tokens to the places of a net. Under certain conditions these tokens can "move" in the net, changing one marking into another.

A marked Petri net $\mathcal{M}$ is a pair $\mathcal{M} = (\mathcal{N}, m_0)$, where:

- $\mathcal{N}$ is a net structure, $\mathcal{N} = (P, T, A)$;
- m_0 is the initial marking function, $m_0 : P \rightarrow \{0, 1, ...\}$ which assigns a nonnegative number of tokens to each place of the net.

Marked nets are also equivalently defined as $\mathcal{M} = (P, T, A, m_0)$.

In a marked net $\mathcal{M}$, a transition t is enabled by a marking m iff:

$$\forall p \in \mathrm{Inp}(t) : m(p) > 0.$$

An enabled transition t can fire (or occur) transforming a marking m into a directly reachable marking m':

$$\forall p \in P : m'(p) = \begin{cases} m(p) - 1, & \text{if } p \in \mathrm{Inp}(t) - \mathrm{Out}(t), \\ m(p) + 1, & \text{if } p \in \mathrm{Out}(t) - \mathrm{Inp}(t), \\ m(p), & \text{otherwise.} \end{cases}$$

A timed Petri net $\mathcal{T}$ is a pair, $\mathcal{T} = (\mathcal{M}, f)$ where:

- $\mathcal{M}$ is a marked net, $\mathcal{M} = (\mathcal{N}, m_0)$;
- f is the firing–time function, $f : T \rightarrow \mathbf{R}^+$, which assigns the (average) firing times (or occurrence times) to transitions of the net.

For performace analysis of timed nets, an additional component is needed to describe random decisions in (nondeterministic) nets. Usually it is a conflict–resolution function, $c : T \rightarrow [0, 1]$, which assigns the probabilities of firings to transitions in free–choice classes of transitions, and relative frequencies of firings to transitions in conflict classes [16], [17]. This function c is not needed for symbolic analysis.

3   Symbolic Analysis

For symbolic analysis, only relations between the firing times of transitions are needed, so the state descriptions can be simpler than for the detailed behavioral analysis [16]. The states can be represented by pairs of functions, current firing function $n : T \rightarrow \{0, 1, ...\}$ and current (residual) marking function $m : P \rightarrow \{0, 1, ...\}$.

For each state $s = (n, m)$, the next states correspond to all possible relations between the durations of currently firing transitions. This is described by all (nonempty) subsets of transitions which finish their firings (and initiate new firings if any transitions become enabled):

$$\mathrm{next}(s) = \bigcup_{T_i \subseteq T_f(s)} \mathrm{Next}(s, T_i)$$

where $T_f(s)$ (or equivalently $T_f(n, m)$ as $s = (n, m)$) is the set of transitions which are firing in state s, i.e., transitions with nonzero entries in n:

$$T_f(n,m) = \{t \in T \mid n(t) > 0\},$$

and $\text{Next}(s, T_i)$ is the set of states which can be reached from s by ending the firings of all transitions in T_i (and then initiating all possible firings).

Finding the set $\text{Next}(s, T_i)$ is done in two steps:

1. Terminating the firings of all transitions in T_i, which creates an intermediate state $s' = (n', m'$ where:

$$\forall t \in T : n'(t) = \begin{cases} n(t) - 1, & \text{if } t \in T_i; \\ n(t), & \text{otherwise}; \end{cases}$$

$$\forall p \in P : m'(p) = m(p) + \sum\nolimits_{t \in Inp(p) \cap T_i} n(t).$$

2. Initiating new firings of transitions which are enabled by $m'i$. These new firings can be described by a set of functions $b_j : T \to \{0, 1, ...\}$ such that:

   - $\forall p \in P : m'(p) - \sum\nolimits_{t \in \text{Out}(p)} b_j(t) \geq 0$, and

   - $\forall t \in T \; \exists p \in \text{Inp}(t) : m'(p) - \sum\nolimits_{t \in \text{Out}(p)} b_j(t) = 0.$

These two conditions guarantee that all transitions which can fire, initiate their firings (free–choice nets and nets with conflicts have more than one function b_j).

The set of states reachable from $s = (n, m)$ by a set of transitions T_i is described by procedure $Next(n, m, T_i)$ which first finds the intermediate state (n', m') and then uses a recursive function $Find$ to find all possible states by firing transitions enabled by m' (and adjusting n' accordingly):

```
proc Next(n[1:k], m[1:ℓ], T₀);
begin
        n' := n;
        m' := m;
        for each tᵢ ∈ T₀ do
            n'[i] := n' − 1;
                for each pⱼ ∈ Inp(tᵢ) do m'[j] := m'[j] + 1 od
        od;
        New := ∅;
        Find(n'ᵢ, m'ᵢ);
        States := States ∪ New
    od
end
```

$States$ is a global variable which stores the set of reachable states.

Recursive function $Find(n, m)$ finds all states which are derived from $s = (n, m)$ by initiating the firings of enabled transitions. It is assumed that n and m are the firing and marking functions, respectively, represented by k-element and ℓ-element vectors (k is the number of transitions and ℓ is the number of places); moreover, $Find$ uses a nonlocal set variable New:

```
proc Find(n[1 : k], m[1 : ℓ]);
begin
    E := ∅;
    for each tᵢ ∈ T do
        check := true;
        for each pⱼ ∈ Inp(tᵢ) do
            if m[j] = 0 then check := false fi;
        od;
        if check then E := E ∪ {tᵢ} fi
    od;
    if E = ∅ then New := New ∪ {(n, m)}
    else
        for each tᵢ ∈ E do
            n' := n;
            n'[i] := n[i] + 1;
            m' := m;
            for each pⱼ ∈ Inp(tᵢ) do m'[j] := m[pⱼ] − 1 od;
            Find(n', m')
        od
    fi
end
```

The initial state (or states) of a marked net is (or are) determined by an invocation $Find(n_0, m_0)$, where n_0 is zero for all $t \in T$, while m_0 is the initial marking function.

The procedures are shown as an illustration of the approach; they can be improved in many ways.

4   Example

A timed Petri net model of a producer–consumer system with an unbouded buffer is shown in Fig.1.

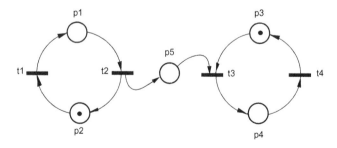

Fig. 1. Model of a producer–consumer system with an unbounded buffer

The two cyclic subnets, (t_1, p_1, t_2, p_2) and (p_3, t_2, p_4, t_4), represent the producer and the consumer, respectively, while place p_5 is the buffer. It is known that for some values of firing timnes the behavior of this system is finite while for others the model becomes unbounded. For example, for $f(t_1) = 3.0$, $f(t_2) = f(t_3) = 0.5$, $f(t_4) = 2.0$, the state transition graph is shown in Fig.2.

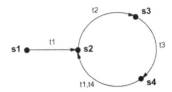

Fig. 2. State transition graph for the net shown in Fig.1 with $f(t_1) = 3.0$, $f(t_2) = f(t_3) = 0.5$, $f(t_4) = 2.0$

Symbolic analysis of the model shown in Fig.1 is presented in Tab.1 where s_i is the current state, n_i and m_i are the two components of s_i, T_i is the set of transitions which terminate their firings in state s_i, b_j is the function describing new firings and s_j is the next state.

It can be traced in Tab.1 that s_{11} is reached from s_{10} by t_2 and that s_{10} is reached from s_8 by t_1. The states s_8 and s_{11} are identical except of marking of p_5. Consequently, if t_1 and t_2 can fire several times before t_3 and t_4 fire, the marking of p_5 can increase arbitrarily, so the model is unbounded.

Similarly, s_{12} is reached from s_9 by t_2, and s_9 is reached from s_7 by t_1.

A timed net is unbounded is there are two states, $s_i = (n_i, m_i)$ and $s_j(n_j, m_j)$ such that s_j is reachable from s_i and s_i is reachable from an initial state, and s_j is componentwise greater or equal to s_i, i.e., for all values of k and ℓ, $n_j[k] \geq n_i[k]$ and $m_j[\ell] \geq m_i[\ell]$.

Moreover, a timed net contains a deadlock if there is a state s_i reachable from an initial state, for which the set of next states is empty.

The conditions for unboundedness and deadlock can be easily recognized during symbolic analysis.

The part of the state transition diagram described in Tab.1 is shown in Fig.3. The regular structure of state transitions can be systematically extended which indicates the unboundedness of the model. Moreover, the behavior shown in Fig.2 can be easily traced in Fig.3 following the same transitions involved in the changes of states:

$$t_1 \rightarrow t_2 \rightarrow t_3 \rightarrow t_1, t_4 \rightarrow t_2 \rightarrow t_3 \rightarrow t_1, t_4 \rightarrow \cdots$$

so the corresponding state transitions (in Fig.3) are:

$$s_1 \rightarrow s_2 \rightarrow s_3 \rightarrow s_5 \rightarrow s_2 \rightarrow s_3 \rightarrow \cdots$$

Table 1. Symbolic analysis of the producer–consumer model

s_i	n_i 1 2 3 4	m_i 1 2 3 4 5	T_i	b_j 1 2 3 4	s_j
s_1	1 0 0 0	0 0 1 0 0	t_1	0 1 0 0	s_2
s_2	0 1 0 0	0 0 1 0 0	t_2	1 0 1 0	s_3
s_3	1 0 1 0	0 0 0 0 0	t_1	0 1 0 0	s_4
			t_3	0 0 0 1	s_5
			t_1,t_3	0 1 0 1	s_6
s_4	0 1 1 0	0 0 0 0 0	t_2	1 0 0 0	s_7
			t_3	0 0 0 1	s_6
			t_2,t_3	1 0 0 1	s_8
s_5	1 0 0 1	0 0 0 0 0	t_1	0 1 0 0	s_6
			t_4	0 0 0 0	s_1
			t_1,t_4	0 1 0 0	s_2
s_6	0 1 0 1	0 0 0 0 0	t_2	1 0 0 0	s_9
			t_4	0 0 0 0	s_2
			t_2,t_4	1 0 1 0	s_3
s_7	1 0 1 0	0 0 0 0 1	t_1	0 1 0 0	s_9
			t_3	0 0 0 1	s_8
			t_1,t_3	0 1 0 1	s_{10}
s_8	1 0 0 1	0 0 0 0 1	t_1	0 1 0 0	s_{10}
			t_4	0 0 1 0	s_3
			t_1,t_4	0 1 1 0	s_4
s_9	0 1 1 0	0 0 0 0 1	t_2	1 0 0 0	s_{12}
			t_3	0 0 0 1	s_{10}
			t_2,t_3	1 0 0 1	s_{11}
s_{10}	0 1 0 1	0 0 0 0 1	t_2	1 0 0 0	s_{11}
			t_4	0 0 1 0	s_4
			t_2,t_4	1 0 1 0	s_7
s_{11}	1 0 0 1	0 0 0 0 2	...		
s_{12}	1 0 1 0	0 0 0 0 2	...		

Structural analysis [18] of the net shown in Fig.1 provides a simple condition for unboundedness for this particular net:

$$f(t_1) + f(t_2) \leq f(t_3) + f(t_4).$$

This condition is clearly not satisfied when $f(t_1) = 1.5$, $f(t_2) = f(t_3) = 0.5$ and $f(t_4) = 3.5$), so net's unbounded behavior is expected in this case. Indeed, the sequence of transitions involved in consecutive state changes is:

$$t_1 \to t_2 \to t_3 \to t_1 \to t_2 \to t_1 \to t_2, t_4 \to t_3 \to t_1 \to t_2 \to t_1 \to t_2, t_4 \to t_3 \cdots$$

with the pattern:

$$t_1 \to t_2 \to t_1 \to t_2, t_4 \to t_3$$

repeated. The sequence of (symbolic) state changes, shown in Fig.4, is more convoluted than in the bounded case:

$$s_1 \to s_2 \to s_3 \to s_5 \to s_6 \to s_8 \to s_{10} \to s_7 \to s_8 \to s_{10} \to s_{11} \to \cdots$$

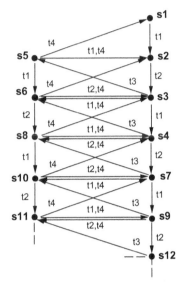

Fig. 3. State transition graph for symbolic analysis of net in Fig.1

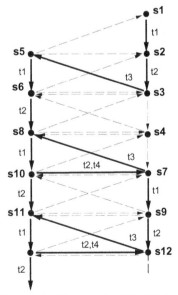

Fig. 4. State transitions for net in Fig.1 with with $f(t_1) = 1.5$, $f(t_2) = f(t_3) = 0.5$, $f(t_4) = 3.5$

5   Concluding Remarks

The behavior of a timed Petri net depends upon the specific values of temporal parameters associated with transitions of a net, and can change in a significant way for even a small chages of these temporal parameters. Symbolic analysis provides general information about the behavior of all nets with the same structure. For example, if symbolic analysis creates a finite space of (symbolic) states, no temporal parameters can result in unbounded behavior. Similarly, if symbolic analysis indicates deadlock freeness, no temporal parameters can create a deadlock in the net.

For large models, symbolic analysis can be quite complex. Therefore analysis of real–life applications is not feasible without efficient software tools. It is expected that such tools will be added to existing software packages for analysis of timed Petri net models.

Symbolic analysis presented in this paper is similar to reachability analysis of marked nets [8], [17]. The obvious difference is that reachability analysis does not consider simultaneous multiple firings. The effects of this difference need to be carefully explored.

Fig.5. shows the initial part of the marking graph for the net in Fig.1.

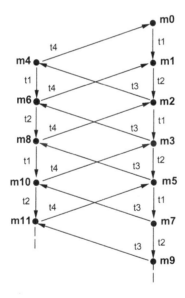

Fig. 5. Marking graph for the net in Fig.1

The similarity of Fig.3 and Fig.5 can be misleading because there is no stright-forward correspondence between the markings (Fig.5) and the states (Fig.3). In particular, there are no changes due to multiple transitions in Fig.5 (like t_1, t_4 or t_2, t_4 in Fig.3).

It is believed that symbolic analysis presented in this paper can be extended to other classes of Petri nets, such as inhibitor Petri nets or high–level Petri nets [6].

Acknowledgement. The Natural Sciences and Engineering Research Council of Canada partially supported this research through grant RGPIN-8222.

References

1. Ajmone Marsan, M., Conte, G., Balbo, G.: A class of generalized stochastic Petri nets for the performance evaluation of multiprocessor systems. ACM Trans. on Computer Systems 2(2), 93–122 (1984)
2. Ajmone Marsan, M., Balbo, G., Conte, G., Donatelli, S., Franceschinis, G.: Modeling with generalized stochastic Petri nets. Wiley and Sons (1995)
3. Bause, F., Kritzinger, P.S.: Stochastic Petri nets – and introduction to the theory, 2nd edn. Vieweg Verlag (2002)
4. Girault, C., Valk, R.: Petri nets for systems engineering. Springer (2002)
5. Haggstrom, O.: Finite Markov chains and algorithmic applications. Cambridge Univ. Press (2003)
6. He, X., Murata, T.: High-level Petri nets - extensions, analysis and applications. In: The Electrical Engineering Handbook, pp. 459–475. Academic Press (2007)
7. Murata, T.: Petri nets: properties, analysis, and applications. Proceedings of the IEEE 77(4), 541–580 (1989)
8. Popova-Zeugmann, L.: Time and Pertri nets. Springer (2013)
9. Proth, J.M., Xie, X.: Petri nets. Wiley & Sons (1996)
10. Reisig, W.: Petri nets – an introduction. EATCS Monographs on Theoretical Computer Science, vol. 4. Springer (1995)
11. Reisig, W.: Understanding Petri nets – modeling techniques, analysis methods, case studies. Springer (2013)
12. Robertazzi, T.A.: Computer networks and systems: queueing theory and performance evaluation, 3rd edn. Springer (2000)
13. Stewart, W.J.: Introduction to the numerical solution of Markov chains. Princeton University Press (1994)
14. Wang, J.: Timed Petri nets. Kluwer Academic Publ. (1998)
15. Zuberek, W.M.: M–timed Petri nets, priorities, preemptions, and performance evaluation of systems. In: Rozenberg, G. (ed.) APN 1985. LNCS, vol. 222, pp. 478–498. Springer, Heidelberg (1986)
16. Zuberek, W.M.: D–timed Petri nets and modelling of timeouts and protocols. Transactions of the Society for Computer Simulation 4(4), 331–357 (1987)
17. Zuberek, W.M.: Timed Petri nets – definitions, properties and applications. Microelectronics and Reliability (Special Issue on Petri Nets and Related Graph Models) 31(4), 627–644 (1991)
18. Zuberek, W.M.: Structural methods in performance analysis of discrete-event systems. In: Proc. 9th IEEE Int. Conf. on Methods and Models in Automation and Robotics, pp. 878–882 (2003)

Author Index